AF761518

PAN-GENOMICS: APPLICATIONS, CHALLENGES, AND FUTURE PROSPECTS

PAN-GENOMICS: APPLICATIONS, CHALLENGES, AND FUTURE PROSPECTS

Edited by

DEBMALYA BARH, PhD

Scientist, Centre for Genomics and Applied Gene Technology, Institute of Integrative Omics and Applied Biotechnology (IIOAB)
Nonakuri, India

SIOMAR SOARES, PhD

Assistant Professor at Department of Immunology, Microbiology and Parasitology, Institute of Biological Sciences and Natural Sciences, Federal University of Triangulo Mineiro (UFTM)
Uberaba, Brazil

SANDEEP TIWARI, PhD

Post-Doctoral Researcher, Laboratory of Cellular and Molecular Genetics, Federal University of Minas Gerais (UFMG)
Belo Horizonte, Brazil

VASCO AZEVEDO, PhD

Senior Professor, Institute of Biological Sciences, Federal University of Minas Gerais (UFMG)
Belo Horizonte, Brazil

ELSEVIER

ACADEMIC PRESS

An imprint of Elsevier

Academic Press is an imprint of Elsevier
125 London Wall, London EC2Y 5AS, United Kingdom
525 B Street, Suite 1650, San Diego, CA 92101, United States
50 Hampshire Street, 5th Floor, Cambridge, MA 02139, United States
The Boulevard, Langford Lane, Kidlington, Oxford OX5 1GB, United Kingdom

© 2020 Elsevier Inc. All rights reserved.

No part of this publication may be reproduced or transmitted in any form or by any means, electronic or mechanical, including photocopying, recording, or any information storage and retrieval system, without permission in writing from the publisher. Details on how to seek permission, further information about the Publisher's permissions policies and our arrangements with organizations such as the Copyright Clearance Center and the Copyright Licensing Agency, can be found at our website: www.elsevier.com/permissions.

This book and the individual contributions contained in it are protected under copyright by the Publisher (other than as may be noted herein).

Notices
Knowledge and best practice in this field are constantly changing. As new research and experience broaden our understanding, changes in research methods, professional practices, or medical treatment may become necessary.

Practitioners and researchers must always rely on their own experience and knowledge in evaluating and using any information, methods, compounds, or experiments described herein. In using such information or methods they should be mindful of their own safety and the safety of others, including parties for whom they have a professional responsibility.

To the fullest extent of the law, neither the Publisher nor the authors, contributors, or editors, assume any liability for any injury and/or damage to persons or property as a matter of products liability, negligence or otherwise, or from any use or operation of any methods, products, instructions, or ideas contained in the material herein.

Library of Congress Cataloging-in-Publication Data
A catalog record for this book is available from the Library of Congress

British Library Cataloguing-in-Publication Data
A catalogue record for this book is available from the British Library

ISBN 978-0-12-817076-2

For information on all Academic Press publications visit
our website at https://www.elsevier.com/books-and-journals

Publisher: Stacy Masucci
Acquisitions Editor: Rafael E. Teixeira
Editorial Project Manager: Sara Pianavilla
Production Project Manager: Maria Bernard
Designer: Greg Harris

Typeset by SPi Global, India

Working together
to grow libraries in
developing countries

www.elsevier.com • www.bookaid.org

Dedication

I dedicate this book to my mother late ***Cacilda de Fátima Soares***, a strong woman who always worked hard to give the best to all her sons.

Dr. Siomar de Castro Soares

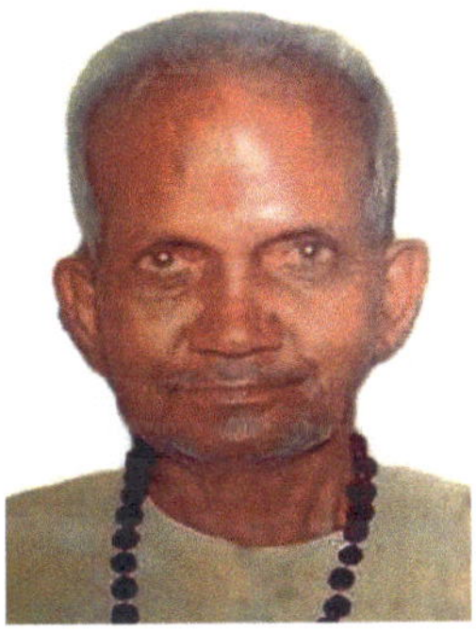

I dedicate this book to my maternal grandfather late ***Prof. Ramawadh Mishra***, a philosopher and guide who had enlightened me to find the meaning of life and shaped my future.

Dr. Sandeep Tiwari

I dedicate this book to my father-in-law late ***Anacer Abi-ackel.*** The Arabic to English translation of his first name and family name says it all. "Anacer" means helps us and "Abi-ackel" is synonyms to father of wisdom. You are the father of wisdom who always helped us.

Prof. Vasco Azevedo

I dedicate this book to ***Nityananda Sarkar,*** signifying the meaning of his name; he is always a happy man whatever the circumstances may be.

Dr. Debmalya Barh

Contents

Contributors

Talita Emile Ribeiro Adelino
Laboratório de Genética Celular e Molecular, ICB, Universidade Federal de Minas Gerais; Fundação Ezequeil Dias (Funed), Belo Horizonte, Brazil

Jamil Ahmad
Research Center for Modeling & Simulation (RCMS), National University of Sciences and Technology (NUST), Islamabad, Pakistan

Shahbaz Ahmed
Atta-ur-Rahman School of Applied Biosciences, National University of Sciences and Technology, Islamabad, Pakistan

Luiz Carlos Junior Alcantara
PG Program in Bioinformatics, Institute of Biological Sciences, Federal University of Minas Gerais (UFMG), Belo Horizonte; Laboratório de Flavivírus, Instituto Oswaldo Cruz, Fundação Oswaldo Cruz, Rio de Janeiro; Laboratório de Genética Celular e Molecular, ICB, Universidade Federal de Minas Gerais, Belo Horizonte, Brazil

Amjad Ali
Department of Plant Biotechnology, Atta-ur-Rahman School of Applied Biosciences (ASAB), National University of Sciences and Technology (NUST), Islamabad, Pakistan

Rabia Amir
Department of Plant Biotechnology, Atta-ur-Rahman School of Applied Biosciences (ASAB), National University of Sciences and Technology (NUST), Islamabad, Pakistan

Fabricio Araujo
Institute of Biological Sciences, Federal University of Pará (UFPA), Belém, Brazil

Muneeba Arveen
Department of Plant Biotechnology, Atta-ur-Rahman School of Applied Biosciences (ASAB), National University of Sciences and Technology (NUST), Islamabad, Pakistan

Vasco Azevedo
PG Program in Bioinformatics, Institute of Biological Sciences, Federal University of Minas Gerais (UFMG), Belo Horizonte, Brazil

Jahanzaib Azhar
Atta-ur-Rahman School of Applied Biosciences, National University of Sciences and Technology, Islamabad, Pakistan

Luciana Balbo
State University of Londrina, Londrina, Brazil

Li Bao
National Clinical Research center for Cancer, Tianjin Medical University Cancer Institute and Hospital; Key Laboratory of Cancer Prevention and Therapy, Tianjin, China

Debmalya Barh
Centre for Genomics and Applied Gene Technology, Institute of Integrative Omics and Applied Biotechnology (IIOAB), Purba Medinipur, India

Fernanda Khouri Barreto
Laboratório de Patologia Experimental, Instituto Gonçalo Moniz, Fiocruz Bahia; Instituto Multidisciplinar em Saúde—IMS, Universidade Federal da Bahia (UFBA), Salvador, Brazil

Attya Bhatti
Atta-ur-Rahman School of Applied Biosciences, National University of Sciences and Technology, Islamabad, Pakistan

Andreas Burkovski
Friedrich-Alexander-Universität Erlangen-Nürnberg, Erlangen, Germany

Roberta Torres Chideroli
State University of Londrina, Londrina, Brazil

Mauricio Corredor
GEBIOMIC Group, FCEN, University of Antioquia, Medellin, Colombia

Kenny da Costa Pinheiro
Institute of Biological Sciences, Federal University of Pará (UFPA), Belém, Brazil

Artur Luiz da Costa Silva
Institute of Biological Sciences, Federal University of Pará (UFPA), Belém, Brazil

Hamza Arshad Dar
Department of Plant Biotechnology, Atta-ur-Rahman School of Applied Biosciences (ASAB), National University of Sciences and Technology (NUST), Islamabad, Pakistan

Letícia de Castro Oliveira
Department of Immunology, Microbiology and Parasitology, Institute of Biological Science and Natural Sciences, Federal University of Triângulo Mineiro (UFTM), Uberaba, Brazil

Siomar de Castro Soares
Department of Immunology, Microbiology and Parasitology, Institute of Biological Science and Natural Sciences, Federal University of Triângulo Mineiro (UFTM), Uberaba, Brazil

Jaqueline Goes de Jesus
Laboratório de Flavivírus, IOC, Fundação Oswaldo Cruz, Rio de Janeiro; Laboratório de Patologia Experimental, Instituto Gonçalo Moniz, Fiocruz Bahia, Salvador, Brazil

Thiago de Jesus Sousa
PG Program in Bioinformatics, Institute of Biological Sciences, Federal University of Minas Gerais (UFMG), Belo Horizonte, Brazil

Tulio de Oliveira
KwaZulu-Natal Research Innovation and Sequencing Platform (KRISP), College of Health Sciences, University of KwaZulu-Natal, Durban, South Africa

Stephane Fraga de Oliveira Tosta
PG Program in Bioinformatics, Institute of Biological Sciences, Federal University of Minas Gerais (UFMG); Laboratório de Genética Celular e Molecular, ICB, Universidade Federal de Minas Gerais, Belo Horizonte, Brazil

Ulisses de Pádua Pereira
State University of Londrina, Londrina, Brazil

Vagner de Souza Fonseca
Laboratório de Flavivírus, IOC, Fundação Oswaldo Cruz, Rio de Janeiro; Laboratório de Genética Celular e Molecular, ICB, Universidade Federal de Minas Gerais, Belo Horizonte, Brazil; KwaZulu-Natal Research Innovation and Sequencing Platform (KRISP), College of Health Sciences, University of KwaZulu-Natal, Durban, South Africa

Dipali Dhawan
Baylor Genetics, Houston, TX, United States

César Toshio Facimoto
State University of Londrina, Londrina, Brazil

Nuno Rodrigues Faria
Department of Zoology, University of Oxford, Oxford, United Kingdom

Nosheen Fatima
Department of Plant Biotechnology, Atta-ur-Rahman School of Applied Biosciences (ASAB), National University of Sciences and Technology (NUST), Islamabad, Pakistan

Henrique César Pereira Figueiredo
AQUACEN, National Reference Laboratory for Aquatic Animal Diseases, Ministry of Fisheries and Aquaculture, Universidade Federal de Minas Gerais, Belo Horizonte, Brazil

Marta Giovanetti
PG Program in Bioinformatics, Institute of Biological Sciences, Federal University of Minas Gerais (UFMG), Belo Horizonte; Laboratório de Flavivírus, IOC, Fundação Oswaldo Cruz, Rio de Janeiro; Laboratório de Genética Celular e Molecular, ICB, Universidade Federal de Minas Gerais, Belo Horizonte, Brazil

Aristóteles Góes-Neto
Molecular and Computational Biology of Fungi Laboratory, Department of Microbiology, Institute of Biological Sciences (ICB), Federal University of Minas Gerais (UFMG), Belo Horizonte, Brazil

Anne Cybelle Pinto Gomide
PG Program in Bioinformatics, Institute of Biological Sciences, Federal University of Minas Gerais (UFMG), Belo Horizonte, Brazil

Luis Carlos Guimarães
Institute of Biological Sciences, Federal University of Pará (UFPA), Belém, Brazil

Raquel Enma Hurtado
PG Program in Bioinformatics, Institute of Biological Sciences, Federal University of Minas Gerais (UFMG), Belo Horizonte, Brazil

Felipe Campos Melo Iani
Laboratório de Genética Celular e Molecular, ICB, Universidade Federal de Minas Gerais; Fundação Ezequeil Dias (Funed), Belo Horizonte, Brazil

Izabela Coimbra Ibraim
PG Program in Bioinformatics, Institute of Biological Sciences, Federal University of Minas Gerais (UFMG), Belo Horizonte, Brazil

Madangchanok Imchen
Department of Genomic Science, School of Biological Sciences, Central University of Kerala, Kasaragod, India

Arun Kumar Jaiswal
Department of Immunology, Microbiology and Parasitology, Institute of Biological Science and Natural Sciences, Federal University of Triângulo Mineiro (UFTM), Uberaba; PG Program in Bioinformatics, Institute of Biological Sciences, Federal University of Minas Gerais (UFMG), Belo Horizonte, Brazil

Syed Babar Jamal
Department of Biological Sciences, National University of Medical Sciences, Rawalpindi, Pakistan

Peter John
Atta-ur-Rahman School of Applied Biosciences, National University of Sciences and Technology, Islamabad, Pakistan

Rodrigo Bentes Kato
Molecular and Computational Biology of Fungi Laboratory, Department of Microbiology, Institute of Biological Sciences (ICB), Federal University of Minas Gerais (UFMG), Belo Horizonte, Brazil

Jaspreet Kaur
University Institute of Engineering and Technology (UIET), Department of Biotechnology, Panjab University, Chandigarh, India

Bineypreet Kaur
University Institute of engineering and Technology (UIET), Department of Biotechnology, Panjab University, Chandigarh, India

Ranjith Kumavath
Department of Genomic Science, School of Biological Sciences, Central University of Kerala, Kasaragod, India

Xiaofeng Liu
National Clinical Research center for Cancer, Tianjin Medical University Cancer Institute and Hospital; Key Laboratory of Cancer Prevention and Therapy, Tianjin, China

Nguyen Thanh Luan
Department of Veterinary Medicine, Institute of Applied Science, Ho Chi Minh City University of Technology—HUTECH, Ho Chi Minh City, Vietnam

Wajahat Maqsood
Department of Plant Biotechnology, Atta-ur-Rahman School of Applied Biosciences (ASAB), National University of Sciences and Technology (NUST), Islamabad, Pakistan

Wanderson Marques da Silva
Institute of Agrobiotechnology and Molecular Biology, INTA-CONICET, Buenos Aires, Argentina

Anupriya Minhas
University Institute of engineering and Technology (UIET), Department of Biotechnology, Panjab University, Chandigarh, India

Faiza Munir
Department of Plant Biotechnology, Atta-ur-Rahman School of Applied Biosciences (ASAB), National University of Sciences and Technology (NUST), Islamabad, Pakistan

Amalia Muñoz-Gómez
GEBIOMIC Group, FCEN, University of Antioquia, Medellin, Colombia

Kanwal Naz
Department of Plant Biotechnology, Atta-ur-Rahman School of Applied Biosciences (ASAB), National University of Sciences and Technology (NUST), Islamabad, Pakistan

Anam Naz
Department of Plant Biotechnology, Atta-ur-Rahman School of Applied Biosciences (ASAB), National University of Sciences and Technology (NUST), Islamabad, Pakistan

Ayesha Obaid
Department of Plant Biotechnology, Atta-ur-Rahman School of Applied Biosciences (ASAB), National University of Sciences and Technology (NUST), Islamabad, Pakistan

Yan Pantoja
Institute of Biological Sciences, Federal University of Pará (UFPA), Belém, Brazil

Rommel Ramos
Institute of Biological Sciences, Federal University of Pará (UFPA), Belém, Brazil

Noor Ul Saba
Department of Plant Biotechnology, Atta-ur-Rahman School of Applied Biosciences (ASAB), National University of Sciences and Technology (NUST), Islamabad, Pakistan

Alvaro Salgado
Laboratório de Genética Celular e Molecular, ICB, Universidade Federal de Minas Gerais, Belo Horizonte, Brazil

Vartul Sangal
Faculty of Health and Life Sciences, Northumbria University, Newcastle upon Tyne, United Kingdom

Qurat-ul-Ain Sani
Department of Plant Biotechnology, Atta-ur-Rahman School of Applied Biosciences (ASAB), National University of Sciences and Technology (NUST), Islamabad, Pakistan

Nubia Seyffert
Biology Institute, Federal University of Bahia, Salvador, Brazil

Faisal Sheraz Shah
Atta-ur-Rahman School of Applied Biosciences, National University of Sciences and Technology, Islamabad, Pakistan

Fatima Shahid
Department of Plant Biotechnology, Atta-ur-Rahman School of Applied Biosciences (ASAB), National University of Sciences and Technology (NUST), Islamabad, Pakistan

Muhammad Shehroz
Department of Plant Biotechnology, Atta-ur-Rahman School of Applied Biosciences (ASAB), National University of Sciences and Technology (NUST), Islamabad, Pakistan

Amnah Siddiqa
Research Center for Modeling & Simulation (RCMS), National University of Sciences and Technology (NUST), Islamabad, Pakistan

Gyan P. Srivastava
Molecular Biology and Genetic Engineering Laboratory, Department of Botany, University of Allahabad, Allahabad, India

Guilherme Campos Tavares
Universidade Nilton Lins, Manaus, Brazil

Hai Ha Pham Thi
Faculty of Biotechnology and Environmental Technology, Nguyen Tat Thanh University, Ho Chi Minh City, Vietnam

Sandeep Tiwari
PG Program in Bioinformatics, Institute of Biological Sciences, Federal University of Minas Gerais (UFMG), Belo Horizonte, Brazil

Basant K. Tiwary
Centre for Bioinformatics, Pondicherry University, Pondicherry, India

Nimat Ullah
Department of Plant Biotechnology, Atta-ur-Rahman School of Applied Biosciences (ASAB), National University of Sciences and Technology (NUST), Islamabad, Pakistan

Ravali Krishna Vennapu
Department of Genomic Science, School of Biological Sciences, Central University of Kerala, Kasaragod, India

Joilson Xavier
Laboratório de Flavivírus, IOC, Fundação Oswaldo Cruz, Rio de Janeiro; Laboratório de Patologia Experimental, Instituto Gonçalo Moniz, Fiocruz Bahia, Salvador, Brazil

Neelam Yadav
Molecular Biology and Genetic Engineering Laboratory, Department of Botany, University of Allahabad, Allahabad, India

Bhupendra N.S. Yadav
Molecular Biology and Genetic Engineering Laboratory, Department of Botany, University of Allahabad, Allahabad, India

Rajiv K. Yadav
Molecular Biology and Genetic Engineering Laboratory, Department of Botany, University of Allahabad, Allahabad, India

Dinesh K. Yadav
Molecular Biology and Genetic Engineering Laboratory, Department of Botany, University of Allahabad, Allahabad, India

Tahreem Zaheer
Department of Plant Biotechnology, Atta-ur-Rahman School of Applied Biosciences (ASAB), National University of Sciences and Technology (NUST), Islamabad, Pakistan

Editors biography

Siomar de Castro Soares holds an MSc in genetics, PhD in genetics and PhD in bioinformatics. He was a senior bioinformatics researcher at the Official Laboratory of the Ministry of Fisheries. Currently, he is working as an assistant professor at Federal University of Triângulo Mineiro—UFTM and is an affiliate member of the Brazilian Academy of Sciences. He has published 79 research publications and 4 book chapters. His areas of expertise include molecular genetics, genomic sequencing, and microbial comparative genomics, mainly focused on pan-genomics, the role of pathogenicity islands and virulence factors in genome plasticity, phylogenomics, molecular epidemiology, reverse vaccinology, and software development.

Sandip Tiwari is a postdoctoral researcher at the Laboratory of Cellular and Molecular Genetics, Institute of Biological Sciences Institute, Federal University of Minas Gerais, Brazil. He earned a bachelor's degree in microbiology in 2009 from Deen Dayal Upadhyaya Gorakhpur University (India), a master's degree in bioinformatics in 2011 from Madhya Pradesh State University (India), and a doctorate in 2017 in bioinformatics from UFMG. He works in the areas of bioinformatics, genomics, transcriptomics, proteomics, and drug target identification against infectious diseases.

Debmalya Barh holds an MSc in applied genetics, MTech in biotechnology, MPhil in biotechnology, PhD in biotechnology, PhD in bioinformatics, Postdoc in bioinformatics, and PGDM in postgraduate in management. He is an honorary scientist at the Institute of Integrative Omics and Applied Biotechnology (IIOAB), India. Dr. Barh is blended with both academic and industrial research for decades and is an expert in integrative omics-based biomarker discovery, molecular diagnosis, and precision medicine in various complex human diseases and traits. He works with over 400 scientists from more than 100 organizations across more than 40 countries. Dr. Barh has published over 150 research publications, more than 32 book chapters, and has edited more than 20 cutting-edge, omics-related reference books published by Taylor & Francis, Elsevier, and Springer. He frequently reviews articles for Nature publications, Elsevier, AACR journals, NAR, BMC journals, PLOS ONE, and Frontiers, to name a few. He has been recognized by *Who's Who in the World* and *Limca Book of Records* for his significant contributions in managing advance scientific research.

Vasco Azevedo is a senior professor of genetics and deputy head of the Department of Genetics, Ecology, and Evolution at Universidade Federal de Minas Gerais, Brazil. He is a member of the Brazilian Academy of Sciences and is a knight of the National Order of Scientific Merit of the Brazilian Ministry of Science, Technology and Innovation. He is also a researcher 1A of the National Council for Scientific and Technological Development (CNPq), which is the highest position. Professor Azevedo is a molecular geneticist who graduated from veterinary school, Federal University of Bahia in 1986. He earned his master's in 1989 and PhD in 1993 in molecular genetics from Institut National Agronomique Paris-Grignon (INAPG) and Institut National de la Recherche Agronomique (INRA), France. He conducted a postdoctoral research at Microbiology Department of Medicine School from University of Pennsylvania, United States, in 1994. In 2017, he earned another PhD in the field of bioinformatics. His total research publications include more than 400 research articles, 3 books, and more than 30 book chapters. Professor Azevedo is a pioneer of genetics of lactic acid bacteria and *Corynebacterium pseudotuberculosis* in Brazil. He has specialized and currently researching on bacterial genetics, genomes, transcriptomes, and proteomes for development of new vaccines and diagnostic against infectious diseases.

Preface

Since the development of the next-generation sequencing technologies, many genomes have been deposited in the databases and, as a result, the term pan-genome was coined in 2005 to describe a new area of genomics analyses that used several strains of the same species to gain insights into the development of bacterial genomes. This area has then expanded, and now other applications have appeared to complement the pan-genomics, creating the pan-omics analyses. This book was conceived to be a compendium of pan-genomics and other pan-omics analyses from different organisms.

The book *Pan-Genomics: Applications, Challenges, and Future Prospects* begins with an introduction on pan-omics focused to Crick's Central Dogma and a brief description of all the chapters of the book (Chapter 1), in which some basic concepts of pan-genomics are introduced. Chapter 2, on the other hand, discusses the use of bioinformatics approaches applied to pan-genomics and their challenges, with a list of software that may be useful in this context. In Chapter 3, Dr. Tiwary and collaborators discuss the use of pan-genomics in evolutionary studies based on gene content and single nucleotide polymorphism.

Next, the chapters explore the pan-genomics of model bacterial organisms and its application such as in the discovery of vaccine and drug targets against bacterial pathogens using reverse vaccinology and drug target analyses (Chapter 16). Chapter 4 describes the pan-genomics analyses of *Corynebacterium diphtheriae* and *Corynebacterium ulcerans*, the causative agents of diphtheria and diphtheria-like diseases. Chapter 5 describes the use of pan-genomics in veterinary pathogens, focusing on the pan-genome analysis of *Corynebacterium pseudotuberculosis*, the causative agent of Caseous lymphadenitis in small ruminants. In Chapter 6, Dr. Amir explores the pan-genomes of plant pathogens, focusing on *Pectobacterium parmentieri*, *Pantoea ananatis*, *Erwinia amylovora*, *Burkholderia*, *Xylella fastidiosa*, *Puccinia graminis,* and *Zymoseptoria tritici*. In Chapter 7, Dr. Pereira explores the pan-genome of food pathogens such as *Escherichia coli*, *Salmonella enterica*, *Clostridium botulinum*, *Clostridium perfringens*, *Listeria monocytogenes,* and *Staphylococcus aureus*. Chapter 8 focuses on the pan-genome of aquatic animals such as *Edwardsiella* and *Aeromonas*. In Chapter 9, Dr. Ali explores the pan-genomes of model bacteria such as *Streptococcus agalactiae*, *Neisseria meningitidis*, *Staphylococcus aureus*, *E. coli*, *Streptococcus pyogenes*, *Haemophilus influenzae,* and *Streptococcus pneumoniae*. Finally, in Chapter 10, the pan-genome of multidrug-resistant human pathogenic bacteria and their resistome are discussed, focusing on bacteria such as *Acinetobacter baumannii* and *Pseudomonas aeruginosa*.

Other chapters focus on virus, plants, algae, fungi, and humans in pan-cancer analyses. Chapter 11 focuses on the pan-genomics of virus and its applications to provide insights into the transmission, biology, and epidemiology of health-care-associated virus pathogens, and also provide a description of software used for this task. In Chapter 12, Góes-Neto performs an intensive literature review and metaanalysis of a customized database to provide insights into fungus pan-genomics, with data on the most studied fungi of the 12 more explored genera. In Chapter 13, Dr. Kaur describes the state of the art in the genomics of algae organisms, from micro to macroalgae. Chapter 14 describes the pan-genome of plants and its applications, focusing on *Brassica rapa, Brassica oleracea, Glycine soja, Oryza sativa,* and *Brachypodium distachyon*. Chapter 15 describes the pan-cancer project, which may be helpful in cancer prevention and in the design of new cancer therapeutics.

Pan-omics analyses are further described in chapters dedicated to pan-proteomics, pan-metagenomics, pan-metabolomics, pan-interactomics, and pan-transcriptomics. Pan-metagenomics is explored in Chapter 17 to better understand the microbiota of a given organism or ecosystem in different conditions and also to explore the commonly shared microorganisms in these conditions. The authors also discuss the importance of pan-metagenomics in pharmacokinetics. Pan-transcriptomics (Chapter 18) and pan-proteomics (Chapter 19) are intended to analyze the dataset of differentially expressed and commonly expressed genes in different conditions in order to give insights into adaptation to these conditions. Chapters 20 and 21 explore pan-metabolomics and pan-interactomics, respectively, which are recent areas of research and may help in elucidating differentially regulated metabolic pathways and protein-protein interactions in different conditions.

A total of 65 experts from 14 countries have contributed to this book to cover wide areas of pan-genomics. We believe this book will provide the readers with the main strategies and their applications utilized so far in pan-genomics.

Editors

Debmalya Barh
Siomar Soares
Sandeep Tiwari
Vasco Azevedo

CHAPTER 1

Pan-omics focused to Crick's central dogma

Arun Kumar Jaiswal*[a,b], Sandeep Tiwari*[a], Guilherme Campos Tavares[h], Wanderson Marques da Silva[d], Letícia de Castro Oliveira[b], Izabela Coimbra Ibraim[a], Luis Carlos Guimarães[e], Anne Cybelle Pinto Gomide[a], Syed Babar Jamal[c], Yan Pantoja[e], Basant K. Tiwary[i], Andreas Burkovski[j], Faiza Munir[k], Hai Ha Pham Thi[l], Nimat Ullah[k], Amjad Ali[k], Marta Giovanetti[a,m], Luiz Carlos Junior Alcantara[a,m], Jaspreet Kaur[n], Dipali Dhawan[o], Madangchanok Imchen[p], Ravali Krishna Vennapu[p], Ranjith Kumavath[p], Mauricio Corredor[q], Henrique César Pereira Figueiredo[g], Debmalya Barh[f], Vasco Azevedo[a], Siomar de Castro Soares[b]

[a]PG Program in Bioinformatics, Institute of Biological Sciences, Federal University of Minas Gerais (UFMG), Belo Horizonte, Brazil
[b]Department of Immunology, Microbiology and Parasitology, Institute of Biological Science and Natural Sciences, Federal University of Triângulo Mineiro (UFTM), Uberaba, Brazil
[c]Department of Biological Sciences, National University of Medical Sciences, Rawalpindi, Pakistan
[d]Institute of Agrobiotechnology and Molecular Biology, INTA-CONICET, Buenos Aires, Argentina
[e]Institute of Biological Sciences, Federal University of Pará (UFPA), Belém, Brazil
[f]Centre for Genomics and Applied Gene Technology, Institute of Integrative Omics and Applied Biotechnology (IIOAB), Purba Medinipur, India
[g]AQUACEN, National Reference Laboratory for Aquatic Animal Diseases, Ministry of Fisheries and Aquaculture, Universidade Federal de Minas Gerais, Belo Horizonte, Brazil
[h]Universidade Nilton Lins, Manaus, Brazil
[i]Centre for Bioinformatics, Pondicherry University, Pondicherry, India
[j]Friedrich-Alexander-Universität Erlangen-Nürnberg, Erlangen, Germany
[k]Department of Plant Biotechnology, Atta-ur-Rahman School of Applied Biosciences (ASAB), National University of Sciences and Technology (NUST), Islamabad, Pakistan
[l]Faculty of Biotechnology and Environmental Technology, Nguyen Tat Thanh University, Ho Chi Minh City, Vietnam
[m]Laboratório de Flavivírus, IOC, Fundação Oswaldo Cruz, Rio de Janeiro, Brazil,
[n]University Institute of Engineering and Technology (UIET), Department of Biotechnology, Panjab University, Chandigarh, India
[o]Baylor Genetics, Houston, TX, United States
[p]Department of Genomic Science, School of Biological Sciences, Central University of Kerala, Kasaragod, India
[q]GEBIOMIC Group, FCEN, University of Antioquia, Medellin, Colombia

1 Introduction

Since the development of the first DNA sequencing technologies, many organisms had their complete DNA repertoire sequenced by Sanger and next-generation sequencing (NGS) technologies, creating the area of genomics, which was originated by the fusion of the words gene and chromosome [1]. In this scenario, a genome is the complete dataset of genes of a given organism. Nowadays, there are more than 200,000 genome projects registered at the Genome Online Database (GOLD), whereas more than 120,000 are

* These authors contributed equally to this work.

Pan-genomics: Applications, Challenges, and Future Prospects
https://doi.org/10.1016/B978-0-12-817076-2.00001-9

© 2020 Elsevier Inc.
All rights reserved.

genomes isolated from bacteria (https://gold.jgi.doe.gov/statistics). Bacteria are widely distributed all over the world and have implications in health, agriculture, industry, and others. Besides, their genomes are small, highly compact, and do not present many repetitions, making them good targets for genome sequencing, once their genomes are easier to sequence than the ones from other organisms. Also, from the genome sequence of bacteria, it is possible to find virulence factors, antibiotic resistance genes, new therapeutic targets for vaccine and drug development, and industrially important genes [2, 3].

Another important point of the development of NGS technologies was the genome sequencing process that has become cheaper and faster, making it possible for small laboratories to use the technology in daily routine. NGS made possible the comparison of several genomes in a multipronged strategy, where phylogenomics, genome plasticity, and whole genome synteny analyses are easier to perform nowadays (Fig. 1). Also, RNA sequencing (RNA-seq) by these platforms and the development of new technologies for sequencing the complete dataset of proteins of an organism created the areas of transcriptomics and proteomics, respectively [4, 5]. Altogether, genomics is responsible for the identification of the complete dataset of genes of a given organism, whereas transcriptomics and proteomics are important for the identification of genes that are differentially expressed

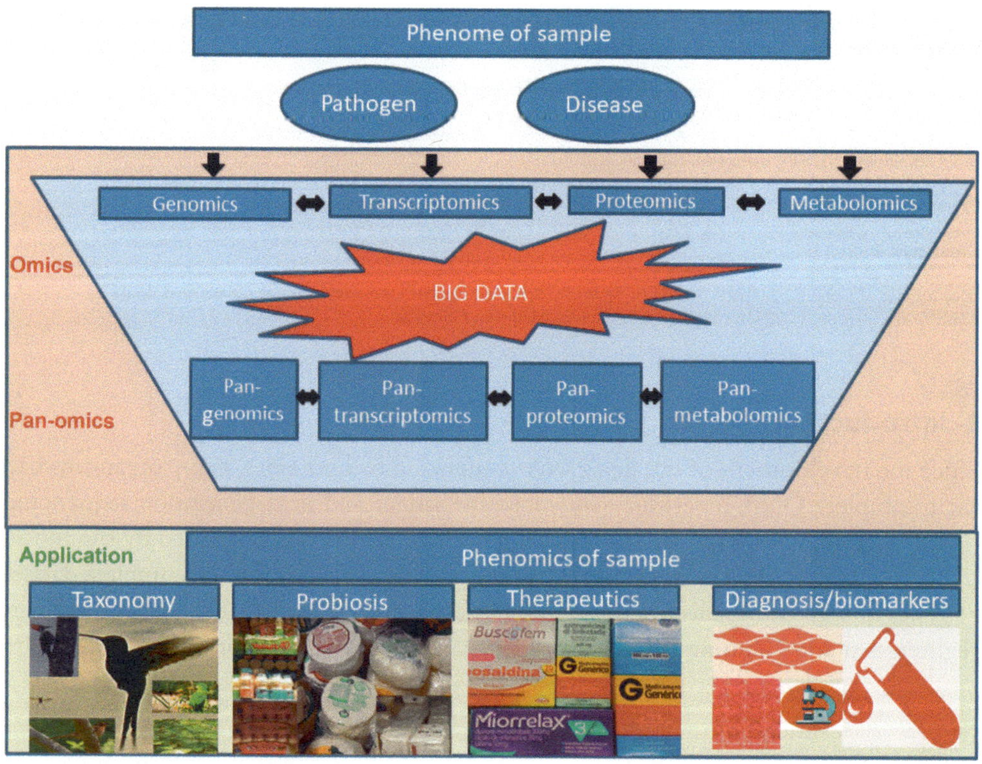

Fig. 1 Pan-omics and its applications.

between strains or species. Finally, the efforts to compare several genomes at once created the area of pan-genomics, which will be further discussed in this book.

1.1 Brief overview of pan-genomics

The term pan-genomics was created by Tettelin and collaborators, in 2005 [6], to describe the complete dataset of genes of a given species through the sequencing of several strains of this species. The pan-genome is composed of the core genome, shared genome, and singletons subsets, whereas the core genome is composed of all the commonly shared genes by all strains of the species; the shared genome contains genes that are present in two or more, but not all strains from a species; and the singletons are strain-specific genes (Fig. 2). From these subsets, one can extrapolate the data to find vaccines and drug targets from the core genome, whereas the shared genes and singletons are responsible for differences between the strains that are normally responsible for the emergence of new pathogens and the adaptation to new traits [6–10].

Normally, the core genome is composed of housekeeping genes and other genes important for metabolism and other important functions of the organism, whereas the shared genes and singletons are the result of genome plasticity. Genome plasticity is the dynamic property of DNA which involves the gain, loss, and rearrangement of genes through plasmids, phages, and genomic islands (GEIs). GEIs are huge blocks of genes acquired through horizontal gene transfer (HGT) that normally share a function in common. They are classified according to the functions of the genes into: pathogenicity islands, harboring virulence factors; metabolic islands, composed of metabolism-related genes; resistance islands, with antibiotic resistant genes; and symbiotic islands, which share in common the presence of symbiotic-related genes [11, 12].

Normally, the subsets of the pan-genome are identified by the use of orthology analyses, which first identify all orthologous genes from the complete dataset using all-vs-all blasts or other alignment search tools. Next, the datasets are classified according to their homology to genes from other strains in the subsets. After the classification, the data is plotted in a chart and mathematical formulas are used to fit the specific curves. Two such

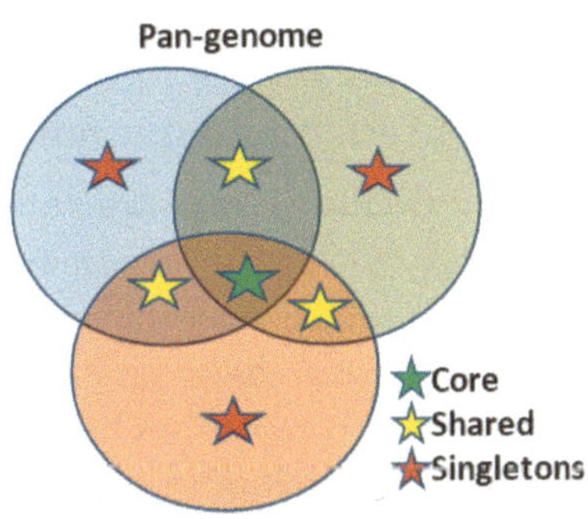

Fig. 2 Schematic representation of the core genome, shared genome, and singleton subsets of pan-genome analysis.

formulas are Heaps' law for the pan-genome development and least-squares fit of the exponential regression decay for the core genome and singleton subsets, which are described respectively as: $n = k \cdot N^{-\alpha}$, where n is the number of genes, N is the number of genomes, and k and α are constants defined by the formula; and $n = k \cdot e^{-x/\tau} + tg\theta$, where n is the number of genes, x is the number of genomes, e is Euler's number, and k, τ, and $tg\theta$ are constants defined by the formula [6, 9].

1.2 Open and closed pan-genomes

According to Heap's law, the α value is representative of the current dynamics of the pan-genome, where an α higher than 1 is representative of a closed pan-genome and an α lower than 1 represents an open pan-genome. A closed pan-genome has all possible genes represented and only few genes will be added to the pan-genome if more genomes are to be sequenced, whereas an open pan-genome is still not fully represented and the sequencing of new genomes will add many genes to the analyses [6, 9]. This definition is controversial, however, once the incorporation of GEIs may change the composition of the pan-genome drastically, even for closed pan-genomes, taking it to be open again. Most important, environmental bacteria and extracellular pathogens normally have open pan-genomes, once they still need to adapt to new traits, whereas obligate intracellular pathogens tend to have closed pan-genomes once they are not in constant contact with other bacteria. Also, intracellular pathogens have lost many genes during evolution, completely adapting to the host organism and, thus, present very compact genomes with a high percentage of essential genes [13].

According to least-squares fit of the exponential regression decay, the $tg\theta$ is representative of the number of genes present in the core genome after stabilization of the core genome curve and, also, of the number of genes that will be added to the analyses after a new genome is sequenced from the singleton development curve. Based on that, researchers may choose the species that need more strains to be sequenced and which do not. Finally, the highest the $tg\theta$ on the singleton development, the lower the α, once a high number of genes will be added to the analyses taking the pan-genome to be more open and the α to be lower (Fig. 3). The opposite is also true, the lower the $tg\theta$, the higher is the α value [6, 7, 10].

1.3 Computational methods used in pan-genomics

Computational methods to find more efficient data structures, algorithms, and statistical methods to perform bioinformatics analyses of pan-genomes have been studied because it is known that in a pan-genome analysis the greater the number of genomes taken to the analysis the greater will be the computational costs, that is, the discovery of a pan-genome content is an NP-hard problem because comparisons between all sets of genes are necessary to solve the task. Furthermore, in an effort to compute standardized pan-genome analysis and minimize computational challenges, several online tools and software suites

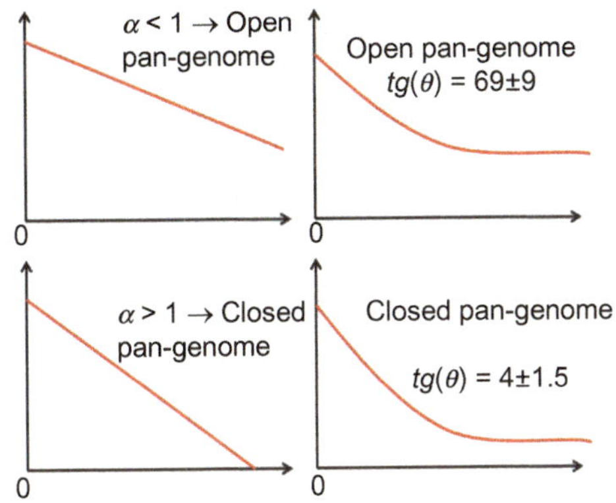

Fig. 3 The concept of open and close pan-genome.

have been developed. Examples of such applications are: PGAP [14], one of the most complete profile available for performing five analysis modules, but the runtime of the analysis grow approximately quadratically with the size of input data and are computationally infeasible with large datasets. The software Roary [15] and BPGA [16] was created to address the computational issues related to performance and execution time. Roary performs a rapid clustering of highly similar sequences, which can reduce the runtime of BLAST. BPGA is an ultrafast computational pipeline with seven functional modules for comprehensive pan-genome studies and downstream analyses. Pan-genome analysis can be applied in many different application domains, such as microbes, metagenomics, viruses, plants, cancer, and others [17]. Nowadays, the processes of similarity search and pan-genome visualization are two of the wide variety of particular computational challenges that need to be considered. For this, novel different computational methods and paradigms are needed over the years, making the computational pangenomics a subarea of research in rapid extension. Furthermore, new technologies that are emerging in rapid development allow to infer the pan-genome with three-dimensional conformation, which means that possibly in the future three-dimensional pan-genomes will not only represent all sequence variation of the species or genus, but will also encode their spatial organization, as well as their mutual relationships in this regard.

1.4 Applications of pan-genomics in evolutionary studies

The manifestation of rich genetic diversity in the form of a pan-genome in a species is an evolutionary puzzle. These three distinct parts of a pan-genome (core, shared, and singletons) of a particular species may undergo different evolutionary trajectories under the differential influence of evolutionary forces. An ideal pan-genome is expected to be very complete, comprehensive, efficient, and stable [18]. The pan-genome of a species has some evolutionary signatures in the form of gene content and single nucleotide polymorphism (SNP). These evolutionary signatures are useful in inferring the phylogenetic relationship among different strains of a species based on the pan-genome.

An evolutionary pan-genomic study of microbes provides a holistic picture of all the genomic variations of a species. These genomic variations endow the bacteria with their unique pathogenic properties and subsequent development of resistance to various antibiotics. Thus, a complete mechanistic detail of the processes involved in the pathogenesis and frequent antibiotic resistance in a bacterium will further pave the way for better detection methods and effective control strategies for the pathogen. In addition, evolutionary pan-genomics of a useful bacterium will help us in exploiting maximally the full potential of the microbe in enhancing industrial productivity. In fact, it will be a boom for the industries actively involved in the production of pharmaceuticals and dairy products using microbial cultures. Eukaryotes including crop plants and farm animals have abundant genomic variations in the form of SNP, copy number variants (CNVs), and presence/absence variants (PAVs). The discovery of SNPs associated with productivity or disease resistance in a crop or a farm animal will be much more efficient with the availability of a complete pan-genome of the species [19].

In a recent past, a work published by Benevides et al. [20] utilized 16S rRNA gene phylogeny, whole-genome multilocus sequence typing (wgMLST), phylogenomics, gene synteny, average nucleotide identity (ANI), and pan-genome to explain the phylogenetic relationships in a better way among strains of *Faecalibacterium*. For this, they used 12 newly sequenced, assembled, and curated genomes of *Faecalibacterium prausnitzii*, which were isolated from the feces of healthy volunteers from France and Australia, and combined these with five strains already published, which were downloaded from public databases. The phylogenetic analysis of the 16S rRNA along with the wgMLST profile and the phylogenetic tree based on the comparison of the similarity of genome supports the grouping of *Faecalibacterium* strains in different genospecies [20].

In another work published by Chen et al. [21], the comparison of whole genome and core genome multilocus sequence typing (MLST) and SNP analyses were carried out to show the maximum biased power achieved by using multiple analyses. It was required to differentiate isolates associated with outbreak from a pulsed-field gel electrophoresis (PFGE)-indistinguishable isolate collected in 2012 from a nonimplicated food source. Whole genome sequencing (WGS) has been proven as a powerful subtyping tool for bacteria like *L. monocytogenes*, a foodborne pathogen [21]. A company produced an environmental isolate that was highly similar to all outbreak isolates. The difference observed between unrelated isolates and outbreak isolates was only 7–14 SNPs; consequently, the minimum spanning tree from the analyses of whole genome, phylogenetic algorithm, and usual variant calling approach for core genome-based analyses could not offer the difference between unrelated isolates. This also suggested that the SNP/allele counts should always be pooled with WGS clustering analysis produced by phylogenetically meaningful algorithms on an adequate number of isolates, and the SNP/allele onset alone does not provide enough evidence to demarcate an outbreak [21]. Hence, it was proposed that the comparison of pan-genome subcategories and their related α value may be utilized as an alternate approach, along with ANI, in the in silico cataloging of new species [20, 22]. We hope that the ever-expanding pan-genome across different species and genera will give impetus to a better data structure of the pan-genome and novel computational methods for a robust evolutionary pan-genomic analysis in near future.

2 Applications of Pan-genomics in Bacteria

2.1 Applications of pan-genomics in model bacteria

Advancement in sequencing technologies and development in sophisticated bioinformatics tools created an overwhelming number of microbial genomic data and allowed the scientific community to estimate the pan-genome of a species. Identification of novel dispensable genes has applications in characterizing novel metabolic pathways, virulence determinants, and molecular fingerprinting targets for epidemiological studies and core genes can be used to predict the evolutionary history of the organism [9]. Therefore, pan-genome analyses are now considered the indispensable and gold standard for bacterial genome comparisons, evolution, and diversity. It is also useful to develop a vaccine against the pathogens of epidemic diseases by filtering different functional genes in the core genome using reverse vaccinology approaches [23].

There are a number of freely accessible tools, pipelines, and web-servers available to estimate the microbial pan-genome including Roary, BPGA, PGAP, PGAPx, Panseq, PanOCT, etc. [16]. A number of model bacterial species pan-genome is determined by researchers and a vast majority of those human pathogens exhibit an open pan-genome, as they colonize multiple environments that facilitate them to exchange genetic materials. These organisms include *Escherichia coli*, *Meningococci*, *Streptococci*, *Salmonellae*, *Helicobacter pylori*, etc. [24]. Therefore, in dealing with such species a reasonable number of genomes is usually required to define the complete gene repertoire of these species. On the other hand, species living in isolated (close) habitats having less possibility to exchange genetic material tend to have closed pan-genome, for example, *Mycobacterium tuberculosis*, *B. anthracis*, and *Chlamydia trachomatis* [25]. Hence, pan-genome analyses serve as a framework to determine and understand the genomic diversity in bacterial species. In Chapter 17, we have discussed the bacterial pan-genome analysis performed till date with specific examples from model organisms along with studying approaches, technical implementations, and their outcome.

2.2 Applications of pan-genomics in *Corynebacterium diphtheriae* and *Corynebacterium ulcerans*

The development of diphtheria toxoid vaccines in the 1920s, the start of mass immunization in the 1940s, and the global introduction of the Expanded Program on

Immunization (EPI) by the World Health Organization (WHO) in 1974 led to a dramatic decrease of diphtheria cases, both in industrialized and developing countries [26]. However, despite this tremendous success story, diphtheria has not been eradicated yet. This has been illustrated dramatically by a diphtheria pandemic connected to the breakdown of the former Union of Socialist Soviet Republics with more than 157,000 cases and more than 5000 deaths reported between 1990 and 1998. Even after the pandemic has finally stopped, local breakouts have been observed constantly during the last years and the reported global cases increased from about 7000 in 2016 to almost 9000 in 2017 with a focus on countries with limited or lacking public health systems, for example India, Indonesia, Nepal, Pakistan, Venezuela, and Yemen. Consequently, *Corynebacterium diphtheriae*, the etiological agent of respiratory and cutaneous diphtheria, is still present on the list of the most important global pathogens [27]. Furthermore, the frequency of human diphtheria-like infections associated with *Corynebacterium ulcerans* appears to be increasing [28]. This species, which was recognized before as a commensal of a large number of animal species, is closely related to *C. diphtheriae* and recognized as an emerging pathogen today [28, 29].

The need of fast and unequivocal identification of especially pathogenic *C. diphtheriae* led to the early development of a number of different methods such as biovar discrimination based on different biochemical reactions, Elek's test to immunologically distinguish between toxigenic and nontoxigenic strains, restriction fragment length polymorphism (RFLP), single-strand conformation polymorphism (SSCP), phage-typing, spoligotyping, ribotyping, MLST and others. This plethora of methods was significantly improved when next-generation sequencing was introduced. The first genome sequence of *C. diphtheriae* was published in 2003 and showed the presence of the *tox* gene on a bacteriophage in addition to a number of other horizontally acquired virulence-associated genes [30]. Subsequent pan-genome studies allowed unraveling the extent of genomic diversity within *C. diphtheriae* and the role of HGT as a source of variation between strains. Furthermore, pan-genomics of *C. ulcerans* helped to estimate the virulence potential of different strains and to verify zoonotic transmission from animals to patients. Today, pan-genomics of *C. diphtheriae* and *C. ulcerans* allow elucidating global transmission traits and local adaptations of pathogenic corynebacteria and, hopefully, a better understanding of population dynamics and strain evolution will help combat diphtheria and other *Corynebacterium*-associated diseases in future.

2.3 Applications of pan-genomics in multidrug-resistant human pathogenic bacteria and pan-resistome

The pan-genome will probably be the largest molecular evolutionary history of the organism ever written. This will integrate all the pan-phenotypes existing on Earth, such as the pan-proteome, the pan-transcriptome, and especially, a portion of pan-genome that has made the organisms successful on Earth: the pan-resistome. The pan-genome represents the set of all current genes in the genomes of a group of organisms. The basic genome common to all bacteria contains about 250 gene families in the extended core, the specific niche adaptive genome of about 8000 gene families in the character gene pool, and the pan-genomic diversity (accessory genes) of more than 139,000 rare gene families scattered throughout the bacterial genomes [31]. The pan-genome analysis, whereby the size of the gene repertoire accessible to any given species is characterized along with an estimate of the number of whole genome sequences required to proper analysis, and currently it is increasing 10 years after Tettelin et al. [6] publication. Different current models for the pan-genome analysis, accuracy, and applicability depend on the case at hand [32]. The NCBI, EMBL, KEEG, PATRIC, MBGD, ENSEMBL, and JGI-IMG/M databases provide complete downloadable genomics information, which can be analyzed for intraspecies diversity, and determine the pan-genome using software tools, currently developed to perform via a personal server [32], or even online resources. The pan-genomics is now a cutting edge of computational genomics field. Pan-genomics is a subarea of computational biology [17]. Therefore, the notion of computational pan-genomics intentionally passes through many other bioinformatics-related disciplines.

The resistome, a term coined by Wright [33], comprises all the genes and their products that contribute to resist whatever environment, substance, or some extreme grow factor. Updated data will close to the metadata available for establishing what part of resistome traits belong both to core-genome as accessory genome inside all bacterial species as well as will offer a broader perspective of bacterial antibiotic resistance. The WHO summarizes antimicrobial resistance (AMR) as the resistance of a microorganism to an antimicrobial drug that was originally effective for the treatment of infections caused by themselves. An adequate approach to solving major questions about the resistome inside of the bacterial genome [34] is to perform a pan-genomics analysis. The updated pan-genome data will be close to the metadata available for establishing the part of resistome traits that belong both to core-genome as accessory genome in bacterial species; as well as a broader perspective of antibiotic resistance in bacteria. The emergent antibiotic-resistant pathogenic bacteria are a current menacing concern. *Pseudomonas aeruginosa*, *Acinetobacter baumannii*, and coliform bacteria are the new emergent antibiotic-resistant bacteria according to the WHO. Pan-genomics has tackled some important concerns, which would be impossible to solve using classical molecular biology or descriptive genomics: it is very important to define the core and accessory genome for establishing the plasticity of resistome. Thousands of unknown bacteria and microorganisms are exposed to manufactured antibiotics, leading us to assume that there are no means to prevent this catastrophe. In opposition, pan-genomics is a powerful approach to prevent such disaster. We must move toward sequencing of known and unknown species, classify them, and establishing its antibiotic-resistant status, their pan-genome, and come out with new alternatives for reducing antibiotic consumption nowadays.

2.4 Applications of pan-genomics in veterinary pathogens

Following the development of NGS, the number of sequenced genomes filed exponentially [35]. Thus, projects aimed at studying groups of organisms became viable, and thus, several studies appeared that are called Omics studies. The studies involving pan-genomes are exposing important information on the differences and similarity between organisms of the same or between species. For concept purposes, we have the Pan-genome as a set of genes in a given group of individuals [10]. This information is being explored and applied by several scientific fronts, for example, in bacteria that infect animals and humans. The main applications of these studies are in the development of prophylactic and diagnostic methods in less time and with less cost, more precise taxonomic studies, studies on genetic variations, and pathogenesis [17]. In this chapter, we describe more recent research involving pan-genomics of the pathogenic bacteria that cause veterinary diseases, including some responsible for zoonoses, they are: *Corynebacterium pseudotuberculosis*; *Corynebacterium ulcerans*; *Streptococcus suis*; *Brachyspira hyodysenteriae*; *Moraxella bovoculi*; *Pasteurella multocida*; *Mannheimia haemolytica*; *Clostridium botulinum*; *Campylobacter*; *Streptococcus agalactiae*; *Francisella tularensis*; *Corynebacterium diphtheriae*; *Brucella* spp. Finally, it is worth highlighting that the influence of the approaches with big data and artificial intelligence are increasing and the influences of these in Pan-genomic studies will bring a new era of studies and discoveries.

2.5 Applications of pan-genomics in aquatic pathogenic bacteria

The sustainability of aquaculture industry is critical both for global food security and economic welfare. However, the massive wealth of pathogenic bacteria poses a key challenge to the development of a sustainable biocontrol method. Recent advances in genome sequencing study combined with pan-genome analysis can be an efficacious management applied to numerous aquatic pathogens [36]. Thus, routine pan genome analyses of genomic-derived aquatic pathogens will deduce the phylogenomic diversity and possible evolutionary trends of aquatic bacterial pathogen strains, elucidate the mechanisms of pathogenesis, as well as estimate patterns of pathogen transmission across epidemiological scales. The whole genome sequencing data is the opportunity to revolutionize the molecular epidemiology of aquaculture pathogens as it has for those pathogens of relevance to public health [37]. Challenges of aquaculture disease management are the biological diversity of pathogens, host-pathogen interactions (e.g., different modes of adaptation and transmission), and shifting environmental pressures, in particular climate change. Hence, analysis of pathogenic phenotype combined with genotype derived from the full potential of genome sequencing data is critical to reconstruct pathogen transmission routes on local and global scales, as well as mitigate disease emergence and spread.

Comparative pan-genome analyses are an effective tool which could possibly be extended to the analysis of aquatic microorganisms and to dynamic characteristics and adaptation to a broad range of their hosts and environmental niches. Conspicuously, our previous pan-genome analysis [38] showed that strain WFLU12 isolated from marine fish exhibited niche-specific characteristics of energy production and conversion, and carbohydrate transport and metabolism by exploring genes in the gene repertoire of strains. Based on the pan-genome categories, the functional annotations of selected genes can be reanalyzed with the Virulence Factors Database (VFDB), Clusters of Orthologous Groups (COG), Kyoto Encyclopedia of Genes and Genomes (KEGG), and Antibiotic Resistance Genes Database (ARDB). Also, comparative pan-genome has advanced to the point when genes are predicted as belonging to cell surface-exposed proteins (SEPs) from important pathogens, including outer membrane proteins, and extracellular proteins. These predicted genes are serving as vaccine candidates in an animal model called Reversed Vaccinology (RV) [39]. In aquaculture, SEPs from pathogens include several important virulence factors that play key roles in bacterial pathogenesis and host immune responses. For example, the expression of *esa1* from *Edwardsiella tarda*, a D15-like surface antigen, in the Japanese flounder model induced the expression of a broad spectrum of genes possibly involved in both innate and adaptive immunity, as well as a high level of fish survival and produced specific serum antibodies [40]. Vaccination using SEPs results in the development of protective effects against *Aeromonas hydrophila* infection, *Flavobacterium columnare* infection, *Pseudomonas putida* infection, and Edwardsiellosis [as in the review of Abdelgayed [41]]. A recent study [42] has successfully implemented a pan-genome analysis to screen SEPs from 17 representative *Leptospira interrogans* strains covering multiepidemic serovars from around the world, and 118 new candidate antigens were identified in addition to several known outer membrane proteins and lipoproteins. We highly consider that the rapid increase in the number of genome sequencing of aquatic pathogens will allow us to develop a rapid-response infection control protocols, but also be a potential trend for studying aquatic pathogenic bacteria to improve the cross-serotype efficacy of vaccines in farmed fish and stem the disease outbreak when implementing pan-genome analysis (using RV strategy). In the chapter "Pan-genomics of aquatic animal pathogens and its applications," we reviewed comparative pan-genome analysis with a particular focus on controlling aquatic diseases and give real-world examples by analyzing genome sequencing data derived from aquatic bacterial isolates.

2.6 Pan-genomics applications for therapeutics

The emergence of bacterial resistance is occurring, threatening the ability of antibiotics that have transformed medicine and saved millions of lives around the globe [43, 44]. The occurrence of bacterial resistance has been identified since the beginning of the antibiotic era but the emergence of most dangerous and easily communicated strains has been reported in past two decades [45, 46]. After several years of the first patient treated with antibiotics, bacterial infections became a threat for society once again. This situation is mainly because of the misuse and/or overuse of antibiotics as well as the inefficiency of pharmaceutical companies for not producing advanced drugs, once economic investments have been reduced [44]. The Centers for Disease Control and Prevention (CDC) has categorized several bacterial strains as an alarming threat that need serious consideration for proper treatment and are already responsible for putting significant burden on the health-care system in the United States (US), ultimately, affecting patients and their families [43, 47, 48]. The infections caused by antibiotic-resistant strains of bacteria are pervasive worldwide [43, 44]. A national survey of infectious-disease specialists led by the IDSA Emerging Infections Network in 2011 found that about two-third (2/3) of the participants had seen a pan-resistant and deadly bacterial infection within the past few years [49]. The rapid emergence of resistant bacteria has been described as a nightmare by several public health organizations that could have disastrous results [50]. The WHO cautioned in 2014 that the disaster of antibiotic resistance is becoming dreadful [51]. Among Gram-positive pathogens, a universal endemic of resistant *S. aureus* and *Enterococcus* species are presently the biggest intimidation [48]. Vancomycin-resistant enterococci (VRE) and additional emergent pathogens are evolving resistance to numerous antibiotics used commonly [43]. The worldwide distribution of common respiratory pathogens includes *Streptococcus pneumoniae* and *Mycobacterium tuberculosis*, which are reported as epidemic [48]. Gram-negative pathogens are in general more troublesome because of the fact that they are becoming more resistant to almost all the available therapeutics, making the conditions evocative to the preantibiotic era [44]. The occurrence of multidrug resistant (MDR) Gram-negative bacilli has outdated all the practice in field of medicine [43]. The most common infections caused by Gram-negative bacteria in health-care settings are usually by *Enterobacteriaceae* (mostly *Klebsiella pneumoniae*), *Acinetobacter*, and *Pseudomonas aeruginosa* [43, 44]. The evolution of bacterial strains and development of antibiotic-resistant genes through HGT make it necessary to look for novel and advanced strategies to cope with the infections [52].

The in silico approaches like pan-genome, pan-modelome, subtractive genomics, and reverse vaccinology are playing vital roles in rapid identification of new therapeutic targets in the postgenomic era [53–55]. Comparative microbial genomics approach along with statistical analysis are useful tools for the identification of essential genetic contents commonly present in all pathogenic isolates, based on sequence similarity. In addition to essential genetic contents, it also helps to identify subset of genes encoding virulence and novel functions as the variable genome [56]. A pan-genome is usually divided into three parts, that is, core genes, accessory genes, and strain-specific genes. In the drug and vaccine discovery process, the very first step is always the identification of a suitable target. Subtractive genomics is a widely used process in this regard. In recent past, working with pathogenic bacteria, using computational approaches, a large number of novel therapeutic targets has been identified, which are either resistant to drugs or no appropriate vaccine is available for these targets [54, 57]. The most popular approach for rapid identification of novel vaccine targets in postgenomic era is reverse vaccinology [54]. Strategies such as comparative genomics, subtractive genomics, and differential genome analyses are being broadly utilized for the identification of targets in several human and animal pathogens (Table 1), that includes *Mycobacterium tuberculosis* [62], *Treponema pallidum* [54], *Corynebacterium diphtheriae* [53, 64], *Hemophilus ducreyi* [52], *Neisseria gonorrhoeae* [59], and *Salmonella typhi* [63]. The basic principle of these approaches is the identification of genes/proteins that are not homologous to gene/protein of the host but are essential for the survival of the pathogen. However, the identified targets might be slightly homologous to host gene/protein but still can be selected for structure-based selective inhibitor development as a supplementary molecular target [54, 64–66].

2.7 Pan-genomics applications for probiotics

The term probiotic has become highlighted in the last few years, but few know that its use is already registered as fermented foods in books such as: the Holy Bible and sacred books of Hinduism [67, 68]. Probiotics are live microorganisms that may provide health to the host [69].

Table 1 Pan-genome studies in bacterial pathogens

Name	Strain/no of strains	No of genes/ proteins	Host	Therapeutic drug/ vaccine targets	References
Treponema pallidum	13	837	Human	15 vaccine/6 drug	[54]
Haemophilus ducreyi	28	1257	Human	13 vaccine/3 drug	[52]
Chlamydia trachomatis	NC_010287.1	934	Human	63 drug	[58]
Neisseria gonorrhoeae	FA 1090		Human	67 drug	[59]
Ureaplasma urealyticum	ATCC 33699	646	Human	2 drug	[60]
Corynebacterium diphtheriae	13	Not mentioned	Animal	8 drug	[53]
Helicobacter pylori	39	59,958	Human	28 vaccine	[61]
Mycobacterium tuberculosis	H37Rv genome	3989	Human	135 drug	[62]
Salmonella typhi		4718	Human	149	[63]

Its importance gained pace in the medical and biotechnological fields with the results found not only related with inflammatory bowel diseases (IBDs) [70, 71], but also with diabetes [72], multiple sclerosis [73], dermatitis [74], and in the production of heterologous proteins [75]. Many species play a role as probiotic and much more are in the process of testing (Table 2).

Table 2 Probiotics and their effects

Name	Strain	Status	Effect	References
Acinetobacter sp.	BR-12	R	Plant phosphate supply	[76]
Acinetobacter sp.	BR-12	R	Plant phosphate supply	[77]
Acinetobacter sp.	WR922	R	Plant growth	[78]
Bacillus amyloliquefaciens	G1	R	Bacterial infections in animals	[79]
Bacillus amyloliquefaciens	SC06	R	Bacterial infections in animals	[80]
Bacillus clausii	UBBC 07	C	Acute diarrhea	[81]
Bacillus coagulans	–	M	Irritable bowel syndrome (IBS)	[82]
Bacillus coagulans	–	C	Antibiotic-induced diarrhea	[83]
Bacillus licheniformis	2336	M	Acute enteric infections	[84]
Bacillus licheniformis	26L-10/3RA	M	Bacterial infections in animals	[85]
Bacillus licheniformis	8-37-0-1	M	Maintenance of aquatic conditions for animals; Heavy metal accumulation	[86]
Bacillus subtilis	E20	M	Immuno-protection for animals	[87]
Bacteroides fragilis	–	R	Autism spectrum disorders (ASD)	[88]
Bifidobacterium animalis subsp. *lactis*	BB-12	M	Reduces the risk of infections in early childhood	[89]
Bifidobacterium animalis subsp. *lactis*	Bb-12	M	*H. pylori* related	[90]
Bifidobacterium animalis subsp. *lactis*	Bb-12	C	Atopic dermatitis	[91]
Enterococcus faecalis (*Streptococcus faecalis*)	SL-5	C	Acne vulgaris	[92]
Enterococcus faecium (*Streptococcus faecium*)	CTC492	R	Antilisteral effect	[93]

Table 2 Probiotics and their effects—cont'd

Name	Strain	Status	Effect	References
Escherichia coli	M-17	R	Pouchitis	[94]
Escherichia coli	Nissle 1917	C	Ulcerative colitis; Crohn's disease; Inflammatory bowel disease (IBD)	[95–97]
Lactobacillus acidophilus	L-92	C	Atopic dermatitis	[98]
Lactobacillus acidophilus	LA-02 (DSM 21717)	C	Vulvovaginal candidiasis	[99]
Lactobacillus brevis	D7	M	Antioxidation process in animals	[100, 101]
Lactobacillus buchneri	P2	R	Cholesterol removal	[102]
Lactobacillus casei	DN-114001	C	Immune modulation	[103]
Lactobacillus casei	F-19	M	Food digestion	[104]
Lactobacillus crispatus	CTV-05	C	Urinary tract infection	[105]
Lactobacillus delbrueckii subsp. *bulgaricus*	OLL1073R-1	C	Reduces the risk of infection in the elderly	[106]
Lactobacillus rhamnosus	CGMCC 1.3724	C	Obesity	[107]
Lactobacillus rhamnosus	JCM1136	M	Immuno-protection for animals	[108]
Lactococcus lactis subsp. *cremoris*	IBB SC1	R	Immunomodulation	[109]
Oxalobacter formigenes	OxCC13	R	Calcium oxalate stone disease	[110]
Propionibacterium freudenreichii subsp. *shermanii*	–	C	Liver cancer	[111]
Streptococcus salivarius	K12	R	Halitosis	[112]
Weissella koreensis	OK1–6	R	Antiobesity	[113, 114]

R = research; C = Clinical trial; M = Marketed.

The Omics studies allowed an advance in the elucidation and characterization of the properties of these organisms, opening a vast field of application, besides providing new ways to access the information about their genomes. Following the pan-genomic approach, the pan-probiosis analysis consists in comparison of two or more strains, aiming to identify some points in the organism genome that differs or presents similarities related with probiotic characteristics, such as genes coding for adhesion.

In comparative genomics, for example, it is possible to retrieve a high number of genome information in silico—an attractive and cheap way [115]. There are some requirements that are important for an organism to be considered as probiotic which is determined through some mechanisms of action, like surviving to gastric acidity and bile salts [116], competing with other organisms via exclusion mechanisms and antimicrobial activity [117], and modulating the immune system [118], and these features may be used to gather the genome information in silico.

A comparative analysis with *L. lactis* subsp. *lactis* NCDO 2118 was performed aiming to find the potential probiotic characteristics of this strain. The authors found, through comparative genomics, phage regions, GEIs (metabolic and symbiotic), bacteriocins of three different classes, bile salts, and acid stress resistance genes found in other *L. lactis*, adhesion-related, and antibiotic-resistant genes. Besides that, comparing in vitro data of the aforementioned strain with another species, already described as nonprobiotic, they could identify genes encoding proteins (secreted and expressed) that are exclusive of NCDO 2118 [119].

Using a pan-genome microarray with probiotic *E. coli* isolates, Willenbrock and coauthors could characterize the pan-genome of 32 species based in two-control strain: *E. coli* K-12 and O157:H7. Despite they observed different sizes of genomes within the species, they believe they achieved the expected results, one of them being the characterization of the core genome with around 1560 essential genes [120].

Pan-genome approach was also used to discover probiotic characteristics of *L. lactis* WFLU12 [38] that showed resistance against streptococcal infection and improved the growth in olive flounder [121]. They identified some data that supported their previous work, like the identification of bacteriocins and genes involved in stress response. Comparing WFLU12 with other *L. lactis*, there are genes and gene clusters for specific niches based on carbohydrate metabolism, defense mechanisms, and envelope biogenesis [38].

Following the idea about niche-specific, Kant and coauthors worked with 13 *Lactobacillus rhamnosus* from different origins with the pan-genomic analysis. They used *L. rhamnosus* GG as reference, focusing in SEPs that may play a role in niche adaptability. The interesting thing was, they could find uncommon information in lactic acid bacteria, a *spaCBA* operon. This operon may be related with the origin of these strains, maybe of a similar microhabitat, for example [122].

Another species used as probiotic was analyzed via pan-genomics in the study by Smokvina and coauthors, in which 34 different *Lactobacillus paracasei* strains were studied using comparative genomics and pan-genomics. They identified 1800 orthologous groups representing the core genome and these genes were related with cell envelope, pili, hydrolases, or the production of branched short-chain fatty acid (SCFAs). About this, they found genes that encode these SCFAs: *bdkABCD*, only found in *Lactobacillus* until this date [123].

Nowadays, we have a lot of information about potential probiotic organisms, beyond those whose are commonly known in the market, but there is no database concentrating all the information about them, like genes related with bile juice and gastric acid resistance, genes coding adhesion, or secret proteins. A database with those information about known probiotic organism could help in future analysis be them in silico, in vivo, and in vitro. Finally, the comparative and pan-genomic analyses have an important role in the most diverse organism analyses and in the case of probiotic ones, it could be very helpful and elucidating in the precision to characterize new potential probiotics. The diversity inside the genomes may be observed and with this information it is possible to have a better idea of how many genomes will be necessary to characterize fully the organisms in these studies.

3 Pan-genomics of virus and its applications

Advances in DNA sequencing technology have ushered in a new era of pan-genomics and genomic surveillance, in which traditional molecular diagnostics and genotyping methods are being enhanced and even replaced by genomic-based methods to aid epidemiologic investigations of communicable diseases [124]. The ability to compare and analyze entire pathogen's genomes has allowed unprecedented resolution into how and why infectious diseases spread. The rapid development of sequencing technologies has made sequencing routine of viral genomes possible [125]. As these genomic-based methods continue to improve regarding speed, costs, and accuracy, they will increasingly be used to inform and guide infection control and public health practices [125a].

There are currently two major ways in which high-throughput sequencing technologies are used in public health and diagnostic applications: (i) to track outbreaks and epidemics in order to call public health responses and (ii) to characterize individual infections to tailor treatment decisions [126, 127].

Focusing on these aims, genome sequencing has been successfully used to describe unique and detailed insights into the transmission, biology, and epidemiology of many health-care-associated viral pathogens. Considering the improvements on portability and quality of sequencing, and the acceleration and standardization of analytical pipelines, the applicable routine of genome sequencing may soon become the common de facto method for infectious diseases control. Using genomic analysis tools to complement existing genotyping and epidemiologic methods, the future of infection control and prevention will lead to more targeted and successful interventions for outbreaks, which will ultimately result in the reduction of infectious diseases impact.

Next-generation sequencing techniques have transformed genomic studies from the analysis of single or few genomes to an ever-increasing amount of genomic data, bringing with it the need to develop novel techniques to treat efficiently, novel tools to assemble, analyze, and derive useful information from overwhelmingly large datasets. The analysis of pan-genomes can uncover significant information regarding the genomes of interest. According to Guimaraes et al. [128], pan-genomic studies can help understand pathogen evolution, niche adaptation, population structure, and host interaction. Furthermore, it can help in vaccine and drug design, as well as in the identification of virulence genes.

In the context of virus investigations, pan-genomics and bioinformatics in general face great challenges. Rapid extraction of genomic features with an evolutionary signal will facilitate evolutionary analyses ranging from the reconstruction of species phylogenies to tracing epidemic outbreaks. Improvements on genome assembly using machine learning techniques are proposed by Padovani De Souza et al. [129]. Finally, in order to better use all the information acquired by high-throughput real-time sequencing and its analysis, text mining and knowledge discovery techniques, integrated with medical and scientific literature and gene family and metabolic pathway databases, could help generate new insights and speed up discoveries. High-throughput real-time next-generation sequencing projects have transformed the field of bioinformatics from single-genome studies to pan-genome analyses. The limiting factor now is no longer data rarity, but immense data availability and dimensionality. In this new context, bottom-up analyses stemming from big data provide great challenges and also great rewards.

4 Pan-genomics of plants and its applications

The plants genomes are highly dynamic as compared to many higher eukaryotes due to the presence of transposable elements and frequent genome duplication events [130]. Thus, the identification of such structural variations and dynamics in plant genomes is a prerequisite for subsequent understanding and their applications based on the sequence-trait associations. Several plant genomes were sequenced during the sequencing initiative in 2000 allowing an assembly of their reference genomes [131]. These reference genomes were mainly used to compare genomes of different plant species and to identify the SNPs across populations [132]. These studies increased our understanding regarding the allelic variations associated with phenotypic outcomes in general. However, such studies were not able to capture fully the diversity of sequence variations in plant genomes being themselves dependent on large genetic variations within strains/species. To this end, the advent of high throughput sequencing has played a major role in examining the genetic variations including SNPs, CNV, and presence/absence variations (PAV) comprehensively. The reduced costs of high-throughput sequencing methods have now revolutionized the ways being used for the analyses of plant genomes previously and for asking relevant biological questions. It has made it possible to easily sequence and compare the whole genomes of many individuals of same plants species and thus capturing the interspecies genetic diversity. Accordingly, the full genome content capturing the interspecies genomic diversity is termed as pan-genome [133]. The pan-genome approach allows to predict the number of additional genome sequences that are necessary to characterize fully the genomic diversity of a species [133]. Analyses of pangenomes of several plants have now revealed the role of structural variations in different plant phenotypes such as flowering times, different stress-resistant mechanisms, etc. [134]. These studies have enhanced our understanding of the diverse applications of these genotypic to phenotypic association such as for increasing the crop production of better varieties in terms of size and flavors, increasing the abiotic stress and pathogens/disease resistances among many others reviewed in this chapter. The pan-genome approach is especially suitable for plant-breeding applications in contrast to the single liner reference genomes because of reduced sampling biases along with the comprehensive representation of genetic diversity [133]. The field of pan-genomics is rapidly evolving based on the underlying sequencing paradigms and the analytical pipelines, tools, and algorithms for sequencing data. The current pangenome assembly approaches can be categorized into a k-mer-based approach, comparative de-novo assembly approach, and iterative assembly approach. One of the challenges associated with the analysis of pan-genome data is related to requiring the increase in precision of the underlying genome assembly approaches. This review chapter aims to describe comprehensively the structural variations in plants genomes, explain the concept of pangenome, and its characterization along with the applications, methods, and approaches to conduct pan-genome analyses for a wide range of plant species.

4.1 Applications of pan-genomics in plant pathogens

The knowledge of plant diseases and host-pathogen interactions is one of the fundamental and active areas of genetic research with a wide array of applications [135]. Previously, linear reference genomes have been widely used for the subsequent analyses of phylogenetic relationships, identification of casual agents, virulence factors, host specificity associations, and pathogenic mechanisms [136]. These studies aided better disease management for economically important crops and plants by counteracting the stress-based resistance factors and better vaccine development. However, there is increasing evidence that the single reference genomes are insufficient in capturing the entire genetic diversity of the strains and subsequent delineation of principles governing the adaptive success of plant pathogens along with the determination of pathogenicity factors [137]. Accordingly, the concept of pan-genome emerged to cater to the interstrain genetic diversity based on different structural variations including CNV, presence/absence variations (PAV), and other allelic transformations. Pan-genome approach is now emerging as an analytical approach for analyzing the genetic diversity of genomes at an unprecedented level of details in contrast to the single reference genome. The strain-specific genome content is especially beneficial for gaining insights into the pathogenic mechanisms of plant pathogens as most of the pathogenic determinants are often strain specific and highly variable. Moreover, the pan-genome analysis allows determining the genome plasticity through studying the evolutionary impact of HGT. As of yet, pan-genome analyses have already been used to identify and detect new strains along with development of vaccines against many plant pathogens [138]. Several computational pipelines based on tools and software especially designed to conduct a pan-genome analysis are available now. These tools can perform several functions including homologous gene clustering, SNPs identification, pan-genomic profiles visualization, phylogenetic analysis based on orthologous genes or gene families based information, pan-genome visualization, curation, and function-based searching. Most of the established pan-genome analysis methods were initially developed to deal with smaller prokaryotic genomes and thus are beneficial in analyzing most of the plant pathogens including bacteria and fungi. However, there are still certain challenges in assembling and analyzing the pangenomes of the species with complex genome structures [32]. Despite this, the pan-genome analyses is emerging as an important research tool to enhance our understanding about host-pathogen interactions and to develop universal vaccines. Since this approach has a potential for organizing pathogenic diversity, integrating pan-genomics with phylogeny and phylogenomics will be an interesting viewpoint for the future. Overall, we have comprehensively reviewed the studies conducted to assemble the pan-genomes of plant pathogens, its applications, available methods, and tools to conduct a pan-genome analysis in our chapter.

5 Genomics of algae and its applications

Genome sequencing unveils the basis of various fundamental processes and origin as well as the evolution of the organism. Advancement in whole-genome sequencing in the field of algal biomass has answered our queries of ecological and economic importance extending from the adaptation of organisms in diverse environments to synthesizing abundant metabolites of vast economical future. WGS of diverse algal genome has been performed using sequencing approaches ranging from shortgun to high throughput. Shortgun approach includes cloning 1–10 kb g-DNA fragments into pUC18 or pBluescript II KS (Stratagene). Plasmids have been sequenced using PE BigDye Terminator/ ET DYEnamic terminator kit. Sequences have been resolved using PE 377 Automated DNA Sequencers and assembled from end sequences using PHRAP (P. Green) and Consed. Primer walking has been used for gap filling. Glimmer, GeneMarks, and Critica have been used to identify ORFs in the genome. High-throughput sequencing technologies include Illumina HiSeq 2000 technology, Illumina GA II x and Solexa Genome Analyzer (Illumina) and paired reads have been assembled using a DeBruijn method or CLC Genomics Workbench tools.

This development has also initiated metagenomics and metatranscriptomics, maneuvering the expression analysis and functional assays to study intraspecies and interspecies variability among nonmodel and complex biological communities of worth.

Comparative genomics is another approach to identify the essential mechanisms of origin and evolution. Genome analysis showed that a cyanobacterium *Synechococcus* sp. strain WH8102 is nutritionally more adaptable as it has acquired more sodium-dependent transporters for the uptake of organic nitrogen and phosphorus. Reduced gene complement in marine cyanobacterium *P. marinus* SS120 is consistent with the fact that the oligotrophic marine environment where it preferentially thrives is much more stable than freshwaters [139]. There are also examples from other algal genome analysis that unveiled the adaptation strategies to thrive under harsh conditions such as *Ostreococcus tauri* that has adapted costly C4 photosynthetic pathway to acquire critical ecological advantage in the CO_2-limiting conditions of phytoplankton blooms, green alga *Chloroidium* sp. UTEX 3007 is able to survive high temperatures in deserts by accumulation of thermostable palmitic acid [140]. Also, an acidophilic green alga *Chlamydomonas eustigma* NIES-2499 has acquired phytochelatin synthase genes providing it tolerance to toxic metal ions such as cadmium [141]. *Galdieria sulphuraria* and *C. merolae* belong to the Cyanidiophyceae group but at the same time possess many contrasting features. The foremost is the ability of *G. sulphuraria* to adapt to extreme acidic thermophilic environments. It is the only alga in this group with an adaptation of the heterotrophic mode of nutrition with multiple substrates, which indicates how it survives in harsh environments [142]. In the process of evolution of ancestral lineages of red algae, the role of HGT is undeniable. This was indicated in the genome of other red algae, *Porphyridium purpureum*. Along with that, several light-harvesting complexes (LHC) were identified. Genomic analysis revealed evidence for sexual reproduction [143]. To cope with ecological stress, the genome of *P. umbilicalis* reveals the presence of genes coding for high-affinity iron transport complex necessary for the iron uptake processes to obtain nutrients during stressful high tides [144]. The study of gene sequences has also thrown light on the conservation of certain key enzymes such as GDP-mannose 6-dehydrogenase (GMD) required in the process of synthesis of alginates in brown algae *Cladosiphon okamuranus*. Also, *C. okamuranus* holds significant commercial importance as it is cultivated for fucoidan, which is a sulfated polysaccharide, a kind of Japanese seaweed [145]. The information on genomics has opened doors to various other research fields like proteomics, expression analysis, structural biology, metabolomics, etc.

6 Pan-metagenomics and human microbiome

Pan-metagenome is the collective study of all or several metagenomes from all possible units belonging to a particular type of ecosystem or host.

In the past decade, most of the metagenomic studies have aimed at understanding the microbial community from a relatively small set of samples. Such studies could miss out important rare taxa. However, the reduction in cost of gigabyte of NGS data has made the NGS application affordable and widespread [146]. This has given rise to an enormous amount of publicly accessible data from various types of samples. The application of pan-metagenome ranges from the mosquito gut microbiome [147] to human gut microbiome [148], including various ecosystems [149, 150]. Pan-metagenome primarily aims to explore and redefine the microbial community at a global scale. This will help to capture all the taxonomical variations between samples and understand the shifts in microbial community on a larger scale.

A pan-metagenome comprising thousands of samples pertaining to an ecosystem or host from multiple locations and studies at global level collaborations could be used as a standard reference. Such a reference-based pan-metagenome could serve as a guideline to answer several questions: What types of ecosystems are most vulnerable to global warming? Are rare taxa distributed based on geography?

7 Pan-proteomics and its applications

In the proteomic approach it is possible to identify and quantify a set of proteins synthetized by a determined cell, tissue, or microorganism [151] when exposed to different experimental conditions (such as temperature, osmolarity, antibiotics, nitric oxide, and others), or different steps of the cell growth, or during infection process [151–153]. At a specific condition, the identified proteins from the complex protein mixtures may be characterized in relation to their expression, cellular localization, structure, biological functions, and interactions with other proteins, posttranslation modifications, and metabolic pathways. In this way, proteomic studies contribute to understanding about cellular adaptation in response to external changes, metabolic stresses, or infection, and this response can vary according to time and environment [154], The proteomic analysis have been considered the most relevant approach to describe a biological system [151].

Proteomic approach in eukaryotic cells is relatively complex due to posttranslational modification, like phosphorylation of proteins, which is involved in protein signaling in different cellular pathways [155]. In humans, datasets from proteome studies have allowed to evaluate the potential methods in diagnosis, prognosis, and treatment for some diseases, including cancer [156]. On the other hand, in prokaryotes the proteomic assays have enabled the investigation of physiological behaviors, mutations, adaptability to different environmental conditions, presence of proteins involved in virulence, and the identification of putative immunogenic proteins [157].

The protein synthesis in eukaryotic and prokaryotic calls can be evaluated by different technologies, such as chromatography-based methods, enzyme-linked immunosorbent assay (ELISA), Western blotting, protein separation using gel-based approaches, especially two-dimensional (2D) polyacrylamide gel electrophoresis, or through the identification and sequencing of polypeptides through mass spectrometry technologies [151]. In chromatography-based techniques, the proteins can be obtained from separation based on their charge nature and charge strength (ion exchange chromatography), molecular size (size exclusion chromatography), or specificity (affinity chromatography) [158]. On the other hand, ELISA uses antibodies or antigens on the solid surface to detect specific peptides or enzymes from the biological sample, forming enzyme-conjugated antibodies which allow to measure the enzyme activity or protein concentration [159]. Last, Western blotting enables the identification of low abundance proteins after electrophoresis separation, transfer onto nitrocellulose membrane, and detection by enzyme-conjugated antibodies [160]. Nevertheless, these three methodologies allow to evaluate few proteins, and they are unable to determine protein expression level [151]. 2D gel electrophoresis is an efficient and widely used technique in proteomic studies to analyze complex protein mixtures extracted especially from bacterial cells. This methodology involves separation of proteins by isoelectric focusing (proteins with different isoelectric points) and by molecular weight (in polyacrylamide gel electrophoresis). Each spot in a 2D matrix corresponds to a single protein in the sample evaluated. In this way, 2D gel electrophoresis allows to obtain information of several proteins simultaneously as apparent molecular weight, isoelectric point, and quantity of each one [161]. And, mass spectrometry can be defined as the study of matter through the formation of ions in the gas phase and their characterization by mass, charge, structure, or physicochemical properties, using mass spectrometer that measures m/z values and abundance of ions [162].

The association between 2D gel electrophoresis and mass spectrometry was already considered the most appropriate method to recognize and identify proteins from pathogenic microorganisms [163] for being a methodology used for the construction of proteomic databases, due to its greater efficiency and high resolution to investigate the complex mixtures of proteins present in cell or tissues [164]. Nevertheless, with the technical advances achieved in recent years, such as solubilization of complex samples, pH gradient, and detection of proteins present in small quantities, the technique of liquid chromatography associated with mass spectrometry (LC-MS) started to be used and allowed the analysis of complex mixtures of proteins by tryptic digestion without prior gel separation [165]. This technique had the advantage of having a low detection limit for peptides and proteins, capability to identify hundreds to thousands of proteins in a simple experiment as well as allowing the study of membrane proteins, poorly accessible by other methods [166].

LC-MS is divided into two approaches: stable isotopic labeling [167] and label-free quantification [168]. In the first, two solutions containing the proteins to be analyzed are labeled with different molecular mass isotopes, and are mixed, trypsin-digested to obtain peptides and submitted to the LC-MS system [169]. The molecular weight difference allows the identification and quantification of peptides of both samples tested [170], but the labeling occurs after the extraction step, which can lead to a reduction in the precision of the quantification method [171]. Alternatively, label-free quantification allows the evaluation of numerous samples at the same time within the LC-MS system, with data-independent acquisition, and the concentration of a given peptide is proportional to its chromatographic area [172].

Among the strategies used in proteomic studies in prokaryotic cells surfome and secretome analyses stand out. The bacterial surface has been considered of great importance for understanding the pathogenesis of an infectious disease. On the surface, it can be found that proteins are associated with mechanisms of defense and virulence factors, which can promote adhesion and cellular invasion, culminating consequently in the appearance of clinical signs in an infected host [173]. Therefore, surfome is a proteome-based method, in which allows the identification of bacterial surface proteins [174]. Apart from surface proteins, extracellular and secreted proteins are important in bacterial pathogenesis, since they also mediate the interaction of the bacterium with the host and by stimulating the immune response. Therefore, the secretome has been associated with adhesion, invasion, immune evasion, and spread of bacterium in host tissues. In addition, these proteins can also be used for the development of antibiotics and vaccines [175]. Besides these two methods, comparative proteome analysis has been used for both prokaryotic and eukaryotic cells. This method has also been used to identify virulence factors and to obtain information on physiological and environmental adaptations in different pathogens [176], as well as to compare cells, tissues, and organs from the eukaryotic host in normal and pathological (inflammation, infection, and cancer) conditions [156].

In this context, pan-proteomics is also an approach with characterizes and compares the qualitative and quantitative proteome; however, the comparison occurs across organisms inside a species, with genetic variation and phenotype [177]. Pan-proteomics can be performed using 2D gel electrophoresis or LC-MS; nevertheless, LC-MS by bottom-up/shotgun techniques, from our expertise, is recommended for this type of study, otherwise, we will always have only part of the proteome and not the whole proteome.

Conceptually, pan-proteome refers to the proteins identified from a whole set of samples/strains tested, which are usually more than two samples, under the same experimental conditions. The analysis of two samples is equivalent to comparative proteomic methodology. Pan-proteome can be divided into core proteome, accessory proteome, and orphan (or unique) proteome [177]. The core proteome represents the subset of identified proteins simultaneously in all samples, whereas accessory proteome represents the detected proteins shared by at least two samples, and orphan proteome represents proteins identified exclusively in a single sample.

In the microbiological field, the genetic variation among isolates has been implicated with virulence factors, drug resistance, and environmental adaptation [178]. In this way, understanding about these mechanisms needs the evaluation of several proteomes and not from single proteome analysis [177]. Thus, pan-proteomic analysis may increase knowledge about the adaptation and pathogenicity of a given microorganism, independent of the genotype. Besides that, this approach can be used to classify bacterial strains in types [179], identify putative vaccine targets from conserved proteins among isolates [178], as well as, to determine drug targets and drug mode of action in analysis with multiple strains [177].

The term pan-proteome and core proteome have been used in different studies of protein identification and quantitation. In this type of study, pan-proteome and core proteome were referenced in the first time from analysis of four epidemic *Salmonella* Paratyphi A strains, with different PFGE types, using 2D gel electrophoresis [180]. From this analysis, the authors verified a high covered (over than 81%) of core proteome among the isolates tested, regardless of the range of pH applied, suggesting a high similarity in protein expression. Proteins involved in metabolic pathways and survival of the bacterium were the most identified within the core proteome. Moreover, the proteome comparison among isolates suggest a geospatial and temporal differentiation of expressed protein profile (spots).

The conserved core proteome was also observed in other works, where this category represented approximately 92% of pan-proteome of five fish-adapted *Streptococcus agalactiae* strains, which belonged to three MLST profiles. This study was performed using a label-free proteomic analysis [178]. The authors suggest that the identified proteins reflect an adaptation to an aquatic environment and fish-pathogen interaction. In addition, in the same study, conserved antigenic proteins were identified and suggested as targets in vaccine design, seeing that the high degree of conservation of these proteins among the isolates would suggest the production of a monovalent vaccine effective against all genetic variants tested.

Another study, despite the conservation of proteins identified simultaneously in avirulent, virulent, and two clinical strains of *Mycobacterium tuberculosis*, the quantitative protein expression profiling revealed a strain-specific variation in proteome patterns of isolates [181]. This study was also performed using label-free analysis, being identified 257 differentially expressed proteins. The differences in virulence among four isolates were suggested to a two-component system, oxidative stress, ribosome biogenesis, energy generation, and transcriptional regulator proteins.

The pan-proteomic analysis of four biotechnological *Lactococcus lactis* strains was performed using label-free analysis and showed a conservation of 52% of core proteome. The identified proteins contribute to physiological adaptation of bacteria, metabolic pathways, microbial metabolism in diverse environment, and proteins involved in post-translational modification, which enable maintenance of cellular integrity and physiological process bacterial during adverse environmental conditions, like temperature and oxidative stress. In this way, the authors suggested that with the results found it would be possible to increase the biotechnological potential of *L. lactis* [182].

On the other hand, in eukaryotic cells, the term pan and core proteome was used in a comparative proteomic analysis of *Gammarus* female reproductive systems (ovaries).

Nevertheless, in this study the authors verified a core proteome relatively low among the three amphipods belonging to *Gammarus* genus [183], identifying proteins involved in cellular process, localization, catalytic activity, and binding. Nevertheless, proteins involved in reproductive process were little found due to the absence of their sequences in the database used.

For the success of pan-proteomic experiments, it is necessary to be attentive as to: sample preparation, being important an optimization of the protocols of protein extraction from the multistrain or multiclinical samples; types of data acquisition from gel-based or gel-free methods; construction of pan-proteome database containing all possible proteins, including the same protein but with sequence variation, to use during searching for peptide identification; and better understand the biological functions of the identified proteins through bioinformatics analysis. All these points were extensively revised in a previous study [177].

8 Pan-transcriptomics and its applications

Transcriptome profiling is a powerful approach to identify and quantify the entire repertoire of transcripts in a cell, including mRNAs, noncoding RNAs, and small RNAs, during specific developmental stages or conditions [184]. Transcriptome analysis has enabled the study of the functional elements of the genome, increasing our understanding of the transcriptional dynamics of biological processes and disease development [185]. Among the various technologies that have been developed for high-throughput transcriptome analyses, microarray and RNA-seq are at the forefront of large-scale genome transcriptome profiling [186]. Microarray is a hybridization-based approach developed in the mid-1990s that measure the abundance of a known set of genes using an array of complementary probes. Microarray is a cost-effective, easy to analyze approach that remains the most extensively used methodology in the scientific community. RNA microarrays are generated using complementary DNA (cDNA) immobilized on a glass slide, where each cDNA fragment represents an individual gene of interest. RNA arrays have been used to identify regulated genes, pathways, networks, biological mechanisms, and processes in a variety of biological conditions [187].

However, since its commercial availability, RNA-seq has been widely applied to identify genes within a genome or to measure the expression of transcripts in an organism in different tissues, conditions, and time points [188]. RNA-seq has many advantages over array-based technology, including a high level of data reproducibility, detection of low abundant transcripts, and identification of isoforms over a wider dynamic range. Moreover, the technology does not depend on existing genome data or annotation, allowing the identification and quantification of novel transcripts [189]. Generating data on RNA transcripts require RNA to be first isolated from the experimental organism, following synthesis of cDNA, PCR amplification of cDNA transcripts, and deep sequencing [188].

Following the increased number of high-throughput RNA data, a wide range of strategies for transcriptome analysis has emerged, ranging from single cell to comparative pan-transcriptomic analysis. The pan-transcriptomics analysis consists of a comparison between complete sets of RNA transcripts, under specific circumstances, aiming to identify genes that are differentially expressed in distinct or related populations, or in response to different treatments to better understand the functional and structural aspect of genes. The integration and collective analysis of transcriptome data has enabled the identification of core and distinct molecular responses that functionally reflect the phenotypical diversity of a specific group or condition including patterns of expression associated with parasitism [190], construction of co-expression networks of differentially expressed genes encoding virulence factors [191], the identification of universal biomarkers of cellular senescence [192], comprehensive analysis of molecular alterations across multiple cancer types [193], and the characterization of tissue-specific expression of long noncoding RNAs (lncRNAs) [194]. Pan-transcriptome analysis is particularly applicable in prokaryotes and has been proven valuable in shedding light on gene expression and transcriptome organization among bacterial groups where the difference in phenotypes cannot be explained by the genome sequences alone [195] (Table 3). Moreover, a comparative approach using high-throughput studies can also show the molecular basis of pathogenicity, orthologous biological features, virulence factors, and signaling pathways responsible for stress tolerance and pathogen resistance of related surrogate bacterial species as well as within larger groups of the bacterial domain (Table 3). In addition, integrated analysis can aid the search for potential targets that can be used in the development of therapeutic strategies against relevant pathogens.

Table 3 Pan-transcriptome studies in prokaryotes

Species	Strains/isolates	Approach	Conditions/remarks	References
Mycobacterium tuberculosis and *Mycobacterium bovis*	Mtb H37Rv Mtb H37Rv Mtb H37Rv Mbovis AF2122/97 Mbovis AF2122/97 Mtb H37Rv	Microarray	Bacterial response to aerobic chemostat, low oxygen chemostat—0.2% DOT, aerobic rolling, batch culture, aerobic chemostat, aerobic rolling batch culture, harvested from macrophages	[196]
Bacillus subtilis	BR16 BR17 16BCE	Microarray	Bacterial stringent response by mimicking isoleucine and leucine starvation	[197]

Continued

Table 3 Pan-transcriptome studies in prokaryotes—cont'd

Species	Strains/isolates	Approach	Conditions/remarks	References
Acinetobacter baumannii		RNA-seq	Dynamics of gene expression in the transcriptomic response of drug resistance multidrug-resistant strains and sensitive strains	[198]
Campylobacter jejuni	NCTC11168 81–176 81,116 RM1221	RNA-seq	Comparative analysis of regulatory elements between four isolates	[195]
Pseudomonas aeruginosa	PA14	RNA-seq	Identification of phenotypic variability among bacteria dependent on gene expression in response to different environments including growth within biofilms, at various temperatures, growth phases, osmolarities, phosphate, and iron concentrations, under anaerobic conditions, attached to a surface, and conditions encountered within the eukaryotic host	[199]
Mycobacterium tuberculosis	TKK-01-0084 TKK-01-0025 TKK-01-0033 TKK-01-0040	RNA-seq	Identification of novel transcriptional mechanisms of drug resistance in Mtb strains	[200]
Escherichia coli	EPEC1 EPEC5 EPEC7	RNA-seq	Investigate the global transcriptional responses of the enteropathogenic *E. coli* (EPEC) and enterotoxigenic *E. coli* (ETEC) using 7 isolates	[201]

9 Pan-cancer analysis and its applications

Pan-cancer analysis has enabled in identifying the molecular aspects underlying cancer thereby benefiting diagnosis, prevention, and therapy for patients. One of the major applications of the pan-cancer data is for drug development by ranking drug targets that can be further exploited to develop targeted therapies for cancer. Further analysis of the data is needed for understanding gene-gene interactions and roles of genetic variants affecting pathways. Extensive research has been done to elucidate the underlying mechanisms of cancer occurrence and progression [202–204]. However, most of the studies are conducted independently on smaller sample sizes, thereby limiting the essence of information that needs to come out of such studies. The numerous projects involved in pan-cancer analysis generated huge volumes of data using various technologies including high-end molecular genetics and cytogenetics techniques. Various web tools have been developed and used to interpret the large amount of data generated by the pan-cancer projects [205]. The International Cancer Genome Consortium hence made a group of researchers conducting such cancer analysis across various tumor types in order to generate a pan-cancer atlas [206]. Data generated through these projects will enable in understanding the molecular aspects of cancer occurrence and further help in cancer prevention and designing cancer therapeutics. There are certain challenges that need to be overcome for the development of clinical trial strategies to connect tumor subsets from diverse tissue types [207].

10 Conclusions

The emergence of NGS technologies and the use of the data generated by these technologies for comparative genomics is a major advancement in understanding the diversity of genomes. There are effective examples of pan-genomic studies in various fields of research. The concept of pan-genomics is so deep that it has been perfectly applied in the studies of several organisms and diseases, for example, in the study of dynamics of biological processes and disease development, identification of therapeutic targets against deadly and emerging pathogens, and in the development of new probiotics. It has great potential, which may bring a closer understanding and help combat prokaryotic and eukaryotic diseases in a better way. Finally, several other fields of research that use pan-genomic idea exist, such as pan-cancer, pan-genomics of plants, virus and fungi, pan-metabolomics, and others. All those fields will be further discussed in the following chapters.

References

[1] J.M. Heather, B. Chain, The sequence of sequencers: the history of sequencing DNA, Genomics 107 (2016) 1–8.

[2] E.S. Donkor, Sequencing of bacterial genomes: principles and insights into pathogenesis and development of antibiotics, Genes (Basel) 4 (2013) 556–572.

[3] M. Land, L. Hauser, S.R. Jun, I. Nookaew, M.R. Leuze, T.H. Ahn, T. Karpinets, O. Lund, G. Kora, T. Wassenaar, S. Poudel, D.W. Ussery, Insights from 20 years of bacterial genome sequencing, Funct. Integr. Genom. 15 (2015) 141–161.

[4] J.W. Prokop, T. May, K. Strong, S.M. Bilinovich, C. Bupp, S. Rajasekaran, E.A. Worthey, J. Lazar, Genome sequencing in the clinic: the past, present, and future of genomic medicine, Physiol. Genom. 50 (2018) 563–579.

[5] J. Zhang, R. Chiodini, A. Badr, G. Zhang, The impact of next-generation sequencing on genomics, J. Genet. Genom. 38 (2011) 95–109.

[6] H. Tettelin, D. Riley, C. Cattuto, D. Medini, Comparative genomics: the bacterial pan-genome, Curr. Opin. Microbiol. 11 (2008) 472–477.

[7] D. Medini, C. Donati, H. Tettelin, V. Masignani, R. Rappuoli, The microbial pan-genome, Curr. Opin. Genet. Dev. 15 (2005) 589–594.

[8] A.J. Van Tonder, S. Mistry, J.E. Bray, D.M.C. Hill, A.J. Cody, C.L. Farmer, K.P. Klugman, A. Von Gottberg, S.D. Bentley, J. Parkhill, K.A. Jolley, M.C.J. Maiden, A.B. Brueggemann, Defining the estimated core genome of bacterial populations using a Bayesian decision model. PLoS Comput. Biol. 10 (8) (2014) e1003788 https://doi.org/10.1371/journal.pcbi.1003788.

[9] L. Rouli, V. Merhej, P.E. Fournier, D. Raoult, The bacterial pangenome as a new tool for analysing pathogenic bacteria, New Microbes New Infect. 7 (2015) 72–85.

[10] H. Tettelin, V. Masignani, M.J. Cieslewicz, C. Donati, D. Medini, N.L. Ward, S.V. Angiuoli, J. Crabtree, A.L. Jones, A.S. Durkin, R.T. Deboy, T.M. Davidsen, M. Mora, M. Scarselli, Y. Margarit, I. Ros, J.D. Peterson, C.R. Hauser, J.P. Sundaram, W.C. Nelson, R. Madupu, L.M. Brinkac, R.J. Dodson, M.J. Rosovitz, S.A. Sullivan, S.C. Daugherty, D.H. Haft, J. Selengut, M.L. Gwinn, L. Zhou, N. Zafar, H. Khouri, D. Radune, G. Dimitrov, K. Watkins, K.J. O'connor, S. Smith, T.R. Utterback, O. White, C.E. Rubens, G. Grandi, L.C. Madoff, D.L. Kasper, J.L. Telford, M.R. Wessels, R. Rappuoli, C.M. Fraser, Genome analysis of multiple pathogenic isolates of Streptococcus agalactiae: implications for the microbial "pan-genome" Proc. Natl. Acad. Sci. U. S. A. 102 (2005) 13950–13955.

[11] S.C. Soares, V.A. Abreu, R.T. Ramos, L. Cerdeira, A. Silva, J. Baumbach, E. Trost, A. Tauch, R. Hirata Jr., A.L. Mattos-Guaraldi, A. Miyoshi, V. Azevedo, PIPS: pathogenicity island prediction software, PLoS ONE 7 (2012) e30848.

[12] S.C. Soares, H. Geyik, R.T. Ramos, P.H. De Sa, E.G. Barbosa, J. Baumbach, H.C. Figueiredo, A. Miyoshi, A. Tauch, A. Silva, V. Azevedo, GIPSy: genomic island prediction software, J. Biotechnol. 232 (2016) 2–11.

[13] M. De Barsy, A. Frandi, G. Panis, L. Theraulaz, T. Pillonel, G. Greub, P.H. Viollier, Regulatory (pan-) genome of an obligate intracellular pathogen in the PVC superphylum, ISME J. 10 (2016) 2129–2144.

[14] Y. Zhao, J. Wu, J. Yang, S. Sun, J. Xiao, J. Yu, PGAP: pan-genomes analysis pipeline, Bioinformatics 28 (2012) 416–418.

[15] A.J. Page, C.A. Cummins, M. Hunt, V.K. Wong, S. Reuter, M.T. Holden, M. Fookes, D. Falush, J.A. Keane, J. Parkhill, Roary: rapid large-scale prokaryote pan genome analysis, Bioinformatics 31 (2015) 3691–3693.

[16] N.M. Chaudhari, V.K. Gupta, C. Dutta, BPGA- an ultra-fast pan-genome analysis pipeline, Sci. Rep. 6 (2016) 24373.

[17] C. Computational Pan-Genomics, Computational pan-genomics: status, promises and challenges, Brief. Bioinform. 19 (2018) 118–135.

[18] Computational Pan-Genomics Consortium, Computational pan-genomics: status, promises and challenges. Brief. Bioinform. 19 (1) (2018) 118–135, https://doi.org/10.1093/bib/bbw089.

[19] B. Hurgobin, D. Edwards, SNP discovery using a pangenome: has the single reference approach become obsolete? Biology (Basel) 6 (1) (2017) pii: E21.

[20] L. Benevides, S. Burman, R. Martin, V. Robert, M. Thomas, S. Miquel, F. Chain, H. Sokol, L.G. Bermudez-Humaran, M. Morrison, P. Langella, V.A. Azevedo, J.M. Chatel, S. Soares, New insights into the diversity of the genus *Faecalibacterium*, Front. Microbiol. 8 (2017) 1790.

[21] Y. Chen, Y. Luo, H. Carleton, R. Timme, D. Melka, T. Muruvanda, C. Wang, G. Kastanis, L.S. Katz, L. Turner, A. Fritzinger, T. Moore, R. Stones, J. Blankenship, M. Salter, M. Parish,

T.S. Hammack, P.S. Evans, C.L. Tarr, M.W. Allard, E.A. Strain, E.W. Brown, Whole genome and core genome multilocus sequence typing and single nucleotide polymorphism analyses of *Listeria monocytogenes* associated with an outbreak linked to cheese, United States, 2013. Appl. Environ. Microbiol. 83 (15) (2017) e00633–17 https://doi.org/10.1128/AEM.00633-17.

[22] G. Vernikos, D. Medini, D.R. Riley, H. Tettelin, Ten years of pan-genome analyses, Curr. Opin. Microbiol. 23 (2015) 148–154.

[23] H. Tettelin, The bacterial pan-genome and reverse vaccinology, Genome Dyn. 6 (2009) 35–47.

[24] O. Lukjancenko, T.M. Wassenaar, D.W. Ussery, Comparison of 61 sequenced *Escherichia coli* genomes, Microb. Ecol. 60 (2010) 708–720.

[25] V. Periwal, A. Patowary, S.K. Vellarikkal, A. Gupta, M. Singh, A. Mittal, S. Jeyapaul, R.K. Chauhan, A.V. Singh, P.K. Singh, P. Garg, V.M. Katoch, K. Katoch, D.S. Chauhan, S. Sivasubbu, V. Scaria, Comparative whole-genome analysis of clinical isolates reveals characteristic architecture of *Mycobacterium tuberculosis* pangenome, PLoS ONE 10 (2015) e0122979.

[26] M.W. Tiwari, Diphtheria toxoid, in: Plotkin's Vaccines, seventh ed., Elsevier, 2017.

[27] M. Hessling, J. Feiertag, K. Hoenes, Pathogens provoking most deaths worldwide: A review, Biosci. Biotechnol. Res. Commun. 10 (2017) 1–7.

[28] E. Hacker, C.A. Antunes, A.L. Mattos-Guaraldi, A. Burkovski, A. Tauch, Corynebacterium ulcerans, an emerging human pathogen, Future Microbiol. 11 (2016) 1191–1208.

[29] A. Burkovski, Pathogenesis of *Corynebacterium diphtheriae* and *Corynebacterium ulcerans*, in: Human Emerging and Re-emerging Infections, Wiley, 2015, pp. 699–709 Print ISBN: 9781118644713, Online ISBN: 9781118644843.

[30] A.M. Cerdeno-Tarraga, A. Efstratiou, L.G. Dover, M.T. Holden, M. Pallen, S.D. Bentley, G.S. Besra, C. Churcher, K.D. James, A. De Zoysa, T. Chillingworth, A. Cronin, L. Dowd, T. Feltwell, N. Hamlin, S. Holroyd, K. Jagels, S. Moule, M.A. Quail, E. Rabbinowitsch, K.M. Rutherford, N.R. Thomson, L. Unwin, S. Whitehead, B.G. Barrell, J. Parkhill, The complete genome sequence and analysis of Corynebacterium diphtheriae NCTC13129, Nucleic Acids Res. 31 (2003) 6516–6523.

[31] P. Lapierre, J.P. Gogarten, Estimating the size of the bacterial pan-genome, Trends Genet. 25 (2009) 107–110.

[32] J. Xiao, Z. Zhang, J. Wu, J. Yu, A brief review of software tools for pangenomics, Genom. Proteom. Bioinform. 13 (2015) 73–76.

[33] G.D. Wright, The antibiotic resistome: the nexus of chemical and genetic diversity, Nat. Rev. Microbiol. 5 (2007) 175–186.

[34] M.R. Gillings, Evolutionary consequences of antibiotic use for the resistome, mobilome and microbial pangenome, Front. Microbiol. 4 (2013) 4.

[35] M.L. Metzker, Sequencing technologies—the next generation, Nat. Rev. Genet. 11 (2010) 31–46.

[36] S. Ghatak, J. Blom, S. Das, R. Sanjukta, K. Puro, M. Mawlong, I. Shakuntala, A. Sen, A. Goesmann, A. Kumar, S.V. Ngachan, Pan-genome analysis of Aeromonas hydrophila, Aeromonas veronii and Aeromonas caviae indicates phylogenomic diversity and greater pathogenic potential for Aeromonas hydrophila, Antonie Van Leeuwenhoek 109 (2016) 945–956.

[37] S.C. Bayliss, D.W. Verner-Jeffreys, K.L. Bartie, D.M. Aanensen, S.K. Sheppard, A. Adams, E.J. Feil, The promise of whole genome pathogen sequencing for the molecular epidemiology of emerging aquaculture pathogens, Front. Microbiol. 8 (2017) 121.

[38] T.L. Nguyen, D.-H. Kim, Genome-wide comparison reveals a probiotic strain *Lactococcus lactis* WFLU12 isolated from the gastrointestinal tract of olive flounder (*Paralichthys olivaceus*) harboring genes supporting probiotic action, Mar. Drugs 16 (5) (2018) pii: E140.

[39] M. Dalsass, A. Brozzi, D. Medini, R. Rappuoli, Comparison of open-source reverse vaccinology programs for bacterial vaccine antigen discovery, Front. Immunol. 10 (2019) 113.

[40] Y. Sun, C.S. Liu, L. Sun, Construction and analysis of the immune effect of an Edwardsiella tarda DNA vaccine encoding a D15-like surface antigen, Fish Shellfish Immunol 30 (2011) 273–279.

[41] M.Y. Abdelgayed, Y.G. Alkhateib, A.M. Laila, S.Z. Mona, DNA-based vaccines against bacterial fish diseases: trials and prospective, Rep. Opinion 9 (2017) 1–16.

[42] L. Zeng, D. Wang, N. Hu, Q. Zhu, K. Chen, K. Dong, Y. Zhang, Y. Yao, X. Guo, Y.F. Chang, Y. Zhu, A novel pan-genome reverse vaccinology approach employing a negative-selection strategy for screening surface-exposed antigens against leptospirosis, Front. Microbiol. 8 (2017) 396.

[43] Z. Golkar, O. Bagasra, D.G. Pace, Bacteriophage therapy: a potential solution for the antibiotic resistance crisis, J. Infect. Dev. Ctries. 8 (2014) 129–136.

[44] C.L. Ventola, The antibiotic resistance crisis: part 1: causes and threats, P T 40 (2015) 277–283.

[45] P.C. Appelbaum, 2012 and beyond: potential for the start of a second pre-antibiotic era? J. Antimicrob. Chemother. 67 (2012) 2062–2068.

[46] R.J. Fair, Y. Tor, Antibiotics and bacterial resistance in the 21st century, Perspect. Medicin. Chem. 6 (2014) 25–64.

[47] B.D. Lushniak, Antibiotic resistance: a public health crisis, Public Health Rep. 129 (2014) 314–316.

[48] G.M. Rossolini, F. Arena, P. Pecile, S. Pollini, Update on the antibiotic resistance crisis, Curr. Opin. Pharmacol. 18 (2014) 56–60.

[49] B. Spellberg, D.N. Gilbert, The future of antibiotics and resistance: a tribute to a career of leadership by John Bartlett, Clin. Infect. Dis. 59 (Suppl 2) (2014) S71–S75.

[50] V.K. Viswanathan, Off-label abuse of antibiotics by bacteria, Gut Microbes 5 (2014) 3–4.

[51] C.A. Michael, D. Dominey-Howes, M. Labbate, The antimicrobial resistance crisis: causes, consequences, and management, Front. Public Health 2 (2014) 145.

[52] A. De Sarom, A. Kumar Jaiswal, S. Tiwari, L. De Castro Oliveira, D. Barh, V. Azevedo, C. Jose Oliveira, S. De Castro Soares, Putative vaccine candidates and drug targets identified by reverse vaccinology and subtractive genomics approaches to control Haemophilus ducreyi, the causative agent of chancroid, J. R. Soc. Interface 15 (142) (2018) 20180032.

[53] S.B. Jamal, S.S. Hassan, S. Tiwari, M.V. Viana, L.J. Benevides, A. Ullah, A.G. Turjanski, D. Barh, P. Ghosh, D.A. Costa, A. Silva, R. Rottger, J. Baumbach, Azevedo, V.a.C., An integrative in-silico approach for therapeutic target identification in the human pathogen Corynebacterium diphtheriae, PLoS ONE 12 (2017) e0186401.

[54] A. Kumar Jaiswal, S. Tiwari, S.B. Jamal, D. Barh, V. Azevedo, S.C. Soares, An in silico identification of common putative vaccine candidates against *Treponema pallidum*: a reverse vaccinology and subtractive genomics based approach, Int. J. Mol. Sci. (2017) 18.

[55] C.D. Rinaudo, J.L. Telford, R. Rappuoli, K.L. Seib, Vaccinology in the genome era, J. Clin. Invest. 119 (2009) 2515–2525.

[56] T. Bhardwaj, P. Somvanshi, Pan-genome analysis of *Clostridium botulinum* reveals unique targets for drug development, Gene 623 (2017) 48–62.

[57] D. Barh, S. Tiwari, N. Jain, A. Ali, A.R. Santos, A.N. Misra, V. Azevedo, A. Kumar, In silico subtractive genomics for target identification in human bacterial pathogens, Drug Dev. Res. 72 (2011) 162–177.

[58] A. Praveena, R. Sindhuja, V. Anuradha, S.K.M. Habeeb, Putative drug target identification for *Chlamydia trachomatis*: an insilico proteome analysis, Int. J. Biomed. Res. 2 (2011) 151–160.

[59] D. Barh, A. Kumar, In silico identification of candidate drug and vaccine targets from various pathways in *Neisseria gonorrhoeae*, In Silico Biol. 9 (2009) 225–231.

[60] S. Madagi, V. Malipatil, Putative drug targets in Ureaplasma urealyticum serovar 10 str. ATCC 33699 by insilico genomics approach and virtual screening, Int. J. Pharma Bio Sci. 4 (2013) 8.

[61] A. Ali, A. Naz, S.C. Soares, M. Bakhtiar, S. Tiwari, S.S. Hassan, F. Hanan, R. Ramos, U. Pereira, D. Barh, H.C.P. Figueiredo, D.W. Ussery, A. Miyoshi, A. Silva, V. Azevedo, Pan-genome analysis of human gastric pathogen *H. pylori*: comparative genomics and pathogenomics approaches to identify regions associated with pathogenicity and prediction of potential core therapeutic targets, Biomed. Res. Int. 2015 (2015) 1–17.

[62] S.M. Asif, A. Asad, A. Faizan, M.S. Anjali, A. Arvind, K. Neelesh, K. Hirdesh, K. Sanjay, Dataset of potential targets for Mycobacterium tuberculosis H37Rv through comparative genome analysis, Bioinformation 4 (2009) 245–248.

[63] B. Rathi, A.N. Sarangi, N. Trivedi, Genome subtraction for novel target definition in *Salmonella typhi*, Bioinformation 4 (2009) 143–150.

[64] S.S. Hassan, S.B. Jamal, L.G. Radusky, S. Tiwari, A. Ullah, J. Ali, Behramand, P. De Carvalho, R. Shams, S. Khan, H.C.P. Figueiredo, D. Barh, P. Ghosh, A. Silva, J. Baumbach, R. Rottger, A.G. Turjanski, V.A.C. Azevedo, The druggable pocketome of *Corynebacterium diphtheriae*: a new approach for in silico putative druggable targets, Front. Genet. 9 (2018) 44.

[65] D. Barh, N. Jain, S. Tiwari, B.P. Parida, V. D'afonseca, L. Li, A. Ali, A.R. Santos, L.C. Guimaraes, S. De Castro Soares, A. Miyoshi, A. Bhattacharjee, A.N. Misra, A. Silva, A. Kumar, V. Azevedo, A novel comparative genomics analysis for common drug and vaccine targets in Corynebacterium pseudotuberculosis and other CMN group of human pathogens, Chem. Biol. Drug Des. 78 (2011) 73–84.

[66] S.S. Hassan, S. Tiwari, L.C. Guimaraes, S.B. Jamal, E. Folador, N.B. Sharma, S. De Castro Soares, S. Almeida, A. Ali, A. Islam, F.D. Povoa, V.A. De Abreu, N. Jain, A. Bhattacharya, L. Juneja, A. Miyoshi, A. Silva, D. Barh, A. Turjanski, V. Azevedo, R.S. Ferreira, Proteome scale comparative modeling for conserved drug and vaccine targets identification in Corynebacterium pseudotuberculosis, BMC Genom. 15 (Suppl 7) (2014) S3.

[67] D.J. Bibel, Elie Metchnikoff's Bacillus of Long Life, ASM News (1988) 661–665.

[68] A. Hosono, Fermented milk in the orient, in: Y. Naga Sawa, A. Hosono (Eds.), Functions of fermented milk. Challenges for the health sciences, 1992. Elsevier Applied Science.

[69] A.W. FAO, Guidelines for the Evaluation of Probiotics in Food, Food and Agriculture Organization of the United Nations, 2002.

[70] R. Bibiloni, R.N. Fedorak, G.W. Tannock, K.L. Madsen, P. Gionchetti, M. Campieri, C. De Simone, R.B. Sartor, VSL#3 probiotic-mixture induces remission in patients with active ulcerative colitis, Am. J. Gastroenterol. 100 (2005) 1539–1546.

[71] A. Tursi, G. Brandimarte, A. Papa, A. Giglio, W. Elisei, G.M. Giorgetti, G. Forti, S. Morini, C. Hassan, M.A. Pistoia, M.E. Modeo, S. Rodino, T. D'amico, L. Sebkova, N. Sacca, E. Di Giulio, F. Luzza, M. Imeneo, T. Larussa, S. Di Rosa, V. Annese, S. Danese, A. Gasbarrini, Treatment of relapsing mild-to-moderate Ulcerative Colitis with the probiotic VSL#3 as adjunctive to a standard pharmaceutical treatment: a double-blind, randomized, Placebo-Controlled Study, Am. J. Gastroenterol. 105 (2010) 2218–2227.

[72] F. Calcinaro, S. Dionisi, M. Marinaro, P. Candeloro, V. Bonato, S. Marzotti, R.B. Corneli, E. Ferretti, A. Gulino, F. Grasso, C. De Simone, U. Di Mario, A. Falorni, M. Boirivant, F. Dotta, Oral probiotic administration induces interleukin-10 production and prevents spontaneous autoimmune diabetes in the non-obese diabetic mouse, Diabetologia 48 (2005) 1565–1575.

[73] D. Unutmaz, S. Lavasani, B. Dzhambazov, M. Nouri, F. Fåk, S. Buske, G. Molin, H. Thorlacius, J. Alenfall, B. Jeppsson, B. Weström, A novel probiotic mixture exerts a therapeutic effect on experimental autoimmune encephalomyelitis mediated by IL-10 producing regulatory T cells, PLoS ONE 5 (2) (2010) e9009.

[74] M. Viljanen, E. Pohjavuori, T. Haahtela, R. Korpela, M. Kuitunen, A. Sarnesto, O. Vaarala, E. Savilahti, Induction of inflammation as a possible mechanism of probiotic effect in atopic eczema–dermatitis syndrome, J. Allergy Clin. Immunol. 115 (2005) 1254–1259.

[75] A. Miyoshi, E. Jamet, J. Commissaire, P. Renault, P. Langella, V. Azevedo, A xylose-inducible expression system for *Lactococcus lactis*, FEMS Microbiol. Lett. 239 (2004) 205–212.

[76] M.T. Islam, A. Deora, Y. Hashidoko, A. Rahman, T. Ito, S. Tahara, Isolation and identification of potential phosphate solubilizing bacteria from the rhizoplane of *Oryza sativa* L. cv. BR29 of Bangladesh, Z. Naturforsch. C 62 (2007) 103–110.

[77] D. Thakuria, N.C. Talukdar, C. Goswami, S. Hazarika, R.C. Boro, M.R. Khan, Characterization and screening of bacteria from rhizosphere of rice grown in acidic soils of Assam, Curr. Sci. 86 (7) (2004) 978–985.

[78] M. Ogut, F. Er, N. Kandemir, Phosphate solubilization potentials of soil *Acinetobacter strains*, Biol. Fertil. Soils 46 (2010) 707–715.

[79] H. Cao, S. He, R. Wei, M. Diong, L. Lu, *Bacillus amyloliquefaciens* G1: a potential antagonistic bacterium against eel-pathogenic *Aeromonas hydrophila*, Evid. Based Complement. Alternat. Med. 2011 (2011) 1–7.

[80] J. Ji, S. Hu, W. Li, Probiotic *Bacillus amyloliquefaciens* SC06 prevents bacterial translocation in weaned mice, Indian J. Microbiol. 53 (2013) 323–328.

[81] M.R. Sudha, S. Bhonagiri, M.A. Kumar, Efficacy of *Bacillus clausii* strain UBBC-07 in the treatment of patients suffering from acute diarrhoea, Benefic. Microbes 4 (2013) 211–216.

[82] H.A. Hong, L.H. Duc, S.M. Cutting, The use of bacterial spore formers as probiotics: Table 1, FEMS Microbiol. Rev. 29 (2005) 813–835.

[83] M. La Rosa, G. Bottaro, N. Gulino, F. Gambuzza, F. Di Forti, G. Ini, E. Tornambe, Prevention of antibiotic-associated diarrhea with Lactobacillus sporogens and fructo-oligosaccharides in children. A multicentric double-blind vs placebo study, Minerva Pediatr. 55 (2003) 447–452.
[84] N.M. Gracheva, A.F. Gavrilov, A.I. Solov'eva, V.V. Smirnov, I.B. Sorokulova, S.R. Reznik, N.V. Chudnovskaia, The efficacy of the new bacterial preparation biosporin in treating acute intestinal infections, Zh. Mikrobiol. Epidemiol. Immunobiol. 1 (1996) 75–77.
[85] P. Pattnaik, S. Grover, V.K. Batish, Effect of environmental factors on production of lichenin, a chromosomally encoded bacteriocin-like compound produced by *Bacillus licheniformis* 26L-10/3RA, Microbiol. Res. 160 (2005) 213–218.
[86] C. Liu, J. Lu, L. Lu, Y. Liu, F. Wang, M. Xiao, Isolation, structural characterization and immunological activity of an exopolysaccharide produced by *Bacillus licheniformis* 8-37-0-1, Bioresour. Technol. 101 (2010) 5528–5533.
[87] D.-Y. Tseng, P.-L. Ho, S.-Y. Huang, S.-C. Cheng, Y.-L. Shiu, C.-S. Chiu, C.-H. Liu, Enhancement of immunity and disease resistance in the white shrimp, *Litopenaeus vannamei*, by the probiotic, *Bacillus subtilis* E20, Fish Shellfish Immunol. 26 (2009) 339–344.
[88] J.A. Gilbert, R. Krajmalnik-Brown, D.L. Porazinska, S.J. Weiss, R. Knight, Toward effective probiotics for autism and other neurodevelopmental disorders, Cell 155 (2013) 1446–1448.
[89] M. Saxelin, S. Tynkkynen, T. Mattila-Sandholm, W.M. De Vos, Probiotic and other functional microbes: from markets to mechanisms, Curr. Opin. Biotechnol. 16 (2005) 204–211.
[90] K.Y. Wang, S.N. Li, C.S. Liu, D.S. Perng, Y.C. Su, D.C. Wu, C.M. Jan, C.H. Lai, T.N. Wang, W.M. Wang, Effects of ingesting Lactobacillus- and Bifidobacterium-containing yogurt in subjects with colonized *Helicobacter pylori*, Am. J. Clin. Nutr. 80 (2004) 737–741.
[91] C.K. Dotterud, O. Storrø, R. Johnsen, T. Øien, Probiotics in pregnant women to prevent allergic disease: a randomized, double-blind trial, Br. J. Dermatol. 163 (2010) 616–623.
[92] B.S. Kang, J.-G. Seo, G.-S. Lee, J.-H. Kim, S.Y. Kim, Y.W. Han, H. Kang, H.O. Kim, J.H. Rhee, M.-J. Chung, Y.M. Park, Antimicrobial activity of enterocins from Enterococcus faecalis SL-5 against Propionibacterium acnes, the causative agent in acne vulgaris, and its therapeutic effect, J. Microbiol. 47 (2009) 101–109.
[93] T. Aymerich, M.G. Artigas, M. Garriga, J.M. Monfort, M. Hugas, Effect of sausage ingredients and additives on the production of enterocin A and B by *Enterococcus faecium* CTC492. Optimization of in vitro production and anti-listerial effect in dry fermented sausages, J. Appl. Microbiol. 88 (2000) 686–694.
[94] B. Olle, Medicines from microbiota, Nat. Biotechnol. 31 (2013) 309–315.
[95] W. Kruis, Maintaining remission of ulcerative colitis with the probiotic *Escherichia coli* Nissle 1917 is as effective as with standard mesalazine, Gut 53 (2004) 1617–1623.
[96] H.A. Malchow, Crohn's disease and *Escherichia coli*. A new approach in therapy to maintain remission of colonic Crohn's disease? J. Clin. Gastroenterol. 25 (1997) 653–658.
[97] A. Sturm, K. Rilling, D.C. Baumgart, K. Gargas, T. Abou-Ghazale, B. Raupach, J. Eckert, R.R. Schumann, C. Enders, U. Sonnenborn, B. Wiedenmann, A.U. Dignass, *Escherichia coli* Nissle 1917 distinctively modulates T-cell cycling and expansion via toll-like receptor 2 signaling, Infect. Immun. 73 (2005) 1452–1465.
[98] Y. Inoue, T. Kambara, N. Murata, J. Komori-Yamaguchi, S. Matsukura, Y. Takahashi, Z. Ikezawa, M. Aihara, Effects of oral administration of *Lactobacillus acidophilus* L-92 on the symptoms and serum cytokines of atopic dermatitis in Japanese adults: a double-blind, randomized, clinical trial, Int. Arch. Allergy Immunol. 165 (2014) 247–254.
[99] F. Murina, A. Graziottin, F. Vicariotto, F. De Seta, Can *Lactobacillus fermentum* LF10 and *Lactobacillus acidophilus* LA02 in a slow-release vaginal product be useful for prevention of recurrent vulvovaginal candidiasis? J. Clin. Gastroenterol. 48 (2014) S102–S105.
[100] Y.-J. Lai, S.-H. Tsai, M.-Y. Lee, Isolation of exopolysaccharide producing Lactobacillus strains from sorghum distillery residues pickled cabbage and their antioxidant properties, Food Sci. Biotechnol. 23 (2014) 1231–1236.
[101] N. Waki, N. Yajima, H. Suganuma, B.M. Buddle, D. Luo, A. Heiser, T. Zheng, Oral administration ofLactobacillus brevisKB290 to mice alleviates clinical symptoms following influenza virus infection, Lett. Appl. Microbiol. 58 (2014) 87–93.

[102] X.Q. Zeng, D.D. Pan, Y.X. Guo, The probiotic properties of *Lactobacillus buchneri* P2. J. Appl. Microbiol. 108 (6) (2010) 2059–2066, https://doi.org/10.1111/j.1365-2672.2009.04608.x.
[103] A. Marcos, J. Wärnberg, E. Nova, S. Gómez, A. Alvarez, R. Alvarez, J.A. Mateos, J.M. Cobo, The effect of milk fermented by yogurt cultures plus *Lactobacillus casei* DN-114001 on the immune response of subjects under academic examination stress, Eur. J. Nutr. 43 (2004) 381–389.
[104] R.J. Siezen, G. Wilson, Probiotics genomics, Microb. Biotechnol. 3 (2010) 1–9.
[105] A.E. Stapleton, M. Au-Yeung, T.M. Hooton, D.N. Fredricks, P.L. Roberts, C.A. Czaja, Y. Yarova-Yarovaya, T. Fiedler, M. Cox, W.E. Stamm, Randomized, placebo-controlled phase 2 trial of a *Lactobacillus crispatus* probiotic given intravaginally for prevention of recurrent urinary tract infection, Clin. Infect. Dis. 52 (2011) 1212–1217.
[106] S. Makino, S. Ikegami, A. Kume, H. Horiuchi, H. Sasaki, N. Orii, Reducing the risk of infection in the elderly by dietary intake of yoghurt fermented with Lactobacillus delbrueckii ssp. bulgaricus OLL1073R-1, Br. J. Nutr. 104 (2010) 998–1006.
[107] M. Sanchez, C. Darimont, V. Drapeau, S. Emady-Azar, M. Lepage, E. Rezzonico, C. Ngom-Bru, B. Berger, L. Philippe, C. Ammon-Zuffrey, P. Leone, G. Chevrier, E. St-Amand, A. Marette, J. Doré, A. Tremblay, Effect of Lactobacillus rhamnosus CGMCC1.3724 supplementation on weight loss and maintenance in obese men and women, Br. J. Nutr. 111 (2013) 1507–1519.
[108] S. Chabot, H.-L. Yu, L. De Léséleuc, D. Cloutier, M.-R. Van Calsteren, M. Lessard, D. Roy, M. Lacroix, D. Oth, Exopolysaccharides from Lactobacillus rhamnosus RW-9595M stimulate TNF, IL-6 and IL-12 in human and mouse cultured immunocompetent cells, and IFN-γ in mouse splenocytes, Lait 81 (2001) 683–697.
[109] J.P. Madej, T. Stefaniak, M. Bednarczyk, Effect ofin ovo-delivered prebiotics and synbiotics on lymphoid-organs' morphology in chickens, Poult. Sci. 94 (2015) 1209–1219.
[110] M.L. Ellis, A.E. Dowell, X. Li, J. Knight, Probiotic properties of Oxalobacter formigenes: an in vitro examination, Arch. Microbiol. 198 (2016) 1019–1026.
[111] H.S. El-Nezami, N.N. Polychronaki, J. Ma, H. Zhu, W. Ling, E.K. Salminen, R.O. Juvonen, S.J. Salminen, T. Poussa, H.M. Mykkänen, Probiotic supplementation reduces a biomarker for increased risk of liver cancer in young men from Southern China, Am. J. Clin. Nutr. 83 (2006) 1199–1203.
[112] J.P. Burton, C.N. Chilcott, J.R. Tagg, The rationale and potential for the reduction of oral malodour using *Streptococcus salivarius* probiotics, Oral Dis. 11 (2005) 29–31.
[113] Y.J. Moon, J.R. Soh, J.J. Yu, H.S. Sohn, Y.S. Cha, S.H. Oh, Intracellular lipid accumulation inhibitory effect of *Weissella koreensis* OK1-6 isolated from Kimchi on differentiating adipocyte, J. Appl. Microbiol. 113 (2012) 652–658.
[114] J.A. Park, P.B. Tirupathi Pichiah, J.J. Yu, S.H. Oh, J.W. Daily, Y.S. Cha, Anti-obesity effect of kimchi fermented withWeissella koreensisOK1-6 as starter in high-fat diet-induced obese C57BL/6J mice, J. Appl. Microbiol. 113 (2012) 1507–1516.
[115] J. Touchman, *Comparative Genomics* [Online], in: Nature Education Knowledge, 2010. Available: https://www.nature.com/scitable/knowledge/library/comparative-genomics-13239404. (Accessed 14 January 2019).
[116] A. Bezkorovainy, Probiotics: determinants of survival and growth in the gut, Am. J. Clin. Nutr. 73 (2001) 399S–405S.
[117] G. Konuray, Z. Erginkaya, Potential use of *Bacillus coagulans* in the food industry, Foods 7 (2018).
[118] B.R. Johnson, T.R. Klaenhammer, Impact of genomics on the field of probiotic research: historical perspectives to modern paradigms, Antonie Van Leeuwenhoek 106 (2014) 141–156.
[119] L.C. Oliveira, T.D. Saraiva, W.M. Silva, U.P. Pereira, B.C. Campos, L.J. Benevides, F.S. Rocha, H. C.P. Figueiredo, V. Azevedo, S.C. Soares, Analyses of the probiotic property and stress resistance-related genes of Lactococcus lactis subsp. lactis NCDO 2118 through comparative genomics and in vitro assays. PLoS ONE 12 (4) (2017) e0175116. https://doi.org/10.1371/journal.pone.0175116.
[120] H. Willenbrock, P.F. Hallin, T.M. Wassenaar, D.W. Ussery, Characterization of probiotic Escherichia coli isolates with a novel pan-genome microarray, Genome Biol. 8 (2007).
[121] T.L. Nguyen, C.-I. Park, D.-H. Kim, Improved growth rate and disease resistance in olive flounder, Paralichthys olivaceus, by probiotic Lactococcus lactis WFLU12 isolated from wild marine fish, Aquaculture 471 (2017) 113–120.

[122] R. Kant, J. Rintahaka, X. Yu, P. Sigvart-Mattila, L. Paulin, J.-P. Mecklin, M. Saarela, A. Palva, I. Von Ossowski, A comparative pan-genome perspective of niche-adaptable cell-surface protein phenotypes in *Lactobacillus rhamnosus*. PLoS ONE 9 (7) (2014) e102762. https://doi.org/10.1371/journal.pone.0102762.

[123] T. Smokvina, M. Wels, J. Polka, C. Chervaux, S. Brisse, J. Boekhorst, J.E. Van Hylckama Vlieg, R.J. Siezen, Lactobacillus paracasei comparative genomics: towards species pan-genome definition and exploitation of diversity, PLoS ONE 8 (2013) e68731.

[124] J.L. Gardy, N.J. Loman, Towards a genomics-informed, real-time, global pathogen surveillance system, Nat. Rev. Genet. 19 (2018) 9–20.

[125] J. Shendure, H. Ji, Next-generation DNA sequencing, Nat. Biotechnol. 26 (2008) 1135–1145.

[125a] J. Quick, N.D. Grubaugh, S.T. Pullan, et al., Multiplex PCR method for MinION and Illumina sequencing of Zika and other virus genomes directly from clinical samples. Nat. Protoc. 12 (6) (2017) 1261–1276, https://doi.org/10.1038/nprot.2017.066.

[126] N.R. Faria, J. Quick, I.M. Claro, J. Theze, J.G. De Jesus, M. Giovanetti, M.U.G. Kraemer, S.C. Hill, A. Black, A.C. Da Costa, L.C. Franco, S.P. Silva, C.H. Wu, J. Raghwani, S. Cauchemez, L. Du Plessis, M.P. Verotti, W.K. De Oliveira, E.H. Carmo, G.E. Coelho, A. Santelli, L.C. Vinhal, C. M. Henriques, J.T. Simpson, M. Loose, K.G. Andersen, N.D. Grubaugh, S. Somasekar, C. Y. Chiu, J.E. Munoz-Medina, C.R. Gonzalez-Bonilla, C.F. Arias, L.L. Lewis-Ximenez, S. A. Baylis, A.O. Chieppe, S.F. Aguiar, C.A. Fernandes, P.S. Lemos, B.L.S. Nascimento, H.A. O. Monteiro, I.C. Siqueira, M.G. De Queiroz, T.R. De Souza, J.F. Bezerra, M.R. Lemos, G. F. Pereira, D. Loudal, L.C. Moura, R. Dhalia, R.F. Franca, T. Magalhaes, E.T. Marques Jr., T. Jaenisch, G.L. Wallau, M.C. De Lima, V. Nascimento, E.M. De Cerqueira, M.M. De Lima, D.L. Mascarenhas, J.P.M. Neto, A.S. Levin, T.R. Tozetto-Mendoza, S.N. Fonseca, M. C. Mendes-Correa, F.P. Milagres, A. Segurado, E.C. Holmes, A. Rambaut, T. Bedford, M.R. T. Nunes, E.C. Sabino, L.C.J. Alcantara, N.J. Loman, O.G. Pybus, Establishment and cryptic transmission of Zika virus in Brazil and the Americas, Nature 546 (2017) 406–410.

[127] J. Theze, T. Li, L. Du Plessis, J. Bouquet, M.U.G. Kraemer, S. Somasekar, G. Yu, M. De Cesare, A. Balmaseda, G. Kuan, E. Harris, C.H. Wu, M.A. Ansari, R. Bowden, N.R. Faria, S. Yagi, S. Messenger, T. Brooks, M. Stone, E.M. Bloch, M. Busch, J.E. Munoz-Medina, C.R. Gonzalez-Bonilla, S. Wolinsky, S. Lopez, C.F. Arias, D. Bonsall, C.Y. Chiu, O.G. Pybus, Genomic epidemiology reconstructs the introduction and spread of zika virus in Central America and Mexico, Cell Host Microbe 23 (855-864) (2018).

[128] L.C. Guimaraes, J. Florczak-Wyspianska, L.B. De Jesus, M.V. Viana, A. Silva, R.T. Ramos, C. Soares Sde, C. Soares Sde, Inside the pan-genome—methods and software overview, Curr. Genom. 16 (2015) 245–252.

[129] K. Padovani De Souza, J.C. Setubal, F. Ponce De Leon, A.C. De Carvalho, G. Oliveira, A. Chateau, R. Alves, Machine learning meets genome assembly, Brief Bioinform. (2018) 1–14.

[130] S.I. Lee, N.S. Kim, Transposable elements and genome size variations in plants, Genom. Inform. 12 (2014) 87–97.

[131] I. Arabidopsis Genome, Analysis of the genome sequence of the flowering plant Arabidopsis thaliana, Nature 408 (2000) 796–815.

[132] K.L. McNally, K.L. Childs, R. Bohnert, R.M. Davidson, K. Zhao, V.J. Ulat, G. Zeller, R.M. Clark, D.R. Hoen, T.E. Bureau, R. Stokowski, D.G. Ballinger, K.A. Frazer, D.R. Cox, B. Padhukasahasram, C.D. Bustamante, D. Weigel, D.J. Mackill, R.M. Bruskiewich, G. Ratsch, C.R. Buell, H. Leung, J.E. Leach, Genomewide SNP variation reveals relationships among landraces and modern varieties of rice, Proc. Natl. Acad. Sci. U. S. A. 106 (2009) 12273–12278.

[133] A.A. Golicz, J. Batley, D. Edwards, Towards plant pangenomics, Plant Biotechnol. J. 14 (2016) 1099–1105.

[134] J.D. Montenegro, A.A. Golicz, P.E. Bayer, B. Hurgobin, H. Lee, C.K. Chan, P. Visendi, K. Lai, J. Dolezel, J. Batley, D. Edwards, The pangenome of hexaploid bread wheat, Plant J. 90 (2017) 1007–1013.

[135] M.G. Milgroom, T.L. Peever, Population biology of plant pathogens: the synthesis of plant disease epidemiology and population genetics, Plant Dis. 87 (2003) 608–617.

[136] B.M. Tyler, S. Tripathy, X. Zhang, P. Dehal, R.H. Jiang, A. Aerts, F.D. Arredondo, L. Baxter, D. Bensasson, J.L. Beynon, J. Chapman, C.M. Damasceno, A.E. Dorrance, D. Dou, A. W. Dickerman, I.L. Dubchak, M. Garbelotto, M. Gijzen, S.G. Gordon, F. Govers, N. J. Grunwald, W. Huang, K.L. Ivors, R.W. Jones, S. Kamoun, K. Krampis, K.H. Lamour, M.K. Lee, W.H. Mcdonald, M. Medina, H.J. Meijer, E.K. Nordberg, D.J. Maclean, M. D. Ospina-Giraldo, P.F. Morris, V. Phuntumart, N.H. Putnam, S. Rash, J.K. Rose, Y. Sakihama, A.A. Salamov, A. Savidor, C.F. Scheuring, B.M. Smith, B.W. Sobral, A. Terry, T.A. Torto-Alalibo, J. Win, Z. Xu, H. Zhang, I.V. Grigoriev, D.S. Rokhsar, J.L. Boore, Phytophthora genome sequences uncover evolutionary origins and mechanisms of pathogenesis, Science 313 (2006) 1261–1266.

[137] J.O. McInerney, A. McNally, M.J. O'Connell, Why prokaryotes have pangenomes, Nat. Microbiol. 2 (2017) 17040.

[138] C. Plissonneau, F.E. Hartmann, D. Croll, Pangenome analyses of the wheat pathogen Zymoseptoria tritici reveal the structural basis of a highly plastic eukaryotic genome, BMC Biol. 16 (2018) 5.

[139] J.C. Meeks, E.L. Campbell, M.L. Summers, F.C. Wong, Cellular differentiation in the cyanobacterium Nostoc punctiforme, Arch. Microbiol. 178 (2002) 395–403.

[140] D.R. Nelson, B. Khraiwesh, W. Fu, S. Alseekh, A. Jaiswal, A. Chaiboonchoe, K.M. Hazzouri, M. J. O'connor, G.L. Butterfoss, N. Drou, J.D. Rowe, J. Harb, A.R. Fernie, K.C. Gunsalus, K. Salehi-Ashtiani, The genome and phenome of the green alga Chloroidium sp. UTEX 3007 reveal adaptive traits for desert acclimatization. Elife 6 (2017) e25783 https://doi.org/10.7554/eLife.25783.

[141] S. Hirooka, Y. Hirose, Y. Kanesaki, S. Higuchi, T. Fujiwara, R. Onuma, A. Era, R. Ohbayashi, A. Uzuka, H. Nozaki, H. Yoshikawa, S.Y. Miyagishima, Acidophilic green algal genome provides insights into adaptation to an acidic environment, Proc. Natl. Acad. Sci. U. S. A. 114 (2017) E8304–E8313.

[142] G. Barbier, C. Oesterhelt, M.D. Larson, R.G. Halgren, C. Wilkerson, R.M. Garavito, C. Benning, A.P. Weber, Comparative genomics of two closely related unicellular thermo-acidophilic red algae, *Galdieria sulphuraria* and Cyanidioschyzon merolae, reveals the molecular basis of the metabolic flexibility of *Galdieria sulphuraria* and significant differences in carbohydrate metabolism of both algae, Plant Physiol. 137 (2005) 460–474.

[143] D. Bhattacharya, D.C. Price, C.X. Chan, H. Qiu, N. Rose, S. Ball, A.P. Weber, M.C. Arias, B. Henrissat, P.M. Coutinho, A. Krishnan, S. Zauner, S. Morath, F. Hilliou, A. Egizi, M.M. Perrineau, H.S. Yoon, Genome of the red alga Porphyridium purpureum, Nat. Commun. 4 (2013) 1941.

[144] S. Bose, S.K. Herbert, D.C. Fork, Fluorescence characteristics of photoinhibition and recovery in a sun and a shade species of the red algal genus porphyra, Plant Physiol. 86 (1988) 946–950.

[145] K. Nishitsuji, A. Arimoto, K. Iwai, Y. Sudo, K. Hisata, M. Fujie, N. Arakaki, T. Kushiro, T. Konishi, C. Shinzato, N. Satoh, E. Shoguchi, A draft genome of the brown alga, *Cladosiphon okamuranus*, S-strain: a platform for future studies of 'mozuku' biology, DNA Res. 23 (2016) 561–570.

[146] A. Sboner, X.J. Mu, D. Greenbaum, R.K. Auerbach, M.B. Gerstein, The real cost of sequencing: higher than you think!, Genome Biol. 12 (2011) 125.

[147] M. Guegan, K. Zouache, C. Demichel, G. Minard, V. Tran Van, P. Potier, P. Mavingui, C. Valiente Moro, The mosquito holobiont: fresh insight into mosquito-microbiota interactions, Microbiome 6 (2018) 49.

[148] D. Aguirre De Carcer, The human gut pan-microbiome presents a compositional core formed by discrete phylogenetic units, Sci. Rep. 8 (2018) 14069.

[149] M.H. Leung, P.K. Lee, The roles of the outdoors and occupants in contributing to a potential pan-microbiome of the built environment: a review, Microbiome 4 (2016) 21.

[150] P. Vandenkoornhuyse, A. Quaiser, M. Duhamel, A. Le Van, A. Dufresne, The importance of the microbiome of the plant holobiont, New Phytol. 206 (2015) 1196–1206.

[151] B. Aslam, M. Basit, M.A. Nisar, M.H. Rasool, M. Khurshid, Proteomics: technologies and their applications, J. Chromatogr. Sci. 55 (2017) 182–196.

[152] W.M. Silva, R.D. Carvalho, S.C. Soares, I.F.S. Bastos, E.L. Folador, G.H.M.F. Souza, Y. Le Loir, A. Miyoshi, A. Silva, V. Azevedo, Label-free proteomic analysis to confirm the predicted proteome of

Corynebacterium pseudotuberculosis under nitrosative stress mediated by nitric oxide, BMC Genom. 15 (2014) 1065.
[153] W.M. Silva, R.D.O. Carvalho, F.A. Dorella, E.L. Folador, G.H.M.F. Souza, A.M.C. Pimenta, H.C. P. Figueiredo, Y. Le Loir, A. Silva, V. Azevedo, Quantitative proteomic analysis reveals changes in the benchmark *Corynebacterium pseudotuberculosis* biovar equi exoproteome after passage in a murine host. Front. Cell. Infect. Microbiol. 7 (2017) 325, https://doi.org/10.3389/fcimb.2017.00325.
[154] T.-C. Chao, N. Hansmeier, The current state of microbial proteomics: where we are and where we want to go, Proteomics 12 (2012) 638–650.
[155] M.A. Moseley, Quantitative proteomics in genomic medicine, in: G.S. Ginsburg, H.F. Willard (Eds.), Genomic and Personalized Medicine, second ed., Academic Press, 2013, pp. 155–165 (Chapter 13).
[156] M.A. Reymond, W. Schlegel, Proteomics in cancer, Adv. Clin. Chem. 44 (2007) 103–142.
[157] M.A. Hussain, F. Huygens, Proteomic and bioinformatics tools to understand virulence mechanisms in *Staphylococcus aureus*, Curr. Proteom. 9 (2012) 2–8.
[158] O. Coskun, Separation techniques: Chromatography, North. Clin. Istanb. 3 (2016) 156–160.
[159] R.M. Lequin, Enzyme immunoassay (EIA)/enzyme-linked immunosorbent assay (ELISA), Clin. Chem. 51 (2005) 2415–2418.
[160] B.T. Kurien, R.H. Scofield, Western blotting: an introduction, in: B.T. Kurien, R.H. Scofield (Eds.), Western Blotting: Methods and Protocols, Springer New York, New York, NY, 2015, pp. 17–30.
[161] M. D'Innocenzo, Identificação das proteínas por meio da eletroforese 2D, in: R. Verlengia, R. Curi, E. Bevilacqua, P. Newsholme (Eds.), Análises de RNA, proteínas e metabólitos: metodologia e procedimentos técnicos, Santos Editora, São Paulo, 2013, pp. 261–280.
[162] R. Vessecchi, N.P. Lopes, F.C. Gozzo, F.A. Dörr, M. Murgu, D.T. Lebre, R. Abreu, O.V. Bustillos, J.M. Riveros, Nomenclaturas de espectrometria de massas em língua portuguesa, Quím. Nova 34 (2011) 1875–1887.
[163] S.J. Cordwell, A.S. Nouwens, B.J. Walsh, Comparative proteomics of bacterial pathogens, Proteomics 1 (2001) 461–472.
[164] P.M. Bisch, Genômica funcional: proteômica, in: L. Mir (Ed.), Genômica, Atheneu, São Paulo, 2004, pp. 139–162.
[165] E.-H. Jeong, B. Vaidya, S.-Y. Cho, M.-A. Park, K. Kaewintajuk, S.R. Kim, M.-J. Oh, J.-S. Choi, J. Kwon, D. Kim, Identification of regulators of the early stage of viral hemorrhagic septicemia virus infection during curcumin treatment, Fish Shellfish Immunol. 45 (2015) 184–193.
[166] N. Solis, S.J. Cordwell, Current methodologies for proteomics of bacterial surface-exposed and cell envelope proteins, Proteomics 11 (2011) 3169–3189.
[167] S. Hölper, A. Ruhs, M. Krüger, Stable isotope labeling for proteomic analysis of tissues in mouse, in: B. Warscheid (Ed.), Stable Isotope Labeling by Amino Acids in Cell Culture (SILAC): Methods and Protocols, Springer New York, New York, NY, 2014, pp. 95–106.
[168] K. Cheng, A. Sloan, S. Mccorrister, L. Peterson, H. Chui, M. Drebot, C. Nadon, J.D. Knox, G. Wang, Quality evaluation of LC-MS/MS-based E. coli H antigen typing (MS-H) through label-free quantitative data analysis in a clinical sample setup, Proteom. Clin. Appl. 8 (2014) 963–970.
[169] S. Kosono, M. Tamura, S. Suzuki, Y. Kawamura, A. Yoshida, M. Nishiyama, M. Yoshida, Changes in the acetylome and succinylome of *Bacillus subtilis* in response to carbon source, PLoS ONE 10 (2015) e0131169.
[170] S.P. Gygi, B. Rist, T.J. Griffin, J. Eng, R. Aebersold, Proteome analysis of low-abundance proteins using multidimensional chromatography and isotope-coded affinity tags, J. Proteome Res. 1 (2002) 47–54.
[171] V.J. Patel, K. Thalassinos, S.E. Slade, J.B. Connolly, A. Crombie, J.C. Murrell, J.H. Scrivens, A comparison of labeling and label-free mass spectrometry-based proteomics approaches, J. Proteome Res. 8 (2009) 3752–3759.
[172] D. Chelius, P.V. Bondarenko, Quantitative profiling of proteins in complex mixtures using liquid chromatography and mass spectrometry, J. Proteome Res. 1 (2002) 317–323.
[173] M.J.G. Hughes, J.C. Moore, J.D. Lane, R. Wilson, P.K. Pribul, Z.N. Younes, R.J. Dobson, P. Everest, A.J. Reason, J.M. Redfern, F.M. Greer, T. Paxton, M. Panico, H.R. Morris, R.

G. Feldman, J.D. Santangelo, Identification of major outer surface proteins of *Streptococcus agalactiae*, Infect. Immun. 70 (2002) 1254–1259.

[174] F. Doro, S. Liberatori, M.J. Rodríguez-Ortega, C.D. Rinaudo, R. Rosini, M. Mora, M. Scarselli, E. Altindis, R. D'aurizio, M. Stella, I. Margarit, D. Maione, J.L. Telford, N. Norais, G. Grandi, Surfome analysis as a fast track to vaccine discovery: identification of a novel protective antigen for group B *Streptococcus hypervirulent* strain COH1, Mol. Cell. Proteom. 8 (2009) 1728–1737.

[175] W.M. Silva, N. Seyffert, A.V. Santos, T.L.P. Castro, L.G.C. Pacheco, A.R. Santos, A. Ciprandi, F.A. Dorella, H.M. Andrade, D. Barh, A.M.C. Pimenta, A. Silva, A. Miyoshi, V. Azevedo, Identification of 11 new exoproteins in *Corynebacterium pseudotuberculosis* by comparative analysis of the exoproteome, Microb. Pathog. 61–62 (2013) 37–42.

[176] W.M. Silva, N. Seyffert, A. Ciprandi, A.V. Santos, T.L.P. Castro, L.G.C. Pacheco, D. Barh, Y. Le Loir, A.M.C. Pimenta, A. Miyoshi, A. Silva, V. Azevedo, Differential exoproteome analysis of two *Corynebacterium pseudotuberculosis* biovar ovis strains isolated from goat (1002) and sheep (C231), Curr. Microbiol. 67 (2013) 460–465.

[177] J.A. Broadbent, D.A. Broszczak, I.U.K. Tennakoon, F. Huygens, Pan-proteomics, a concept for unifying quantitative proteome measurements when comparing closely-related bacterial strains, Expert Rev. Proteom. 13 (2016) 355–365.

[178] G.C. Tavares, F.L. Pereira, G.M. Barony, C.P. Rezende, W.M. Da Silva, G.H.M.F. De Souza, T. Verano-Braga, V.A. De Carvalho Azevedo, Leal, C.a.G., and Figueiredo, H.C.P., Delineation of the pan-proteome of fish-pathogenic Streptococcus agalactiae strains using a label-free shotgun approach, BMC Genom. 20 (2019) 11.

[179] J. Rothen, J.F. Pothier, F. Foucault, J. Blom, D. Nanayakkara, C. Li, M. Ip, M. Tanner, G. Vogel, V. Pflüger, C.A. Daubenberger, Subspecies typing of *Streptococcus agalactiae* based on ribosomal subunit protein mass variation by MALDI-TOF MS, Front. Microbiol. 10 (2019) 471.

[180] L. Zhang, D. Xiao, B. Pang, Q. Zhang, H. Zhou, L. Zhang, J. Zhang, B. Kan, The core proteome and pan proteome of *Salmonella* Paratyphi A epidemic strains, PLoS ONE 9 (2014) e89197.

[181] G.D. Jhingan, S. Kumari, S.V. Jamwal, H. Kalam, D. Arora, N. Jain, L.K. Kumaar, A. Samal, K.V.S. Rao, D. Kumar, V.K. Nandicoori, Comparative proteomic analyses of avirulent, virulent, and clinical strains of *Mycobacterium tuberculosis* identify strain-specific patterns, J. Biol. Chem. 291 (2016) 14257–14273.

[182] W.M. Silva, C.S. Sousa, L.C. Oliveira, S.C. Soares, G. Souza, G.C. Tavares, C.P. Resende, E.L. Folador, F.L. Pereira, H. Figueiredo, V. Azevedo, Comparative proteomic analysis of four biotechnological strains *Lactococcus lactis* through label-free quantitative proteomics, Microb. Biotechnol. 12 (2019) 265–274.

[183] J. Trapp, C. Almunia, J.-C. Gaillard, O. Pible, A. Chaumot, O. Geffard, J. Armengaud, Proteogenomic insights into the core-proteome of female reproductive tissues from crustacean amphipods, J. Proteome 135 (2016) 51–61.

[184] Z. Wang, M. Gerstein, M. Snyder, RNA-Seq: a revolutionary tool for transcriptomics, Nat. Rev. Genet. 10 (2009) 57–63.

[185] I. Korf, Genomics: the state of the art in RNA-seq analysis, Nat. Methods 10 (2013) 1165–1166.

[186] M.F. Rai, E.D. Tycksen, L.J. Sandell, R.H. Brophy, Advantages of RNA-seq compared to RNA microarrays for transcriptome profiling of anterior cruciate ligament tears, J. Orthop. Res. 36 (2018) 484–497.

[187] S.C. Sealfon, T.T. Chu, RNA and DNA microarrays, Methods Mol. Biol. 671 (2011) 3–34.

[188] R. Lowe, N. Shirley, M. Bleackley, S. Dolan, T. Shafee, Transcriptomics technologies, PLoS Comput. Biol. 13 (2017) e1005457.

[189] S. Zhao, W.P. Fung-Leung, A. Bittner, K. Ngo, X. Liu, Comparison of RNA-Seq and microarray in transcriptome profiling of activated T cells, PLoS ONE 9 (2014) e78644.

[190] M. Blaxter, S. Kumar, G. Kaur, G. Koutsovoulos, B. Elsworth, Genomics and transcriptomics across the diversity of the Nematoda, Parasite Immunol. 34 (2012) 108–120.

[191] M.S. Kim, H. Zhang, H. Yan, B.J. Yoon, W.B. Shim, Characterizing co-expression networks underpinning maize stalk rot virulence in Fusarium verticillioides through computational subnetwork module analyses, Sci. Rep. 8 (2018) 8310.

[192] Z. Wei, H. Guo, J. Qin, S. Lu, Q. Liu, X. Zhang, Y. Zou, Y. Gong, C. Shao, Pan-senescence transcriptome analysis identified RRAD as a marker and negative regulator of cellular senescence, Free Radic. Biol. Med. 130 (2019) 267–277.

[193] X. Ma, Y. Liu, Y. Liu, L.B. Alexandrov, M.N. Edmonson, C. Gawad, X. Zhou, Y. Li, M.C. Rusch, J. Easton, R. Huether, V. Gonzalez-Pena, M.R. Wilkinson, L.C. Hermida, S. Davis, E. Sioson, S. Pounds, X. Cao, R.E. Ries, Z. Wang, X. Chen, L. Dong, S.J. Diskin, M.A. Smith, J.M. Guidry Auvil, P.S. Meltzer, C.C. Lau, E.J. Perlman, J.M. Maris, S. Meshinchi, S.P. Hunger, D.S. Gerhard, J. Zhang, Pan-cancer genome and transcriptome analyses of 1,699 paediatric leukaemias and solid tumours, Nature 555 (2018) 371–376.

[194] C.R. Cabanski, N.M. White, H.X. Dang, J.M. Silva-Fisher, C.E. Rauck, D. Cicka, C.A. Maher, Pan-cancer transcriptome analysis reveals long noncoding RNAs with conserved function, RNA Biol. 12 (2015) 628–642.

[195] G. Dugar, A. Herbig, K.U. Forstner, N. Heidrich, R. Reinhardt, K. Nieselt, C.M. Sharma, High-resolution transcriptome maps reveal strain-specific regulatory features of multiple Campylobacter jejuni isolates, PLoS Genet. 9 (2013) e1003495.

[196] B. Sidders, M. Withers, S.L. Kendall, J. Bacon, S.J. Waddell, J. Hinds, P. Golby, F. Movahedzadeh, R. A. Cox, R. Frita, A.M. Ten Bokum, L. Wernisch, N.G. Stoker, Quantification of global transcription patterns in prokaryotes using spotted microarrays, Genome Biol. 8 (2007) R265.

[197] C. Eymann, G. Homuth, C. Scharf, M. Hecker, Bacillus subtilis functional genomics: global characterization of the stringent response by proteome and transcriptome analysis, J. Bacteriol. 184 (2002) 2500–2520.

[198] H. Qin, N.W. Lo, J.F. Loo, X. Lin, A.K. Yim, S.K. Tsui, T.C. Lau, M. Ip, T.F. Chan, Comparative transcriptomics of multidrug-resistant *Acinetobacter baumannii* in response to antibiotic treatments, Sci. Rep. 8 (2018) 3515.

[199] A. Dotsch, M. Schniederjans, A. Khaledi, K. Hornischer, S. Schulz, A. Bielecka, D. Eckweiler, S. Pohl, S. Haussler, The *Pseudomonas aeruginosa* transcriptional landscape is shaped by environmental heterogeneity and genetic variation, MBio 6 (2015) e00749.

[200] L. De Welzen, V. Eldholm, K. Maharaj, A.L. Manson, A.M. Earl, A.S. Pym, Whole-transcriptome and -genome analysis of extensively drug-resistant *Mycobacterium tuberculosis* clinical isolates identifies downregulation of etha as a mechanism of ethionamide resistance, Antimicrob. Agents Chemother. 61 (2017).

[201] T.H. Hazen, J. Michalski, Q. Luo, A.C. Shetty, S.C. Daugherty, J.M. Fleckenstein, D.A. Rasko, Comparative genomics and transcriptomics of *Escherichia coli* isolates carrying virulence factors of both enteropathogenic and enterotoxigenic *E. coli*, Sci. Rep. 7 (2017) 3513.

[202] P.A. Northcott, C. Lee, T. Zichner, A.M. Stutz, S. Erkek, D. Kawauchi, D.J. Shih, V. Hovestadt, M. Zapatka, D. Sturm, D.T. Jones, M. Kool, M. Remke, F.M. Cavalli, S. Zuyderduyn, G.D. Bader, S. Vandenberg, L.A. Esparza, M. Ryzhova, W. Wang, A. Wittmann, S. Stark, L. Sieber, H. Seker-Cin, L. Linke, F. Kratochwil, N. Jager, I. Buchhalter, C.D. Imbusch, G. Zipprich, B. Raeder, S. Schmidt, N. Diessl, S. Wolf, S. Wiemann, B. Brors, C. Lawerenz, J. Eils, H.J. Warnatz, T. Risch, M.L. Yaspo, U.D. Weber, C.C. Bartholomae, C. Von Kalle, E. Turanyi, P. Hauser, E. Sanden, A. Darabi, P. Siesjo, J. Sterba, K. Zitterbart, D. Sumerauer, P. Van Sluis, R. Versteeg, R. Volckmann, J. Koster, M.U. Schuhmann, M. Ebinger, H.L. Grimes, G.W. Robinson, A. Gajjar, M. Mynarek, K. Von Hoff, S. Rutkowski, T. Pietsch, W. Scheurlen, J. Felsberg, G. Reifenberger, A.E. Kulozik, A. Von Deimling, O. Witt, R. Eils, R.J. Gilbertson, A. Korshunov, M.D. Taylor, P. Lichter, J.O. Korbel, R.J. Wechsler-Reya, S.M. Pfister, Enhancer hijacking activates GFI1 family oncogenes in medulloblastoma, Nature 511 (2014) 428–434.

[203] E. Papaemmanuil, M. Cazzola, J. Boultwood, L. Malcovati, P. Vyas, D. Bowen, A. Pellagatti, J. S. Wainscoat, E. Hellstrom-Lindberg, C. Gambacorti-Passerini, A.L. Godfrey, I. Rapado, A. Cvejic, R. Rance, C. Mcgee, P. Ellis, L.J. Mudie, P.J. Stephens, S. Mclaren, C.E. Massie, P. S. Tarpey, I. Varela, S. Nik-Zainal, H.R. Davies, A. Shlien, D. Jones, K. Raine, J. Hinton, A. P. Butler, J.W. Teague, E.J. Baxter, J. Score, A. Galli, M.G. Della Porta, E. Travaglino, M. Groves, S. Tauro, N.C. Munshi, K.C. Anderson, A. El-Naggar, A. Fischer, V. Mustonen, A.

J. Warren, N.C. Cross, A.R. Green, P.A. Futreal, M.R. Stratton, P.J. Campbell, Chronic Myeloid Disorders Working Group of the International Cancer Genome Consortium, Somatic SF3B1 mutation in myelodysplasia with ring sideroblasts, N. Engl. J. Med. 365 (2011) 1384–1395.

[204] X.S. Puente, M. Pinyol, V. Quesada, L. Conde, G.R. Ordonez, N. Villamor, G. Escaramis, P. Jares, S. Bea, M. Gonzalez-Diaz, L. Bassaganyas, T. Baumann, M. Juan, M. Lopez-Guerra, D. Colomer, J. M. Tubio, C. Lopez, A. Navarro, C. Tornador, M. Aymerich, M. Rozman, J.M. Hernandez, D. A. Puente, J.M. Freije, G. Velasco, A. Gutierrez-Fernandez, D. Costa, A. Carrio, S. Guijarro, A. Enjuanes, L. Hernandez, J. Yague, P. Nicolas, C.M. Romeo-Casabona, H. Himmelbauer, E. Castillo, J.C. Dohm, S. De Sanjose, M.A. Piris, E. De Alava, J. San Miguel, R. Royo, J. L. Gelpi, D. Torrents, M. Orozco, D.G. Pisano, A. Valencia, R. Guigo, M. Bayes, S. Heath, M. Gut, P. Klatt, J. Marshall, K. Raine, L.A. Stebbings, P.A. Futreal, M.R. Stratton, P. J. Campbell, I. Gut, A. Lopez-Guillermo, X. Estivill, E. Montserrat, C. Lopez-Otin, E. Campo, Whole-genome sequencing identifies recurrent mutations in chronic lymphocytic leukaemia, Nature 475 (2011) 101–105.

[205] Z. Liu, S. Zhang, Toward a systematic understanding of cancers: a survey of the pan-cancer study, Front. Genet. 5 (2014) 194.

[206] T.J. Hudson, W. Anderson, A. Artez, A.D. Barker, C. Bell, R.R. Bernabe, M.K. Bhan, F. Calvo, I. Eerola, D.S. Gerhard, A. Guttmacher, M. Guyer, F.M. Hemsley, J.L. Jennings, D. Kerr, P. Klatt, P. Kolar, J. Kusada, D.P. Lane, F. Laplace, L. Youyong, G. Nettekoven, B. Ozenberger, J. Peterson, T.S. Rao, J. Remacle, A.J. Schafer, T. Shibata, M.R. Stratton, J.G. Vockley, K. Watanabe, H. Yang, M.M. Yuen, B.M. Knoppers, M. Bobrow, A. Cambon-Thomsen, L. G. Dressler, S.O. Dyke, Y. Joly, K. Kato, K.L. Kennedy, P. Nicolas, M.J. Parker, E. Rial-Sebbag, C.M. Romeo-Casabona, K.M. Shaw, S. Wallace, G.L. Wiesner, N. Zeps, P. Lichter, A. V. Biankin, C. Chabannon, L. Chin, B. Clement, E. De Alava, F. Degos, M.L. Ferguson, P. Geary, D.N. Hayes, T.J. Hudson, A.L. Johns, A. Kasprzyk, H. Nakagawa, R. Penny, M. A. Piris, R. Sarin, A. Scarpa, T. Shibata, M. Van De Vijver, P.A. Futreal, H. Aburatani, M. Bayes, D.D. Botwell, P.J. Campbell, X. Estivill, D.S. Gerhard, S.M. Grimmond, I. Gut, M. Hirst, C. Lopez-Otin, P. Majumder, M. Marra, J.D. Mcpherson, H. Nakagawa, Z. Ning, X. S. Puente, Y. Ruan, T. Shibata, M.R. Stratton, H.G. Stunnenberg, H. Swerdlow, V. E. Velculescu, R.K. Wilson, H.H. Xue, L. Yang, P.T. Spellman, G.D. Bader, P.C. Boutros, P. J. Campbell, P. Flicek, et al., International network of cancer genome projects, Nature 464 (2010) 993–998.

[207] D.A. Levine, Integrated genomic characterization of endometrial carcinoma, Nature 497 (2013) 67–73.

CHAPTER 2

Bioinformatics approaches applied in pan-genomics and their challenges

Yan Pantoja, Kenny da Costa Pinheiro, Fabricio Araujo, Artur Luiz da Costa Silva, Rommel Ramos
Institute of Biological Sciences, Federal University of Pará (UFPA), Belem, Brazil

1 Introduction

Since the advent of next-generation sequencing (NGS), it became possible to evaluate an increasing number of genomes and, consequently, genetically related organisms [1]. Currently, it is known that there are a great number of genomic variations within a particular bacterial population or species. Thus, the functional annotation of such variants is now possible as well as the analysis of different strains that constitute a particular bacterial species. And the trend is that this scenario will be even bigger and more complex in the future [2].

As the number of genomes available in biological databases increased due to NGS technologies, it became necessary to rethink the idea of a "reference" genome that represents a particular species and aids in research [3]. This reference genome can be shaped in many forms, including:

- the genome of a single individual selected;
- a consensus from an entire population;
- a "functional" genome (without disabling mutations of any gene); and
- a maximum genome that captures every sequence of a given species already detected.

Depending on the context, each one of these options might be best suited for a particular research approach. However, many initial reference sequences did not contain any of the previously mentioned characteristics [3].

In this context, in order to take the most advantage of the data produced by NGS platforms, using a reference, it was necessary to do a paradigm shift: instead of focusing only on a single reference genome, use a "pan-genome," that is, a representation of the entire gene repertoire of a particular species or phylogenetic clade [3].

A decade after the beginning of the genomic era, identifying the number of genomes that could describe a bacterial species became the target of the major questions. Understanding the genomic versatility has become particularly relevant for the study of disease-causing bacteria, which frequently have a large number of variable genes [4].

Pan-genomics: Applications, Challenges, and Future Prospects
https://doi.org/10.1016/B978-0-12-817076-2.00002-0
© 2020 Elsevier Inc.
All rights reserved.

However, species classification was never simple. Since the first use of the term in a biological context by the English naturalist John Ray in the 17th century [5], the definition of species has been repeated several times, based on different criteria; from shared physical characteristics or ability to produce viable descendants until a shared pattern, niche, or evolutionary history. But regardless of the used definition, the frontier between one taxonomic group and the next is not always clear. While a reproductive definition effectively organizes most multicellular animals into distinct taxonomic groups, the bacteriologists community has not yet been able to establish a uniformly accepted definition for bacterial species due to the fact that these microorganisms possess high levels of genomic diversity and because of their complexity in terms of cultivability, in addition to the high level of horizontal transfer observed [6].

Facing such complexity, some researchers are developing a more subtle view. In prokaryotes, where the lines between taxonomic units are more diffuse, pan-genome analysis (which divides the genome into core and variable genes depending on their presence or absence among species) could offer a more effective way to distinguish closely related organisms when compared to the traditional alternative approaches. While most current methods compare the sequences of only one or a few genes (such as the 16S rRNA gene, or housekeeping genes in the case of multilocus sequence typing) to determine relationships between organisms, pan-genome analysis compare and contrast whole genomes of several individuals, providing an expanded view of similarities and differences between organisms [7, 8].

2 Pan-genome analysis

The pan-genome analysis in the last decade has allowed researchers to develop universal vaccines that could be effective against all strains of one species, or even against several related species. In 2005, the work of Tettelin and colleagues on *Streptococcus agalactiae* (or group B Streptococcus [GBS]) led to the creation of a potentially universal vaccine based on the combination of four bacterial surface proteins [4]. And in June of 2016, researchers at the University of California, San Diego, published a study on methicilin-resistant hospital superbug *Staphylococcus aureus* (MRSA). This study started with 64 strains as a starting point for the development of a vaccine that is widely effective against MRSA [9].

Now that pan-genome approach is widely accepted as a useful way of organizing bacterial diversity, efforts are concentrated on incorporating such studies into phylogenetics, taxonomy, and even into metagenomics, in a more recent metapangenome area [10].

As pan-genome research in microbiology continues to increase, observed intraspecific variation also influences the genomic descriptions of other taxons. As an example, it can be mentioned the eukaryotic species, where the horizontal transfer is even more complex when compared to the prokaryotes. It is also noted that the sequencing of multiple individuals of the same species begins to reveal an extensive genomic diversity that goes far beyond the small differences observed between genes. Besides, the horizontal transfer events can also occur between prokaryotes and eukaryotes, increasing the diversity of these taxons [11, 12].

Researchers from San Marcos, California State University, realized the importance of such genomic variation a few years ago, shortly after the assembly of a reference genome for the eukaryotic phytoplankton *Emiliania huxleyi*. This species can be found in several ocean sites all over the world. Suspecting that the organism's ability to adapt to varied conditions may depend on single-nucleotide polymorphisms (SNPs) within genes, scientists started to work on the sequencing of more isolate organisms [13].

After sequencing 13 distinct strains, researchers were surprised to find that the size of the genome, originally estimated at about 30,000 genes, varied widely among the analyzed strains, with some strains losing more than 2000 genes. When they performed a pan-genome analysis, the researchers found that only two-thirds of the genes they had identified initially were shared by all sequenced isolates. In particular, there was a high degree of variability in genes encoding metal-binding proteins—key components in the adaptation of *E. huxleyi* to the environment [13].

Given the lack of evidence for horizontal gene transfer in *E. huxleyi*, it is unlikely that the availability of the total genetic pool for each individual is similar to that of prokaryotes. But it is believed that the bigger pan-genome in relation to the central genome of an individual supports the adaptability of this unicellular eukaryote [13].

Emiliania huxleyi hardly is the only one to have this diversity in its DNA. Large-scale sequencing projects were applied to thousands of whole genomes of model eukaryotic organisms, such as *Saccharomyces cerevisiae* and *Arabidopsis thaliana*. These also revealed significant numbers of duplicate new genes. And in cultivated plants, whose genomes often contain large duplicate regions, some studies already support the correlation between the presence or the absence of "variable" genes, disease resistance, metabolite production, and stress responses, showing that the genetic difference has a great impact [13].

2.1 Pan-genome approaches

Computational methods to find more efficient data structures, algorithms, and statistical methods to perform bioinformatic analyses of pan-genomes give rise to a new area known as "computational pan-genomics." This field has desirable characteristics [3]:

- *Completeness*: The presence of all functional elements.
- *Stability*: To present unique identifiable characteristics that can be studied.
- *Comprehensibility*: Understanding the complexity of the genome structure from many species.
- *Efficiency*: Organization of data in a way that accelerates downstream analysis.

The main objective of pan-genome analysis is to determine the genomic diversity of the available dataset, and to predict, via extrapolation, how many genomic sequences would be necessary to characterize the whole pan-genome or repertoire of genes [14]. Most of the pan-genome projects that emerged after 2005 had as their main differences: the number of genomes/strains analyzed, the phylogenetic resolution, the mathematical prediction model used, the threshold of orthology definition, the algorithm used for alignment and search beside the parameters of percentage of alignment, and completeness of the product [8].

The approach to estimate the pan-genome size, the core genome, and the novel gene discovery rate was started by Tettelin and colleagues; intuitively, starting from a small pan-genome model (i.e., two genomes) and adding more genomes to it, a large number of new genes will be found, since the repertoire of the starting genes were small; conversely, the size of the central genome will decrease, since genes will be less likely to be shared by all genomes. The higher the number of genomes added, the greater the pan-genome and the lower the number of new genes that will be revealed. In parallel, the size of the core genome will decrease. It is possible that a point of "saturation" will be reached, in the sense that the addition of new genomes will not increase the size of the core genome, while the ratio of new genes will be asymptotically stabilized at a given value. For a closed pan-genome, this value is higher than 1 and the pan-genome size can be estimated; for an open pan-genome, this value is lower than 1, and the size of the pan-genome cannot be estimated (i.e., it will probably grow "indefinitely"). Since the number of shared genes and the number of specific genes for a pan-genome depends on how many strains are taken into account, the approach used by Tettelin and colleagues was to use eight genomes of pathogenic strains of *S. agalactiae* and to compute all possible comparisons among n genomes (i.e., eight possible combinations for pan-genome of $n = 2$ genomes) [15].

Plotting the number of shared genes and the number of new genes for each comparison as a function of the n strains considered, Tettelin and colleagues were able to fit exponential decaying function curves over the data which asymptotically reached the values of 1806 shared genes and 33 novel genes, corresponding to the estimate of core-genome size and novel gene discovery rate. The latter value was used for extrapolating the *S. agalactiae* pan-genome size [15].

Users interested in pan-genome analysis have the option of implementing methods such as alignment of multiple nucleotide sequences (complete genomes) to improve sensitivity, for comparisons of high resolution in the species/subspecies or at strain level. They may also use amino acid similarity, protein grouping, structural alignment, and metabolic pathway information at higher levels to reduce noise and eliminate artifacts resulting from nucleotide sequence alignment [8].

The original implementation of the algorithm or workflow pipeline for pan-genome analysis, while conceptually intuitive, has several potential technical pitfalls, some of which are essential enough to impact the conclusions drawn. Issues include the prediction of an open versus closed pan-genome, a rapid or slow pan-genome growing (the rate at which new genes identified from additional genomes expands the pan-genome), genes that are assigned to the core genome versus accessory genome (the choice of parameters affects whether genes are considered shared/core or noncore), and determining the size of the core genome (the asymptote for the extrapolation of the core genome tends to decrease as more genomes are added to the analysis) [8].

In addition, there is the combinatorial aspect of this approach, where all possible permutations when adding a genome to a set of previously analyzed genomes is considered. The number of comparisons (n) used to calculate the number of new genes, genes belonging to the core, and genes shared in the nth genome can be modeled with the following function, where C is the total number of combinations and N is the total number of genomes in the analysis [8]:

$$C = \frac{N!}{(n-1)!^{*}(N-n)!} \tag{1}$$

These combinations can be represented in the form of a boxplot that can be drawn for both pan- and core-genomes. The combinations from 1 to the total number of samples are placed in the x-axis of the graph, being that in combination 1, the number of genes found in each individual genome is determined. In the combination 2, all possible combinations of 2 × 2 genomes are observed. In the combination 3, all possible combinations of 3 × 3 genomes are observed and so on, until reaching the maximum combination that corresponds to the set of all samples [8] (Figs. 1 and 2).

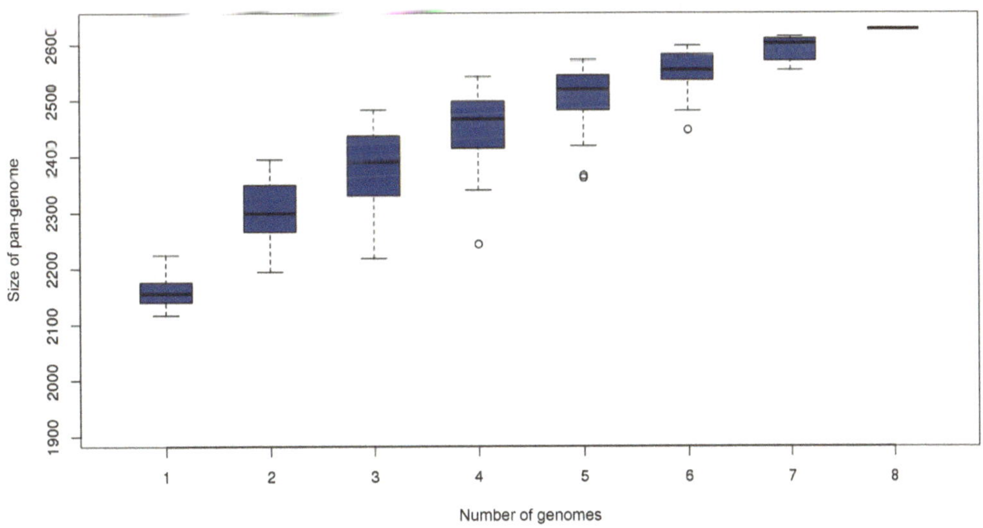

Fig. 1 Pan-genome being displayed graphically. Combinations 1–8 are presented as boxplot (*blue*). It is possible to note that as the number of samples inserted in the combinations increases, the pan-genome also increases.

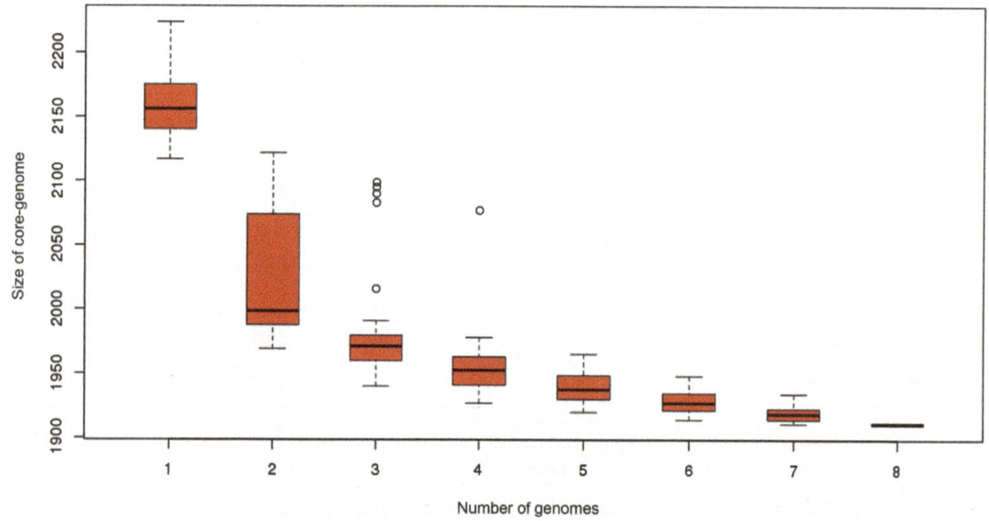

Fig. 2 Core genome being displayed graphically. Combinations 1–8 are shown as box distributions (*red*). It is possible to note that as the number of samples inserted in the combinations increases, the core genome decreases.

2.2 Mathematical model: Heaps' law

It is common to adjust the regression curves of box charts using a power law model (Heaps' law) rather than an exponential decay. Heaps' law is an empirical law that describes the number of distinct words in a document (or set of documents) as a function of document length, and is represented by the formula [15]:

$$n = k^* N^{-\alpha} \tag{2}$$

where n is the expected number of genes for a given set of genomes and N is the number of genomes in a given analysis. K and α are the free coefficients of the regression.

Heaps' law is used in pan-genome analysis to determine whether a given pan-genome is open or closed. This is done after adjusting the regression curve where it is possible to get the values of the alpha coefficient (α). This way it can be inferred that a certain pan-genome is open when the value of α is less than 1. On the other hand, we have that a pan-genome is considered closed when the observed value of α is greater than 1 [15] (Fig. 3).

To obtain the complete gene repertoire of a given microbial species, it is necessary to identify how many extra genes can be added to each new genome sequenced. If each new genome sequenced increases the amount of new genes inserted considerably, we say this pan-genome is open. Generally, open pan-genomes can be observed in species that undergo frequent horizontal gene transfer and colonize multiple environments. In contrast, microorganisms that are more conserved and that live in more isolated niches and consequently have a low capacity to acquire new genes have greater tendency to have a

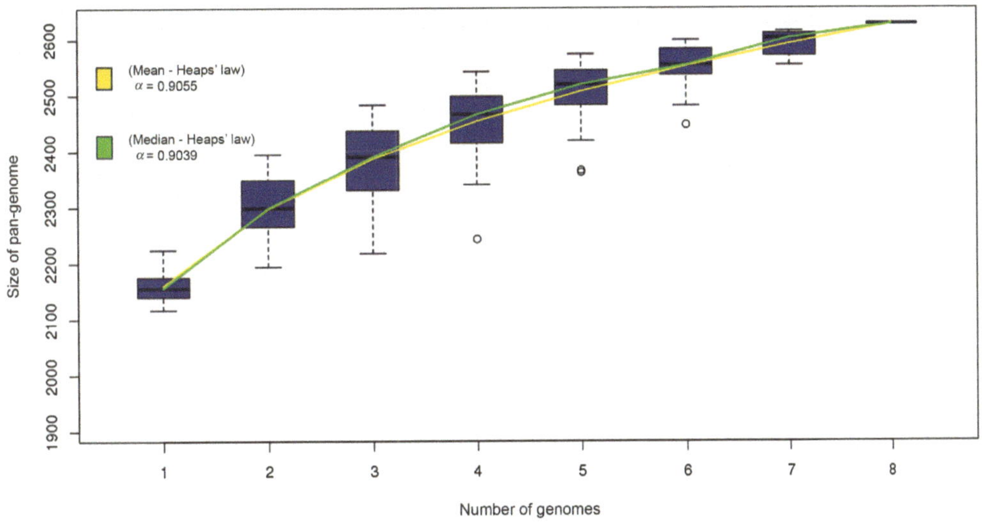

Fig. 3 Pan-genome being plotted along with the regression curves. The curves are adjusted for both the median (*green*) and the mean (*yellow*) values of each distribution. It can be observed in the figure that the values of α (alpha) are close to 0.9 considering a pan-genome near to being closed.

closed pan-genome [7]. It is important to note that a closed pan-genome is not always synonymous with the same phenotype for all the bacterial strains analyzed, because different SNPs can confer different characteristics to different strains [4].

2.3 Software packages and tools

Existing software packages and tools responsible for performing pan-genome analysis have some common functions, such as the search and identification of orthologous and paralogous genes, calculation of the pan-genome profile, and definition of the core genome, accessory genome, and strain-specific genes [7].

2.3.1 Composition and annotation

In order to evaluate the composition and later annotation, the search for orthologs is performed in order to estimate the composition of the pan-genome (core genes, accessory genes, and unique genes). This search is made with tools and algorithms most often used in bioinformatics such as BLAST [16] or OrthoMCL [17]. OrthoMCL uses the Markov clustering algorithm, a method based on a graph flow theory that determines the transition probabilities among the nodes in the graphs, eventually producing clusters of nodes representing groups of orthologous proteins between two or more species [17]. In the later steps, to characterize the sequences found (annotation), tools such as COG (Cluster of Orthologous Groups), InterPro, and KEGG (Kyoto Encyclopedia of Genes and Genomes) are used to obtain data on how the function of the genes is distributed within the core and accessory genome as well as assessing the metabolic pathways found [7].

Another important factor is the study of the regulation of protein expression and related transcription factors, since the identification of these elements in one or more isolates may help to explain some characteristics that distinguish the different strains. A very useful online tool for this purpose is P2RP (Predicted Prokaryotic Regulatory Proteins), which was developed to make this type of search feasible for all researchers and not only for bioinformaticians, since it has a user friendly interface and is simple, fast, and effective [18].

In addition to the regulatory elements, another important factor is the definition of homology relations between genes belonging to different genomes. Basically, there are two types of situations: when genes descend from an event of speciation (orthologs) and when the genes come from a duplication event (paralogs) from a common ancestor. To find these two groups, it is often used alignment and sequence comparison tools. Homologous genes are conceptualized as corresponding genes in different species. The approach used to find such sequences (genes or proteins) is based on similarity and on the assumption that they are more similar to each other than in any other genome sequence, or they are bidirectional best hits (BBHs). Thus, it is common to assume that BBHs are composed of orthologs that serve to identify families of genes. However, this approach does not take into account the duplication events that may have occurred after a speciation event, since it captures only one-to-one orthological relationships. To overcome this problem, other approaches can be used as COGs proteins and InParanoid/MultiParanoid, which are, respectively, used to call orthologs in pairwise comparison and multiple genome comparison [18a].

InParanoid [19] was initially designed to find orthologous sequences in pairwise genome analysis. Subsequently, the algorithm called MultiParanoid [20] was created to complement and extend the InParanoid approach by taking as input the pairwise orthologous clusters and thus producing clusters of orthologous genes. The comparison of the results obtained using these two different methods showed that there are only small differences in performance between them (Fondi, 2015).

There are several bioinformatics tools capable of predicting microbial genes from genomic sequences. Among them, we can cite GeneMarkHMM [21], Glimmer [22], or Prodigal [23], which depend on statistical methods of learning such as the hidden Markov model to accomplish this task. Tools that use unsupervised learning (Prodigal) are simpler to use since they do not require a trained data set and are able to infer algorithm parameters from the provided genomic sequence.

In global alignment, MAUVE can be used [24], or it can be possible to try a multiple alignment [25] to perform the phylogeny. The MEGA [26] or MAFFT [27] tools are recommended for the reconstruction of trees in the study of phylogeny, and the algorithms most used for this purpose are: neighbor joining and maximum parsimony.

The search for SNPs in the core genome can be used to estimate the age of the species of interest. However, it is necessary that the genomes of the analyzed species are very close in order to study in detail the mutational events that led to the separation in two distinct species. As an example we can mention the work that was carried out in *Yersinia pestis*, in which a comparative analysis was performed with *Yersinia pseudotuberculosis* and *Yersinia enterocolitica* [7].

2.3.2 Pan-genome tools

In an effort to compute standardized pan-genome analysis, several online tools and software suites have been developed. Among the early-developed packages, Panseq [28] and PanCGHweb [29] were published in 2010, followed by Prokaryotic-genome Analysis Tool (PGAT) [30] in 2011. Panseq is a software suite that supports core/dispensable gene mapping and classification of a collection of genome sequences. This tool defines the core and accessory genome based on the sequence identity and segmentation length and not on the predicted proteins. For this purpose, the Novel Region Finder (NRF) module was developed. The module first splits the genome sequence into fragments with predefined sizes, and then the MUMmer alignment program [31] identifies the sequences and contiguous regions that are present or absent in the database [28].

Subsequently, a second module called Core and Accessory Genome Finder (CAGF) is executed and through it a comparison of a single sequence file is performed against all other sequences. The sequence will be added to the pan-genome if it fits in with the predefined parameters, and then, the newly added to fragment sequence is used for subsequent comparisons, and the looping continues until all of the fragment sequences have been tested [28]. PanCGHweb is a web tool for pan-genome microarray analysis based on PanCGH algorithm [32]. It enables users to group genes into orthologs and to construct gene-based phylogenies of related strains and isolates. However, this tool is rather specific to analyze microarray data but not RNA-seq data. The package PGAT integrates several functions, such as identifying SNPs among orthologs and syntenic regions, plotting the presence and the absence of genes among members of a pan-genome, comparing gene orders among different strains and isolates, providing KEGG pathway analysis tools, and searching for genes through different annotations such as the COGs of proteins, PSORT, SignalP, the transmembrane hidden Markov model, and Pfam. However, PGAT is just a database with a limited number of species curated and it cannot perform analysis for new sequencing data from users [33].

GET_HOMOLOGUES [34] is a customizable and detailed pan-genome analysis platform for microorganisms addressed to nonbioinformaticians that was written in Perl and R and can be installed on personal machines. The program starts using BLAST [16] and HMMER [35] to build clusters of orthologous groups. Then, the sequences, features, and intergenes are extracted, sorted, and indexed. Next, the genomes are classified by size being the smallest used as a reference, and then the paralogous genes that arose by duplication after the speciation process are identified, this whole process is performed through the bidirectional best hit (BBH) algorithm. Subsequently, new genomes are added and compared with the reference genome, and their BBHs are annotated; in the last step, clusters that comprise at least one sequence per genome are conserved [34]. Concomitantly, the results are submitted to OrthoMCL [36] and COGtriangles [37]. Another software that performs pan-genome analysis is called PanGP [38] that implements two sampling algorithms totally random and distance guide on combinations of N strains and generates pan-genome, core genome, and new gene graphs similar to Tettelin and colleagues [4]. The basic difference between the totally random and distance guide algorithms consists of estimating the sample size, where the totally random algorithm repeats randomly the samples in nonredundant combinations for all possible combinations, and the distance guide algorithm has a variable amplification coefficient, which controls the sample size for evaluating the genome diversity of all of the combinations. Tests performed by the authors showed that the distance guide algorithm has better efficiency [38].

PanOCT [39] and PGAP [40] perform scalable pan-genome analyses and require an all-against-all comparison using BLAST, with the running time growing approximately quadratically with the size of input data and are computationally infeasible with large datasets. They also have quadratic memory requirements, quickly exceeding the RAM available in high-performance servers for large datasets. PanOCT is a graph-based ortholog clustering tool for pan-genome analysis of closely related prokaryotic genomes exploiting conserved gene neighborhood information to separate recently diverged paralogs into distinct clusters of orthologs [39]. PGAP executes five analysis modules: cluster analysis of functional genes (the core module), pan-genome profile analysis, genetic variation analysis of functional genes, species evolution analysis, and function enrichment analysis of gene clusters. The software uses two methods to calculate all of the analyses: (i) the GF method to detect homologous genes and (ii) the MP method to detect orthologous genes. The GF method is based on the protein BLAST and MCL algorithms. All of the protein sequences are brought together, and protein BLAST is performed; the results are filtered and clustered using the MCL algorithm [16, 41]. The MP method is based on two algorithms: (i) Inparanoid to search orthologous and parologous genes using BLAST. Then, the pairwise ortholog clusters are moved to (ii) MultiParanoid, which was specifically developed to search for gene clusters among multiple strains [20, 42]. Large-scale BLAST score ratio (LS-BSR) introduces a preclustering step that makes it an order of magnitude faster than PGAP; however, it is less sensitive [43].

The software Roary [44] and BPGA [45] were created to address the computational issues related to performance and execution time. Roary performs a rapid clustering of highly similar sequences, which can reduce the running time of BLAST [16] substantially, and carefully manage RAM usage so that it increases linearly, both of which make it possible to analyze datasets with thousands of samples using commonly available computing hardware without compromising on the accuracy of results [44]. The Bacterial Pan Genome Analysis tool (BPGA) is written in perl programming language but complied in executable files for both Windows and Linux so that no module installation is required. The tool is an ultrafast computational pipeline with seven functional modules for comprehensive pan-genome studies and downstream analyses, these include (i) pan-genome profile analysis, (ii) pan-genome sequence extraction, (iii) exclusive gene family analysis, (iv) atypical GC content analysis, (v) pan-genome functional analysis, (vi) species phylogenetic analysis, and (vii) subset analysis. Other notable features include user friendly command-line interface and high-quality graphics outputs [45].

In the work of Page et al., an accuracy study was performed between four similar stand-alone pan-genome applications. They accurately analyzed the clustering quality of the programs by performing simulated data analysis based on *Salmonella enterica* serovar Typhi (*S. typhi*) CT18 (accession no. AL513382) and they used a single processor (AMD Opteron 6272) and provided 60 GB of RAM. For the study, 12 genomes with 994 identical central genes and 23 accessory genes in various combinations were created and they concluded that all the applications created clusters that are within 1% of the expected results and that the overlap of clusters is almost identical among all applications, except LS-BSR, as shown in Table 1 [44].

The tools and software packages shown so far are the main and best-known available in the scientific community. Although these tools perform different approaches in their pan-genome analysis process, most have common features and functions. Table 2 shows, briefly, each step performed by the cited tools [33, 45].

It is known that in a pan-genome analysis the greater the amount of genomes taken to the analysis the greater will be the computational costs, that is, the discovery of a pan-genome content is an NP-hard problem because comparisons between all sets of genes are necessary to solve the task [46]. The task of recognizing homologous genes becomes even more difficult in the presence of phylogenetically distant genomes, due to the variability introduced in duplication and gene transmission. This research field has the challenge of designing similarity measures that are fast and adaptive, in order to find an adequate homology pan-genome structure [46]. Therefore, in the study of Bonnici

Table 1 Accuracy of each pan-genome application on a dataset of simulated data [44]

	Core genes	Total genes	Incorrect merge
Expected	994	1017	0
PGAP	991	1012	4
PanOCT	993	1015	1
LS-BSR	974	994	23
Roary	994	1017	0

Table 2 Features of each pan-genome application

Name tools	Link	Platform	Main features
BPGA	http://iicb.res.in/bpga/index.html	Windows Linux	a, b, c, d, e, f, g, h
PGAP	https://sourceforge.net/projects/pgap/	Linux	b, c, d, e, f
PGAT	http://nwrce.org/pgat/	Online	b, h
LS-BSR	https://github.com/jasonsahl/LS-BSR	Linux	b
Roary	https://sanger-pathogens.github.io/Roary/	Linux	b, c, d, e
Panseq	https://lfz.corefacility.ca/panseq/	Online Windows Linux	b, e
GET_HOMOLOGUES	http://github.com/eead-csic-compbio/get/_homologues/	MacOS Linux	b, d, e
PanCGHweb	http://bamics2.cmbi.ru.nl/websoftware/pancgh/	Online	b
PanOCT	http://bamics2.cmbi.ru.nl/websoftware/pancgh/	Online	b
PanGP	https://pangp.zhaopage.com/	Windows Linux	c, d

Notes: The main features are represented by letters: (a) Preparation step; (b) clustering; (c) matrix generation (pan-matrix); (d) pan-genome profile analysis; (e) phylogeny construction; (f) function and pathway analysis; (g) pan-genome statistics; and (h) atypical GC content analysis.
Source: (a) From N. Chaudhari, V. Gupta, C. Dutta, BPGA—an ultra-fast pan-genome analysis pipeline, Sci. Rep. 6 (2016) 24373.

et al. [46], a computational tool called PanDelos was developed with the purpose of minimizing these challenges. It is an autonomous dictionary-based tool for the discovery of pan-genome contents among distant genomes phylogenetically.

Pan-genome analysis can be applied in many different application domains. Table 3 summarizes the main fields.

The approaches to pan-genome content discovery need to take into account that duplication and gene transmission may introduce sequence changes [30, 45]. These variations hamper the task of recognizing homologous genes, especially when ancestral genomes are no longer available. The sequences present in the core genome are transferred almost without any change, since the genes present in the core genome are often under strong evolutionary selection. The process is different for the genes present in the accessory genome because these dispensable genes have a number of inconstant and varied variations, and depending on the phylogenetic distance, the similarity between the homologous sequences tends to decrease. Organisms very close phylogenetically, when

Table 3 Description of pan-genome applications [3]

Application	Description
Microbes	Important to understand the functional and evolutionary repertoire of microbial genomes, which opens possibilities for the development of therapies and engineering applications
Metagenomics	In the metagenome, there is the possibility of revealing common adaptations to the environment, as well as the coevolution of the interactions through the pan-genome
Viruses	One of the goals of pan-genomics, both in virology and in medical microbiology, will be to fight infectious disease
Plants	A pan-genome available for a certain crop that includes its wild relatives provides a unique coordinate system to anchor all known phenotype and variation information, and will allow the identification of new genes from the available germplasm that are not present in the genome of reference(s)
Human genetic diseases	Pan-genome data structures are able to handle combinations of genomic variants with comprehensive functional annotations—for example, epigenomic datasets or gene expression
Cancer	A pan-genome of somatic cancer, representing variability in the inferred rate of change throughout the genome, would increase the identification of disease-related genomic changes based on their recurrence among individuals
Phylogenomics	The pan-genome extracts genomic features with an evolutionary signal, such as gene content tables, alignments of shared marker gene sequences, genomic SNPs, or transcribed internal spacer sequences, depending on the level of kinship of the included organisms

they are analyzed, reasonable thresholds are applied in the similarity of the sequences so that recognition of gene families occurs [46].

The Roary and EDGAR tools [47] are based on sequence alignment; however, some alternative strategies can be used to retrieve domain architecture between homologous genes [48] or for the detection of horizontal gene transfer [49], through the exploration of free alignment techniques.

PanDelos uses a different strategy, the tool seeks to discover pan-genome content in phylogenetically distant organisms based on the information theory and network analysis. The use of parameters is not a requirement of the software and the limits are automatically deduced from the context. PanDelos avoids sequence alignment by introducing a measure of similarity based on k-mers multiplicity, rather than the simple presence/absence of mers. Strategy confidence is supported by a nonempirical choice of the most appropriate k-mer length. In addition, when two sequences are identified as homologous, the selection of the least similarity between them is based on the knowledge from the mapping of the readings that were used in the sequence sequencing and reconstruction processes [46].

To infer thresholds for the discovery of paralogs, the best results from the 1vs1 comparison of the genome that was made previously, aiming at the discovery of orthologous genes, are used. The homology relationships between organisms are incorporated and form part of a global network and the groups of homologous genes used in the analysis are extracted from that network using applications with detection algorithms. According to Bonnici et al. [46], the PanDelos tool overcomes existing tools such as Roary and EDGAR in terms of execution time and accuracy of analysis, both in real applications and in synthetic analysis with simulated data.

2.3.3 Machine learning applied to pan-genome

Machine learning techniques have been widely used in the field of bioinformatics [50]. Techniques such as supervised classification, grouping, and probabilistic graphical models for discovery of knowledge, as well as deterministic and stochastic heuristics for optimization [50]. The rapidly growing data diversity, produced by modern molecular biology and made available in public databases, has stimulated the need for accurate classification and prediction algorithms [51]. With this exponential growth in the amount of biological data, computational problems arise such as the proper storage and management of this astronomical amount of information being generated, as well as problems for extracting useful information from such data. The second problem is one of the main challenges of computational biology [52]. Therefore, there is a need in the development of methods and tools capable of transforming all this heterogeneous data into biological knowledge about the fundamental mechanisms. These tools and methods should allow us to provide knowledge in the form of testable models and not just describe the content present in those data. By means of this simplifying abstraction that constitutes a model, we can obtain predictions from the system [52].

Machine learning techniques basically consist of developing algorithms for computers to optimize one performance criteria using example data or past experience. The optimized criteria can be the precision provided by a predictive model—in a modeling problem—and the value of a function of adequacy or evaluation—in an optimization problem [52].

The techniques and computational methods of machine learning are applied in several biological fields such as genomics, proteomics, microarrays, systems biology, evolution, text mining [52], and even pan-genome analysis because researchers face challenges such as processing and maintaining large datasets, while providing accurate and efficient analysis approaches. Genomics is one of the most important fields of bioinformatics, mainly because of the exponential increase in the number of available sequences that need to be processed. The initial step is to obtain and extract the location and structure of the genes, either by prediction or genomic annotation, from genome sequences [50]. In addition, it is possible to further identify regulatory elements and RNA noncoding genes present in intergenic regions.

In the field of proteomics, the main application of computational methods is the prediction of protein structure. Proteins are very complex macromolecules and therefore, the number of possible structures is enormous. This makes the prediction of protein structure a very complicated combinatorial problem, where optimization techniques are required [52].

The management of the large amount of complex experimental data is another application in which computational methods of machine learning can be used [52]. The microarray assays are one of the best known, but not the only, fields where this type of data is collected. Complex experimental data raise two different problems: first, the data need to go through a preprocessing step, that is, they need to be formatted to be used properly by machine learning algorithms. The second problem would be the analysis of the data itself, which will depend on what it is searched. In the case of microarray data, the most typical applications are identification of patterns of expression, classification, and induction of genetic networks [52]. Systems biology is another field in which biology and machine learning work very well together as it is very complex to model the life processes that occur within the cell. Thus, computer learning techniques are extremely useful in the modeling of biological networks, especially genetic networks, signal transduction networks, and metabolic pathways.

Not very different, the analysis of evolution and, especially, the reconstruction of phylogenetic trees is also used of the techniques of machine learning. Phylogenetic trees are schematic representations of organisms' evolution [52]. Generally, they were constructed according to different characteristics of the organisms (morphological characteristics, metabolic characteristics, etc.) but, nowadays, with the great amount of biological sequences available in public databases, phylogenetic tree-building algorithms are based on comparison between different genomes [50]. This comparison is made through the alignment of multiple sequences, where optimization techniques, used with machine learning algorithms, are very useful.

In the paper by Her et al. [53], a machine learning approach based on pan-genome was developed to predict antimicrobial resistance (AMR) activities in *Escherichia coli* strains. Machine learning approaches were applied in the pan-genome to better define and predict AMR. According to the authors, AMR is becoming a major problem in the developed and developing countries, and the identification of resistant or susceptible strains of certain antibiotics is essential in the fight against antibiotic-resistant pathogens [53].

Antimicrobial-resistant pathogens (AMR) have an ultrarapid mutation rate which renders most of the existing drugs against superbugs unavoidable, and existing classes of antibiotics are probably the best there will ever be [54]. Another study published in 2013 also identified that additional economic costs due to AMR could reach $55 billion and that trivial bacterial infections, such as hip replacements, for example, could increase the mortality rate from approximately 0% to 30% [55].

Pan-genome was also used in the analysis of diversity, virulence, and AMR phenotypes in the organism *Klebsiella pneumoniae* [56]. In this study, they found that *K. pneumoniae* can be divided into three distinct groups, and that certain branches in all three groups may be hypervirulent or resistant to multiple drugs [56]. In addition, in another study a computational approach, called Scoary, was developed to make an association between the genetic components found in the pan-genome with the observed phenotypic traits and to identify the gene pools that were associated with activities of high level of AMR, such as resistance to linezolid in *Staphylococcus epidermidis* [57]. These examples have suggested that the pan-genome idea can be very useful in defining genetic components that can contribute to the phenotypes of living organisms.

The PATRIC database is known as one of the most comprehensive antibiotic resistance databases that collects genes, proteins, and genomic information related to the resistance or susceptibility of pathogens to various antibiotics [58]. PATRIC has a collection of more than 80,000 bacterial genomes available in its database allowing scientists to understand the mechanisms of AMR in terms of genes, proteins, and genomes.

Thus, it was developed a pan-genome-based approach to characterize strains that are resistant to antibiotics and strains of *E. coli* were used as a model in which 59 strains of *E. coli* from the PATRIC database were selected [58]. By using machine learning techniques through genetic algorithms (GA), it was obtained better predictive performance than the sets of genes established in the literature, suggesting that gene sets selected by GA may justify a more in-depth analysis in investigating more details on how *E. coli* fights against antibiotics.

3 Challenges

The data analyzed in a pan-genome study have characteristics of Big Data such as volume, variety, speed, and veracity. These studies presented great challenges to algorithm and software developers, especially due to the size of the data generated by the new-generation sequencers, the data heterogeneity, and their complex interaction [3].

The International Cancer Genome Consortium has accumulated a dataset of more than two petabytes in just 5 years, resulting in the need to store data in clouds, providing a scalable, dynamic and parallel way of processing data in an inexpensive, flexible, reliable, and safe manner. Currently, there are large providers with a complex computing infrastructure and large public repositories (e.g., National Center for Biotechnology Information, European Bioinformatics Institute, and DNA Data Bank of Japan) that assist both researchers who choose to download/upload data for analysis, but also provide a secure and reliable storage environment for this large set of information. Distributed and parallel computing has also been used as a resource to deal with the considerable volume of data stored in public databases [3].

Pan-genome has also introduced new challenges for data visualization. As the relationships between several genomes can be highly complex and the homology relations can vary widely with each dataset studied, it became necessary to obtain new ways of visualizing these relations in their total complexity without loss of information. In general, mathematical approaches to comparing sets are used to evaluate homology relationships such as Venn and Flower Plots diagrams [3].

New data visualization packages for pan-genome are developed to facilitate the research and generate a better visualization of the relations of homology existing in the genomes. As an example we can mention the UpSetR package that has provided users with an improved alternative to the Venn chart; while a normal Venn graph accepts up to five data sets at most (five genomes), the visualization offered by UpSetR does not have limit to the data set analyzed [59].

3.1 Pan-genome analysis with draft genomes

Pan-genome analysis are usually performed using complete genomes to analyze the complete gene repertoire. However, depositing a complete genome of an organism in a public database is not an easy task, the finalization of this process is directly linked to a number of variables, and therefore, the number of drafts deposited genomes increases exponentially, thus increasing the number of projects that use this type of genome in pan-genome analysis. According to the Genomes OnLine Database [60], the number of complete and draft genomes deposited in public databases in 2017 reached 4311 and 31,332, respectively. Bacteria have a greater number of reports of genomes being deposited, due to their compact nature, being relatively less complex in the sequencing process, and due to the importance of their application in various fields, such as biotechnology, agriculture, medicine, etc. [61].

Working with draft genomes in any type of analysis, even in pan-genome analysis, brings a series of challenges and requires greater attention precisely because it is not yet finalized, that is, the genomic repertoire of this genome is not yet totally represented. In addition, draft genomes may contain a number of errors, such as broken products or frameshifts. Several factors may explain the reason why a given genome was not yet fully finalized, such as errors in sequencing, assembly, or even genomic annotation errors. In this case, there may be a lot that has not yet been represented, such as important products and functions for the bacteria, which may imply errors in the final result of a given analysis, such as pan-genome. Therefore, an important step before using a draft genome in any type of analysis is to seek to represent its gene repertoire as much as possible. In the study by Veras et al. [62], for example, a computational tool was developed in JAVA programming language, called Pan4draft, especially to work with drafts genomes in pan-genome analysis. Pan4draft uses the PGAP software pipeline to perform the pan-genome analysis, but performs a series of previous steps, automatically integrating several tools, responsible for seeking a better representation of the gene repertoire of these genomes drafts, thus increasing the accuracy of the pan-genome analysis [62].

3.2 Perspectives for pan-genome applied to the human genome

The human genome project was founded in 1990, and after numerous surveys carried out in several centers, it is now known that *Homo sapiens* cannot be described only by a single reference sequence. Although the variation occurring in the human genome is inferior in comparison to microbes and plants, the first attempt to construct a human pan-genome in 2009 (based on the human reference genome and other two genomes) estimated that up to 40 megabases of sequence including the coding regions of proteins, were absent from the reference genome [63]. Still in 2009, researchers estimated that gene counts ranged from 73 to 87 genes found in two randomly selected individuals [64].

Such observed differences are increasingly associated with genetic disorders such as autism, Parkinson's disease, and Alzheimer's, causing research to turn even further to the study of these variations observed in our species [65, 66]. Researchers at the Case Western Reserve University have identified that more than 300 small sequences absent from the reference genome were present in at least 1% of the human population, leading to a reconsideration of the whole concept of the reference genome used not only for prokaryotes but also for eukaryotes [67].

In this way, it is possible to evaluate that we can still improve in many aspects the approaches and methodologies used in pan-genomic studies. The main objective in overcoming such challenges is to find a more complete scenario that presents all the desired characteristics when analyzing certain species of both prokaryotes and eukaryotes.

4 Conclusion and future direction

With the development of sequencing technologies, thousands of biological data have become accessible in the past years. In this context, in order to take the most advantage of the data produced by NGS platforms, using a reference, it was necessary to do a paradigm shift: instead of focusing only on a single reference genome, use a pan-genome, that is, a representation of the entire gene repertoire of a particular species or phylogenetic clade. Thus, life sciences have entered the era of pan-genomics, which is known to represent "all" major genetic variation of a collection of genomes of interest. The search for sequence similarity is the important step in the pan-genome analysis and in comparative genomics in general.

Nowadays, the process of similarity search and pan-genome visualization are two of the wide variety of particular computational challenges that need to be considered. For this, novel different computational methods and paradigms are needed over the years, making the computational pan-genomics a subarea of research in rapid extension.

A current pan-genome analysis can be considered a "one-dimensional" approach by mainly working with genomes only as sequences and thus concentrating on storing and analyzing sequences and relations between certain parts of subsequences, such as variant alleles and their interconnections, genes, and/or transcriptomes.

However, new technologies that are emerging in rapid development allow to infer the pan-genome with three-dimensional conformation, that is, in the medium term, one can expect to be able to raise the pan-genome in up to three dimensions. This will mean that future three-dimensional pan-genomes will not only represent all sequence variation of the species or genus, but also will encode their spatial organization, as well as their mutual relationships in this regard.

References

[1] B. Hall, G. Ehrlich, F. Hu, Pan-genome analysis provides much higher strain typing resolution than multi-locus sequence typing, Microbiology 156 (2010) 1060–1068.

[2] M. Pallen, B. Wren, Bacterial pathogenomics, Nature 449 (2007) 835.

[3] Computational Pan-Genomics Consortium, Computational pan-genomics: status, promises and challenges, Brief. Bioinf. 19 (2016) 118–135.

[4] H. Tettelin, V. Masignani, M. Cieslewicz, C. Donati, D. Medini, N. Ward, S. Angiuoli, J. Crabtree, A. Jones, A. Durkin, Genome analysis of multiple pathogenic isolates of *Streptococcus agalactiae*: implications for the microbial "pan-genome", Proc. Natl. Acad. Sci. USA 102 (2005) 13950–13955.

[5] I. Stevenson, John Ray and his contributions to plant and animal classification, J. Hist. Med. Allied Sci. 2 (1947) 250–261.

[6] L. Olendzenski, J. Gogarten, M. Gogarten, J. Gogarten, L. Olendzenski, Horizontal Gene Transfer: Genomes in Flux, Humana Press, Totowa, NJ, 2009.

[7] L. Rouli, V. Merhej, P. Fournier, D. Raoult, The bacterial pangenome as a new tool for analysing pathogenic bacteria, New Microbes New Infect. 7 (2015) 72–85.

[8] G. Vernikos, D. Medini, D. Riley, H. Tettelin, Ten years of pan-genome analyses, Curr. Opin. Microbiol. 23 (2015) 148–154.

[9] E. Bosi, J. Monk, R. Aziz, M. Fondi, V. Nizet, B. Palsson, Comparative genome-scale modelling of *Staphylococcus aureus* strains identifies strain-specific metabolic capabilities linked to pathogenicity, Proc. Natl. Acad. Sci. USA 113 (2016) E3801–E3809.

[10] T. Delmont, A. Eren, Linking pangenomes and metagenomes: the *Prochlorococcus metapangenome*, PeerJ 6 (2018) e4320.

[11] K. Sieber, R. Bromley, J. Hotopp, Lateral gene transfer between prokaryotes and eukaryotes, Exp. Cell Res. 358 (2017) 421–426.

[12] J. Huang, Horizontal gene transfer in eukaryotes: the weak-link model, Bioessays 35 (2013) 868–875.

[13] B. Read, J. Kegel, M. Klute, A. Kuo, S. Lefebvre, F. Maumus, C. Mayer, J. Miller, A. Monier, A. Salamov, Pan genome of the phytoplankton *Emiliania underpins* its global distribution, Nature 499 (2013) 209.

[14] P. Lapierre, J. Gogarten, Estimating the size of the bacterial pan-genome, Trends Genet. 25 (2009) 107–110.

[15] H. Tettelin, D. Riley, C. Cattuto, D. Medini, Comparative genomics: the bacterial pan-genome, Curr. Opin. Microbiol. 11 (2008) 472–477.

[16] S. Altschul, W. Gish, W. Miller, E. Myers, D. Lipman, Basic local alignment search tool, J. Mol. Biol. 215 (1990) 403–410.

[17] F. Chen, A. Mackey, C. Stoeckertjr, D. Roos, OrthoMCL-DB: querying a comprehensive multi-species collection of ortholog groups, Nucleic Acids Res. 34 (2006) D363–D368.
[18] M. Barakat, P. Ortet, D. Whitworth, P2RP: a web-based framework for the identification and analysis of regulatory proteins in prokaryotic genomes, BMC Genomics 14 (2013) 269.
[18a] F. Del Chierico, M. Ancora, M. Marcacci, C. Cammà, L. Putignani, S. Conti, Bacterial pangenomics [Internet], in: A. Mengoni, M. Galardini, M. Fondi (Eds.), Methods in Molecular Biology, Springer, New York, NY, 2015, pp. 31–47. Available from: http://link.springer.com/10.1007/978-1-4939-1720-4.
[19] K. O'brien, M. Remm, E. Sonnhammer, Inparanoid: a comprehensive database of eukaryotic orthologs, Nucleic Acids Res. 33 (2005) D476–D480.
[20] A. Alexeyenko, I. Tamas, G. Liu, E. Sonnhammer, Automatic clustering of orthologs and inparalogs shared by multiple proteomes, Bioinformatics 22 (2006) e9–e15.
[21] J. Besemer, M. Borodovsky, GeneMark: web software for gene finding in prokaryotes, eukaryotes and viruses, Nucleic Acids Res. 33 (2005) W451–W454.
[22] A. Delcher, K. Bratke, E. Powers, S. Salzberg, Identifying bacterial genes and endosymbiont DNA with Glimmer, Bioinformatics 23 (2007) 673–679.
[23] D. Hyatt, G.L. Chen, P.F. Locascio, M.L. Land, F.W. Larimer, L.J. Hauser, Prodigal: prokaryotic gene recognition and translation initiation site identification, BMC Bioinf. 11 (2010) 119.
[24] A. Darling, B. Mau, N. Perna, progressiveMauve: multiple genome alignment with gene gain, loss and rearrangement, PLoS ONE 5 (2010) e11147.
[25] A. Jacobsen, R. Hendriksen, F. Aaresturp, D. Ussery, C. Friis, The *Salmonella enterica* pan-genome, Microb. Ecol. 62 (2011) 487.
[26] K. Tamura, D. Peterson, N. Peterson, G. Stecher, M. Nei, S. Kumar, MEGA5: molecular evolutionary genetics analysis using maximum likelihood, evolutionary distance, and maximum parsimony methods, Mol. Biol. Evol. 28 (2011) 2731–2739.
[27] K. Katoh, K. Misawa, K. Kuma, T. Miyata, MAFFT: a novel method for rapid multiple sequence alignment based on fast Fourier transform, Nucleic Acids Res. 30 (2002) 3059–3066.
[28] C. Laing, C. Buchanan, E. Taboada, Y. Zhang, A. Kropinski, A. Villegas, J. Thomas, V. Gannon, Pan-genome sequence analysis using Panseq: an online tool for the rapid analysis of core and accessory genomic regions, BMC Bioinf. 11 (2010) 461.
[29] J. Bayjanov, R. Siezen, S. Vanhijum, PanCGHweb: a web tool for genotype calling in pangenome CGH data, Bioinformatics 26 (2010) 1256–1257.
[30] M. Brittnacher, C. Fong, H. Hayden, M. Jacobs, M. Radey, L. Rohmer, PGAT: a multistrain analysis resource for microbial genomes, Bioinformatics 27 (2011) 2429–2430.
[31] S. Kurtz, A. Phillippy, A. Delcher, M. Smoot, M. Shumway, C. Antonescu, S. Salzberg, Versatile and open software for comparing large genomes, Genome Biol. 5 (2004) R12.
[32] J. Bayjanov, M. Wels, M. Starrenburg, J. Vanhylckamavlieg, R. Siezen, D. Molenaar, PanCGH: a genotype-calling algorithm for pangenome CGH data, Bioinformatics 25 (2009) 309–314.
[33] J. Xiao, Z. Zhang, J. Wu, J. Yu, A brief review of software tools for pangenomics, Genom. Proteom. Bioinform. 13 (2015) 73–76.
[34] B. Contreras-moreira, P. Vinuesa, GET_HOMOLOGUES, a versatile software package for scalable and robust microbial pan-genome analysis, Appl. Environ. Microbiol. 79 (2013) 7696–7701.
[35] R. Finn, J. Clements, S. Eddy, HMMER web server: interactive sequence similarity searching, Nucleic Acids Res. 39 (2011) W29–W37.
[36] L. Li, C. Stoeckert, D. Roos, OrthoMCL: identification of ortholog groups for eukaryotic genomes, Genome Res. 13 (2003) 2178–2189.
[37] D. Kristensen, L. Kannan, M. Coleman, Y. Wolf, A. Sorokin, E. Koonin, A. Mushegian, A low-polynomial algorithm for assembling clusters of orthologous groups from intergenomic symmetric best matches, Bioinformatics 26 (2010) 1481–1487.
[38] Y. Zhao, X. Jia, J. Yang, Y. Ling, Z. Zhang, J. Yu, J. Wu, J. Xiao, PanGP: a tool for quickly analyzing bacterial pan-genome profile, Bioinformatics 30 (2014) 1297–1299.
[39] D. Fouts, L. Brinkac, E. Beck, J. Inman, G. Sutton, PanOCT: automated clustering of orthologs using conserved gene neighborhood for pan-genomic analysis of bacterial strains and closely related species, Nucleic Acids Res. 40 (2012) e172.

[40] Y. Zhao, J. Wu, J. Yang, S. Sun, J. Xiao, J. Yu, PGAP: pan-genomes analysis pipeline, Bioinformatics 28 (2011) 416–418.
[41] A. Enright, S. Vandongen, C. Ouzounis, An efficient algorithm for large-scale detection of protein families, Nucleic Acids Res. 30 (2002) 1575–1584.
[42] G. Ostlund, T. Schmitt, K. Forslund, T. Köstler, D. Messina, S. Roopra, O. Frings, E. Sonnhammer, InParanoid 7: new algorithms and tools for eukaryotic orthology analysis, Nucleic Acids Res. 38 (2009) D196–D203.
[43] J. Sahl, J. Caporaso, D. Rasko, P. Keim, The large-scale blast score ratio (LS-BSR) pipeline: a method to rapidly compare genetic content between bacterial genomes, PeerJ 2 (2014) e332.
[44] A. Page, C. Cummins, M. Hunt, V. Wong, S. Reuter, M. Holden, M. Fookes, D. Falush, J. Keane, J. Parkhill, Roary: rapid large-scale prokaryote pan genome analysis, Bioinformatics 31 (2015) 3691–3693.
[45] N. Chaudhari, V. Gupta, C. Dutta, BPGA—an ultra-fast pan-genome analysis pipeline, Sci. Rep. 6 (2016) 24373.
[46] V. Bonnici, R. Giugno, V. Manca, PanDelos: a dictionary-based method for pan-genome content discovery, BMC Bioinf. 19 (2018) 437.
[47] J. Blom, J. Kreis, S. Spänig, T. Juhre, C. Bertelli, C. Ernst, A. Goesmann, EDGAR 2.0: an enhanced software platform for comparative gene content analyses, Nucleic Acids Res. 44 (2016) W22–W28.
[48] D. Syamaladevi, A. Joshi, R. Sowdhamini, An alignment-free domain architecture similarity search (ADASS) algorithm for inferring homology between multi-domain proteins, Bioinformation 9 (2013) 491.
[49] G. Bernard, C. Chan, Y. Chan, X. Chua, Y. Cong, J. Hogan, S. Maetschke, M. Ragan, Alignment-free inference of hierarchical and reticulate phylogenomic relationships, Brief. Bioinform. 20 (2019) 426–435.
[50] P. Baldi, S. Brunak, F. Bach, Bioinformatics: The Machine Learning Approach, MIT Press, Cambridge, 2001.
[51] H. Bhaskar, D. Hoyle, S. Singh, Machine learning in bioinformatics: a brief survey and recommendations for practitioners, Comput. Biol. Med. 36 (2006) 1104–1125.
[52] P. Larranaga, B. Calvo, R. Santana, C. Bielza, J. Galdiano, I. Inza, J. Lozano, R. Armañanzas, G. Santafé, A. Pérez, Machine learning in bioinformatics, Brief. Bioinform. 7 (2006) 86–112.
[53] H. Her, Y. Wu, A pan-genome-based machine learning approach for predicting antimicrobial resistance activities of the *Escherichia coli* strains, Bioinformatics 34 (2018) i89–i95.
[54] M. Cormican, A. Vellinga, Existing classes of antibiotics are probably the best we will ever have, Br. Med. J. (Online) 344 (2012).
[55] R. Smith, J. Coast, The true cost of antimicrobial resistance, BMJ 346 (2013) f1493.
[56] K. Holt, H. Wertheim, R. Zadoks, S. Baker, C. Whitehouse, D. Dance, A. Jenney, T. Connor, L. Hsu, J. Severin, Genomic analysis of diversity, population structure, virulence, and antimicrobial resistance in *Klebsiella pneumoniae*, an urgent threat to public health, Proc. Natl. Acad. Sci. USA 112 (2015) E3574–E3581.
[57] O. Brynildsrud, J. Bohlin, L. Scheffer, V. Eldholm, Rapid scoring of genes in microbial pan-genome-wide association studies with Scoary, Genome Biol. 17 (2016) 238.
[58] A. Wattam, J. Davis, R. Assaf, S. Boisvert, T. Brettin, C. Bun, N. Conrad, E. Dietrich, T. Disz, J. Gabbard, Improvements to PATRIC, the all-bacterial bioinformatics database and analysis resource center, Nucleic Acids Res. 45 (2016) D535–D542.
[59] J. Conway, A. Lex, N. Gehlenborg, UpSetR: an R package for the visualization of intersecting sets and their properties, Bioinformatics 33 (2017) 2938–2940.
[60] S. Mukherjee, D. Stamatis, J. Bertsch, G. Ovchinnikova, O. Verezemska, M. Isbandi, A. Thomas, R. Ali, K. Sharma, N. Kyrpides, Genomes OnLine Database (GOLD) v. 6: data updates and feature enhancements, Nucleic Acids Res. 45 (2016) D446–D456. D1.
[61] V. Wanchai, P. Patumcharoenpol, I. Nookaew, D. Ussery, dBBQs: dataBase of bacterial quality scores, BMC Bioinf. 18 (2017) 483.
[62] A. Veras, F. Araujo, K. Pinheiro, L. Guimarães, V. Azevedo, S. Soares, A. Dasilva, R. Ramos, Pan4-Draft: a computational tool to improve the accuracy of pan-genomic analysis using draft genomes, Sci. Rep. 8 (2018) 9670.

[63] R. Li, Y. Li, H. Zheng, R. Luo, H. Zhu, Q. Li, W. Qian, Y. Ren, G. Tian, J. Li, Building the sequence map of the human pan-genome, Nat. Biotechnol. 28 (2010) 57.
[64] C. Alkan, J. Kidd, T. Marques-Bonet, G. Aksay, F. Antonacci, F. Hormozdiari, J. Kitzman, C. Baker, M. Malig, O. Mutlu, Personalized copy number and segmental duplication maps using next-generation sequencing, Nat. Genet. 41 (2009) 1061.
[65] H. Yoo, Genetics of autism spectrum disorder: current status and possible clinical applications, Exp. Neurobiol. 24 (2015) 257–272.
[66] C. Klein, A. Westenberger, Genetics of Parkinson's disease, Cold Spring Harb. Perspect. Med. 2 (2012) a008888.
[67] Y. Liu, M. Koyutürk, S. Maxwell, M. Xiang, M. Veigl, R. Cooper, B. Tayo, L. Li, T. Laframboise, Z. Wang, Discovery of common sequences absent in the human reference genome using pooled samples from next generation sequencing, BMC Genomics 15 (2014) 685.

Further reading

[68] D. Andersson, B. Levin, The biological cost of antibiotic resistance, Curr. Opin. Microbiol. 2 (1999) 489–493.
[69] J. Bower, H. Bolouri, Computational Modeling of Genetic and Biochemical Networks, MIT Press, Cambridge, 2004.
[70] J. Hogg, F. Hu, B. Janto, R. Boissy, J. Hayes, R. Keefe, J. Post, G. Ehrlich, Characterization and modeling of the *Haemophilus influenzae* core and supragenomes based on the complete genomic sequences of Rd and 12 clinical nontypeable strains, Genome Biol. 8 (2007) R103.
[71] G. Kettler, A. Martiny, K. Huang, J. Zucker, M. Coleman, S. Rodrigue, F. Chen, A. Lapidus, S. Ferriera, J. Johnson, Patterns and implications of gene gain and loss in the evolution of *Prochlorococcus*, PLoS Genet. 3 (2007) e231.
[72] M. Krallinger, R. Erhardt, A. Valencia, Text-mining approaches in molecular biology and biomedicine, Drug Discov. Today 10 (2005) 439–445.

CHAPTER 3

Evolutionary pan-genomics and applications

Basant K. Tiwary
Centre for Bioinformatics, Pondicherry University, Pondicherry, India

1 Introduction

The human genome was completely sequenced and assembled in the form of a reference sequence in the year 2001 [1]. The advent of next-generation sequencing methods has paved the way for the resequencing of entire populations of a particular species or a phylogenetic clade in a short span of time with minimum cost [2]. Thus, there was a paradigm shift in the concept of genome from a single reference genome to pan-genome after this technological revolution. The pan-genome represents a full set of genes in a particular species consisting of three major categories, a core genome which is present in all individuals of a species, accessory genome which is present in some individuals of a species, and singleton or unique genome restricted to one individual only (Figs. 1 and 2) [3]. The genes present in the core genome participate in the basic metabolic functions of the cell like housekeeping and conferring antibiotic resistance in bacteria. In addition, the core genome is treated as a conserved genomic unit to infer evolutionary relationships among different strains of bacteria. On the other hand, accessory genes frequently undergo gene gain/loss events and are often subjected to horizontal gene transfer to facilitate adaptations in a novel ecological niche.

The first ever concept of the pan-genome was developed by Tettelin et al. [3] during their study on a bacterial species, *Streptococcus agalactiae*. Since then, the research work on pan-genomes was extended to many prokaryotic species followed by some work on eukaryotic species. The pan-genome may also be defined as a combined analysis of a collection of genomic sequences treated as reference for particular species [4]. A pan-genome analysis can generate three types of new information; the size of the core genome, size of the accessory genome, and gene gain/loss events with addition of new samples. A successful study regarding a pan-genome is based on the quality of the reference assembly, quality of annotation, and the selection of appropriate individuals for study. In prokaryotes, the core part is associated with vertical transmission and homologous recombination whereas the variable part is related to horizontal gene transfer and site-specific recombination. Even the core part and accessory part may follow different evolutionary trajectories in a particular species. Generally, the core part provides a stable

Pan-genomics: Applications, Challenges, and Future Prospects
https://doi.org/10.1016/B978-0-12-817076-2.00003-2
© 2020 Elsevier Inc.
All rights reserved.

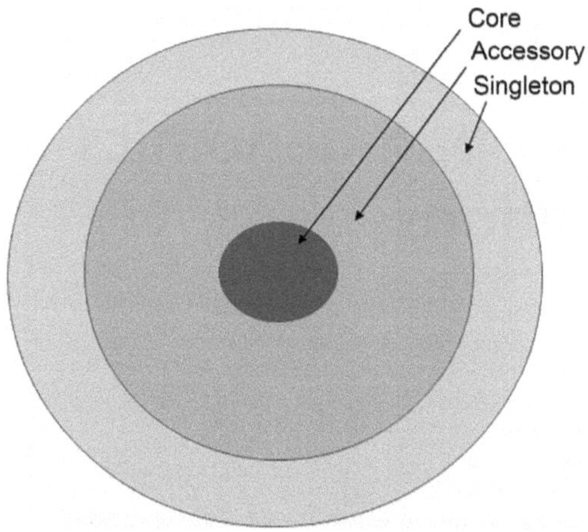

Fig. 1 A pan-genome can be classified as the core, accessory, and singleton parts.

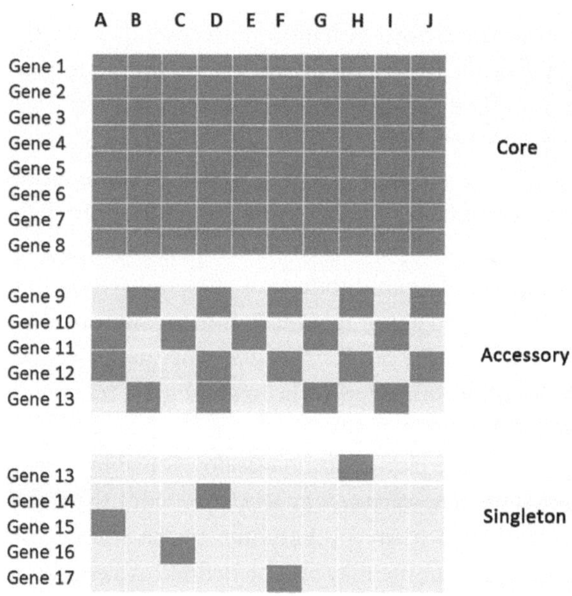

Fig. 2 Distribution of individual genes as core genes, accessory genes, and singleton genes in the pan-genome of 10 strains (A–J) of a species.

metabolic and genomic support to the species and the variable part, on the other hand, is responsible for high diversity among individuals in a population [5]. The majority of this variable part is restricted to the flexible genomic islands having size more than 10 kb [6]. Therefore, the desirable features of an ideal pan-genome are completeness (i.e., includes all functional elements), stability (i.e., unique characteristic features), comprehensibility (i.e., includes genomic information of all individuals or species), and efficiency (i.e., organized data structures) [4]. The evolutionary history of a species can be reconstructed using their genome sequences. The evolutionary signals in the genome in the form of gene content, shared marker gene or single-nucleotide polymorphisms (SNPs) across the genome may provide useful information during phylogenetic reconstruction for inferring evolutionary relationships among strains or species.

2 Computational methods in evolutionary pan-genomics

Pan-genomes are constructed from various many available resources such as the reference sequence and its variants, raw reads and haplotype reference panels. The data structure of a pan-genome is represented by a coordinate system with explicit information on all genetic variants (Fig. 3).

The simplest form of a pan-genome is a set of unaligned sequences which does not provide much useful information. A better representation of the pan-genome is multiple sequence alignment, which provides a coordinate system with many columns specifying the particular location of genes on the pan-genome [7]. However, it is only suitable for small genomic segments and does not demonstrate major genomic rearrangements like inversions and translocations. More efficiently k-mers, which are sequences with length k, provide a better representation of the pan-genome in form of de Bruijn graph (DBG) [8]. DBG is widely used as an algorithm for assembly of short reads. Further, the colored DBG suits better for the pan-genome and provides a promising method for representing the pan-genome [9]. The color of each k-mer is assigned as per the input sample in a colored DBG. A graph structure with nodes and edges can also represent a pan-genome with individual genomes as edges and coordinate system as nodes. The sequence graph may be cyclic or acyclic in nature. Even there are haplotype-centric models, where each haplotype denotes a sequence of fixed length. The positional Burrows-Wheeler Transform (PBWT) is an efficient data structure to represent a haplotype panel with compression facility [10]. Another widely used haplotype-centric model is the Li-Stephens model, which is a hidden Markov model with a matrix of states with rows indicating haplotypes and columns indicating each variant [11].

There are many popular software packages available for evolutionary pan-genome analysis (Table 1). They are primarily used for identifications of SNPs, orthologous genes, reconstruction of phylogenetic tree and profiling of different parts of pan-genome. Pan-seq is the first online and most popular tool for identification of core and variable parts of the genome along with SNPs associated with the core genomic region [12]. However, functional enrichment analysis to understand the functional role of each element of the genomic region is not available in this tool. The PanCGHweb is another online tool to perform pangenomic microarray analysis for the classification of orthologs and phylogenetic reconstruction among related strains [13]. The major limitation of this algorithm is not to facilitate RNA-Seq data analysis. The CAMBer can identify multigene families

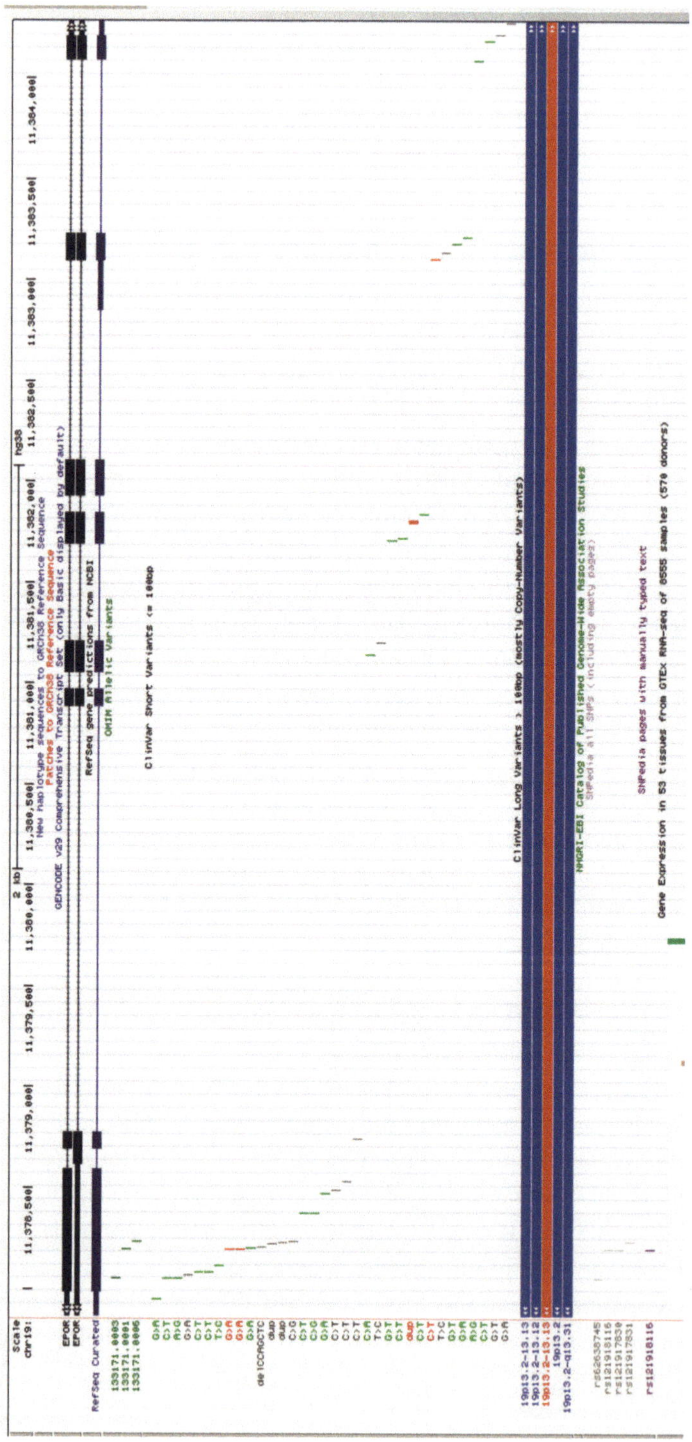

Fig. 3 A sequence coordinate graph generated using the UCSC browser showing genetic variants in the form of single-nucleotide polymorphism and copy number variants of the erythropoietin receptor gene in human.

Table 1 Popular software for evolutionary pangenomics

Name	Authors	Reference
Panseq	Laing et al. (2010)	[12]
PanCGHweb	Bayjanov et al. (2010)	[13]
CAMBer	Wozniak et al. (2011)	[14]
PGAT	Brittnacher et al. (2011)	[15]
PGAP	Zhao et al. (2012)	[16]
GET_HOMOLOGUES	Contreras-Moreira and Vinuesa (2013)	[17]
GET_HOMOLOGUES-EST	Contreras-Moreira et al. (2017)	[18]
PanTools	Sheikhizadeh et al. (2016)	[19]
EDGAR 2.0	Blom et al. (2016)	[20]
PanX	Ding et al. (2018)	[21]
Micropan	Snipen and Liland (2015)	[22]
FindMyFriends	Pedersen (2015)	[23]
Piggy	Thorpe et al. (2018)	[24]
PanViz	Pedersen et al. (2017)	[25]

and mutations in a variety of bacterial strains but does not provide evolutionary analysis of these strains [14]. The prokaryotic genome analysis tool (PGAT) is a web-based database tool with multiple functions for limited number of species [15]. The functions of this tool include identification of SNPs, comparison of gene order across the strains, association with the KEGG pathway and Cluster of Orthologous Groups of proteins (COG). PGAP is another package with standalone facility for creating a pan-genomic profile, and evolutionary analysis of different species along with functional enrichment of strains of a particular pan-genome [16]. GET_HOMOLOGUES is a standalone program that can perform a variety of tasks such as identification of homologues, profiling of pangenome with graphics and reconstruction of the phylogenetic tree of bacterial species [17]. An improved version of this program, GET_HOMOLOGUES-EST was developed for the evolutionary analysis of intraspecific eukaryotic pan-genomes [18]. PanTools is a java application-based tool both for prokaryotes and eukaryotes using de Bruijn graph algorithm for constructing, annotating, and grouping the homologous genes of the pan-genome [19]. The current version of the web server, EDGAR 2.0 provides very powerful phylogenetic analysis features such as average amino acid identity and average nucleotide identity among microbial genomes [20]. Recently, PanX was developed for evolutionary analysis of microbial pan-genomes with capability to display alignment, reconstruct the phylogenetic tree, infer gene gain/loss, and map mutations on the core genome [21]. Micropan is an R-package available in the R language and environment [26] for computing various properties of microbial pan-genome such as pan-genome size, openness or closeness of pan-genome, genomic fluidity, and pan-genome phylogenetic tree [22]. Another R-package FindMyFriends has a broader scope than the Micropan in the sense that it does alignment-free sequence-guided comparison following cosine similarity of k-mer vectors instead of depending on a tedious all-vs-all BLAST process [23]. Piggy detects highly divergent intergenic regions upstream of coding sequences in microbial pan-genomes [24]. PanViz is an interactive visualization tool for pangenomes written in JavaScript but can be accessed in the R environment using a package, PanVizGenerator [25].

3 Evolutionary pan-genomics of prokaryotes

Microbes are most widely studied organisms due to their small genome size and their clinical importance. An evolutionary study of the pan-genome may open up new avenues for diagnosis and therapy of microbial infections. Therefore, due to the availability of a large number of sequences of different strains of a particular microbe, a complete pan-genome of a microbial species can be created with full information regarding individual variations across strains. Microbes provide an extremely variable genome generated by point mutations in the form of SNPs and subsequently fixed in the population under the influence of evolutionary forces such as natural selection and genetic drift. The pan-genomic studies on various microbes have been conducted and core genome size varies widely across bacterial species (Table 2) [27–44]. The highest core genome size in the terms of number genes (3972) was observed in the pathogen for anthrax (*Bacillus anthracis*) whereas the minimum core genome (746) was found in *Gardnerella vaginalis*.

Majority of bacterial species have demonstrated an open pan-genome that needs a large number of additional genomes to further expand the pan-genome of the species. For example, the *E. coli* genome is an open genome and expanding further with the discovery of a new strain. On the other hand, the pan-genome of a species is fully saturated and characterized in the closed pan-genome. *Bacillus anthracis* is the best example of a closed genome because it became fully saturated after the sequencing of the first four genomes. The Heaps law model provides a metric called the alpha parameter to measure the openness or closeness of a pan-genome [43]. The alpha value is always more than 1 in the case of a closed pan-genome but it is less than 1 for the open pan-genome. Horizontal gene transfer (HGT) is another vital evolutionary force in microbial evolution for adaptation to ever-changing environments. HGT is a predominant force of microbial evolution supplemented by a lesser contribution of gene duplication in the evolutionary process [45]. Considering the fast pace of sequencing of microbial genomes, the size of the accessory genome is expanding with increasing number of samples whereas the size of the core genome is concomitantly shrinking with more number of sequenced samples. McInerney et al. opined that the effective population size and tendency to occupy novel ecological niches are two major factors regulating the pan-genome size in microbes [45].

Table 2 Pan-genomic features of bacterial species

Species	Core genome size (No. of genes)	Authors	Reference
Streptococcus agalactiae	1806	Tettelin et al. (2005)	[3]
Streptococcus pyogenes	1376	Lefebure et al. (2007)	[27]
Haemophilus influenzae	1450	Hogg et al. (2007)	[28]
Streptococcus pneumoniae	1400	Hiller et al. (2007)	[29]
Escherichia coli	2344	Rasko et al. (2008)	[30]
Neisseria meningitidis	1337	Schoen et al. (2008)	[31]
Enterococcus faecium	2172	van Schaik et al. (2010)	[32]
Yersinia pestis	3668	Eppinger et al. (2010)	[33]
Clostridium difficile	1033	Scaria et al. (2010)	[34]
Lactobacillus casei	1715	Broadbent et al. (2012)	[35]
Gardnerella vaginalis	746	Ahmed et al. (2012)	[36]
Borrelia burgdoferi	1200	Mongodin et al. (2013)	[37]
Lactobacillus paracasei	1800	Smokvina et al. (2013)	[38]
Campylobacter jejuni	1042	Meric et al. (2014)	[39]
Campylobacter coli	947	Meric et al. (2014)	[39]
Moritella viscosa	3737	Karlsen et al. (2017)	[44]
Pseudoalteromonas	1571	Bosi et al. (2017)	[40]
Bacillus amyloliquefaciens	2870	Kim et al. (2017)	[41]
Bacillus anthracis	3972	Kim et al. (2017)	[41]
Bacillus cereus	1656	Kim et al. (2017)	[41]
Bacillus subtilis	1022	Kim et al. (2017)	[41]
Bacillus thuringiensis	2299	Kim et al. (2017)	[41]
Lactobacillus plantarum	2144	Inglin et al. (2018)	[42]

4 Evolutionary pan-genomics of eukaryotes

The evolution of the eukaryote pan-genome is different from prokaryotes due to the fact that gene duplication is a predominant process in eukaryotes in contrast to HGT in prokaryotes. The genomic variations in eukaryotes are manifested in the form of SNPs, copy number variants (CNVs) (i.e., variable number of copies of a sequence in individuals), and presence or absence of variants (PAVs) (i.e., presence or absence of a sequence in individuals). Pan-genome studies on crop plants using quantitative trait loci (QTL), genome-wide association mapping and phylogenetic analysis may decipher the SNPs associated with crop productivity. Most of the genomic SNPs are not selected by natural selection and fixed by random genetic drift in the population and thus are selectively neutral. The presence of a nonsynonymous SNP changes the encoded amino acid and thereby alters the overall protein structure and function. On the other hand, a synonymous SNP does not change the encoded amino acid and contributes in retaining the

Table 3 Pan-genomic features in eukaryotes

Species	Core genome size (No. of genes)	Authors	Reference
Zymoseptoria tritici	9149	Plissonneau et al. (2018)	[48]
Glycine soja	28712	Li et al. (2014)	[49]
Oryza sativa	23914	Sun et al. (2017)	[47]
Brassica oleracia	49895	Golicz et al. (2016)	[47]

overall stability of native protein structure. The availability of a pan-genome instead of a single reference sequence may improve the efficiency of SNP discovery in crops. Further, it will discriminate SNPs located in the core and variable regions of the pan-genome. A phylogenetic study on the variable and conserved sites in different individuals of a pan-genome will provide an insight into evolutionary trend in a population. The concept of molecular clock can be implicated using SNPs to estimate the divergence time of species. Pan-genome-based phylogenetic studies have been performed in the few species of plants (Table 3) [45–48]. The crop plant *Brassica oleracia* has the maximum size of the core genome (49,895) among eukaryotes studies till date but the core genome size is comparatively smaller for a wheat plant fungal pathogen (*Zymoseptoria tritici*). More pan-genomic studies are expected for a new species of crops and their pathogen in the near future.

5 Orthology prediction and genomic plasticity in pan-genomics

Orthologous gene detection is a prerequisite evolutionary method to create the pan-genome of a species. It is useful in inferring phylogenetic trees, annotating a genome, and predicting the function of a gene. Orthologous genes are homologous genes derived from a common ancestor through the speciation process whereas paralogous genes are products of gene duplication events [49]. Orthologous genes have a common biological function but paralogous genes tend to have distinct biological functions even within a particular species. As per ortholog conjecture, orthologs are likely to have closely related function due to constant selection pressure unlike paralogs [50]. The orthology detection methods can be benchmarked using some functional similarity measures such as conservation of a protein domain or coexpression levels of genes [51]. A web-based facility is also developed to benchmark all available orthology detection tools on a large-scale basis [52]. There are several computational methods for orthologous gene detection using both graph-based and tree-based approaches. Graph-based methods heuristically search a sequence similarity score for a large number of sequences. OrthoMCL is the most popular algorithm among graph-based methods for the automated classification of eukaryotic orthologous groups [53]. First, it constructs a similarity score matrix in the form of a graph with protein sequences as nodes and relationship among protein sequences as edges.

Several subgraphs representing orthologous clusters are created from this graph using the Markov clustering algorithm (MCL). The MCL algorithm simulates random walks on a graph using Markov matrices to obtain transition probabilities among the nodes [54]. Although this algorithm is computationally efficient, it does not consider evolutionary information available on the sequences. Thus, orthology detection using this algorithm is prone to error in clustering, especially when there is a differential gene loss in the lineages under study [55]. Tree-based method is a better approach of orthology prediction, which looks for congruency between the gene tree and the species tree to infer orthologs and paralogs [56, 57]. First, a gene phylogeny is reconstructed from multiple sequence alignment of a certain gene and a particular gene phylogeny is then compared to overall species phylogeny using maximum parsimony in order to distinguish speciation and duplication processes [58]. The maximum parsimony is based on the notion that the evolutionary path showing the minimum number of mutations is the most probable path of evolution. Tree-based is although based on a powerful evolutionary concept of maximum parsimony but suffers from two disadvantages; the species phylogeny of many species is not yet resolved and large-scale phylogenetic analysis is not possible due high computational cost of this approach. However, there are some hybrid methods such as Ortholuge [59], EnsemblCompara [60], and HomoloGene [61], etc. combining the merits of both graph-based and tree-based methods.

A microbial genome can be visualized as a dynamic entity undergoing recurrent gene gain and loss processes. The genomic plasticity in a microbial species is the result of various events in which horizontal gene transfer is of primary importance [62]. Horizontal gene transfer facilitates in acquiring blocks of genes known as genomic islands in a species resulting in accelerated rate of evolution. The core genes in a microbe represent the conserved nature of evolution under high selective constraints. In fact, Koonin has advocated that these core genes provide a strong backbone structure for remaining part of the genome [63]. Although character genes constitute a major part of the bacterial genome (64%), the number of gene families represented by them is very small (~7900) [64]. However, these genes are flexible enough to adapt to novel functions in a short span of time. Although these genes show similarity at the sequence level but exhibit great diversity in specificity to different substrates. Thus, it appears that nature does not opt for creating a new gene *de novo* whenever necessity arises. Instead, new biological solutions are obtained from the existing number of gene families, although limited in number, through two evolutionary processes: gene mutations and gene duplications [65–68]. For example, ABC transporters exhibit some wide substrate specificities due to gene substitutions. In contrast, accessory genes are not strongly linked to any particular lineage and are not highly conserved unlike core genome. They are also not subjected to strong evolutionary pressure unlike core genes [69] and have high turnover rates in microbial genomes [70]. The majority of accessory genes are involved in the process of gene creation, generally leading to loss of a gene from the genome. Rarely do they get adaptive advantage during the gene creation process and ultimately transformed as a character gene in the genome.

6 Phylogenomics and genomic epidemiology in pan-genomics

A phylogenetic tree based on the genome (Phylogenomics) is reconstructed using a set of genes in the genome rather than a single gene. A species or genus can be characterized based on a pan-genomic study on all available strains. This diversity within a genome across different strains can be visualized in the form of a tree. There are two major approaches, namely sequence based and gene content based for reconstructing phylogenomic trees using whole genome data [71]. In a sequence-based tree approach, we first align the sequences using multiple sequence alignment and a phylogenomic tree is reconstructed based on evolutionary distances. On the other hand, we use binary data of presence and absence of a gene in different genomes in a gene content-based tree and then a phylogenomic tree is reconstructed using a derived distance matrix from the data. Two types of distances between pan-genome profiles are commonly used in the pan-genomic tree reconstruction: Manhattan distance and Jaccard distance. Manhattan distance is defined as the sum of the differences between each element of two genomes. Jaccard distance between two genomes, on the other hand, measures the degree of similarity between two genomes in each element with respect to the presence or absence of a gene cluster. Genomic fluidity is another measure of a similar kind but it computes the population diversity of the whole population by taking the average of each pair [72]. A pan-genomic tree can be reconstructed based on hierarchical clustering using distance-based UPGMA or neighbor joining methods on these distances (Fig. 4). Such a tree will demonstrate the differences in gene content between genomes. Different gene family weights are necessary for core, accessory, and singleton genes due to wide variation in the degree of their conservation. For example, core genes are highly conserved across the pan-genome and provide no signal for differences between genomes. Therefore, zero weights are assigned to the core genes. Similarly, genes present in a single genome (singleton or ORFans) are often doubtful and therefore, given zero weights as well. The R package micropan is commonly used for reconstructing the pan-tree from the central genome after partitioning into the medoide genome [22]. The bcgTree is an automatic pipeline for reconstruction of the pan-tree both from genomic databases or in-house generated sequences in the laboratory [73]. It retrieved automatically 107 single copy bacterial core genes using hidden Markov models and subsequently reconstructed a pan-tree using partitioned maximum likelihood analysis.

Genome-based molecular epidemiology or genomic epidemiology is a powerful tool of public health investigations of bacterial infections [74]. Alternatively, different subtypes of pathogenic bacteria were identified using some common laboratory techniques like pulse-field gel electrophoresis and multi-sequence typing. These techniques are

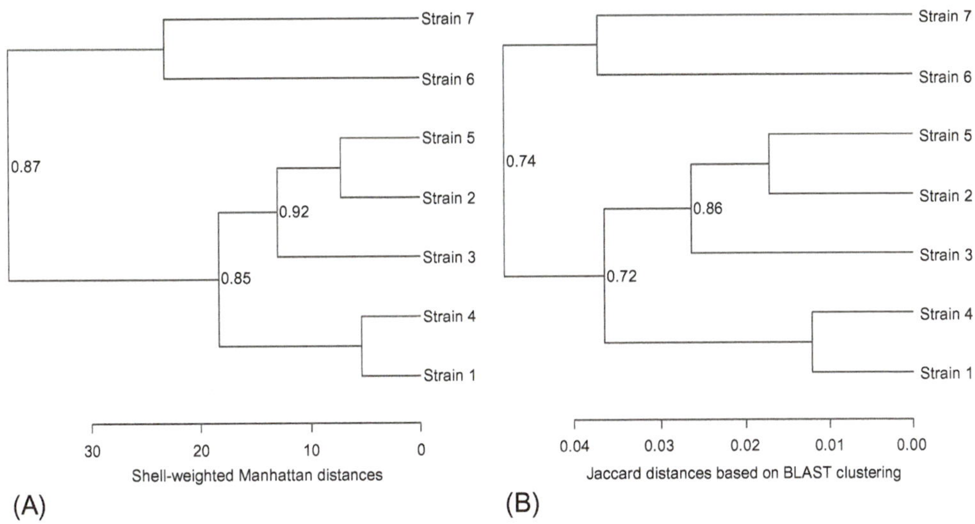

Fig. 4 A pan-tree showing evolutionary relationship between seven strains of a bacterial species based on Manhattan distances (A) and Jaccard distances (B). The values at the node indicate bootstrap values for each clade.

although tedious and time consuming generate limited genetic information regarding the pathogen. However, next-generation whole genome sequencing methods can uncover all SNPs spanning the genome present in different strains of a pathogen within a short span of time. Different strains of *Legionella* were classified into outbreak and non-outbreak groups based whole genome sequencing [75]. It was found that the persistence and virulence of *Legionella pneumophila* were encoded by the core genes [76]. However, some pathogens such as *Yersinia pestis* and *Bacillus anthracis* are found in the soil in dormant state and becomes active and proliferates only in the host. Thus, they do not get an opportunity to exchange genes, and therefore have a closed genome. In fact, the core/pan-genome ratio reaches to an extreme value of 99% in the *B. anthracis* [77]. Therefore, pan-genomic study on a pathogen in an environmental sample will reveal the genomic details of different strains of a pathogen and thereby further help us controlling the outbreak of any epidemic disease.

7 Future directions

There are successful examples of pan-genomic evolutionary studies in various species of prokaryotes and eukaryotes. Concomitantly, appropriate data structures and suitable computational algorithms are being developed for better data analysis of the pan-genome across genera and species. However, there is an urgent need to develop qualitatively better data structure and new computational methods to analyze the fast expanding pan-genomic data. Another major challenge in this area is a better annotation of the pan-genome with relevant functional and phenotypic information. Biochemical modifications on the sequences such as hyper-methylated regions will be a useful additional feature of future pan-genomes. Some additional features like SNPs, non-coding RNA, and indels need special attention in future. There is also a significant development of orthology prediction methods till date. A statistically robust method is needed to discriminate the orthologs and the paralogs with minimal false positives. The evolutionary mechanism regulating genomic plasticity is not yet clear and needs further investigation. Distance-based phylogenomic analysis is a powerful tool to infer evolutionary relationship between different taxa. Character-based methods need more emphasis in their implementation in phylogenomic analysis of the pan-genome for better results.

8 Conclusion

In summary, the emergence of evolutionary pan-genomics is a major advance in understanding the diversity of genomes and inferring the full picture of their variability. I expect that with the development of new computational tools and techniques, we will have some better insights into the regulatory mechanisms generating and governing biodiversity in nature under multiple evolutionary forces in action. Future evolutionary studies are all poised to be focussed on the ever expanding pan-genome instead of a single genome sequencing representing a taxon.

References

[1] International Human Genome Sequencing Consortium, Initial sequencing and analysis of the human genome, Nature 409 (2001) 860–921.

[2] H.P.J. Buermans, J.T. den Dunnen, Next generation sequencing technology: advances and applications, Biochim. Biophys. Acta 1842 (2014) 1932–1941.

[3] H. Tettelin, V. Masignani, M.J. Cieslewicz, C. Donati, D. Medini, N.L. Ward, S.V. Angiuoli, J. Crabtree, A.L. Jones, A.S. Durkin, et al., Genome analysis of multiple pathogenic isolates of Streptococcus agalactiae: implications for the microbial "pangenome", Proc. Natl. Acad. Sci. U.S.A. 102 (39) (2005) 13950–13955.

[4] The Computational Pan-Genomics Consortium, Computational pan-genomics: status, promises and challenges, Brief. Bioinform. 19 (1) (2016) 118–135.

[5] F. Rodriguez-Valera, D.W. Ussery, Is the pan-genome also a pan-selectome? F1000Res. 1 (2012) 16.

[6] M. López-Pérez, F. Rodriguez-Valera, Pangenome evolution in the marine bacterium Alteromonas, Genome Biol. Evol. 8 (5) (2016) 1556–1570.

[7] C. Notredame, Recent evolutions of multiple sequence alignment algorithms, PLoS Comput. Biol. 3 (8) (2007) e123.

[8] J.R. Miller, S. Koren, G. Sutton, Assembly algorithms for next generation sequencing data, Genomics 95 (6) (2010) 315–327.

[9] Z. Iqbal, M. Caccamo, I. Turner, P. Flicek, G. McVean, De novo assembly and genotyping of variants using colored de Bruijn graphs, Nat. Genet. 44 (2) (2012) 226–232.

[10] R. Durbin, Efficient haplotype matching and storage using the positional Burrows–Wheeler transform (PBWT), Bioinformatics 30 (9) (2014) 1266–1272.

[11] N. Li, M. Stephens, Modeling linkage disequilibrium and identifying recombination hotspots using single-nucleotide polymorphism data, Genetics 165 (4) (2003) 2213–2233.
[12] C. Laing, C. Buchanan, E.N. Taboada, Y.X. Zhang, A. Kropinski, A. Villegas, J.E. Thomas, V. P. Gannon, Pan-genome sequence analysis using Panseq: an online tool for the rapid analysis of core and accessory genomic regions, BMC Bioinform. 11 (2010) 461.
[13] J.R. Bayjanov, R.J. Siezen, S.A. van Hijum, PanCGHweb: a web tool for genotype calling in pangenome CGH data, Bioinformatics 26 (9) (2010) 1256–1257.
[14] M. Wozniak, L. Wong, J. Tiuryn, CAMBer: an approach to support comparative analysis of multiple bacterial strains, BMC Genomics 12 (2011) S6.
[15] M.J. Brittnacher, C. Fong, H.S. Hayden, M.A. Jacobs, M. Radey, L. Rohmer, PGAT: a multistrain analysis resource for microbial genomes, Bioinformatics 27 (17) (2011) 2429–2430.
[16] Y. Zhao, J. Wu, J. Yang, S. Sun, J. Xiao, J. Yu, PGAP: pan-genomes analysis pipeline, Bioinformatics 28 (3) (2012) 416–418.
[17] B. Contreras-Moreira, P. Vinuesa, GET_HOMOLOGUES, a versatile software package for scalable and robust microbial pangenome analysis, Appl. Environ. Microbiol. 79 (24) (2013) 7696–7701.
[18] B. Contreras-Moreira, C.P. Cantalapiedra, M.J. García-Pereira, S.P. Gordon, J.P. Vogel, E. Igartua, A.M. Casas, P. Vinuesa, Analysis of plant pan-genomes and transcriptomes with GET_HOMOLOGUES-EST, a clustering solution for sequences of the same species, Front. Plant Sci. (2017), https://doi.org/10.3389/fpls.2017.00184.
[19] S. Sheikhizadeh, M.E. Schranz, M. Akdel, D. De Ridder, S. Smit, PanTools: representation, storage and exploration of pan-genomic data, Bioinformatics 32 (17) (2016) i487–i493.
[20] J. Blom, J. Kreis, S. Spänig, T. Juhre, C. Bertelli, C. Ernst, A. Goesmann, EDGAR 2.0: an enhanced software platform for comparative gene content analyses, Nucleic Acids Res. 44 (W1) (2016) W22–W28.
[21] W. Ding, F. Baumdicker, R.A. Neher, panX: pan-genome analysis and exploration, Nucleic Acids Res. 46 (1) (2018) e5.
[22] L. Snipen, K.H. Liland, micropan: an R-package for microbial pan-genomics, BMC Bioinform. 16 (2015) 79.
[23] T.L. Pedersen, FindMyFriends: Microbial Comparative Genomics in R, R package version 1.12.0, http://bioconductor.org/packages/FindMyFriends, 2015.
[24] H.A. Thorpe, S.C. Bayliss, S.K. Sheppard, E.J. Feil, Piggy: a rapid, large-scale pan-genome analysis tool for intergenic regions in bacteria, Gigascience 7 (4) (2018) 1–11.
[25] T.L. Pedersen, I. Nookaew, D.W. Ussery, M. Månsson, PanViz: interactive visualization of the structure of functionally annotated pangenomes, Bioinformatics 33 (7) (2017) 1081–1082.
[26] R Core Team, R: A Language and Environment for Statistical Computing, version 3.5, second ed., R Foundation for Statistical Computing, Vienna, Austria, 2018.
[27] T. Lefebure, M.J. Stanhope, Evolution of the core and pangenome of Streptococcus: positive selection, recombination, and genome composition, Genome Biol. (5) (2007) R71.
[28] J.S. Hogg, F.Z. Hu, B. Janto, R. Boissy, J. Hayes, R. Keefe, J.C. Post, G.D. Ehrlich, Characterization and modeling of the *Haemophilus influenzae* core and supragenomes based on the complete genomic sequences of Rd and 12 clinical nontypeable strains, Genome Biol. 8 (6) (2007) R103.
[29] N.L. Hiller, B. Janto, J.S. Hogg, R. Boissy, S. Yu, E. Powell, R. Keefe, N.E. Ehrlich, K. Shen, J. Hayes, et al., Comparative genomic analyses of seventeen *Streptococcus pneumoniae* strains:insights into the pneumococcal supragenome, J. Bacteriol. 189 (22) (2007) 8186–8195.
[30] D.A. Rasko, M.J. Rosovitz, G.S. Myers, E.F. Mongodin, W.F. Fricke, P. Gajer, J. Crabtree, M. Sebaihia, N.R. Thomson, R. Chaudhuri, et al., The pangenome structure of *Escherichia coli*: comparative genomic analysis of *E. coli* commensal and pathogenic isolates, J. Bacteriol. 190 (20) (2008) 6881–6893.
[31] C. Schoen, J. Blom, H. Claus, A. Schramm-Gluck, P. Brandt, T. Muller, A. Goesmann, B. Joseph, S. Konietzny, O. Kurzai, et al., Whole genome comparison of disease and carriage strains provides insights into virulence evolution in *Neisseria meningitidis*, Proc. Natl. Acad. Sci. U.S.A. 105 (9) (2008) 3473–3478.
[32] W. van Schaik, J. Top, D.R. Riley, J. Boekhorst, J.E. Vrijenhoek, C.M. Schapendonk, A. P. Hendrickx, I.J. Nijman, M.J. Bonten, H. Tettelin, et al., Pyrosequencing-based comparative

genome analysis of the nosocomial pathogen *Enterococcus faecium* and identification of a large transferable pathogenicity island, BMC Genomics 11 (2010) 239.

[33] M. Eppinger, P.L. Worsham, M.P. Nikolich, D.R. Riley, Y. Sebastian, S. Mou, M. Achtman, L. E. Lindler, J. Ravel, Genome sequence of the deep-rooted *Yersinia pestis* strain Angola reveals new insights into the evolution and pangenome of the plague bacterium, J. Bacteriol. 192 (6) (2010) 1685–1699.

[34] J. Scaria, L. Ponnala, T. Janvilisri, W. Yan, L.A. Mueller, Y.F. Chang, Analysis of ultra low genome conservation in *Clostridium difficile*, PLoS One 5 (12) (2010).

[35] J.R. Broadbent, E.C. Neeno-Eckwall, B. Stahl, K. Tandee, H. Cai, W. Morovic, P. Horvath, J. Heidenreich, N.T. Perna, R. Barrangou, et al., Analysis of the *Lactobacillus casei* supragenome and its influence in species evolution and lifestyle adaptation, BMC Genomics 13 (2012) 533.

[36] A. Ahmed, J. Earl, A. Retchless, S.L. Hillier, L.K. Rabe, T.L. Cherpes, E. Powell, B. Janto, R. Eutsey, N.L. Hiller, et al., Comparative genomic analyses of 17 clinical isolates of *Gardnerella vaginalis* provide evidence of multiple genetically isolated clades consistent with subspeciation into genovars, J. Bacteriol. 194 (15) (2012) 3922–3939.

[37] E.F. Mongodin, S.R. Casjens, J.F. Bruno, Y. Xu, E.F. Drabek, D.R. Riley, B.L. Cantarel, P. E. Pagan, Y.A. Hernandez, L.C. Vargas, et al., Inter- and intra-specific pan-genomes of *Borrelia burgdorferi sensu lato*: genome stability and adaptive radiation, BMC Genomics 14 (2013) 693.

[38] T. Smokvina, M. Wels, J. Polka, C. Chervaux, S. Brisse, J. Boekhorst, J.E. van Hylckama Vlieg, R. J. Siezen, *Lactobacillus paracasei* comparative genomics: towards species pan-genome definition and exploitation of diversity, PLoS One 8 (7) (2013).

[39] G. Meric, K. Yahara, L. Mageiros, B. Pascoe, M.C. Maiden, K.A. Jolley, S.K. Sheppard, A reference pan-genome approach to comparative bacterial genomics: identification of novel epidemiological markers in pathogenic campylobacter, PLoS One 9 (3) (2014).

[40] E. Bosi, M. Fondi, V. Orlandini, E. Perrin, I. Maida, D. de Pascale, M.L. Tutino, E. Parrilli, A. Lo Giudice, A. Filloux, R. Fani, The pangenome of (Antarctic) Pseudoalteromonas bacteria: evolutionary and functional insights, BMC Genomics 18 (2017) 93.

[41] Y. Kim, I. Koh, L.M. Young, W.H. Chung, M. Rho, Pan-genome analysis of Bacillus for microbiome profiling, Sci. Rep. 7 (1) (2017).

[42] R.C. Inglin, L. Meile, M.J.A. Stevens, Clustering of pan- and core-genome of lactobacillus provides novel evolutionary insights for differentiation, BMC Genomics 19 (1) (2018) 284.

[43] H. Tettelin, D. Riley, C. Cattuto, D. Medini, Comparative genomics: the bacterial pan-genome, Curr. Opin. Microbiol. 12 (2008) 472–477.

[44] C.R. Karlsen, E. Hjerde, T. Klemetsen, N.P. Willassen, Pan genome and CRISPR analyses of the bacterial fish pathogen *Moritella viscosa*, BMC Genomics 18 (2017) 313.

[45] J.O. McInerney, A. McNally, M.J. O'Connell, Why prokaryotes have pangenomes, Nat. Microbiol. 2 (2017) 17040.

[46] C. Sun, Z. Hu, T. Zheng, K. Lu, Y. Zhao, W. Wang, J. Shi, C. Wang, J. Lu, D. Zhang, Z. Li, C. Wei, RPAN: rice pan-genome browser for ∼3000 rice genomes, Nucleic Acids Res. 45 (2) (2017) 597–605.

[47] A.A. Golicz, P.E. Bayer, G.C. Barker, P.P. Edger, H. Kim, P.A. Martinez, C.K. Chan, A. Severn-Ellis, W.R. McCombie, I.A. Parkin, A.H. Paterson, J.C. Pires, A.G. Sharpe, H. Tang, G. R. Teakle, C.D. Town, J. Batley, D. Edwards, The pangenome of an agronomically important crop plant *Brassica oleracea*, Nat. Commun. 7 (2016).

[48] C. Plissonneau, F.E. Hartmann, D. Croll, Pangenome analyses of the wheat pathogen *Zymoseptoria tritici* reveal the structural basis of a highly plastic eukaryotic genome, BMC Biol. 16 (1) (2018) 5.

[49] Y.H. Li, G. Zhou, J. Ma, W. Jiang, L.G. Jin, Z. Zhang, Y. Guo, J. Zhang, Y. Sui, L. Zheng, S.S. Zhang, Q. Zuo, X.H. Shi, Y.F. Li, W.K. Zhang, Y. Hu, G. Kong, H.L. Hong, B. Tan, J. Song, Z.X. Liu, Y. Wang, H. Ruan, C.K. Yeung, J. Liu, H. Wang, L.J. Zhang, R.X. Guan, K.J. Wang, W.B. Li, S.Y. Chen, R.Z. Chang, Z. Jiang, S.A. Jackson, R. Li, L.J. Qiu, De novo assembly of soybean wild relatives for pan-genome analysis of diversity and agronomic traits, Nat. Biotechnol. 32 (10) (2014) 1045–1052.

[50] E.V. Koonin, Orthologs, paralogs, and evolutionary genomics, Annu. Rev. Genet. 39 (2005) 309–338.

[51] A.M. Altenhoff, R.A. Studer, M. Robinson-Rechavi, C. Dessimoz, Resolving the ortholog conjecture: orthologs tend to be weakly, but significantly, more similar in function than paralogs, PLoS Comput. Biol. 8 (2012).
[52] T. Hulsen, M.A. Huynen, J. de Vlieg, P.M. Groenen, Benchmarking ortholog identification methods using functional genomics data, Genome Biol. 7 (2006) R31.
[53] A. Altenhoff, B. Boeckmann, S. Capella-Gutierrez, D.A. Dalquen, T. DeLuca, K. Forslund, J. Huerta-Cepas, B. Linard, C. Pereira, L.P. Pryszcz, et al., Standardized benchmarking in the quest for orthologs, Nat. Methods 13 (2016) 425–430.
[54] L. Li, C.J. Stoeckert, D.S. Roos, Orthomcl: identification of ortholog groups for eukaryotic genomes, Genome Res. 13 (9) (2003) 2178–2189.
[55] A.J. Enright, S.V. Dongen, C.A. Ouzounis, An efficient algorithm for large-scale detection of protein families, Nucleic Acids Res. 30 (7) (2002) 1575–1584.
[56] D.R. Scannell, K.P. Byrne, J.L. Gordon, S. Wong, K.H. Wolfe, Multiple rounds of speciation associated with reciprocal gene loss in polyploidy yeasts, Nature 440 (7082) (2006) 341–345.
[57] B. Mirkin, I. Muchnik, T.F. Smith, A biologically consistent model for comparing molecular phylogenies, J. Comput. Biol. 2 (4) (1995) 493–507.
[58] R.D.M. Page, M.A. Charleston, From gene to organismal phylogeny: reconciled trees and the gene tree/species tree problem, Mol. Phylogenet. Evol. 7 (2) (1997) 231–240.
[59] M. Goodman, J. Czelusniak, G.W. Moore, A.E. Romero-Herrera, G. Matsuda, Fitting the gene lineage into its species lineage, a parsimony strategy illustrated by cladograms constructed from globin sequences, Syst. Biol. 28 (2) (1979) 132–163.
[60] D.L. Fulton, Y.Y. Li, M.R. Laird, B.G.S. Horsman, F.M. Roche, F.S.L. Brinkman, Improving the specificity of high-throughput ortholog prediction, BMC Bioinform. 7 (1) (2006) 270.
[61] A.J. Vilella, J. Severin, A. Ureta-Vidal, L. Heng, R. Durbin, E. Birney, EnsemblcomparaGeneTrees: complete, duplication-aware phylogenetic trees in vertebrates, Genome Res. 19 (2) (2009) 327–335.
[62] D.L. Wheeler, T. Barrett, D.A. Benson, S.H. Bryant, K. Canese, V. Chetvernin, D. M. Church, M. DiCuccio, R. Edgar, S. Federhen, et al., Database resources of the national center for biotechnology information, Nucleic Acids Res. 36 (Suppl 1) (2007) D13–D21.
[63] H. Schmidt, M. Hensel, Pathogenicity islands in bacterial pathogenesis, Clin. Microbiol. Rev. 17 (2004) 14–56.
[64] E.V. Koonin, Comparative genomics, minimal gene-sets and the last universal common ancestor, Nat. Rev. Microbiol. 1 (2003) 127–136.
[65] P. Lapierre, J.P. Gogarten, Estimating the size of the bacterial pan-genome, Trends Genet. 25 (3) (2009) 107–110.
[66] A.L. Davidson, J. Chen, ATP-binding cassette transporters in bacteria, Annu. Rev. Biochem. 73 (2004) 241–268.
[67] D.M. Nanavati, T.N. Nguyen, K.M. Noll, Substrate specificities and expression patterns reflect the evolutionary divergence of maltose ABC transporters in *Thermotoga maritima*, J. Bacteriol. 187 (6) (2005) 2002–2009.
[68] K. Fukami-Kobayashi, Y. Tateno, K. Nishikawa, Parallel evolution of ligand specificity between LacI/GalR family repressors and periplasmic sugar-binding proteins, Mol. Biol. Evol. 20 (2003) 267–277.
[69] V. Daubin, H. Ochman, Start-up entities in the origin of new genes, Curr. Opin. Genet. Dev. 14 (2004) 616–619.
[70] J.P. Gogarten, J.P. Townsend, Horizontal gene transfer, genome innovation and evolution, Nat. Rev. Microbiol. 3 (2005) 679–687.
[71] J.G. Lawrence, H. Ochman, Amelioration of bacterial genomes: rates of change and exchange, J. Mol. Evol. 44 (1997) 383–397.
[72] A.O. Kislyuk, B. Haegeman, N.H. Bergman, J.S. Weitz, Genomic fluidity: an integrative view of gene diversity within microbial populations, BMC Genomics 12 (2011) 32.
[73] M.J. Ankenbrand, A. Keller, bcgTree: automatized phylogenetic tree building from bacterial core genomes, Genome 59 (10) (2016) 783–791.
[74] M.W. Gilmour, M. Graham, A. Reimer, G. Van Domselaar, Public health genomics and the new molecular epidemiology of bacterial pathogens, Public Health Genomics 16 (2013) 25–30.

[75] S. Reuter, T.G. Harrison, C.U. Koser, M.J. Ellington, G.P. Smith, J. Parkhill, A pilot study of rapid whole-genome sequencing for the investigation of a Legionella outbreak, BMJ Open 3 (2013).
[76] G. D'Auria, N. Jimenez-Hernandez, F. Peris-Bondia, A. Moya, A. Latorre, *Legionella pneumophila* pangenome reveals strain-specific virulence factors, BMC Genomics 11 (2010) 181.
[77] L. Rouli, V. Merhej, P.E. Fournier, D. Raoult, The bacterial pangenome as a new tool for analysing pathogenic bacteria, New Microbes New Infect. 7 (2015) 72–85.

Further reading

[78] L. Snipen, D.W. Ussery, Standard operating procedure for computing pangenome trees, Stand. Genomic Sci. 2 (1) (2010) 135–141.

CHAPTER 4

Insights into old and new foes: Pan-genomics of *Corynebacterium diphtheriae* and *Corynebacterium ulcerans*

Vartul Sangal[a], Andreas Burkovski[b]
[a]Faculty of Health and Life Sciences, Northumbria University, Newcastle upon Tyne, United Kingdom
[b]Friedrich-Alexander-Universität Erlangen-Nürnberg, Erlangen, Germany

1 *Corynebacterium diphtheriae* and *Corynebacterium ulcerans*

The genus *Corynebacterium* was first described by Lehmann and Neumann in 1896 as a taxonomic group of bacteria showing morphological similarities to the diphtheroid bacillus. At the time of writing, 132 species and 11 subspecies have been published and assigned to the genus, including corynebacteria of biotechnological importance, commensals of humans and animals as well as pathogenic bacteria such as *Corynebacterium diphtheriae*, *Corynebacterium ulcerans*, and *Corynebacterium pseudotuberculosis* [1, 2].

C. diphtheriae is the most prominent member and the type species of the taxon and forms together with its close taxonomic relatives *C. ulcerans* and *C. pseudotuberculosis* the group of toxigenic corynebacteria, based on the fact that these species can be lysogenized by *tox* gene-carrying corynephages [3]. In this chapter, the historical background and pan-genomic insights on *C. diphtheriae* and *C. ulcerans* are discussed, while the pan-genomics of *C. pseudotuberculosis* is covered in Chapter 6.

C. diphtheriae was isolated by Klebs and Löffler and identified as etiological agent of diphtheria [4–6]. As an old foe of mankind [7, 8] diphtheria is known since ancient times with large number of reported cases during industrialization and a major cause of child death. The development of toxoid vaccines and the introduction of immunization programs reduced the number of cases dramatically. A major epidemic occurred in the 1940s due to the miserable health situation during World War II. Thereafter, only local and relatively small epidemics were observed, a positive development, which changed dramatically with the breakdown of the former Union of Socialist Soviet Republics. In 1990, a large-scale outbreak started with the Russian Federation and Ukraine as centers of this epidemic [9–11]. The outbreak spread quickly to neighboring countries and diphtheria infections were observed in Azerbaijan, Belarus, Estonia, Finland, Kazakhstan, Latvia, Lithuania, Poland, Tajikistan, Turkey, and Uzbekistan. Between 1990 and 1998

Pan-genomics: Applications, Challenges, and Future Prospects
https://doi.org/10.1016/B978-0-12-817076-2.00004-4
© 2020 Elsevier Inc.
All rights reserved.

more than 157,000 cases and 5000 deaths were reported [12–14]. The mass immunization started in 1993 effectively controlled the pandemic and today, diphtheria is again uncommon in developed countries; however, it continues to cause a significant morbidity and mortality in many countries. For instance, more than 65,000 cases of diphtheria were reported to the World Health Organization between 2011 and 2015 from India. The most recent major outbreaks were reported from Rohingya refugee camps in Bangladesh and from Venezuela [15, 16].

In addition to diphtheria, increasing numbers of systemic infections are caused by nontoxigenic *C. diphtheriae* strains. The rise in the numbers of nontoxigenic isolates may indicate a shift in the bacterial population [17–19].

C. ulcerans was first described in 1927 by Gilbert and Stewart who isolated this organism from the throat of a patient with respiratory diphtheria-like illness [20]. The bacterium was primarily known as causative agent of mastitis in cattle. Human infections were rare and have traditionally been reported among rural populations with direct contact to domestic livestock or who consumed raw milk and other unpasteurized dairy products [21, 22]. However, during the last 20 years, the frequency and severity of human infections associated with *C. ulcerans* appear to be increasing [23–25] and can most often be ascribed to zoonotic transmission [26, 27] (for recent reviews, see Ref. [6, 28]). The range of hosts that may serve as a reservoir for *C. ulcerans* is extremely broad and includes a plethora of animals such as camels, cats, cows, dogs, ferrets, goats, ground squirrels, monkeys, otters, owls, pigs, roe deer, shrew-moles, water rats, whales, wild boars, and others (for review, see Ref. [28]).

2 Phenotypic and genotypic separation of strains—A historical retrospective

Evolution and adaptation to different environments or ecological niches often introduce genetic and/or phenotypic diversity among bacterial strains. In case of *C. diphtheriae*, four distinct biovars, that is, mitis, gravis, intermedius, and belfanti, may be distinguished based on different biochemical reactions. *C. diphtheriae* biovar gravis is hemolytic, although some strains may show only weak hemolysis activity, and positive for nitrate reduction, starch, and glycogen utilization. Biovar mitis is weakly hemolytic, nitrate reduction-positive and strains can rarely use starch but not glycogen as the carbon source. Strains of biovar intermedius are lipophilic and nonhemolytic, nitrate reduction-positive and may utilize glycogen and starch. The hemolytic properties of biovar belfanti are not clear in the literature. Strains of this biovar are not able to reduce nitrate and utilize starch or glycogen [29, 30].

In addition to these biochemical tests, the Elek test allowed an immunological differentiation between toxin-producing and nonproducing strains [31]. Introduction of PCR allowed the separation of tox^+ and tox^- strains depending on the presence or absence of the *tox* gene that is borne by corynephages [32] and in combination with Elek's test also nontoxigenic *tox*-gene-bearing (NTTB) strains could be distinguished. NTTB *C. diphtheriae* strains possess the *tox* gene but do not produce toxin due to a frame-shift mutation in the nucleotide sequence [33].

Triggered by the need to unravel transmission routes and understand the genetic diversity among *C. diphtheriae* strains, a number of methods were developed in the pre-genome era including restriction fragment length polymorphism (RFLP), single-strand conformation polymorphism (SSCP), phage typing, spoligotyping, and others (reviewed in Ref. [34]). The most efficient methods in this respect were ribotyping and MLST.

Ribotyping has been extensively used for genotyping *C. diphtheriae* that allows strain differentiation based on the nucleotide diversity within rRNA gene operons [35, 36]. Each profile was allocated an arbitrary ribotype name/code until an international nomenclature was published in 2004, where each ribotype was assigned a name based on the place of isolation [35]. Some ribotypes showed geographic association, for example, ribotypes C1 and C5 were commonly isolated in Russia and Moldova while ribotypes C3 and C7 were prevalent in Romania [37]. Majority of epidemic-associated strains in Belarus were ribotypes D1 (Sankt-Peterburg) and D4 (Rossija) [38]. However, a shift was observed in the distribution of these ribotypes during the period from 2001 to 2005 with a significant decrease in the number of D1 (Sankt-Peterburg) isolate and an increase in isolates of ribotypes D4 (Rossija) and D10 (Cluj). Interestingly, the infections caused by toxigenic ribotypes decreased in this period, potentially due to an improved vaccination strategy [7, 39].

CRISPR-based spoligotyping (spacer oligonucleotide typing) was also developed for genotyping *C. diphtheriae* isolates [40]. This approach defines spoligotypes based on the variation in the macroarray-based reverse hybridization patterns at two direct repeat loci, named DRA and DRB [40, 41]. This technique was able to discriminate between strains within each ribotype, for example, 45 distinct spoligotypes were identified among 156 strains of ribotypes Sankt-Peterburg and Rossija from Russia [42]. Similarly, three spoligotypes were identified among 20 isolates of ribotype Rossija from Belarus [41]. The high resolution of this typing scheme has been useful for characterization of the outbreak-associated strains from the former Soviet Union [40, 41].

An MLST scheme based on the sequencing the fragments of *atpA*, *dnaE*, *dnaK*, *fusA*, *leuA*, *odhA*, and *rpoB* genes was developed in 2010 for *C. diphtheriae* [43]. An eBURST group of four sequence types, ST8, ST12, ST52, and ST66 was found to be associated with the epidemic in the former Soviet Union [43]. Consistent with the ribotyping, most isolates of ribotypes (Sankt-Peterburg and Rossija) were ST8. *C. diphtheriae* isolates of ST31 were mainly responsible for the outbreak in Haiti and the Dominican Republic [43]. A post-epidemic prevalence of ST8 isolates in Poland replacing other pre-epidemic strains was recently reported [44]. However, the correlation between STs and biovars or the severity of the disease has been reported as poor [19, 43]. MLST proved to be an efficient method to uncover genetic diversity of *C. diphtheriae* by analyzing sequence variations with more than 11 major clonal groups (eBURST groups; [7]). At the time of writing this article, 580 MLST profiles were listed at PubMLST for *C. diphtheriae* [45, 46].

In summary, a considerable variability of *C. diphtheriae* was already recognized, which can be characterized in much more detail by pan-genomics analyses. Compared to *C. diphtheriae*, less data are available for *C. ulcerans*, for example, no biovars were described. However, the plethora of hosts, including besides humans, pet animals, cattle, platypus, orcas, and many more, strongly hint to a certain genetic variability. In fact, ribotyping and MLST revealed the presence of different lineages of *C. ulcerans* strains [23, 26, 47], which was confirmed by pan-genomics studies presented below.

3 Beginning of the genome era

The first genome of *C. diphtheriae* was sequenced in 2003. The corresponding strain, NCTC 13129, was isolated during the then ongoing outbreak in Eastern Europe from a tourist returned to the United Kingdom from a Baltic cruise [48]. This study showed the presence of the *tox* gene on a bacteriophage that encoded the toxin and was responsible for diphtheria. In addition, a number of other horizontally acquired virulence-associated genes including those involved in the uptake of iron, adhesins, and fimbrial proteins were identified [48]. This study and a subsequent re-annotation approach [49] helped understand the basic genetics behind the pathogenicity of *C. diphtheriae*. Further sequencing of more strains was performed almost a decade later after next-generation sequencers become a regular tabletop laboratory instrument. These studies unraveled the mechanisms of the virulence in greater detail and variation in the degree of pathogenicity between different strains [30, 50–52], which are discussed in more detail in the section on pan-genomics of *C. diphtheriae*.

The first set of two *C. ulcerans* strains, one isolated from a human host (strain 809) and the other one from a canine host (BR-AD22), was also sequenced in the beginning of NGS era [53]. This study revealed that the size (approximately 2.5 Mb) and the GC content (approximately 53.5 mol%) of *C. ulcerans* genomes are similar to *C. diphtheriae* with high genomic synteny. However, prophages introduced some diversity between the *C. ulcerans* strains [53]. While none of these strains carried diphtheria-like *tox* gene, a number of other virulence-associated genes encoding phospholipase D (Pld), neuraminidase H (NanH), corynebacterial protease (CP40), venom serine protease (Vsp1 and Vsp2), ribosomal-binding protein (Rbp, similar to Shiga-like toxin), and adhesive surface pili were reported. Rbp and Vsp2 were only present in the human isolate, potentially contribution to enhanced virulence capacities of the strain 809 [53]. The *rbp* gene encodes a ribosome-binding protein with structural similarity to Shiga-like toxins SLT-1 and

SLT-2 from *Escherichia coli* that may be responsible for multiple organ failure in the patient infected by strain 809 [53, 54].

4 Pan-genomics of *C. diphtheriae*

The extent of genomic diversity within *C. diphtheriae* begins to unravel as more genome sequencing was performed since the year 2012 [30, 50–52, 55–57]. Comparative genomic analysis of 117 *C. diphtheriae* isolates revealed a conservation of more than 50% of the coding sequences (1267 genes [52]). It also showed that horizontal gene transfer is the key source of variation between these strains as most of the diversity is borne on the pathogenicity islands [50, 51]. A phylogenetic tree from the core genome separated the strains into two distinct lineages, Lineage 1 encompassed 116 isolates of all four biovars and Lineage 2 with a single ST106 isolate belonging to biovar belfanti [52] (Fig. 1). These comparative genomic studies helped understanding the genetics behind phenotypic and virulent characteristics of different *C. diphtheriae* strains as discussed below.

4.1 Biochemical subdivision of *C. diphtheriae* into biovars

As mentioned before, *C. diphtheriae* strains are biochemically subdivided into biovars. However, this process is quite complex and unreliable which is reflected by significant misidentification of these biovars by several reference laboratories across Europe [30, 58, 59]. A comparison of the representative genomes from the four biovars revealed an absence or loss of functions due to frameshift mutations in four genes that are involved in carbohydrate metabolism (DIP0660: putative propionyl-CoA carboxylase beta-subunit; DIP1011: putative aldose 1-epimerase; DIP1302: putative ribose-5-phosphate isomerase; DIP1639: putative dihydrolipoamide acetyltransferase) in biovar intermedius [30]. The strains of this biovar are lipophilic and need lipids for optimal growth, probably due to compromised abilities to use carbohydrates as the major energy source [7, 30].

Strains of biovar belfanti are characterized by their inability to reduce nitrate; however, the genomic analyses reveal the presence of the nitrate reductase gene cluster, *narIJHGK* (DIP0497-DIP0502), in belfanti strain INCA 402 [50]. These genes are likely to be iron regulated with a DtxR-binding site upstream to the cluster, which is depleted due to integration of an insertion sequence [50]. The DtxR-binding site upstream to the cluster is also depleted in strains of other biovars including VA01 of biovar gravis and C7(β) that is derived from a mitis strain [60, 61]. Therefore, it may have limited impact on the ability of strains to assimilate nitrogen. Some strains identified as belfanti belong to a distinct lineage (Lineage 2; [43, 52]). The strains of this lineage lack the *narIJHGK*operon [62] and potentially represent true biovar befanti. Apart from these differences, genetic basis of other biochemical characteristics are poorly understood and genome-based phylogeny and pan-genomic analyses do not support the biochemical separation of *C. diphtheriae* isolates into biovars [30, 52].

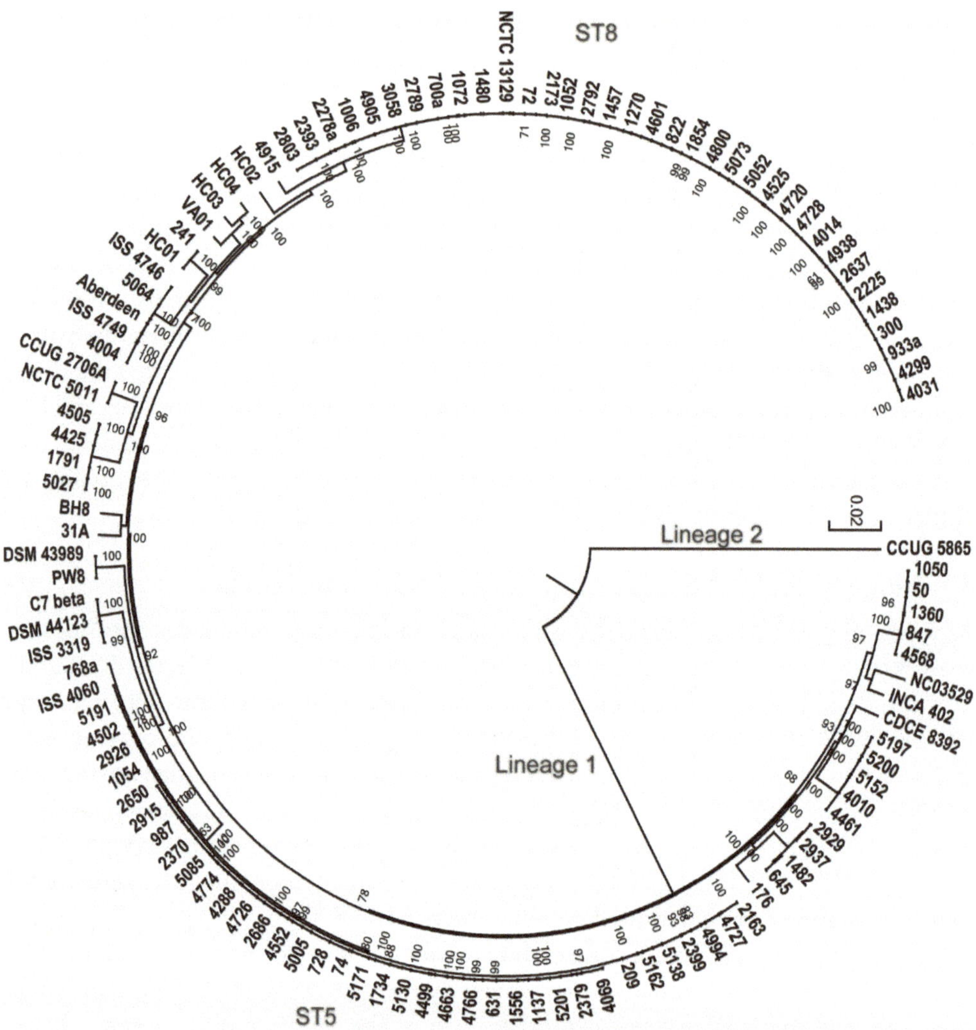

Fig. 1 A maximum-likelihood tree from concatenated nucleotide sequenced alignment of the core genome of 117 *C. diphtheriae* isolates. The scale bar represents nucleotide substitutions per nucleotide site. Lineages and major STs (ST5 and ST8) are labeled. *(Adapted from S. Grosse-Kock, V. Kolodkina, E.C. Schwalbe, J. Blom, A. Burkovski, P.A. Hoskisson, S. Brisse, D. Smith, I.C. Sutcliffe, L. Titov, V. Sangal, Genomic analysis of endemic clones of toxigenic and non-toxigenic* Corynebacterium diphtheriae *in Belarus during and after the major epidemic in 1990s. BMC Genomics 18 (2017) 873).*

4.2 Virulence characteristics and the variation in the degree of pathogenesis

Diphtheria is a toxin-mediated disease of upper respiratory tract in humans. The toxin is encoded by the *tox* gene, which is present on a β-corynephage integrated between duplicated arginine tRNA genes in *C. diphtheriae* genomes [32, 50]. The toxin is produced under low iron conditions and induces apoptosis by catalyzing NAD^+-dependent ADP-ribosylation of elongation factor 2, resulting in cell death. Iron is crucial for several cellular activities such as respiration and catalase activity and the toxin production in low iron conditions may help pathogens to liberate iron from the host cells or compete with the host for the available iron [50]. A number of genes are involved in iron uptake and transport including Irp6A-C (DIP0108-DIP0110), DIP0582-0586, HmuT-V (DIP0626-0628), and DIP1059-1062 and uptake of hemoglobin-haptoglobin complexes, ChtC-CirA (DIP0522-DIP0523), ChtAB (DIP1519-DIP1520), and HtaA-C (DIP0624, DIP0625, and DIP0629) [63]. While majority of iron uptake and transport genes are conserved, the presence of hemoglobin-haptoglobin complex uptake genes is variable [52] that may affect the ability of the strain to acquire iron from the host cells and hence, the strain's fitness or survival.

Recently, nontoxigenic strains lacking the *tox* genes are emerging as a major cause of invasive infections [17–19]. These strains can vary in their abilities to adhere to host cells, to survive intracellularly, and to induce cytokine production by the host immune system that will likely influence the severity of infection [64–66]. Three pilus gene clusters, *spaA*, *spaD*, and *spaH*, are present in *C. diphtheriae* (Fig. 2); however, presence or absence and loss or gain of the gene function within these operons influence the interaction of bacteria with the host cells [50, 51]. A *spaA* gene cluster with disrupted *spaC* gene and a degenerated form of SpaD pilus gene cluster with several intact and disrupted genes encoding two sortases SrtB and SrtE, and SpaD, SpaE, and SpaF pilins are present in the strain Park-Williams no. 8 (PW8) [50]. SpaA pili are responsible for adhesion to the pharyngeal epithelial cells whereas SpaD and SpaH interact with the laryngeal and lung epithelial cells [67].

Two *C. diphtheriae* isolates, ISS 4746 and ISS 4749 showed higher adhesion to pharyngeal D562 cell lines [51, 68] possessed all three pilus gene clusters. However, *spaF* gene encoding a surface-anchored fimbrial subunit for SpaD-type pili was a pseudogene.

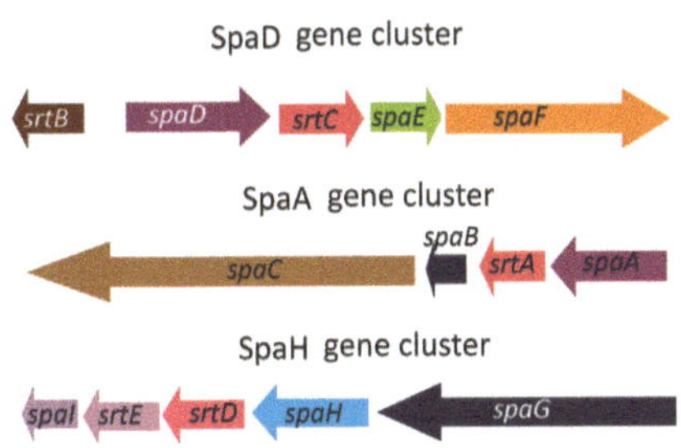

Fig. 2 General organization of the pilus gene clusters in *C. diphtheriae*. The schematic representation is not to scale. Some gene functions may be gained or lost in these operons and the orientation may be different depending on the strains. *(Adapted from V. Sangal, J. Blom, I.C. Sutcliffe, C. von Hunolstein, A. Burkovski, P.A. Hoskisson, Adherence and invasive properties of* Corynebacterium diphtheriae *strains correlates with the predicted membrane-associated and secreted proteome. BMC Genomics 16 (2015) 765).*

SpaA and SpaH clusters are intact in strain ISS 4749 which exhibited highest number of surface pili and highest adhesion to the cell lines in comparison to other strains [50, 51, 64]. SpaA gene cluster in strain ISS 4746 also had a pseudogene, *spaB*, which encodes pilus base subunit. Therefore, SpaA pili may be defective that may be secreted extracellularly by the strain [51]. Strains ISS 3319 and ISS 4060 showed lower adhesion to the cell lines [68] and only possessed SpaD and SpaH-type pili with some gain or loss of gene functions [51]. The *spaG* gene which encodes minor pilins of SpaH-type pilus was a pseudogene in ISS 3319. The *srtB* gene in SpaD cluster of ISS 4060 was a pseudogene [51], which encodes for a sortase responsible for incorporating SpaE into SpaD subunit, suggesting that these pili may not be expressed on the cell surface. Overall, these results show a strong correlation between the presence of pilus gene clusters and gain or loss of the gene functions with the adhesive properties of *C. diphtheriae* strains.

Nontoxigenic *C. diphtheriae* strains are also shown to induce different levels of cytokine production and severity of arthritis in mice model [65, 69]. Genomic analyses revealed a variation in the accessory proteins that include a number of membrane-associated and secreted proteins between these strains that may be associated with the degree of pathogenesis and severity of the infection [51]. However, most of these are hypothetical proteins, indicating a need of molecular characterization of these proteins in the laboratory to understand their roles in cellular and virulence characteristics [7].

4.3 Genomic characterization of outbreak strains

Despite a high vaccine coverage communicated to the WHO, a number of recent diphtheria outbreaks have been reported globally [7, 15, 16, 70–72]. In the 1990s diphtheria epidemic in the former Soviet Union resulted in more than 157,000 cases and approximately 5000 deaths [13]. Whole genome analysis of 93 *C. diphtheriae* strains collected during and after this outbreak from 1996 to 2014 in the former Soviet State Belarus identified two major clones, toxigenic ST8 and nontoxigenic ST5 [52]. In addition to possessing the *tox* gene, the majority of ST8 isolates carried all three pilus gene clusters, *spaA*, *spaD*, and *spaH* as well as genes encoding SapA (surface-anchored pilus protein) and a Sdr-like adhesin that helps the pathogen in adhering and invading the host cells [66, 67]. ST8 isolates were associated with the outbreak [43] and no cases have been reported since 2011 due to an improved vaccination strategy. However, nontoxigenic ST5 isolates were present during the outbreak and continue to infect the population in Belarus [52]. These isolates do not possess a *tox* gene, *sapA* and the Sdr-like adhesion-encoding gene *dip2093*. In addition, a subgroup also lacked the SpaA gene cluster. Therefore, ST8 isolates have greater virulence capacities in comparison to ST5 isolates. ST5 isolates acquired more regions through horizontal gene transfer than ST8 isolates which may reflect the influence of vaccine induced host immune response on the evolutionary dynamics of these clones [52]. Both ST5 and ST8 isolates were able to colonize some individuals asymptomatically and cause disease in others, regardless of their virulence potential. These asymptomatic carriers may serve as a reservoir and disseminate the pathogen to the wider community.

C. diphtheriae strains are known to cause cutaneous infections that is mostly associated with the travel to endemic regions [73–76]. An outbreak of cutaneous diphtheria was observed among the refugees from Northeast Africa and Syria in Germany and Switzerland [70]. Genomic analyses revealed three phylogenetic groups, two containing toxigenic isolates and one nontoxigenic *C. diphtheriae*. Two pilus gene clusters, SpaA and SpaD, were present in one toxigenic (Cluster-1) and the nontoxigenic (Cluster-3) group but were missing toxigenic Cluster-2 [70]. Also, prophages carrying the *tox* genes were different between Cluster-1 and Cluster-2 that were similar to β^{tox+} and ω^{tox+}, respectively [70]. An analysis of the single nucleotide polymorphism within each cluster suggested the recent transmission of *C. diphtheriae* strains between patients before migration to Europe.

An outbreak of respiratory diphtheria was reported from the KwaZulu-Natal province in South Africa that was caused by toxigenic ST378; however, nontoxigenic strains of ST395 were also isolated from some patients. Both these clones were not reported elsewhere and may have other genomic variations. For example, ST378 isolates possessed type I-E-a CRISPR-Cas system while ST395 possessed both type I-E-a and type I-E-b systems [71].

Outbreaks of nontoxigenic strains are also not uncommon [77, 78]. A recent outbreak was caused by nontoxigenic strains belonging to ST8 in Northern Germany among homeless and drug/alcohol abuser people. The core-genome multilocus sequence typing (cg-MLST) analyses divided these strains into four clusters with some geographic distribution. Cluster 1 was relatively homogenous and was concentrated around Hamburg whereas Clusters 2 and 3 were relatively heterogeneous in diversity with most isolates from Berlin region. Cluster 4 only encompassed two isolates, both from Hanover. There were minor exceptions, for example, three isolates from Berlin, Bremen, and Hanover were genomically identical in Cluster 2 which may reflect travel-associated transmission from Berlin region [78].

5 Genomics of *C. ulcerans*

Since the publication of two *C. ulcerans* genomes in 2011 [53], a relatively smaller number of *C. ulcerans* genomes have been sequenced thus far when compared to *C. diphtheriae* [54]. In our recent study, the genome sequence of 19 *C. ulcerans* strains were compared [54]. *C. ulcerans* pan-genome encompasses 4120 genes including 1405 core genes. Similar to *C. diphtheriae*, the core genome phylogeny separated *C. ulcerans* into two distinct lineages; Lineage 1 included 13 strains belonging to eBurst groups (eBG)-325 and eBG332, sequence types ST329, ST338, ST339, and ST349 and three strains that were not assigned any ST designations (Fig. 3) [7, 26, 54]. Six isolates of ST335, ST344 and

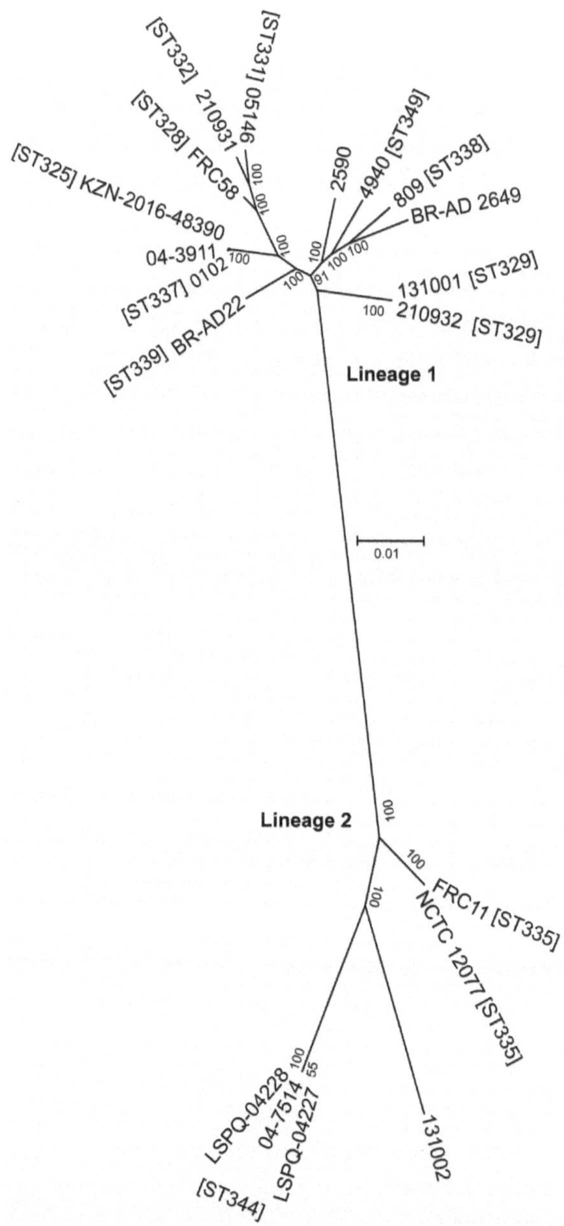

Fig. 3 A maximum-likelihood tree from concatenated nucleotide-sequenced alignment of the core genome of *C. ulcerans* isolates. Scale bar represents nucleotide substitutions per nucleotide site. *(Adapted from R. Subedi, V. Kolodkina, I.C. Sutcliffe, L. Titov, A. Burkovski, L. Simpson Louredo, R. Hirata Jr., A.L. Mattos-Guaraldi, V. Sangal, Genomic analyses reveal two distinct lineages of* Corynebacterium ulcerans *strains. New Microbes New Infect. 25 (2018) 7-13).*

one strain without an ST designation clustered in Lineage 2. Lineage 1 appears to be globally prevalent with isolates from Belarus, Brazil, France, Japan, and South Africa. In contrast, most isolates in Lineage 2 were isolated in France and Canada [54]. However, this observation is based on a smaller number of isolates and more genomes should be analyzed to confirm this finding.

5.1 Virulence potential of *C. ulcerans* strains

Pathogen interaction with the host cells is crucial for successful invasion. Key structures for recognition and attachment are pili on the bacterial cell surface [67]. *C. ulcerans* strains possess two pilus gene clusters namely, *spaDEF* and *spaBC* [53]. These gene clusters were present in all strains analyzed so far [54] with minor exceptions. Major pilin subunit, minor subunit and the tip protein in the *spaDEF* cluster are encoded by *spaD*, *spaE*, and *spaF*, respectively, which are assembled by SrtB and SrtC sortases (Fig. 4). However, *spaD* was absent in one strain (04-3911) while *spaD* and *spaF* genes were present as two smaller genes in another strain [54]. This disruption of *spaD* and *spaF* genes may result in the secretion of smaller protein with the N-terminal domain while the wall anchor protein (C-terminal domain) is encoded separately, potentially resulting in defective pilus that may comprise the ability to interact with the host cells. The *spaBC* cluster lacks gene encoding major pilin subunit; *spaB* and *spaC* genes encode minor pilin subunit and the tip protein, respectively that are assembled by SrtA sortase. Trost and coworkers suggested the potential interaction of these pili with pharyngeal epithelial cells through homodimeric or heterodimeric SpaB/SpaC proteins [53].

Other putative virulence genes including *cpp*, *pld*, *cwlH*, *nanH*, *rpfI*, *tspA*, and *vsp1* were present in all strains while *vsp2* was absent in five isolates [53, 79]. The *cpp* gene encodes a corynebacterial protease CP40 that was identified as an antigen in *C. pseudotuberculosis* [80]. Vaccination of sheep with this antigen provided protection from *C. pseudotuberculosis* infection [80, 81]. Phospholipase D is encoded by the gene *pld* and is involved in persistence and spread of the pathogen within the host cells [82]. CwlH (DIP1621), a cell wall-associated hydrolase and RpfI (DIP1281), a putative secreted protein, is involved in adhesion and internalization of the bacterial cells [53, 83, 84]. NanH,

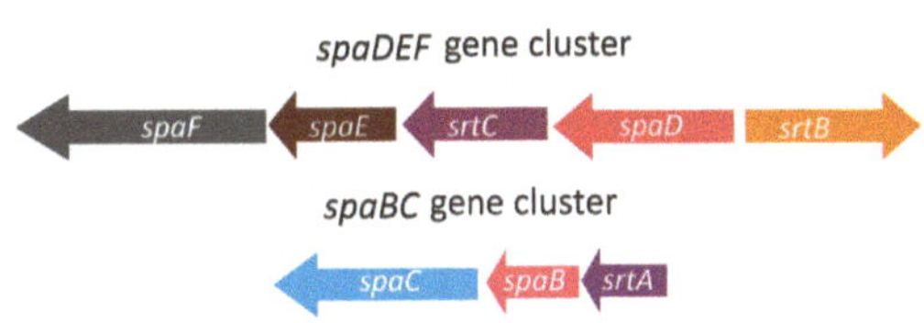

Fig. 4 General organization of the pilus gene clusters in *C. ulcerans*. The schematic representation is not to scale. Some gene functions may be gained or lost in these operons and the orientation may be different depending on the strains. *(Adapted from R. Subedi, V. Kolodkina, I.C. Sutcliffe, L. Titov, A. Burkovski, L. Simpson Louredo, R. Hirata Jr., A.L. Mattos-Guaraldi, V. Sangal, Genomic analyses reveal two distinct lineages of* Corynebacterium ulcerans *strains. New Microbes New Infect. 25 (2018) 7-13).*

an extracellular neuraminidase, may alter the host cell response to bacterial infections by modifying the acceptor molecules on the host cell surface [53, 85]. TspA, Vsp1, and Vsp2 are secreted serine protease type proteins with multiple potential pathogenic functions including interactions with host defense mechanisms and tissue components and may help in host-pathogen interaction and intracellular survival.

Some *C. ulcerans* isolates possess a *tox* gene encoding diphtheria-like toxin and often result in fatal outcomes [53, 86]. However, *C. ulcerans* strain 809 carries a Shiga-like toxin gene (*rbp*), the sole isolate known to possess this gene [54]. This protein has structural similarities to Shiga-like toxins SLT-1 and SLT-2 [53] that are known to cause severe damage to human organs [87, 88]. This gene is flanked by phage-associated genes and has lower G+C content (45.1 mol%) than the average G+C content of the genome (53.3 mol%) [53, 54]. Therefore, this region is potentially acquired via horizontal gene transfer.

An analysis of the accessory genome identified variation in the number of transmembrane, lipoprotein and secreted proteins between different *C. ulcerans* strains [54]. These proteins play a key role in host-pathogen interaction and virulence [51] and hence, may be responsible for some variation in virulence characteristics of these strains.

5.2 Genomic plasticity

Prophages play the key role in introducing diversity in *C. ulcerans* [53, 54, 89, 90]. A prophage of approximately 42 kb in size was identified in the genome of strain 809 whereas three additional prophages with sizes between 14 and 45 kb were detected in the genome of canine isolate BR-AD22 [53]. ΦCULC22I of strain BR-AD22 and ΦCULC809I from strain 809 are similar to each other in size and are located at the same genomic position in these strains [53]. These regions are quite similar with minor variations, for example, the former comprises 42 genes while the latter contains 45 genes; however, 36 of the corresponding proteins show more than 98% amino acid sequence identity [53]. The second prophage region in strain BR-AD22, ΦCULC22II is 44.9 kb in size and includes 60 genes [53]. This prophage has integrated into a gene encoding a hypothetical protein, resulting into two pseudogenes (CULC22_01663 and CULC22_01724) flanking the prophage. The third prophage (ΦCULC22III) is approximately 14 kb is size with 19 genes and has integrated into the direct repeats adjacent to a tRNALys gene [53]. This prophage is relatively smaller than other prophage regions and is likely to be incomplete or a remnant of a former larger corynephage. The fourth prophage, ΦCULC22IV, is approximately 41 kb and is present adjacent to a tRNAThr gene [53]. As mentioned before, some strain may possess the *tox* gene that encodes diphtheria-like toxin and is present on a corynephage. Similar to the corynephages in *C. diphtheriae*, ΦCULC0102-I prophage carrying the *tox* gene has integrated into tRNAArg site in *C. ulcerans* strain 0102 [89]. However, the corynephage in *C. ulcerans* was quite distinct from the one observed in C. *diphtheriae* strain NCTC 13129 [89]. Two additional prophages, ΦCULC0102-II and ΦCULC0102-III, were also present in this strain.

While the *tox* gene is commonly present on corynephages, a novel pathogenicity island was identified in some strains that encompassed the *tox* gene at the same tRNA-Arg locus. This PAI is 7571 bp in size with a G+C content of approximately 48 mol% and carries eight protein-encoding genes [90].

C. ulcerans strain 4940 appears to lack prophages except for a potential incomplete prophage of 8.6 kb in size with a G+C content of 55.42 mol% [54]. This region is also present in strains 809 and BR-AD22 but not identified as prophage associated. Some of the 11 genes in this region show similarities with genes on previously reported phages in other species and may represent remnants of a prophage that was not detected previously. Similarly, one of the detected prophages in strain 2590 was also present in strains 809 and BR-AD 22 (99% sequence similarity) but has not been identified as prophage-associated before. The second prophage in this strain is approximately 31 kb in size that has integrated between *attL* and *attR* sites in the genome while the third predicted prophage is similar to ΦCULC22IV of strain BR-AD 22 [54].

Five prophages-associated regions were identified in the genome of the canine isolate BR-AD 2649; however, two of those regions (contigs 1 and 14) were similar to the prophages ΦCULC809I and ΦCULC22I and likely represent a single prophage [54]. These regions may have been separated due to gaps in the draft assembly of the genome. Second predicted prophage (contig 2) is similar to the one detected in strain 4940, both in size and G+C content. The third predicted on contig 6 is 16.7 kb in size and showed significant similarity with the genome of another *C. ulcerans* strain FRC58 [91]. The fourth putative novel prophage on contig 7 is 8.8 kb in size with a G+C content of 50.36 mol%. Therefore, prophage-like sequences are responsible for the genome in *C. ulcerans* [53, 89, 90].

5.3 Zoonotic transmission

C. ulcerans infections are zoonotic in nature and are often associated with close animal contacts. The genome sequences of isolates from patient-animal (cat, dog, and pigs) pairs indicated the transmission of the pathogen from animals to humans [27, 90]. The strain pairs from patients and their pet and farm animals showed only zero to two SNPs confirming the zoonotic transmission [27, 90]. The number of SNPs between individual strains from different groups were significantly higher and varied from 5,000 to 20,000 SNPs [90].

Similarly, both the lineages identified from the core genome based phylogenetic analyses include strains from canine and human hosts, suggesting that the *C. ulcerans* strains are similar in from animals and human hosts and further supports the zoonotic nature of *C. ulcerans* infections [54].

6 Toxin variation and diphtheria toxoid vaccine

The basic principle of diphtheria vaccine production is the purification of diphtheria toxin and its inactivation by formaldehyde cross-linking. This converts the potentially fatal toxin in a completely harmless protein aggregate, which is still immunogenic and induces antibody production in the vaccinated person. For a broad and optimal protection, it is crucial that the toxin used for vaccine production is to the greatest possible extent identical to the toxin synthesized by strains distributed among the population.

Today, almost all companies use derivatives of the PW8 strain for toxin production [92]. When the variability of PW8 strains was studied by comparative genomic hybridization and PCR analyses, a great heterogeneity in respect to genome organization and pathogenicity was found [93]. However, when the heterogeneity of the *tox* gene was analyzed for 72 strains from Russia and Ukraine by direct sequencing, 28 sequences were identical to the PW8 *tox* sequence, while in the remaining 40 strains only four point mutations were found in the *tox* gene. Based on these results the authors concluded that changes in the efficacy of current vaccines are unlikely to occur [94]. This idea was further supported by a pan-genomic study of mainly Brazilian strains isolated from cases of classical diphtheria, endocarditis, and pneumonia. All *tox* genes detected in this study showed a perfect nucleotide sequence identity, with the exception of a single nucleotide exchange in one of the strains [50]. In our study of 93 *C. diphtheriae* strains collected during and after the diphtheria outbreak in the former Soviet State Belarus, 54 isolates carried the *tox* gene. Eight synonymous single nucleotide polymorphisms were observed between the *tox* genes of the vaccine strain PW8 and other toxigenic strains [52]. However, a single base deletion in the *tox* gene of ST40 isolates introduced a frameshift mutation, converting them into NTTB strains [52].

The first two *C. ulcerans* strains sequenced were not lysogenized by *tox* gene-bearing corynephages [53]. However, other studies showed that toxigenic *C. ulcerans* outnumber toxigenic *C. diphtheriae* in infections analyzed in the United Kingdom [24] and in Germany [90]. In our recent study, the available genome sequences of 19 strains were analyzed and 11 of these were found to be toxigenic [54] further emphasizing the importance of this virulence factor not only for *C. diphtheriae* but also for *C. ulcerans* infections.

The presence of the diphtheria toxin gene in a different species and the description of a new horizontal gene transfer mechanism in *C. ulcerans* [90] give rise to the question, whether diphtheria toxin from *C. diphtheriae* and *C. ulcerans* differ from each other or not. When PCR-amplified *tox* genes from 19 *C. ulcerans* isolates from the CDC's collection were analyzed, mismatches in the toxin gene sequences of seven strains were observed [95]. Later, sequencing of *tox*-specific PCR products derived from 12 toxigenic *C. ulcerans* isolates from Germany revealed only one *C. ulcerans*-specific nucleotide substitution [96].

Based on the current information, variations among *C. diphtheriae* strains and between *C. diphtheriae* and *C. ulcerans* toxin-encoding genes are detectable. However, these seem to have no major influence on toxin detection by antitoxin or vaccination-induced human antibodies. These results suggest that the diphtheria toxoid vaccine may protect against the *C. ulcerans* toxin as well. In fact, Möller and coworkers demonstrated the efficacy of diphtheria vaccine against the toxin from three different *C. ulcerans* strains recently (Möller and coworkers, unpublished observation).

7 Conclusions and future directions

The introduction of next generation sequencing has generated a wealth of information, which allow different levels of analysis from evolutionary and epidemiological traits to identification of important virulence-associated genes. Recent comparative genomic analyses led to an improved biochemical identification scheme for different *Corynebacterium* species including *C. diphtheriae* and *C. ulcerans* [97] and a proposal to assign subspecies designations to the two *C. diphtheriae* lineages namely, *C. diphtheriae* ssp. *diphtheriae* and *C. diphtheriae* ssp. *lausannense* [62]. The genome sequencing has only been applied to a smaller number of strains (STs) and more efforts are required to characterize the global diversity for both *C. diphtheriae* and *C. ulcerans*. We believe that future pan-genomics studies will help improve the current understanding of global transmission and local adaptation of these pathogens and will also help in developing an effective vaccine to protect from toxigenic and nontoxigenic infections by these pathogens.

References

[1] A. Tauch, J. Sandbote, The family Corynebacteriaceae, in: E. Rosenberg, E. Delong, S. Lory, E. Stackebrandt, F. Thompson (Eds.), The Prokaryotes, Springer, Berlin, Heidelberg, Germany, 2014, pp. 239–277.

[2] Bacterio. www.bacterio.net/corynebacterium.html, 2018 (Accessed 16 October 2018).

[3] P. Riegel, R. Ruimy, D. De Briel, G. Prévost, F. Jehl, R. Christen, H. Monteil, Taxonomy of *Corynebacterium diphtheriae* and related taxa, with recognition of *Corynebacterium ulcerans* sp. nov. nom. rev., FEMS Microbiol. Lett. 126 (1995) 271–276.

[4] A. Burkovski, Diphtheria, in: E. Rosenberg, E.F. DeLong, S. Lory, E. Stackebrandt, F. Thompson (Eds.), The prokaryotes, fourth ed., Human Microbiology, vol. 5, Springer, New York, USA, 2013, pp. 237–246.

[5] A. Burkovski, Diphtheria and its etiological agents, in: A. Burkovski (Ed.), *Corynebacterium diphtheriae* and Related Toxigenic Species, Springer, Dordrecht, The Netherlands, 2014, pp. 1–14.

[6] A. Burkovski, Pathogenesis of *Corynebacterium diphtheriae* and *Corynebacterium ulcerans*, in: S.K. Singh (Ed.), Human Emerging and Re-Emerging Infections, vol. 2, John Wiley & Sons/Wiley Blackwell Press, Hoboken, New Jersey, USA, 2016, pp. 697–708.

[7] V. Sangal, P.A. Hoskisson, Evolution, epidemiology and diversity of *Corynebacterium diphtheriae*: new perspectives on an old foe, Infect. Genet. Evol. 43 (2016) 364–370.

[8] P.A. Hoskisson, Microbe Profile: *Corynebacterium diphtheriae*—an old foe always ready to seize opportunity, Microbiology 164 (2018) 865–867.

[9] A.M. Galazka, S.E. Robertson, G.P. Oblapenko, Resurgence of diphtheria, Eur. J. Epidemiol. 11 (1995) 95–105.
[10] J. Eskola, J. Lumio, J. Vuopio-Varkila, Resurgent diphtheria—are we safe? Br. Med. Bull. 54 (1998) 635–645.
[11] T. Popovic, I.K. Mazurova, A. Efstratiou, J. Vuopio-Varkila, M.W. Reeves, A. De Zoysa, T. Glushkevich, P. Grimont, Molecular epidemiology of diphtheria, J. Infect. Dis. 181 (Suppl 1) (2000) S168–S177.
[12] C.R. Vitek, M. Wharton, Diphtheria in the former Soviet Union: reemergence of a pandemic disease, Emerg. Infect. Dis. 4 (1998) 539–550.
[13] S. Dittmann, M. Wharton, C. Vitek, M. Ciotti, A. Galazka, S. Guichard, I. Hardy, U. Kartoglu, S. Koyama, J. Kreysler, M. Martin, D. Mercer, T. Ronne, C. Roure, R. Steinglass, P. Strebel, R. Sutter, M. Trostle, Successful control of epidemic diphtheria in the states of the Former Union of Soviet Socialist Republics: lessons learned, J. Infect. Dis. 181 (Suppl 1) (2000) S10–S22.
[14] S.S. Markina, N.M. Maksimova, C.R. Vitek, E.Y. Bogatyreva, A.A. Monisov, Diphtheria in the Russian Federation in the 1990s, J. Infect. Dis. 181 (2000) S27–S34.
[15] R. Matsuyama, A.R. Akhmetzhanov, A. Endo, H. Lee, T. Yamaguchi, S. Tsuzuki, H. Nishiura, Uncertainty and sensitivity analysis of the basic reproduction number of diphtheria: a case study of a Rohingya refugee camp in Bangladesh, November-December 2017, PeerJ. 6 (2018).
[16] A. Lodeiro-Colatosti, U. Reischl, T. Holzmann, C.E. Hernandez-Pereira, A. Risquez, A.E. Paniz-Mondolfi, Diphtheria outbreak in Amerindian communities, Wonken, Venezuela, 2016-2017, Emerg. Infect. Dis. 24 (2018) 1340–1344.
[17] M.G. Romney, D.L. Roscoe, K. Bernard, S. Lai, A. Efstratiou, A.M. Clarke, Emergence of an invasive clone of nontoxigenic *Corynebacterium diphtheriae* in the urban poor population of Vancouver, Canada, J. Clin. Microbiol. 44 (2006) 1625–1629.
[18] B. Edwards, A.C. Hunt, P.A. Hoskisson, Recent cases of non-toxigenic *Corynebacterium diphtheriae* in Scotland: justification for continued surveillance, J. Med. Microbiol. 60 (2011) 561–562.
[19] E. Farfour, E. Badell, A. Zasada, H. Hotzel, H. Tomaso, S. Guillot, N. Guiso, Characterization and comparison of invasive *Corynebacterium diphtheriae* isolates from France and Poland, J. Clin. Microbiol. 50 (2012) 173–175.
[20] R. Gilbert, F.C. Stewart, *Corynebacterium ulcerans*; a pathogenic microorganism resembling *Corynebacterium diphtheriae*, J. Lab. Clin. Med. 12 (1927) 756–761.
[21] R.J. Hart, *Corynebacterium ulcerans* in humans and cattle in North Devon, J. Hyg. (Lond.) 92 (1984) 161–164.
[22] A.D. Bostock, F.R. Gilbert, D. Lewis, D.C. Smith, *Corynebacterium ulcerans* infection associated with untreated milk, J. Infect. 9 (1984) 286–288.
[23] A. De Zoysa, P.M. Hawkey, K. Engler, R. George, G. Mann, W. Reilly, D. Taylor, A. Efstratiou, Characterization of toxigenic *Corynebacterium ulcerans* strains isolated from humans and domestic cats in the United Kingdom, J. Clin. Microbiol. 43 (2005) 4377–4378.
[24] K.S. Wagner, J.M. White, N.S. Crowcroft, S. De Martin, G. Mann, A. Efstratiou, Diphtheria in the United Kingdom, 1986-2008: the increasing role of *Corynebacterium ulcerans*, Epidemiol. Infect. 138 (2010) 1519–1530.
[25] K.S. Wagner, J.M. White, I. Lucenko, D. Mercer, N.S. Crowcroft, S. Neal, A. Efstratiou, Diphtheria Surveillance Network, Diphtheria in the postepidemic period, Europe, 2000-2009, Emerg. Infect. Dis. 18 (2012) 217–225.
[26] C. König, D.M. Meinel, G. Margos, R. Konrad, A. Sing, Multilocus sequence typing of *Corynebacterium ulcerans* provides evidence for zoonotic transmission and for increased prevalence of certain sequence types among toxigenic strains, J. Clin. Microbiol. 52 (2014) 318–4324.
[27] D.M. Meinel, R. Konrad, A. Berger, C. König, T. Schmidt-Wieland, M. Hogardt, H. Bischoff, N. Ackermann, S. Hörmansdorfer, S. Krebs, H. Blum, G. Margos, A. Sing, Zoonotic transmission of toxigenic *Corynebacterium ulcerans* strain, Germany, 2012, Emerg. Infect. Dis. 21 (2015) 356–358.
[28] E. Hacker, C. Azevedo Antunes, A.-L. Mattos-Guaraldi, A. Burkovski, A. Tauch, *Corynebacterium ulcerans*—an emerging human pathogen, Future Microbiol. 11 (2016) 1191–1208.

[29] M. Goodfellow, P. Kaempfer, H.-J. Busse, M.E. Trujillo, K.-I. Suzuki, W. Ludwig, W.B. Whitman, Bergey's Manual of Systematic Bacteriology, second ed., Springer, London, UK, 2012.
[30] V. Sangal, A. Burkovski, A.C. Hunt, B. Edwards, J. Blom, P.A. Hoskisson, A lack of genetic basis of biovar differentiation in clinically important *Corynebacterium diphtheriae* from whole genome sequencing, Infect. Genet. Evol. 21 (2014) 54–57.
[31] S.D. Elek, The plate virulence test for diphtheria, J. Clin. Pathol. 2 (1949) 250–258.
[32] V. Sangal, P.A. Hoskisson, Corynephages: infections of the infectors, in: A. Burkovski (Ed.), *Corynebacterium diphtheriae* and Related Toxigenic Species, Springer, Heidelberg, Germany, 2014, pp. 67–82.
[33] K. Zakikhany, S. Neal, A. Efstratiou, Emergence and molecular characterisation of non-toxigenic *tox* gene-bearing *Corynebacterium diphtheriae* biovar mitis in the United Kingdom, 2003-2012, Euro Surveill. 19 (2014) 22.
[34] S.K. Rajamani Sekar, B. Veeraraghavan, S. Anandan, N.K. Devanga Ragupathi, L. Sangal, S. Joshi, Strengthening the laboratory diagnosis of pathogenic *Corynebacterium* species in the vaccine era, Lett. Appl. Microbiol. 65 (2017) 354–365.
[35] P.A. Grimont, F. Grimont, A. Efstratiou, A. De Zoysa, I. Mazurova, C. Ruckly, M. Lejay-Collin, S. Martin-Delautre, B. Regnault, European Laboratory Working Group on Diphtheria, International nomenclature for *Corynebacterium diphtheriae* ribotypes, Res. Microbiol. 155 (2004) 162–166.
[36] A. De Zoysa, P. Hawkey, A. Charlett, A. Efstratiou, Comparison of four molecular typing methods for characterization of *Corynebacterium diphtheriae* and determination of transcontinental spread of *C. diphtheriae* based on BstEII rRNA gene profiles, J. Clin. Microbiol. 46 (2008) 3626–3635.
[37] M. Damian, F. Grimont, O. Narvskaya, M. Straut, M. Surdeanu, R. Cojocaru, I. Mokrousov, A. Diaconescu, C. Andronescu, A. Melnic, L. Mutoi, P.A. Grimont, Study of *Corynebacterium diphtheriae* strains isolated in Romania, northwestern Russia and the Republic of Moldova, Res. Microbiol. 153 (2002) 99–106.
[38] L. Titov, V. Kolodkina, A. Dronina, F. Grimont, P.A. Grimont, M. Lejay-Collin, A. De Zoysa, C. Andronescu, A. Diaconescu, B. Marin, A. Efstratiou, Genotypic and phenotypic characteristics of *Corynebacterium diphtheriae* strains isolated from patients in Belarus during an epidemic period, J. Clin. Microbiol. 41 (2003) 1285–1288.
[39] V. Kolodkina, L. Titov, T. Sharapa, F. Grimont, P.A. Grimont, A. Efstratiou, Molecular epidemiology of *C. diphtheriae* strains during different phases of the diphtheria epidemic in Belarus, BMC Infect. Dis. 6 (2006) 129.
[40] I. Mokrousov, O. Narvskaya, E. Limeschenko, A. Vyazovaya, Efficient discrimination within a *Corynebacterium diphtheriae* epidemic clonal group by a novel macroarray-based method, J. Clin. Microbiol. 43 (2005) 1662–1668.
[41] I. Mokrousov, A. Vyazovaya, V. Kolodkina, E. Limeschenko, L. Titov, O. Narvskaya, Novel macroarray-based method of *Corynebacterium diphtheriae* genotyping: evaluation in a field study in Belarus, Eur. J. Clin. Microbiol. Infect. Dis. 28 (2009) 701–703.
[42] I. Mokrousov, E. Limeschenko, A. Vyazovaya, O. Narvskaya, *Corynebacterium diphtheriae* spoligotyping based on combined use of two CRISPR loci, Biotechnol. J. 2 (2007) 901–906.
[43] F. Bolt, P. Cassiday, M.L. Tondella, A. De Zoysa, A. Efstratiou, A. Sing, A. Zasada, K. Bernard, N. Guiso, E. Badell, M.L. Rosso, A. Baldwin, C. Dowson, Multilocus sequence typing identifies evidence for recombination and two distinct lineages of *Corynebacterium diphtheriae*, J. Clin. Microbiol. 48 (2010) 4177–4185.
[44] U. Czajka, A. Wiatrzyk, E. Mosiej, K. Formińska, A.A. Zasada, Changes in MLST profiles and biotypes of *Corynebacterium diphtheriae* isolates from the diphtheria outbreak period to the period of invasive infections caused by nontoxigenic strains in Poland (1950-2016), BMC Infect. Dis. 18 (2018) 121.
[45] K.A. Jolley, M.C. Maiden, BIGSdb: scalable analysis of bacterial genome variation at the population level, BMC Bioinformatics 11 (2010) 595.
[46] Pubmlst. https://pubmlst.org/cdiphtheriae, 2018 (Accessed 16 October 2018).
[47] T. Komiya, Y. Seto, A. De Zoysa, M. Iwaki, A. Hatanaka, A. Tsunoda, Y. Arakawa, S. Kozaki, M. Takahashi, Two Japanese *Corynebacterium ulcerans* isolates from the same hospital: ribotype, toxigenicity and serum antitoxin titre, J. Med. Microbiol. 59 (2010) 1497–1504.

[48] A.M. Cerdeno-Tarraga, A. Efstratiou, L.G. Dover, M.T. Holden, M. Pallen, S.D. Bentley, G.S. Besra, C. Churcher, K.D. James, A. De Zoysa, T. Chillingworth, A. Cronin, L. Dowd, T. Feltwell, N. Hamlin, S. Holroyd, K. Jagels, S. Moule, M.A. Quail, E. Rabbinowitsch, K.M. Rutherford, N. R. Thomson, L. Unwin, S. Whitehead, B.G. Barrell, J. Parkhill, The complete genome sequence and analysis of *Corynebacterium diphtheriae* NCTC13129, Nucleic Acids Res. 31 (2003) 6516–6523.

[49] S.C. Santos, V. D'Alfonseca, A. Ali, A.R. Santos, A.C. Pinto, A.A.C. Magalhaes, C.J. Faria, E. Barbosa, L.C. Guimaraes, M. Eslabao, S.S. Almeida, V.A.C. Abreu, A.Z. Neto, A.R. Carneiro, L.T. Cerdeira, R.T.J. Ramos, R. Hirata Jr., A.L. Mattos-Guaraldi, E. Trost, A. Tauch, A. Silva, M.P. Schneider, A. Miyoshi, V. Azevedo, Reannotation of the *Corynebacterium diphtheriae* NCTC13129 genome as a new approach to studying gene targets connected to virulence and pathogenicity in diphtheria, Open Access Bioinform. 3 (2011) 1–13.

[50] E. Trost, J. Blom, S. de Castro Soares, I.H. Huang, A. Al-Dilaimi, J. Schröder, S. Jaenicke, F.A. Dorella, F.S. Rocha, A. Miyoshi, V. Azevedo, M.P. Schneider, A. Silva, T.C. Camello, P.S. Sabbadini, S.C. Santos, L.S. Santos, R. Hirata Jr., A.L. Mattos-Guaraldi, A. Efstratiou, M.P. Schmitt, H. Ton-That, A. Tauch, Pangenomic study of *Corynebacterium diphtheriae* that provides insights into the genomic diversity of pathogenic isolates from cases of classical diphtheria, endocarditis, and pneumonia, J. Bacteriol. 194 (2012) 3199–3215.

[51] V. Sangal, J. Blom, I.C. Sutcliffe, C. von Hunolstein, A. Burkovski, P.A. Hoskisson, Adherence and invasive properties of *Corynebacterium diphtheriae* strains correlates with the predicted membrane-associated and secreted proteome, BMC Genomics 16 (2015) 765.

[52] S. Grosse-Kock, V. Kolodkina, E.C. Schwalbe, J. Blom, A. Burkovski, P.A. Hoskisson, S. Brisse, D. Smith, I.C. Sutcliffe, L. Titov, V. Sangal, Genomic analysis of endemic clones of toxigenic and non-toxigenic *Corynebacterium diphtheriae* in Belarus during and after the major epidemic in 1990s, BMC Genomics 18 (2017) 873.

[53] E. Trost, A. Al-Dilaimi, P. Papavasiliou, J. Schneider, P. Viehoever, A. Burkovski, S. de Castro Soares, S.S. Almeida, F. Alves Dorella, A. Miyoshi, V. Azevedo, M.P. Cruz Schneider, A. Silva, C.S. Santos, P. Sabbadini, A.A. Dias, R. Hirata Jr., A.L. Mattos-Guaraldi, A. Tauch, Comparative analysis of two complete *Corynebacterium ulcerans* genomes and detection of candidate virulence factors, BMC Genomics 12 (2011) 383.

[54] R. Subedi, V. Kolodkina, I.C. Sutcliffe, L. Titov, A. Burkovski, L. Simpson Louredo, R. Hirata Jr., A.L. Mattos-Guaraldi, V. Sangal, Genomic analyses reveal two distinct lineages of *Corynebacterium ulcerans* strains, New Microbes New Infect. 25 (2018) 7–13.

[55] V. Sangal, N.P. Tucker, A. Burkovski, P.A. Hoskisson, The draft genome sequence of *Corynebacterium diphtheriae mitis* NCTC 3529 reveals significant diversity between the primary disease causing biovars, J. Bacteriol. 194 (2012) 3269.

[56] V. Sangal, N.P. Tucker, A. Burkovski, P.A. Hoskisson, The genome of *Corynebacterium diphtheriae* biovar *intermedius* NCTC 5011, J. Bacteriol. 194 (2012) 4738.

[57] C. Azevedo Antunes, E.J. Richardson, J. Quick, P. Fuentes Utrilla, G.L. Isom, E. Godall, J. Möller, P.A. Hoskisson, A.L. Mattos-Guaraldi, A.F. Cunningham, N.J. Loman, V. Sangal, A. Burkovski, I.R. Henderson, Complete closed genome sequence of non-toxigenic invasive *Corynebacterium diphtheriae* bv. mitis strain ISS-3319, Genome Announc. 6 (2018). e01566-17.

[58] S.E. Neal, A. Efstratiou, International external quality assurance for laboratory diagnosis of diphtheria, J. Clin. Microbiol. 47 (2009) 4037–4042.

[59] L. Both, S.E. Neal, A. De Zoysa, G. Mann, I. Czumbel, A. Efstratiou, Members of the European Diphtheria Surveillance, External quality assessments for microbiologic diagnosis of Diphtheria in Europe, J. Clin. Microbiol. 52 (2014) 4381–4384.

[60] V.J. Freeman, Studies on the virulence of bacteriophage-infected strains of *Corynebacterium diphtheriae*, J. Bacteriol. 61 (1951) 675–688.

[61] W.L. Barksdale, A.M. Pappenheimer Jr., Phage-host relationships in nontoxigenic and toxigenic diphtheria bacilli, J. Bacteriol. 67 (1954) 220–232.

[62] F. Tagini, T. Pillonel, A. Croxatto, C. Bertelli, A. Koutsokera, A. Lovis, G. Greub, Distinct genomic features characterise two clades of *Corynebacterium diphtheriae*: proposal of *Corynebacterium diphtheriae* subsp. *diphtheriae* subsp. nov. and *Corynebacterium diphtheriae* subsp. *lausannense* subsp. nov., Front. Microbiol. 9 (2018) 1743.

[63] C.E. Allen, M.P. Schmitt, Utilization of host iron sources by *Corynebacterium diphtheriae*: multiple hemoglobin-binding proteins are essential for the use of iron from the hemoglobin-haptoglobin complex, J. Bacteriol. 197 (2015) 553–562.
[64] L. Bertuccini, L. Baldassarri, C. von Hunolstein, Internalization of non-toxigenic *Corynebacterium diphtheriae* by cultured human respiratory epithelial cells, Microb. Pathog. 37 (2004) 111–118.
[65] M. Puliti, C. von Hunolstein, M. Marangi, F. Bistoni, L. Tissi, Experimental model of infection with non-toxigenic strains of *Corynebacterium diphtheriae* and development of septic arthritis, J. Med. Microbiol. 55 (2006) 229–235.
[66] R.S. Peixoto, G.A. Pereira, L. Sanches Dos Santos, C.M. Rocha-De-Souza, D.L. Gomes, C. Silva Dos Santos, L.M. Werneck, A.A. Dias, R. Hirata Jr., P.E. Nagao, A.L. Mattos-Guaraldi, Invasion of endothelial cells and arthritogenic potential of endocarditis-associated *Corynebacterium diphtheriae*, Microbiology 160 (2014) 537–546.
[67] M.E. Reardon-Robinson, H. Ton-That, Assembly and function of *Corynebacterium diphtheriae* pili, in: A. Burkovski (Ed.), *Corynebacterium diphtheriae* and Related Toxigenic Species, Springer, Heidelberg, Germany, 2014, pp. 123–141.
[68] L. Ott, M. Höller, J. Rheinlaender, T.E. Schaffer, M. Hensel, A. Burkovski, Strain-specific differences in pili formation and the interaction of *Corynebacterium diphtheriae* with host cells, BMC Microbiol. 10 (2010) 257.
[69] R. Stavracakis Peixoto, C. Azevedo Antunes, D. Weerasekera, L. Simpson Louredo, V. Goncalves Viana, C. Silva dos Santos, J.F. Ribeiro da Silva, R. Hirata Jr., E. Hacker, A.L. Mattos-Guaraldi, A. Burkovski, Functional characterization of the putative adhesin DIP2093 and its influence on the arthritogenic potential of *Corynebacterium diphtheriae*, Microbiology 163 (2017) 692–701.
[70] D.M. Meinel, R. Kuehl, R. Zbinden, V. Boskova, C. Garzoni, D. Fadini, M. Dolina, B. Blümel, T. Weibel, S. Tschudin-Sutter, A.F. Widmer, J.A. Bielicki, A. Dierig, U. Heininger, R. Konrad, A. Berger, V. Hinic, D. Goldenberger, A. Blaich, T. Stadler, M. Battegay, A. Sing, A. Egli, Outbreak investigation for toxigenic *Corynebacterium diphtheriae* wound infections in refugees from Northeast Africa and Syria in Switzerland and Germany by whole genome sequencing, Clin. Microbiol. Infect. 22 (2016) 1003.e1–1003.e8.
[71] M. du Plessis, N. Wolter, M. Allam, L. De Gouveia, F. Moosa, G. Ntshoe, L. Blumberg, C. Cohen, M. Smith, P. Mutevedzi, J. Thomas, V. Horne, P. Moodley, M. Archary, Y. Mahabeer, S. Mahomed, W. Kuhn, K. Mlisana, K. Mccarthy, A. von Gottberg, Molecular characterization of *Corynebacterium diphtheriae* outbreak isolates, South Africa, March-June 2015, Emerg. Infect. Dis. 23 (2017) 1308–1315.
[72] L. Sangal, S. Joshi, S. Anandan, V. Balaji, J. Johnson, A. Satapathy, P. Haldar, R. Rayru, S. Ramamurthy, A. Raghavan, P. Bhatnagar, Resurgence of diphtheria in North Kerala, India, 2016: laboratory supported case-based surveillance outcomes, Front. Public Health 5 (2017) 218.
[73] C.L. Gordon, P. Fagan, J. Hennessy, R. Baird, Characterization of *Corynebacterium diphtheriae* isolates from infected skin lesions in the Northern Territory of Australia, J. Clin. Microbiol. 49 (2011) 3960–3962.
[74] N. Cassir, D. Bagneres, P.E. Fournier, P. Berbis, P. Brouqui, P.M. Rossi, Cutaneous diphtheria: easy to be overlooked, Int. J. Infect. Dis. 33 (2015) 104–105.
[75] R.P. Fitzgerald, A.J. Rosser, D.N. Perera, Non-toxigenic penicillin-resistant cutaneous *C. diphtheriae* infection: a case report and review of the literature, J. Infect. Public Health 8 (2015) 98–100.
[76] T.G. Nelson, C.D. Mitchell, G.M. Sega-Hall, R.J. Porter, Cutaneous ulcers in a returning traveller: a rare case of imported diphtheria in the UK, Clin. Exp. Dermatol. 41 (2016) 57–59.
[77] J. Gubler, C. Huber-Schneider, E. Gruner, M. Altwegg, An outbreak of nontoxigenic *Corynebacterium diphtheriae* infection: single bacterial clone causing invasive infection among Swiss drug users, Clin. Infect. Dis. 27 (1998) 1295–1298.
[78] A. Dangel, A. Berger, R. Konrad, H. Bischoff, A. Sing, Geographically diverse clusters of non-toxigenic *Corynebacterium diphtheriae* infection, Germany, 2016-2017, Emerg. Infect. Dis. 24 (2018) 1239–1245.
[79] V. Sangal, L. Nieminen, B. Weinhardt, J. Raeside, N.P. Tucker, C.D. Florea, K.G. Pollock, P.A. Hoskisson, Diphtheria-like disease caused by toxigenic *Corynebacterium ulcerans*strain, Emerg. Infect. Dis. 20 (2014) 1257–1258.

[80] J. Walker, H.J. Jackson, D.G. Eggleton, E.N. Meeusen, M.J. Wilson, M.R. Brandon, Identification of a novel antigen from *Corynebacterium pseudotuberculosis* that protects sheep against caseous lymphadenitis, Infect. Immun. 62 (1994) 2562–2567.
[81] F.A. Dorella, L.G. Pacheco, N. Seyffert, R.W. Portela, R. Meyer, A. Miyoshi, V. Azevedo, Antigens of *Corynebacterium pseudotuberculosis* and prospects for vaccine development, Expert Rev. Vaccines 8 (2009) 205–213.
[82] S.C. McKean, J.K. Davies, R.J. Moore, Expression of phospholipase D, the major virulence factor of *Corynebacterium pseudotuberculosis*, is regulated by multiple environmental factors and plays a role in macrophage death, Microbiology 153 (2007) 2203–2211.
[83] L. Ott, M. Höller, R.G. Gerlach, M. Hensel, J. Rheinlaender, T.E. Schaffer, A. Burkovski, *Corynebacterium diphtheriae* invasion-associated protein (DIP1281) is involved in cell surface organization, adhesion and internalization in epithelial cells, BMC Microbiol. 10 (2010) 2.
[84] V. Kolodkina, T. Denisevich, L. Titov, Identification of *Corynebacterium diphtheriae* gene involved in adherence to epithelial cells, Infect. Genet. Evol. 11 (2011) 518–521.
[85] S. Kim, D.B. Oh, O. Kwon, H.A. Kang, Identification and functional characterization of the NanH extracellular sialidase from *Corynebacterium diphtheriae*, J. Biochem. 147 (2010) 523–533.
[86] K. Otsuji, K. Fukuda, T. Endo, S. Shimizu, N. Harayama, M. Ogawa, A. Yamamoto, K. Umeda, T. Umata, H. Seki, M. Iwaki, M. Kamochi, M. Saito, The first fatal case of *Corynebacterium ulcerans* infection in Japan, JMM Case Rep. 4 (2017).
[87] V.L. Tesh, A.D. O'Brien, The pathogenic mechanisms of Shiga toxin and the Shiga-like toxins, Mol. Microbiol. 5 (1991) 1817–1822.
[88] Y.S. Chan, T.B. Ng, Shiga toxins: from structure and mechanism to applications, Appl. Microbiol. Biotechnol. 100 (2016) 1597–1610.
[89] T. Sekizuka, A. Yamamoto, T. Komiya, T. Kenri, F. Takeuchi, K. Shibayama, M. Takahashi, M. Kuroda, M. Iwaki, *Corynebacterium ulcerans* 0102 carries the gene encoding diphtheria toxin on a prophage different from the *C. diphtheriae* NCTC 13129 prophage, BMC Microbiol. 12 (2012) 72.
[90] D.M. Meinel, G. Margos, R. Konrad, S. Krebs, H. Blum, A. Sing, Next generation sequencing analysis of nine *Corynebacterium ulcerans* isolates reveals zoonotic transmission and a novel putative diphtheria toxin-encoding pathogenicity island, Genome Med. 6 (2014) 113.
[91] S. Silva Ado, R.A. Barauna, P.C. De Sa, D.A. Das Gracas, A.R. Carneiro, M. Thouvenin, V. Azevedo, E. Badell, N. Guiso, A.L. Da Silva, R.T. Ramos, Draft genome sequence of *Corynebacterium ulcerans* FRC58, isolated from the bronchitic aspiration of a patient in France, Genome Announc. 2 (2014).
[92] J.B. Milstien, B.G. Gellin, M. Kane, J.L. di Fabio, A. Homma, Global DTP manufacturing capacity and capability. Status report: January 1995, Vaccine 14 (1996) 313–320.
[93] M. Iwaki, T. Komiya, A. Yamamoto, A. Ishiwa, N. Nagata, Y. Arakawa, M. Takahashi, Genome organization and pathogenicity of *Corynebacterium diphtheriae* C7(-) and PW8 strains, Infect. Immun. 78 (2010) 3791–3800.
[94] H. Nakao, I.K. Mazurova, T. Glushkevich, T. Popovic, Analysis of heterogeneity of *Corynebacterium diphtheriae* toxin gene, *tox*, and its regulatory element, *dtxR*, by direct sequencing, Res. Microbiol. 148 (1997) 45–54.
[95] P.K. Cassiday, L.C. Pawloski, T. Tiwari, G.N. Sanden, P.P. Wilkins, Analysis of toxigenic *Corynebacterium ulcerans* strains revealing potential for false-negative real-time PCR results, J. Clin. Microbiol. 46 (2008) 331–333.
[96] A. Sing, A. Berger, W. Schneider-Brachert, T. Holzmann, U. Reischl, Rapid detection and molecular differentiation of toxigenic *Corynebacterium diphtheriae* and *Corynebacterium ulcerans* strains by LightCycler PCR, J. Clin. Microbiol. 49 (2011) 2485–2489.
[97] A.S. Santos, R.T. Ramos, A. Silva, R. Hirata Jr., A.L. Mattos-Guaraldi, R. Meyer, V. Azevedo, L. Felicori, L.G.C. Pacheco, Searching whole genome sequences for biochemical identification features of emerging and reemerging pathogenic *Corynebacterium* species, Funct. Integr. Genomics (2018), https://doi.org/10.1007/s10142-018-0610-3. epub ahead of print.

CHAPTER 5

Pan-genomics of veterinary pathogens and its applications

Thiago de Jesus Sousa[a], Arun Kumar Jaiswal[a,b], Raquel Enma Hurtado[a], Stephane Fraga de Oliveira Tosta[a], Siomar de Castro Soares[b], Anne Cybelle Pinto Gomide[a], Luiz Carlos Junior Alcantara[d], Debmalya Barh[c], Vasco Azevedo[a], Sandeep Tiwari[a]

[a]PG Program in Bioinformatics, Institute of Biological Sciences, Federal University of Minas Gerais (UFMG), Belo Horizonte, Brazil

[b]Department of Immunology, Microbiology and Parasitology, Institute of Biological Science and Natural Sciences, Federal University of Triângulo Mineiro (UFTM), Uberaba, Brazil

[c]Centre for Genomics and Applied Gene Technology, Institute of Integrative Omics and Applied Biotechnology (IIOAB), Purba Medinipur, India

[d]Laboratório de Flavivírus, Instituto Oswaldo Cruz, FIOCRUZ, Rio de Janeiro, Brazil

1 Introduction

Pan-genome is an approach that contributes to the research of bacterial pathogenesis. This terminology was proposed in 2005 in research with the bacterium *Streptococcus agalactiae*, by the researcher Tettelin and collaborators [1]. In this work, they define the pan-genome as a set of genes in a given study group, considering core genome, the genes present in all strains in the group of study; dispensable genes as absent genes in one or more strains; and, genes that are considered unique in each lineage of the study group. Pan-genome can be considered open or closed, depending on the bacterial ability to acquire exogenous regions (DNA) [1] and the lifestyle that will determine this issue [2]. From the sequencing, one can thoroughly study each region of the genome, contributing with unpublished information. Since 2005, with the era of new sequencers, the speed, ease, and reliability of data have been increasing and with them the number of bacterial genomes deposited in public databases [3]. Pan-genome studies can be applied with different goals, such as taxonomy, reverse vaccinology, gene variation, pathogenesis [4], among others. This chapter is foucused on the pan-genomics studies carried out on pathogenic bacteria that cause veterinary diseases, including the ones responsible for zoonotic diseases. From the genetic repertoire studies, the key points (genes) supposedly involved in the spread of disease, bacterial resistance, infection, adhesion, can be detected, leading to practical solutions against the disease being studied. An important fact is an identification, from taxonomic studies among the lineages, of horizontal gene transfer, which in addition to contributing to evolutionary information, may be used to infer possibly emerging pathogens, once the previously harmless pathogen may become pathogenic. Horizontal gene transfer causes a considerable impact on genomic plasticity,

Pan-genomics: Applications, Challenges, and Future Prospects
https://doi.org/10.1016/B978-0-12-817076-2.00005-6
© 2020 Elsevier Inc.
All rights reserved.

Table 1 Pathogenic bacterial species of veterinary and human importance

Bacterial species causing animal infection	Bacterial species causing animal and human infection
Corynebacterium pseudotuberculosis	*Brucella*
Corynebacterium ulcerans	*Corynebacterium diphtheriae*
Streptococcus suis	*Francisella tularensis*
Brachyspira hyodysenteriae	*Campylobacter*
Moraxella bovoculi	*Clostridium botulinum*
Mannheimia haemolytica	*Streptococcus agalactiae*
Pasteurella multocida	

bacterial evolution, and adaptation, and leads to an inquiry into species determination [5]. The strains of the same species can differ considerably in the gene repertoire, which confers a versatile adaptation to a wide range of environments [6]. From the Pan-genome studies, one can perform this thorough analysis between different genomes leading to an understanding of evolutionary strategies, acquisition of resistance, hereditary variation leading to its evolutionary adaptation and, in some situations, the results can lead to even a proposal of species redefinition [2, 3].

Table 1 shows the list of pathogenic bacterial species of veterinary and human importance that already have Pan-genome studies. The studies have a high impact in the diagnosis, prophylaxis, and verification of the genetic variation among the strains. Thus, these studies could provide effective solutions to fight against the diseases that cause significant damage to the agribusiness or a constant public health problem.

2 Pan-genomics studies of pathogenic bacteria causing veterinary and zoonotic diseases

2.1 *Corynebacterium pseudotuberculosis*

Corynebacterium pseudotuberculosis is the agent of Caseous lymphadenitis (CLA), but may also cause other chronic diseases such as ulcerative lymphangitis. *C. pseudotuberculosis* has as host small and large ruminants, causing significant economic losses, and there are already some cases in the literature of transmission in humans as well. One way to contribute to the health of these host is the study of the genomes of *C. pseudotuberculosis* strains, which brings good discussions about the evolutionary understanding of the species, adaptation, and interaction with the host. Thus, it is possible to elucidate inferences about genes or virulence factors, and consequently more efficient and cheaper vaccines, drugs, and diagnostics. An example of this is reverse vaccinology, which aims to identify targets for vaccines and/or drugs by computational means and thereby reducing in vivo and in vitro tests [7].

In 2011, Ruiz et al. compared two genomes of *C. pseudotuberculosis*, strains 1002 and C231, which were first complete genomes deposited in National Center for Biotechnology Information (NCBI). These two strains are very similar, with approximately 95% similarity from the amino acid sequences of the predicted protein pool. The two strains are also very similar concerning genomic composition, G+C content values, gene size, operon composition, and gene density. However, significant differences are observed for genome size, a number of pseudogenes and lineage-specific genes. As expected, the strains including *C. pseudotuberculosis* 1002 and C231 showed high conservation in the genus, with approximately 97% of their genes presenting conservation in the gene order [8].

In 2013, Soares et al. did an antigenic target prediction study with the *C. pseudotuberculosis* 258 strain genome for the prediction of biotechnology vaccines. Then, by reverse vaccinology, 49 possible proteins were identified as vaccine target candidates, where one target was present on a pathogenicity island [9]. In the same year, Soares et al. made a pan-genomic analysis with 15 genomes of *C. pseudotuberculosis*, characterizing this species with an open pan-genome, in which approximately 19 new protein coding sequences were to be added for each new genome. The core genome consists of 1504 sequences encoding proteins. More detailed analyses of the pan-genome revealed differences between *ovis* and *equi* biovar strains, where the biovar *ovis* showed a more clonal behavior than the biovar *equi* strains [10].

But in the last quarter of 2018, the number of complete genomes has increased to 72. This genome data was made with new sequencing platforms and methodologies. In addition, many other works that corrected errors present in the assembly of the deposited genomes were published, suggesting an update in these pan-genomic studies [11].

2.2 *Corynebacterium ulcerans*

Corynebacterium ulcerans has emerged as a relevant zoonotic pathogen. An increasing number of cases of *C. ulcerans* infection have been reported from many countries including Brazil. *C. ulcerans* has a wide range of animal hosts [12].

Pan-genomic studies of *C. ulcerans* showed that the main virulence factor in this species is the *tox* gene, mainly present in *Corynebacterium diphtheriae*. The *tox* gene is found in lysogenic corynephages, but also on a pathogenicity island. In some strains the function of *tox* gene found inactivated due to frameshift mutation. However, several other genes encoding virulence-associated proteins, such as phospholipase D (*Pld*), neuraminidase H (*NanH*), corynebacterial protease (*CP40*), venom serine protease (*Vsp1* and *Vsp2*), ribosomal-binding protein (*Rbp*, similar to Shiga-like toxin), and adhesive surface pili are present in different *C. ulcerans* strains [13].

Pan-genomic studies have identified the presence of multiple prophages that are an important source of genomic plasticity. Surface pili are responsible for adhesion and invasion of host cells, which play an essential role in the virulence of pathogenic bacteria. A study with 19 strains of *C. ulcerans* published in 2018 by Subedi et al. identified 4120 genes, including 1405 core genes and 2715 accessory genes. Among the proteins of the core genome, there were 351 proteins with transmembrane domains, 3 with additional signal peptides, 2 cell wall-anchored proteins, and 82 secreted proteins, of which 46 were identified as putative lipoproteins. The accessory genome included 611 membrane-associated proteins, 65 with additional signal peptide features and 46 with an LPXTG motif. A total of 116 accessory proteins were secreted via sec-dependent secretory pathways. Membrane-associated and secreted proteins are essential for host-pathogen interactions and virulence [13]. Therefore, in addition to the variation in the virulence genes, the number of transmembrane, lipoprotein, and secreted proteins may be responsible for the variation in their virulence characteristics. Indeed, a variation in the ability to cause arthritis in a mice model by different *C. ulcerans* strains was previously reported. As mentioned earlier, prophages are the primary source of diversity among these strains.

2.3 *Streptococcus suis*

Streptococcus suis is a Gram-positive bacterium considered one of the essential bacterial pathogens in the swine industry in the world, mainly in China. In addition, *S. suis* is also an emerging zoonotic pathogen. It is classified into 33 serotypes, where serotypes 1, 2, 3, 7, 9, and 1/2 are the most prevalent in swine, and strains that cause human infections were also found among these serotypes. In 2018, there are 42 complete genomes deposited in the NCBI, with a single chromosome of approximately 2 Mb [14].

A study in 2011 by Zhang et al. [14], with 13 complete genomes found 2374 orthologous genes and 1211 unique genes, a core genome with 1343 genes, and the observed pan-genome shared by the 13 strains consisted of 3585 genes. In this pan-genomic analysis, they estimated that for each newly sequenced genome, 82 genes are added, characterizing that the species has an open pan-genome. This is consistent with an earlier study on the core and pan-genome of *Streptococcus*, which indicated that *S. suis* was the ancestor with the highest number of genetic gains and losses [14].

2.4 *Brachyspira hyodysenteriae*

Brachyspira spp. is found colonizing intestines of some species of mammals and birds, and shows different degrees of enteropatogenicity. *B. hyodysenteriae* is an important swine pathogen, which causes dysentery in these animals. It has three complete genomes deposited in NCBI, and its genome size consists of a ~3-Mb chromosome and a ~36-kb plasmid. This plasmid is conserved among several strains, but it is not found in any non-virulent isolated strain in the field, suggesting that it may be an essential virulence factor for the species [15].

Genomic studies between *Brachyspira pilosicoli*, *Brachyspira intermedia*, *Brachyspira hyodysenteriae*, and *Brachyspira murdochii*, suggest *B. pilosicoli* lost many transport-related proteins, which might reflect its adaptation to a more specialized ecological niche. The highest level of reductive evolution in *B. pilosicoli* suggests that it is a pathogen older than *B. hyodysenteriae*. The pathogenicity of the younger *B. hyodysenteriae* may be related to the acquisition of the 32 kb plasmid [15]. In general, recent studies suggest that *B. hyodysenteriae* and *B. pilosicoli* are more specialized pathogens and have less genetic material and diversity. These strains have undergone specialization process independently, which is suggested by the little genetic material that is shared only between them. In addition, studies suggest that there was a reductive evolution with *B. hyodysenteriae* and *B. pilosicoli* since they have the two smaller genomes. Reductive evolution may be involved in the loss of genes, especially transport proteins [15]

2.5 *Moraxella bovoculi*

Infectious bovine keratoconjunctivitis (IBK) affects cattle, causing pain, blindness in severe cases, and reduced weight gain in animals. In addition to concern about animal health and welfare, IBK's economic impact may be significant, with estimates exceeding US$ 150 million in direct and indirect economic losses. As microbiological characteristics, they are coccobacillus and Gram-negative. *Moraxella bovoculi* has been extensively associated with IBK in the absence of *Moraxella bovis* since its initial description in 2007 [16].

Genomic studies with this species are scarce. Studies in the literature have shown that the diversity of single nucleotide polymorphisms (SNPs) in *M. bovoculi* is high, with 81,284 SNPs identified in eight genomes (being seven complete genomes). Two distinct genotypes are represented, isolated from IBK (genotype 1) and the nasopharynx of cattle without clinical IBK signs (genotype 2). Only in genotype 1, it found repeats-in-toxin (RTX) putative pathogenesis factor and 10 putative antibiotic resistance genes carried within a genomic island (GI). Due to very high recombination, genotype 1 subtypes cannot be distinguished at the SNPs level, although these subtypes may vary in their virulence potential. Interspecific recombination with *M. bovis* indicates that, for at least two loci, these species share a common genetic set. Because of this, future work as the development of IBK vaccines may benefit from the identification and characterization of conserved outer membrane proteins shared by both *Moraxella* species [16].

2.6 *Pasteurella multocida*

Pasteurella multocida is a Gram-negative commensal and bacterial pathogen causing economically important diseases of veterinarian interest as hemorrhagic septicemia, fowl cholera, atrophic rhinitis, and pneumonia in a broad range of animal species, likewise it is a zoonotic agent to humans through bites infections [17]. A last pangenomic study on 109 *P. multocida* isolates describes a pan-genome with 4256 repertoire genes, 1806 core genes (42.43%), 1841 dispensable genes (43.25%), and 609 strain-specific genes (14.3%) [18]. Similar results describe the accessory genome with 52.91% and dispensable genes with 33.47%, showing an open pangenome to species [19]. The dispensable genes content assigned to COG categories belong to carbohydrate transport and metabolism (9.54%), transcription (4.85%), replication, recombination and repair (3.08%), inorganic ion transport, and metabolism (4.6%) [19]. The presence of these highlighted functional categories could be associated with its environmental fitness [20, 21], whereas 46.35% and 49% of unique and dispensable genes are assigned to unknown function, revealing a large number of noncharacterized proteins involved in diversification process [19]. Association studies of the accessory genome would show the presence of specific genes in a specific disease [19, 22, 23] but not a predilection to a host [18]. Complementary comparative genomic analysis show the accessory genome belonged to prophages, ICE, GI and plasmids, as well as the presence of a unique large integrative conjugative element, ICEPmu1, containing 88 genes of which 12 genes encoding resistance to antibiotics [24]. Likewise, pathogenomics analysis among virulent avian *P. multocida* strains (P1059 and/or X73) against an avirulent strain Pm70 identified 336 genes of which 61 genes present unknown function [22]. Other studies corroborated the presence of a cluster of genes involved in the transport and modification of citrate, galactitol-specific phosphotransferases, transport and utilization of L-fucose shared by at least two fowl cholera strains X73, F216, P1059, and F218 [19, 22]. The presence of these cluster of genes related to metabolism and adhesion could provide the capacity of adaptation and virulence to avian host [22]. Also, the genomic comparison among Hemorragic Septicemia-associated strains and strains not associated with the disease show two unique intact prophages present on all HS strains [23]. Additionally, phylogenomic and comparative genomics analysis based on the accessory genome shows the clustering of some *P. multocida* strains by disease [19, 22, 23], which supports the SNPs phylogenetic clustering [19]. Population phylogenies based on core genes show a relationship with the predilection to a host and geographical association [25] or MLST distribution [19]. These studies showed a great diversity at the gene level; likewise, this reflects the associations of genetic groups that present determinate mobile genetic element that could be involved with the capacity to infect. All the studies so far allow us to show the importance of accessory genome in the genetic diversification process and evolutionary adaptation of *P. multocida* species [19, 25, 26].

2.7 *Mannheimia haemolytica*

Mannheimia haemolytica is a hemolytic, Gram-negative coccobacillus, commensal of the upper respiratory tract and nasopharynx, and causal agent of respiratory disease on ruminants, mainly associated with the bovine respiratory disease with economic losses to the cattle industry worldwide [27, 28]. Pan-genome analysis of 21 *M. haemolytica* isolates identified 9507 orthologous groups of genes, 1333 core genes (14%), and 6350 dispensable genes (66.8%) [29]. The pan-genome of all 21 *M. haemolytica* strains is open and the accessory genome is composed of 66.8% and 81.8% of dispensable and unique genes, respectively, containing uncharacterized or hypothetical proteins [29]. The virulence and etiology of *M. haemolytica* is strongly associated with serotypes, being serotype 1 and 6 responsible for pneumonia in bovine and serotype 2 responsible for pneumonia in sheep and prevalent as commensal among healthy cattle [29, 30]. Comparative pathogenomic studies found differences between S1, S6 bovine strains with the presence of more integrative conjugative elements and prophages than S2 strain and also differences of spacer sequences on CRISPR arrays. Likewise, the presence of antimicrobial-resistant (AMR) contained in conjugable element (ICE) is more prevalent in S2 than S1 and S6 strains. The AMR may be removed in SA and S6 through effective antimicrobial therapies in diseases animal compared with healthy animals. However, little is known about how genetic differences among serotypes contribute to pathogenesis in this species [29, 30]. The identification of variable mobile genetic elements as prophages and ICEs would be implied in the genetic diversification process, pathogenicity, and evolutionary adaptation [29–31]. First comparative genomic analysis between three strains of *M. haemolytica* from bovines and ovines found a high percentage of hypothetical proteins in the content of unique genes (57%) and phage related genes (20% and 29% from A1 and B strain, respectively), where the authors correlated the variable gene pool with specific phenotypes (strain virulence, species specificity, etc.) [30]. From the analysis of 11 bovine isolates, 14 prophage clusters were identified, which contain toxin-antitoxin systems and multiple virulence-associated genes involved in virulence and antimicrobial resistance [29]. It was detected a CRISPR-Cas that play a role in immune evasion or adhesion during infection [29]. Integrative conjugative elements were found in nine strains, playing a role in the survival through the multidrug resistances [32, 33], and regulating their dissemination through toxin-antitoxin and entry exclusion systems [29]. Comparative genomic analyses of pathogenic strains would allow a better comprehension of the pathogenicity and the prediction of resistance mechanisms. Likewise, pan-genome analysis allows the discovery of all spectrum of genes represented, which are implicated in the genetic diversity and evolution of the species (Table 2).

2.8 *Clostridium botulinum*

Clostridium botulinum is an anaerobic, Gram-positive, and spore-forming pathogen in charge of the rising of food contamination cases over the world. The transmission of the disease from *C. botulinum* is resonating, by the unexpected hospital outbreaks and expanded obstruction against multiple drugs [38]. *C. botulinum* is able to produce botulinum toxins and these toxins (BoNT) are considered to be the most toxic substances

Table 2 An overview of Pan-genome studies in veterinary infection related bacteria

Name of the bacteria	Disease	Host	Genome size (Mb)	Pan genome analysis	References
Brucella spp.	Brucellosis	Human, Bovine and small ruminants	3.3	To get insights of the survival mechanism	[34–36]
Brachyspira hyodysenteriae	Swine dysentery	Mammals and birds	3.052	Reduction of many transport-related proteins	[15]
Corynebacterium ulcerans	Diphtheria-like infection and extrapharyngeal infections	Animal/ Human	2.497	In core genome, 351 were transmembrane domains, 3 with additional signal peptides, and 2 were cell wall-anchored proteins, 82 were predicted to be secreted, of which 46 were identified as putative lipoproteins.	[13]
Corynebacterium diphtheriaediphtheria	Diphtheria	Humans and animal	2.444	57 genomics islands, most of them pathogenicity islands and associated with adhesive pili, responsible for the adhesion	[37]
Corynebacterium pseudotuberculosis	Caseous lymphadenitis/ ulcerative lymphangitis	Animal	2.337	Revealed differences between ovis and equi biovar strains	[10]
Clostridium botulinum	Botulism	Human and animal	3.917	Open pangenome, the study was to study symptoms related to this	[38]

				bacteria with respect to the wide range of hosts	
Campylobacter spp.	Campylobacteriosis	Human and animal	1.818		
Francisella tularensis	Tularaemia	Lagomorphs and humans	1.825	The presence of point mutations, insertion elements and small indels resulting in gene deactivation in the process of differentiation from the nonpathogenic strain into the human pathogenic strains	[39]
Moraxella bovoculi	Infectious Bovine Keratoconjunctivitis (IBK)	Cattle	2.214	81,284 SNPs identified in eight genomes	[16]
Mannheimia haemolytica	Respiratory disease	Cattle	2.635	Open and the accessory genome is composed of 66.8% and 81.8% of dispensable and unique genes, respectively	[29]
Pasteurella multocida	Hemorrhagic septicemia, fowl cholera, atrophic rhinitis and pneumonia	Animals	2.305	The importance of accessory genome in the genetic diversification process and evolutionary adaptation	[19, 25, 26]
Streptococcus agalactiae	Meningoencephalitis, Septicemia, Meningitis, Neonatal sepsis and pneumonia	Cattle, Fish and Human	2.081	Vaccine targets identification, 36 antigenic proteins as possible vaccine targets	[40]
Streptococcus suis	Meningitis, septicaemia	Swine and Human	2.096	Each newly sequenced genome, 82 genes were added, Open pan-genome	[14]

occurring in nature [41]. Botulism is a perilous flaccid paralytic disease caused by eight different neuroparalytic toxin subtypes (A–H) [42]. Toxin subtypes A, B, E, and F are rarely and recently discovered, and serotype H is mainly responsible for human botulism, whereas toxin types C and D are involved in animal botulism around the world [42, 43]. The instances of Botulism infection are exceptionally normal in wild and local creatures and happen sporadically just as hugely everywhere throughout the world. The cattle and birds are extremely affected species of animals, despite the fact that botulism cases likewise are typically found among horses, sheep, and goats. The bacteria produce botulinum neurotoxins that act on the nerve endings, blocking acetylcholine discharge [44–46]. *C. botulinum* is the third most infectious agent worldwide to human and animal health. Botulism cases are exceptionally critical in ruminants, common in birds and dogs, and have additionally been reported in other species, specifically dogs, pigs, horses, and wild mammals in Brazil [47]. The first Botulism disease was reported in Brazil in 1960s in the state of Piauí in cattle, and was later identified in other species, such as sheep, goats, and buffaloes in all Brazilian regions [47]. The strain A2 of *C. botulinum* was recognized as resistant to metronidazole and penicillin [48]. A pan-genome work was published by Bhardwaj et al. [38], to comprehend the symptoms related to this bacteria with respect to the wide range of hosts. The successive calculation and characterization of the core and pan-genome subset disclosed the identification of more specific targets for drug designing and vaccine development [38]. In this study, 13 genomes of *C. botulinum* were used for pan-genome analysis and they identified 889 genes as core genome and 287 strain-specific genes. The reported open pan-genome in their analysis, which indicates unique genes, suggests that new genes could be added with every newly added genome sequence. Core, unique, and accessory genes were further categorized, in which most of core genes belong to metabolism and genetic information processing. Core-genome calculation exposes high level of genomic similarity among the genomes with low variation in GC content. The persistence of singleton genes shows the capacity to get novel virulence traits. The identification and analysis of GIs helped characterize potential drugs and vaccine targets [38].

2.9 *Campylobacter*

The *Campylobacter* species constitute a highly biological diverse group of organisms, some of which are widely known causative agents of clinical illness in animals and humans [49]. The disease Campylobacteriosis is an aggregate depiction for infectious diseases, caused by members of the bacterial genus *Campylobacter*. The infection is present in animals such as poultry, cattle, pigs, wild birds, and wild mammals. *Campylobacter* bacterium is one of the greatest agents of foodborne diarrheal illness in humans, and in addition, commonly causes gastroenteritis worldwide [50–52] and affects 9 million people each year, costing around €2.4 billion [53, 54]. Generally, infections are not extreme, being the most critical symptom the gastroenteritis; however, they can also cause extraintestinal manifestations such as reactive arthritis, inflammatory bowel disease (IBD), Guillain-Barré syndrome (GBS), and in some cases, infection lead to death. Infections in Human are fundamentally connected with taking care of and additionally devouring poultry meat [54, 55]. The related subspecies *C. fetus* subsp. fetus and *C. fetus* subsp. *venerealis* of *Campylobacter fetus* are well-known pathogens of reproductive failures in ruminants [56]. The *C. fetus* subsp. *fetus* shows a wide ranging of host distribution, colonizes the gastrointestinal tract, and is generally linked with sheep and cattle abortion, while *C. fetus* subsp. *venerealis* has low host range, is restricted to the bovine genital tract, and the primary cause of venereally transmitted infectious, infertility, and embryonic mortality in cattle [49, 57]. In addition to *C. fetus* subsp. *fetus*, *Campylobacter jejuni* subsp. *jejuni* is also a major pathogen of *Campylobacter* species related with sheep abortion [49, 57, 58]. *C. fetus* subsp. *venerealis*, infections is also known as bovine genital campylobacteriosis (BGC), bovine venereal campylobacteriosis, or vibriosis, which is characterized by infertility and early embryonic deaths [57, 59, 60]. Rather than its public health importance, the ecological and evolutionary aspects of the *Campylobacter* are still poorly understood. Nevertheless, they could have an intense effect on transmission and human infection and it is not explained properly how *Campylobacter coli* and *C. jejuni*, which have similar host niches and frequently exchange genetic material, show differences in their disease epidemiology [61]. Throughout the decades, antibiotics have been arbitrarily used in animal production to control, prevent, and treat infections and to increase animal growth [62]. The primary cause of rise and spread of antibiotic resistance among *Campylobacter* spp. is the use of unregulated antimicrobial agents in food animal production, which has led to the development of antibiotic resistance in campylobacter subspecies [63–65]. *Campylobacter* antibiotics resistance is emerging globally and has already been described by several authors earlier and also acknowledged by the WHO, as a problem of public health importance [63, 65–68]. Antibiotics, generally tetracycline, macrolides, and (fluoro) quinolones, are used for more severe cases. Nevertheless, the growth of resistance to tetracycline, erythromycin, and (fluoro) quinolones of *C. coli* and *C. jejuni* strains might compromise the efficacy of this treatment [65]. Work published by Lefébure et al., in 2010, used 42 strains of *C. coli* and 43 strains of *C. jejuni*, where the pan-genome of both species combined reaches approximately 3000 genes [69]. In another study published in 2014 by Méric et al., seven strains of *C. jejuni* and *C. coli* genomes were used for pan-genome analysis. They identified 3933 genes as pan-genome, a core genome of 1035, and the accessory genome contained 2792 genes [61].

2.10 *Streptococcus agalactiae*

S. agalactiae is a bacterium that causes illnesses in cattle, fish, and human [40]. In human, it is frequently associated with meningitis, neonatal sepsis, pneumonia, and pregnant women [40, 70]. This bacterium is associated with typical gut flora and genital tract, moreover, it is also found colonizing 10%–40% of pregnant women [71, 72]. A notable number of newborn infant infections from *S. agalactiae* have been identified, making it necessary to investigate it in view of its substantial morbidity and mortality [73, 74]. In dairy cattle, *S. agalactiae* (Lancefield group B; GBS) is additionally a noteworthy pathogen of clinical and subclinical mastitis, which affects quality and production of milk [70]. *S. agalactiae* is an evolving pathogen in fish, which causes meningoencephalitis and septicemia. The pathogen has been accounted with high mortality in wild and cultured species worldwide [40, 75, 76]. *S. agalactiae* developed phenotypic and genotypic antibiotic resistance patterns in China, being isolated from cows with mastitis [77]. Bolukaoto et al. [71] isolated an antibiotic resistant strain of *S. agalactiae* from pregnant women in Garankuwa, South Africa. *In silico* techniques like Pan-genome, Pan-modelome, Subtractive genomics, and Reverse vaccinology are playing a key role in quick and rapid identification of new therapeutic targets in the post-genomic era [78]. In 2013 Pereira et al., published research article for vaccine targets against *S. agalactiae* where they used 15 genomic strains from different isolates (10 from human isolates, 4 from fish and 1 from cow). Their pan-genome analysis identified 5143 genes on the pan-genome and 1111 genes as part of the core-genome, shared by all genomes. They identified 36 antigenic proteins as possible vaccine targets, which were conserved in all 15 strains and, in future, will be used as vaccine candidates [40].

2.11 *Francisella tularensis*

F. tularensis is a highly infectious, Gram-negative, facultative, and intracellular bacterium, which presents rod-shaped or coccoid cells and is also aerobic and nonmotile [79]. *F. tularensis* is the etiological agent of tularaemia—a zoonotic disease that has been described in animals, predominantly in rodents, lagomorphs, and humans [80]. In this group, six clinical manifestations are characterized by the form of entrance of bacteria: ulceroglandular, glandular, oropharyngeal, oculoglandular, pneumonic tularaemia, and typhoidal tularaemia forms [81]. The occurrence of tularemia is equally influenced by the host and the different subspecies [80] as the four proposed subspecies of *F. tularensis* subspecies *tularensis, holarctica, novicida,* and *mediasiatica* differ in virulence and geographical range.

Rohmer and collaborators (2007) compared two pathogenic subspecies in humans; *F. tularensis* subspecies *tularensis* and *holarctica* against *F. tularensis* subspecies *novicida* U112, described as nonpathogenic in humans but reproducing in mice a tularaemia-like disease [82]. The comparison revealed the presence of point mutations, insertion elements, and small indels resulting in gene deactivation in the process of differentiation from the nonpathogenic strain into the human pathogenic strains [39].

In order to investigate adaptations within the genus *Francisella*, in 2009, Larsson and collaborators compared 13 *F. tularensis* isolates from different subspecies to the genomes of 3 isolates of *Francisella novicida* and 1 isolate of *Francisella philomiragia*. Although *F. novicida* and *F. tularensis* present an average nucleotide identity of >97%, *F. novicida* is less virulent in mammals, with rare descriptions of human infections and seems to have a less specialized cycle. This increased host association were related to events of random insertions like the duplication of the *Francisella* Pathogenicity Island [83].

2.12 *Corynebacterium diphtheriae*

C. diphtheriae is the etiological agent of diphtheria, an acute disease localized in the upper respiratory tract leading to ulceras at the mucosa, and formation of an inflammatory pseudomembrane [84]. *C. diphtheriae* strains can be divided into toxigenic strains, which carry the *tox* structural gene and nontoxigenic strains, which do not carry the tox gene. Nontoxigenic *C. diphtheriae* strains are related with severe pharyngitis and tonsillitis, endocarditis, osteomyelitis, splenic abscesses, and septic arthritis [85]. In order to explore the genetic basis of different interactions with host tissues and clinical manifestations of infection by a variety of *C. diphtheriae* strains, several studies have been developed to provide information if a group of genes can be related with a clinical manifestation of *C. diphtheriae* infection.

Trost and collaborators performed a pan-genome study of *C. diphtheriae* comparing 13 genomes of strains isolated from patients with classical diphtheria, pneumonia, endocarditis, and the strain *C. diphtheriae* NCTC 13129 as a reference. It was demonstrated a high synteny level and a core genome consisting of 1632 conserved genes and, on average, 65 unique genes per strain. The analysis of genome-wide motif searches of *tox*-controlling regulator DtxR showed that the DtxR regulons presented differences due to gene variation on those sites responsible for interactions with DtxR. One important finding was the identification of 57 genomics islands, most of them are pathogenicity islands and associated with adhesive pili, responsible for the adhesion of *C. diphtheriae* to different host tissues [37]. Other study performed with 48 *C. diphtheriae* isolates from Australia over a 12-year period. The pan-genome analysis revealed 22 genes from gene group I significantly associated with respiratory infection [86].

Although the detection and isolation of *C. diphtheriae* in animals is poorly described at the literature, it is extremely relevant to try to understand the role of animals at transmission of *C. diphtheriae*, once the majority of isolated strains from animals had direct contact with humans [87].

C. diphtheriae had been characterized from four different animal species (dog, cat, cow, and horse) showing nontoxigenic, toxigenic, and nontoxigenic tox-bearing (NTTB) *C. diphtheriae* strains. All these reports had different clinical manifestations from pharyngitis, parotitis, otitis, chronic active dermatitis, draining wound infection to nonhealing pyogenic stake wound [88–93], which may contribute to the poor investigations of injuries as a result of *C. diphtheriae* infection in other animal species.

As described by Sing et al. [87], in the first finding of a nontoxigenic *C. diphtheriae* biovar *belfanti* in a wild red fox with no human contact, *C. diphtheriae* was accompanied by *Streptococcus canis*, an opportunistic pathogen of this species. Even though the contributions of lesions cannot be attributed just to *C. diphtheriae* [87], this case brings forward the possibility of *C. diphtheriae* infections being not detected as pathogenic bacteria in humans and animal infections [87].

2.13 *Brucella* spp.

The genus *Brucella* is composed by seven species, being them *Brucella neotomae*, *Brucella melitensis*, *Brucella abortus*, *Brucella suis*, *Brucella ovis*, *Brucella canis*, and *Brucella maris*. They are Gram-negative facultative intracellular and coccobacilli nonmotile bacteria [34]. Brucellosis is a zoonotic disease caused by *Brucella* spp. affecting mainly mammals, such as cattle, goats, camels, sheeps, pigs, dogs, which can lead to sterility or abortion, and humans, which causes serious, debilitating illness [34].

More than one species as the etiological agent of brucellosis has fomented several pan-genomic studies in order to identify different contributions of each agent.

Yang and collaborators performed pan-genomics analysis with 42 *Brucella* complete genomes in order to get insights on the survival mechanism of *Brucella* spp. in vivo. From the genes analyzed, the core genome contains 1710 clusters, 1182 clusters were strain-specific genes, and 2477 clusters were accessory genome. The core functions were mainly related with conservation, amino acid metabolism, and energy [36].

Although many studies look for genomic characteristics that can be distinguishable as a host adaptation, a comparative genomics study identified clonal isolates of *B. melitensis* Biovar 3 with no signature of host adaptation, investigating strains of a same outbreak from three different species (human, bovine, small ruminants) [35].

3 Conclusions

The infectious diseases that can be naturally transmitted between animals and humans are known as zoonoses. The causative agent of zoonoses includes wide range of pathogens such as viruses, bacteria, fungi, and parasites. Due to the advancement of the sequencing technology, there are multiple genome data of these pathogens available. Using bioinformatics and comparative genomics approaches can help in better understanding the dynamics of the pathogenies. Such as in the identification of common virulence factors in pathogenicity islands, which have a direct impact in the shared and singletons genes. Also, they may help in finding new vaccine and drug targets through the use of core genome information. Other omics analyses may also be performed like, pan-transcriptomics and pan-proteomics to discover the different patterns of gene expression of these organisms in different hosts, shedding a light on their adaptability. Finally, pan-genomics may contribute in the search for efficient new solutions against these diseases that cause several animal and human losses worldwide in agriculture and health systems.

References

[1] H. Tettelin, V. Masignani, M.J. Cieslewicz, C. Donati, D. Medini, N.L. Ward, et al., Genome analysis of multiple pathogenic isolates of Streptococcus agalactiae: implications for the microbial "pan-genome", Proc. Natl. Acad. Sci. U. S. A. 102 (39) (2005) 13950–13955.

[2] L. Rouli, V. Merhej, P.E. Fournier, D. Raoult, The bacterial pangenome as a new tool for analysing pathogenic bacteria, New Microbes New Infect. 7 (2015) 72–85.

[3] X. Zhang, X. Liu, F. Yang, L. Chen, Pan-genome analysis links the hereditary variation of Leptospirillum ferriphilum with its evolutionary adaptation, Front. Microbiol. 9 (2018) 577.

[4] P.-G.C. Computational, Computational pan-genomics: status, promises and challenges, Brief. Bioinform. 19 (1) (2018) 118–135.

[5] V. Daubin, G.J. Szollosi, Horizontal gene transfer and the history of life, Cold Spring Harb. Perspect. Biol. 8 (4) (2016).

[6] A. Mira, A.B. Martin-Cuadrado, G. D'Auria, F. Rodriguez-Valera, The bacterial pan-genome:a new paradigm in microbiology, Int. Microbiol. 13 (2) (2010) 45–57.

[7] L.C. Guimaraes, J. Florczak-Wyspianska, L.B. de Jesus, M.V. Viana, A. Silva, R.T. Ramos, et al., Inside the pan-genome—methods and software overview, Curr. Genomics 16 (4) (2015) 245–252.

[8] J.C. Ruiz, V. D'Afonseca, A. Silva, A. Ali, A.C. Pinto, A.R. Santos, et al., Evidence for reductive genome evolution and lateral acquisition of virulence functions in two Corynebacterium pseudotuberculosis strains, PLoS One 6 (4) (2011).

[9] S.C. Soares, E. Trost, R.T. Ramos, A.R. Carneiro, A.R. Santos, A.C. Pinto, et al., Genome sequence of Corynebacterium pseudotuberculosis biovar equi strain 258 and prediction of antigenic targets to improve biotechnological vaccine production, J. Biotechnol. 167 (2) (2013) 135–141.

[10] S.C. Soares, A. Silva, E. Trost, J. Blom, R. Ramos, A. Carneiro, et al., The pan-genome of the animal pathogen Corynebacterium pseudotuberculosis reveals differences in genome plasticity between the biovar ovis and equi strains, PLoS One 8 (1) (2013).

[11] D.C. Mariano, J. Sousa Tde, F.L. Pereira, F. Aburjaile, D. Barh, F. Rocha, et al., Whole-genome optical mapping reveals a mis-assembly between two rRNA operons of Corynebacterium pseudotuberculosis strain 1002, BMC Genomics 17 (2016) 315.

[12] W.B. Whitman, F. Rainey, P. Kämpfer, M. Trujillo, J. Chun, P. DeVos, et al., Bergey's Manual of Systematics of Archaea and Bacteria, 2015.

[13] R. Subedi, V. Kolodkina, I.C. Sutcliffe, L. Simpson-Louredo, R. Hirata Jr., L. Titov, et al., Genomic analyses reveal two distinct lineages of Corynebacterium ulcerans strains, New Microbes New Infect. 25 (2018) 7–13.

[14] A. Zhang, M. Yang, P. Hu, J. Wu, B. Chen, Y. Hua, et al., Comparative genomic analysis of Streptococcus suis reveals significant genomic diversity among different serotypes, BMC Genomics 12 (2011) 523.

[15] T. Hafstrom, D.S. Jansson, B. Segerman, Complete genome sequence of Brachyspira intermedia reveals unique genomic features in Brachyspira species and phage-mediated horizontal gene transfer, BMC Genomics 12 (2011) 395.

[16] A.M. Dickey, G. Schuller, J.D. Loy, M.L. Clawson, Whole genome sequencing of Moraxella bovoculi reveals high genetic diversity and evidence for interspecies recombination at multiple loci, PLoS One 13 (12) (2018).

[17] B.A. Wilson, M. Ho, Pasteurella multocida: from zoonosis to cellular microbiology, Clin. Microbiol. Rev. 26 (3) (2013) 631–655.

[18] Z. Peng, W. Liang, F. Wang, Z. Xu, Z. Xie, Z. Lian, et al., Genetic and phylogenetic characteristics of Pasteurella multocida isolates from different host species, Front. Microbiol. 9 (2018) 1408.

[19] R. Hurtado, D. Carhuaricra, S. Soares, M.V.C. Viana, V. Azevedo, L. Maturrano, et al., Pan-genomic approach shows insight of genetic divergence and pathogenic-adaptation of Pasteurella multocida, Gene 670 (2018) 193–206.

[20] A.N. Brooks, S. Turkarslan, K.D. Beer, F.Y. Lo, N.S. Baliga, Adaptation of cells to new environments, Wiley Interdiscip. Rev. Syst. Biol. Med. 3 (5) (2011) 544–561.

[21] C. Simon, A. Wiezer, A.W. Strittmatter, R. Daniel, Phylogenetic diversity and metabolic potential revealed in a glacier ice metagenome, Appl. Environ. Microbiol. 75 (23) (2009) 7519–7526.

[22] T.J. Johnson, J.E. Abrahante, S.S. Hunter, M. Hauglund, F.M. Tatum, S.K. Maheswaran, et al., Comparative genome analysis of an avirulent and two virulent strains of avian Pasteurella multocida reveals candidate genes involved in fitness and pathogenicity, BMC Microbiol. 13 (2013) 106.
[23] A.M. Moustafa, T. Seemann, S. Gladman, B. Adler, M. Harper, J.D. Boyce, et al., Comparative genomic analysis of asian haemorrhagic septicaemia-associated strains of Pasteurella multocida identifies more than 90 haemorrhagic septicaemia-specific genes, PLoS One 10 (7) (2015).
[24] G.B. Michael, K. Kadlec, M.T. Sweeney, E. Brzuszkiewicz, H. Liesegang, R. Daniel, et al., ICEPmu1, an integrative conjugative element (ICE) of Pasteurella multocida: structure and transfer, J. Antimicrob. Chemother. 67 (1) (2012) 91–100.
[25] D. Zhu, J. He, Z. Yang, M. Wang, R. Jia, S. Chen, et al., Comparative analysis reveals the Genomic Islands in Pasteurella multocida population genetics: on symbiosis and adaptability, BMC Genomics 20 (1) (2019).
[26] J.D. Boyce, T. Seemann, B. Adler, M. Harper, Pathogenomics of Pasteurella multocida, Curr. Top. Microbiol. Immunol. 361 (2012) 23–38.
[27] G.H. Frank, Pasteurellosis of cattle, in: C. Adlam, J.M. Rutter (Eds.), Pasteurella and Pasteurellosis, Academic Press, New York, 1989, pp. 197–221.
[28] M.R. Ackermann, K.A. Brogden, Response of the ruminant respiratory tract to Mannheimia (Pasteurella) haemolytica, Microbes Infect. 2 (9) (2000) 1079–1088.
[29] C.L. Klima, S.R. Cook, R. Zaheer, C. Laing, V.P. Gannon, Y. Xu, et al., Comparative genomic analysis of Mannheimia haemolytica from bovine sources, PLoS One 11 (2) (2016).
[30] P.K. Lawrence, W. Kittichotirat, J.E. McDermott, R.E. Bumgarner, A three-way comparative genomic analysis of Mannheimia haemolytica isolates, BMC Genomics 11 (2010) 535.
[31] E.C. Keen, Paradigms of pathogenesis: targeting the mobile genetic elements of disease, Front. Cell. Infect. Microbiol. 2 (2012) 161.
[32] C. Eidam, A. Poehlein, A. Leimbach, G.B. Michael, K. Kadlec, H. Liesegang, et al., Analysis and comparative genomics of ICEMh1, a novel integrative and conjugative element (ICE) of Mannheimia haemolytica, J. Antimicrob. Chemother. 70 (1) (2015) 93–97.
[33] M.L. Clawson, R.W. Murray, M.T. Sweeney, M.D. Apley, K.D. DeDonder, S.F. Capik, et al., Genomic signatures of Mannheimia haemolytica that associate with the lungs of cattle with respiratory disease, an integrative conjugative element, and antibiotic resistance genes, BMC Genomics 17 (1) (2016) 982.
[34] K.L. Cosford, Brucella canis: an update on research and clinical management, Can. Vet. J. 59 (1) (2018) 74–81.
[35] M. Holzapfel, G. Girault, A. Keriel, C. Ponsart, D. O'Callaghan, V. Mick, Comparative genomics and in vitro infection of field clonal isolates of Brucella melitensis biovar 3 did not identify signature of host adaptation, Front. Microbiol. 9 (2018) 2505.
[36] X. Yang, Y. Li, J. Zang, Y. Li, P. Bie, Y. Lu, et al., Analysis of pan-genome to identify the core genes and essential genes of Brucella spp, Mol. Gen. Genomics. 291 (2) (2016) 905–912.
[37] E. Trost, J. Blom, S. de Castro Soares, I.H. Huang, A. Al-Dilaimi, J. Schroder, et al., Pangenomic study of Corynebacterium diphtheriae that provides insights into the genomic diversity of pathogenic isolates from cases of classical diphtheria, endocarditis, and pneumonia, J. Bacteriol. 194 (12) (2012) 3199–3215.
[38] T. Bhardwaj, P. Somvanshi, Pan-genome analysis of Clostridium botulinum reveals unique targets for drug development, Gene 623 (2017) 48–62.
[39] L. Rohmer, C. Fong, S. Abmayr, M. Wasnick, T. Larson Freeman, M. Radey, et al., Comparison of Francisella tularensis genomes reveals evolutionary events associated with the emergence of human pathogenic strains, Genome Biol. 8 (6) (2007).
[40] U.P. Pereira, S.C. Soares, J. Blom, C.A.G. Leal, R.T.J. Ramos, L.C. Guimarães, et al., In silico prediction of conserved vaccine targets in Streptococcus agalactiae strains isolated from fish, cattle, and human samples, Genet. Mol. Res. 12 (3) (2013) 2902–2912.
[41] C. Rasetti-Escargueil, E. Lemichez, M. Popoff, Variability of botulinum toxins: challenges and opportunities for the future, Toxins 10 (9) (2018).
[42] M.R. Popoff, Ecology of neurotoxigenic strains of clostridia, Curr. Top. Microbiol. Immunol. 195 (1995) 1–29.

[43] J.R. Barash, S.S. Arnon, A novel strain of Clostridium botulinum that produces type B and type H botulinum toxins, J. Infect. Dis. 209 (2) (2014) 183–191.
[44] M. Kruger, M. Skau, A.A. Shehata, W. Schrodl, Efficacy of Clostridium botulinum types C and D toxoid vaccination in Danish cows, Anaerobe 23 (2013) 97–101.
[45] K. Oguma, T. Yamaguchi, K. Sudou, N. Yokosawa, Y. Fujikawa, Biochemical classification of Clostridium botulinum type C and D strains and their nontoxigenic derivatives, Appl. Environ. Microbiol. 51 (2) (1986) 256–260.
[46] E.L. Ortolani, L.A. Brito, C.S. Mori, U. Schalch, J. Pacheco, L. Baldacci, Botulism outbreak associated with poultry litter consumption in three Brazilian cattle herds, Vet. Hum. Toxicol. 39 (2) (1997) 89–92.
[47] R.O.S. Silva, C. Oliveira, L.A. Gonçalves, F.C.F. Lobato, Botulism in ruminants in Brazil, Ciência Rural 46 (8) (2016).
[48] C. Mazuet, E.J. Yoon, S. Boyer, S. Pignier, T. Blanc, I. Doehring, et al., A penicillin- and metronidazole-resistant Clostridium botulinum strain responsible for an infant botulism case, Clin. Microbiol. Infect. 22 (7) (2016). 644.e7-e12.
[49] O. Sahin, M. Yaeger, Z. Wu, Q. Zhang, Campylobacter-associated diseases in animals, Annu Rev Anim Biosci. 5 (2017) 21–42.
[50] R. Jain, S. Singh, V. SK, A. Jain, Genome-wide prediction of potential vaccine candidates for Campylobacter jejuni using reverse vaccinology, Interdiscip. Sci. 11 (2019) 337–347.
[51] A.H.M. van Vliet, J.M. Ketley, Pathogenesis of enteric Campylobacter infection, J. Appl. Microbiol. 90 (S6) (2001) 45S–56S.
[52] J.I. Dasti, A.M. Tareen, R. Lugert, A.E. Zautner, U. Groß, Campylobacter jejuni: a brief overview on pathogenicity-associated factors and disease-mediating mechanisms, Int. J. Med. Microbiol. 300 (4) (2010) 205–211.
[53] I.A. Gillespie, S.J. O'Brien, J.A. Frost, G.K. Adak, P. Horby, A.V. Swan, et al., A case-case comparison of Campylobacter coli and Campylobacter jejuni infection: a tool for generating hypotheses, Emerg. Infect. Dis. 8 (9) (2002) 937–942.
[54] M. Meunier, M. Guyard-Nicodème, E. Hirchaud, A. Parra, M. Chemaly, D. Dory, Identification of novel vaccine candidates against campylobacterthrough reverse vaccinology, J Immunol Res 2016 (2016) 1–9.
[55] R. Janssen, K.A. Krogfelt, S.A. Cawthraw, W. van Pelt, J.A. Wagenaar, R.J. Owen, Host pathogen interactions in Campylobacter infections: the host perspective, Clin. Microbiol. Rev. 21 (3) (2008) 505–518.
[56] M.A. van Bergen, K.E. Dingle, M.C. Maiden, D.G. Newell, L. van der Graaf-Van Bloois, J.P. van Putten, et al., Clonal nature of Campylobacter fetus as defined by multilocus sequence typing, J. Clin. Microbiol. 43 (12) (2005) 5888–5898.
[57] M.B. Skirrow, Diseases due to Campylobacter, Helicobacter and related bacteria, J. Comp. Pathol. 111 (2) (1994) 113–149.
[58] O. Sahin, C. Fitzgerald, S. Stroika, S. Zhao, R.J. Sippy, P. Kwan, et al., Molecular evidence for zoonotic transmission of an emergent, highly pathogenic Campylobacter jejuni clone in the United States, J. Clin. Microbiol. 50 (3) (2012) 680–687.
[59] S. Hum, Bovine abortion due to Campylobacter fetus, Aust. Vet. J. 64 (10) (1987) 319–320.
[60] C.A. Kirkbride, Etiologic agents detected in a 10-year study of bovine abortions and stillbirths, J. Vet. Diagn. Investig. 4 (2) (1992) 175–180.
[61] G. Méric, K. Yahara, L. Mageiros, B. Pascoe, M.C.J. Maiden, K.A. Jolley, et al., A reference pan-genome approach to comparative bacterial genomics: identification of novel epidemiological markers in pathogenic Campylobacter, PLoS One 9 (3) (2014).
[62] E. Rozynek, K. Dzierzanowska-Fangrat, B. Szczepanska, S. Wardak, J. Szych, P. Konieczny, et al., Trends in antimicrobial susceptibility of Campylobacter isolates in Poland (2000-2007), Pol. J. Microbiol. 58 (2) (2009) 111–115.
[63] J. Takkinen, A. Ammon, O. Robstad, T. Breuer, Campylobacter Working Group, European survey on Campylobacter surveillance and diagnosis 2001, Euro Surveill. 8 (11) (2003) 207–213.
[64] J.L. Smith, P.M. Fratamico, Fluoroquinolone resistance in campylobacter, J. Food Prot. 73 (6) (2010) 1141–1152.

[65] J. Silva, D. Leite, M. Fernandes, C. Mena, P.A. Gibbs, P. Teixeira, Campylobacter spp. as a foodborne pathogen: a review, Front. Microbiol. 2 (2011) 200.

[66] J.R. Greig, Quinolone resistance in Campylobacter, J. Antimicrob. Chemother. 51 (3) (2003) 740–742.

[67] P.F. McDermott, S.M. Bodeis-Jones, T.R. Fritsche, R.N. Jones, R.D. Walker, Broth microdilution susceptibility testing of Campylobacter jejuni and the determination of quality control ranges for fourteen antimicrobial agents, J. Clin. Microbiol. 43 (12) (2005) 6136–6138.

[68] J.E. Moore, M.D. Barton, I.S. Blair, D. Corcoran, J.S. Dooley, S. Fanning, et al., The epidemiology of antibiotic resistance in Campylobacter, Microbes Infect. 8 (7) (2006) 1955–1966.

[69] T. Lefébure, P.D. Pavinski Bitar, H. Suzuki, M.J. Stanhope, Evolutionary dynamics of complete Campylobacter pan-genomes and the bacterial species concept, Genome Biol. Evol. 2 (2010) 646–655.

[70] V.P. Richards, P. Lang, P.D. Bitar, T. Lefebure, Y.H. Schukken, R.N. Zadoks, et al., Comparative genomics and the role of lateral gene transfer in the evolution of bovine adapted Streptococcus agalactiae, Infect. Genet. Evol. 11 (6) (2011) 1263–1275.

[71] J.Y. Bolukaoto, C.M. Monyama, M.O. Chukwu, S.M. Lekala, M. Nchabeleng, M.R. Maloba, et al., Antibiotic resistance of Streptococcus agalactiae isolated from pregnant women in Garankuwa, South Africa, BMC Res. Notes 8 (2015) 364.

[72] S.D. Manning, Molecular epidemiology of Streptococcus agalactiae (group B Streptococcus), Front. Biosci. 8 (2003) s1–18.

[73] C.J. Baker, Group B streptococcal infections, Clin. Perinatol. 24 (1) (1997) 59–70.

[74] A. Schuchat, Epidemiology of group B streptococcal disease in the United States: shifting paradigms, Clin. Microbiol. Rev. 11 (3) (1998) 497–513.

[75] G.F. Mian, D.T. Godoy, C.A. Leal, T.Y. Yuhara, G.M. Costa, H.C. Figueiredo, Aspects of the natural history and virulence of S. agalactiae infection in Nile tilapia, Vet. Microbiol. 136 (1-2) (2009) 180–183.

[76] M. Chen, R. Wang, L.P. Li, W.W. Liang, J. Li, Y. Huang, et al., Screening vaccine candidate strains against Streptococcus agalactiae of tilapia based on PFGE genotype, Vaccine 30 (42) (2012) 6088–6092.

[77] J. Gao, F.Q. Yu, L.P. Luo, J.Z. He, R.G. Hou, H.Q. Zhang, et al., Antibiotic resistance of Streptococcus agalactiae from cows with mastitis, Vet. J. 194 (3) (2012) 423–424.

[78] S.B. Jamal, S.S. Hassan, S. Tiwari, M.V. Viana, L.J. Benevides, A. Ullah, et al., An integrative in-silico approach for therapeutic target identification in the human pathogen Corynebacterium diphtheriae, PLoS One 12 (10) (2017).

[79] A.B. Sjöstedt, Francisella, Bergey's Manual of Systematics of Archaea and Bacteria, John Wiley & Sons, Ltd, 2015.

[80] D.J. Brenner, N.R. Krieg, J.T. Staley, G.M. Garrity, D.R. Boone, P. De Vos, et al., Bergey's Manual® of Systematic Bacteriology, Springer-Verlag, 2005.

[81] M. Maurin, M. Gyuranecz, Tularaemia: clinical aspects in Europe, Lancet Infect. Dis. 16 (1) (2016) 113–124.

[82] M. Santic, M. Molmeret, Y. Abu Kwaik, Modulation of biogenesis of the Francisella tularensis subsp. novicida-containing phagosome in quiescent human macrophages and its maturation into a phagolysosome upon activation by IFN-gamma, Cell. Microbiol. 7 (7) (2005) 957–967.

[83] P. Larsson, D. Elfsmark, K. Svensson, P. Wikström, M. Forsman, T. Brettin, et al., Molecular evolutionary consequences of niche restriction in Francisella tularensis, a facultative intracellular pathogen, PLoS Pathog. 5 (6) (2009).

[84] L. Hadfield Ted, P. McEvoy, Y. Polotsky, V.A. Tzinserling, A.A. Yakovlev, The pathology of diphtheria, J. Infect. Dis. 181 (s1) (2000) S116–S120.

[85] V. Sangal, P.A. Hoskisson, Evolution, epidemiology and diversity of Corynebacterium diphtheriae: new perspectives on an old foe, Infect. Genet. Evol. 43 (2016) 364–370.

[86] V.J. Timms, T. Nguyen, T. Crighton, M. Yuen, V. Sintchenko, Genome-wide comparison of Corynebacterium diphtheriae isolates from Australia identifies differences in the Pan-genomes between respiratory and cutaneous strains, BMC Genomics 19 (1) (2018).

[87] A. Sing, R. Konrad, D.M. Meinel, N. Mauder, I. Schwabe, R. Sting, Corynebacterium diphtheriae in a free-roaming red fox: case report and historical review on diphtheria in animals, Infection 44 (4) (2016) 441–445.

[88] L. Corboz, R. Thoma, U. Braun, R. Zbinden, Isolation of Corynebacterium diphtheriae subsp. belfanti from a cow with chronic active dermatitis, Schweiz. Arch. Tierheilkd. 138 (12) (1996) 596–599.
[89] A. Kraszewska, Z. Anusz, Appearance in domestic animals of Corynebacterium diphtheriae and other Corynebacterium strains pathogenic for man, Przegl. Epidemiol. 33 (2) (1979) 269–276.
[90] L. Detemmerman, D. Rousseaux, A. Efstratiou, C. Schirvel, K. Emmerechts, I. Wybo, et al., Toxigenic Corynebacterium ulcerans in human and non-toxigenic Corynebacterium diphtheriae in cat, New Microbes New Infect. 1 (1) (2013) 18–19.
[91] B.A. Leggett, A. De Zoysa, Y.E. Abbott, N. Leonard, B. Markey, A. Efstratiou, Toxigenic Corynebacterium diphtheriae isolated from a wound in a horse, Vet. Rec. 166 (21) (2010) 656–657.
[92] B. Henricson, M. Segarra, J. Garvin, J. Burns, S. Jenkins, C. Kim, et al., Toxigenic Corynebacterium diphtheriae associated with an equine wound infection, J. Vet. Diagn. Investig. 12 (3) (2000) 253–257.
[93] K. Zakikhany, S. Neal, A. Efstratiou, Emergence and molecular characterisation of non-toxigenic tox gene-bearing Corynebacterium diphtheriae biovar mitis in the United Kingdom, 2003-2012, Euro Surveill. 19 (22) (2014).

CHAPTER 6

Pan-genomics of plant pathogens and its applications

Rabia Amir[a], Qurat-ul-Ain Sani[a], Wajahat Maqsood[a], Faiza Munir[a], Nosheen Fatima[a], Amnah Siddiqa[b], Jamil Ahmad[b]
[a]Department of Plant Biotechnology, Atta-ur-Rahman School of Applied Biosciences (ASAB), National University of Sciences and Technology (NUST), Islamabad, Pakistan
[b]Research Center for Modeling & Simulation (RCMS), National University of Sciences and Technology (NUST), Islamabad, Pakistan

1 Introduction

Last century has witnessed a huge transition in genomics analyses from the sequencing of single or few genomes to hundreds and thousands of genomes simultaneously [1]. The emergence and development of sophisticated ultrahigh throughput next-generation sequencing (NGS) technologies and subsequent genomes sequencing projects unleashed whole-genome sequences of many strains of different plant pathogenic species easily accessible than previously. These sequences are called the "reference genomes," which served as a basis of many genomics studies. Particularly, with reference to plant pathogens genomes, the analysis aided in analyzing the evolutionary relationships, population genetics, identification of casual agents, virulence factors, host specificity associations, and pathogenic mechanisms [2–5]. In order to bring together useful potential encrypted inside the specific organism, "tried-and-true" annotation schemes were usually applied by means of HMM, or BLASTX searches of protein groups (e.g., TIGRfam, Pfam, COG, etc.) and that genome solely reflected the dynamic potential of that organism. Overall, the sequencing projects aided in immense characterization of plants pathogen effector sequences involved in pathogenic mechanisms which laid the basis of rational improvements in drug design for controlling infectious plant diseases by plants pathogens and economically important crop plants disease management [6–8].

However, the availability of vast number of genomes particularly increased the potential number of comparative genomics-based studies which further yielded advanced data. The classical use of comparative genomics remained mainly associated with comparisons of whole genomes of different organisms which aided in gaining biological insights into functional characterization, evolutionary relationships and genomes plasticity [3, 9]. Comparative genomic analysis among the plants pathogens resulted in identifying the distinctive operations forming the basis of pathogenicity [10, 11], variations between the pathogenic and non-or-less pathogenesis causing strains [12], hotspots for horizontal

Pan-genomics: Applications, Challenges, and Future Prospects
https://doi.org/10.1016/B978-0-12-817076-2.00006-8
© 2020 Elsevier Inc.
All rights reserved.

gene transfers among strains [9], the pathogenicity determinants [13], and mechanistic principles for their adaptive success [14, 15]. Such studies additionally played a role in the management of different types of stress-resistant crop varieties in plants through identification of less-or-no pathogen-resistant varieties, accelerated development of genetic-based diagnostic tools for plant pathogens identification, increased knowledge of the effector and pathogenic sequences associated with pathogenesis thus facilitating revamped vaccine design [16–19].

However, it is now increasingly being accepted that sequence of a single reference genome does not reflect the genetic variability of an organism giving rise to pan-genome concept which has now emerged as a new method for analyzing and characterizing the genomes from a broader heterogeneity viewpoint. Relative genomic analysis between multiple strains of single species has disclosed highly diverse genomic content within the species. Several types of structural variations including copy number variations (CNVs), present absent variations (PAVs), and other allelic transformations are among some of the demonstrated factors behind these variations [20–23]. The presence of CNVs, PAVs, repeat-driven expansions, and other structural variations due to transposable elements, epigenetic processes, gene conversion, mitotic recombination, and horizontal gene/chromosome transfer have been demonstrated in many plants pathogens already [11, 24–32].

The pan-genome analysis was pioneered by Tettlin while comparing several full-length genomes of *Streptococcus agalactiae*, followed by the studies on *Haemophilus influenzae* genomes carried out by Hogg with the aim to analyze the intraspecies genomic diversity [33]. The main findings of both of these studies included the determination of a core genome that consisted of genes shared by all the strains and that a huge percentage of every genomic sequence was particular to each strain. Thus, the addition of every newly sequenced genome-based comparison supplemented the pan-genome with numerous new genes not characterized hitherto which represented a nonredundant set of genes/genome identified in different strains of the same species. The pan-genome concept that was brought forward included representation of nonredundant whole-genome repertoire of a species based on different strains where the sets of genes among different strains could be categorized into three different categories including core genome, accessory (dispensable) genome, and species-/strain-specific genome [34] (graphically illustrated in Fig. 1A). A core genome comprises of mutual genes of all the strains (or individuals or samples) studied and is associated with the roles associated with basics homeostatic processes and phenotypical appearance of the species. The genes that are part of the core genome undergo selective pressure and therefore are not drastically changed. The accessory (or dispensable) genome comprises of the subset of genes which are present in more than one but not all of the studied strains (or individuals or samples) of the same species. The accessory genome often has been associated with survival- and lifestyle-related functional genes. Therefore, it is highly interesting to

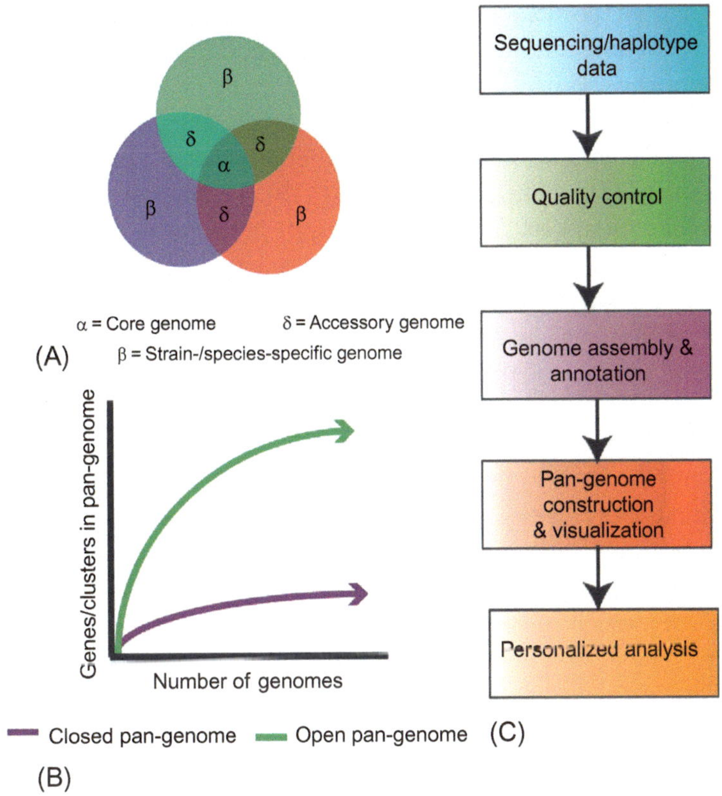

Fig. 1 (A) The pan-genome components are illustrated. Three different categories including core genome, accessory genome, and strain-/species-specific genomes comprise a pan-genome. (B) Closed and open pan-genomes illustration. (C) Flow chart illustrating the steps followed for a pan-genome analysis.

examine the genetic features of the core genome that is responsible for all probable lifestyles in a species. The species-/strain-specific genome is present in only one of the strains/species being studied. The genes that are part of this subset of pan-genome are often associated with species-/strain-specific virulence, pathogenesis, or adaptive evolution.

The pan-genomes may spread due to modifications in structural features that emerge as a result of genomic rearrangements, instead of simply contracting to gene content. The phenotype of pathogens is affected by alterations in genome construction at the species level, even if it is in the same gene repertoire. The reason is the location of a gene that influences vital physiological mechanisms such as expression level or protein dosage. In addition to that the change in location of moveable components may place other genes in such a way that they interact with regulatory regions to stimulate gene expression. Cell fitness is overwhelmed by the genomic alterations as a result of these evolving features and such alterations possibly have a biological meaning. Therefore, the structural pan-genome is not insignificant, as various genomic constructs can affect important features like rate of development or pathogenicity of strain.

The pan-genome of a species is also mathematically extrapolated to be "closed" or "open" based on the number of new sequences/genes added to a pan-genome with every additional genome added to the comparison (Fig. 1B). Tettlin et al. demonstrated that the number of unique genes kept on increasing despite hundreds of sequenced genomes of *S. agalactiae* pan-genome analyzed. Thus the pan-genome of *S. agalactiae* was classified as "open" with a likelihood of getting new unique genes with every new genome comparison. Such types of pan-genomes are specifically reported with highly flexible genomic composition. On the other hand, the pan-genomes may also be classified as closed for the species which does not show genetic variations and their genomes are not usually expanding because of isolated lifestyles. The status of a pan-genome as open or closed is calculated based on Heaps Law [1, 34]. One important factor in estimating the pan-genome is to use a sufficient number of genomes for comparisons and interpretation of results.

Capturing the strain-specific insights is especially relevant for studying the biological mechanisms of plants pathogens because of the fact that the major determinants of pathogenicity are often strain-specific and highly variable. Besides, the pan-genome study also offers characterizing the strains by their individual gene set and to study the evolutionary impact of horizontal gene transfer. Thus, the pan-genomic analyses studies would act as two sword edge by capturing the genome plasticity along with the identification of unique and novel determinants of pathogenicity. Up till now, pan-genome analysis has already been used for identification, detection, and tracking of new strains in metagenomics samples and developing vaccines against many plant pathogenic strains [35–37]. Moreover, exploring strain diversity in environmental population genomics are some other benefits of extracting the pan-genome. Pan-genome analysis serves as an outline to assess the genomic variability of all the data at hand and forecasting additional whole-genome sequences that would be required to entirely depict that diversity. Here, we review the evolution of plant-pathogen pan-genomics, its impact in understanding pathogen evolution and physiology and the opportunities and applications presented by pan-genomics as applied to pathogen genome comparisons.

2 Pan-genomics of plant pathogens

The diversity within a pathogen genome poses difficulties in the identification of the genes that are linked within all the strains of that pathogen [1]. The genes associated with a single pathogen are thought to be unlimited, however, many groups are attempting to devise a practical value for it [2]. Therefore it was crucial to introduce the idea of pangenomes and core genomes [3]. The pan-genome sizes of several bacteria and fungi are illustrated in Table 1 for a coarse grain view.

Table 1 Pan-genome size of several bacteria and fungi

Organism	Family	Host plants	Pan-genome size	Core genome size	Accessory genome size	Open/closed	Chromosome size
Pan-genome of bacteria							
Pectobacteria [38]	Enterobacteriaceae	Potato and ornamental plants					
Pectobacterium parmentieri [39]	Enterobacteriaceae	Potato		3706 core genes	1468 accessory genes	Open	
Pantoea ananatis [40]	Enterobacteriaceae	Maize, onion, Eucalyptus, Sudan grass, honey mildew	4225–4415 CDSs				4.39–4.61 Mb
Erwinia amylovora [41]	Enterobacteriaceae	Malus, Pyrus, Crataegus, Sorbus, raspberries, blackberries	5751 CDS	3414 CDS		Open	3.8 Mb
Burkholderia [42]	Burkholderiaceae	Rice	86,000–88,000 genes	587 genes		Open [43]	
Xylella fastidiosa [12, 44]	Xanthomonadaceae	Grapes, citrus fruits, almonds					2.679 Mb
Pan-genome of fungi							
Puccinia graminis [45, 46]	Pucciniaceae	Wheat, barley, triticale	92 Mb		13 Mb		
Zymoseptoria tritici [47]	Mycosphaerellaceae	Wheat		13 chromosomes	8 chromosomes		

2.1 Pan-genomics of plant pathogenic bacteria

2.1.1 Pectobacteria

Pectobacteria are vital plant pathogenic enterobacteria which produce a range of disease symptoms such as soft rot, wilt, and blackleg in potato and ornamental plants [38]. The genus Pectobacterium includes *Pectobacterium atrosepticum*, *P. wasabiae*, *P. carotovorum*, and *P. betavasculorum* [48]. A fifth species, phylogenetically distinct from other four is *P. brasiliensis* [49]. Comparison of their genomes unveiled core genome (signifying nearly 80% of the nucleotides per species) carrying varied sequences. Unique islands were rich in regulatory genes and the genes for proteins of DNA replication, mostly of phage origin [38]. Arrangement of the Pectobacterium genomes by means of Mauve has shown that almost 77% of the complete *P. atrosepticum* chromosome exists in *P. brasiliensis* and *P. carotovorum* [50]. Among the genome of *Pectobacterium* strains and their subgroups, the variable segment of the pan-genome is uniformly scattered. *P. brasiliensis* and *P. carotovorum* sequences show nearly 5.4% similarity with each other but do not match with *P. atrosepticum*, supporting a close association between *P. brasiliensis* and *P. carotovorum* [51].

2.1.2 Pectobacterium parmentieri

P. parmentieri is a recently recognized species in family *Pectobacteriaceae*. Such bacteria are highly harmful to economically vital crops including potato in diverse environmental habitats. There are certain virulence elements including cell wall damaging enzymes that may cause severe disease symptoms. Prominent differences in the phenotypes of *P. parmentieri* isolates have been observed concerning virulence elements construction and their capabilities to deliquesce plants.

The pan-genome of *P. parmentieri* is composed of 3706 core genes, 1468 accessory genes, and a high number of distinctive 1847 genes. Several genes that encode virulence elements in the core genome segment, but others were positioned indispensable genome. Many significant differences in phenotypes are likely because of virulence-related gene duplications, a higher fraction of horizontally transferred genes together with various CRISPR assortments.

Thus it is hypothesized that a significant mass of the genes in the accessory genome and major genomic variability among *P. parmentieri* strains might be the source of the broad host range and extensive dispersal of *P. parmentieri*. The information regarding gene content and structure of *P. parmentieri* strains enables to find the significance of great genomic plasticity for *P. parmentieri* acclimatization to various environmental niches [39].

2.1.3 Pantoea ananatis

P. ananatis belongs to the family of Enterobacteriaceae, familiarized by its universality in nature, and repeated link with both the plant as well as animal hosts. *P. ananatis* is often isolated from leaves, roots, and stems [40] of onion, honeydew melons, maize, Sudan grass, and Eucalyptus [52].

The genome of *P. ananatis* comprises of a chromosome that is 4.39–4.61 Mb in size with an average of 53.7% G + C content and a large plasmid pPANA1 of 281–353 kb in size with an average of ~52% G + C content. Almost 4225–4415 CDSs are encoded on the chromosome and pPANA1 plasmid collectively [52].

The assessment of protein complements encoded by genomes of *P. ananatis* strains were undertaken through Reciprocal Best BlastP Hit analysis [53], showed an average amino acid identity of 99.4%. These outcomes proposed a comprehensive and highly conserved core genome that encompasses the bulk of proteins encoded by an individual genome [52].

2.1.4 *Erwinia amylovora*

E. amylovora is a causative agent of fire blight disease [54], which is divided into two host-specific groups; strains that infect a wide range of hosts among Spiraeoideae including Malus, Pyrus, Sorbus, and Crataegus and the strains which infect Rubus including blackberries and raspberries.

The pan-genome consists of 5751 coding sequences of which 3414 CDS were identified as core and is well conserved (>99% amino acid similarity between all strains) as compared to other plant pathogenic bacteria. Study of the aligned sequences has shown that approximately 86% of *E. amylovora* genome is comprised of coding sequences of about one per kb density.

The chromosomes are nearly 3.8 Mb. The highly infective strains of spiraeoideae family have homogeneous chromosomes with 53.6% G + C content, whereas a large genetic diversity was detected among Spiraeoideae and Rubus infecting strains and between individual Rubus infecting strains with chromosomes having 53.3%–53.4% G + C content [41].

It has been predicted that *E. amylovora* has moderately low genetic diversity as compared to other phytopathogens such as *P. syringae* because it experiences narrow genetic recombination with constricted ecological habitat. However, strains that infect Spiraeoideae are exposed to restricted selection pressures due to pome fruit breeding schemes that favors high-value varieties which are susceptible to fire blight [55, 56].

Based on EDGAR analysis [57] that utilizes 2 whole-genome sequences and 10 draft sequences of the genome of *E. amylovora*, the pan-genome is projected to be open [41].

2.1.5 *Burkholderia*

The genus *Burkholderia* consists of *Burkholderia glumae* [58], *B. gladioli*, and *B. plantarii* that inhabit diverse ecological niches. These representative species cause seedling blight, grain rot, and sheath rot, which may result in adverse losses in the production of rice [42].

The pan-genome of *Burkholderia* is open with the saturation between 86,000 and 88,000 genes. *Burkholderia* genomes are unusual due to their multichromosomal organization [59]. Their genomes are comprised of two or three chromosomes [43].

Pan-genome analysis of *Burkholderia* has revealed several genomic characteristics of pathogenic species of *Burkholderia* compared to a wider variety of *Burkholderia* strains, comprising both animal/human pathogens together with individuals from environmental niches. Overall pan-genome comprised of 78,782 orthologs, of which 587 genes were highly common among *Burkholderia* genomes, thus forming the core genome.

A better understanding of the specificities and variability in *Burkholderia* individuals may give an insight of their ability to acclimatize to various environments, in addition to their distinctive interactions with host species during the course of pathogenesis [60].

2.1.6 *Xylella fastidiosa*

Xylella is a plant pathogenic bacterium that is responsible for various economical yield losses in crops such as citrus fruits, grapes, almonds, and many plant hosts [12]. The pathogen causes Pierce's disease in wineries and citrus variegated chlorosis in citrus [61].

Genomic variations among closely related strains offer not only an understanding of functional and evolutionary processes but also points toward describing the extent of pathogenicity of one strain than others [62].

Pan-genome analysis of *X. fastidiosa* has revealed 2680 protein clusters in the "shell" category of proteins dispersed in 3–24 genomes tracked by a "cloud" group of 2668 protein clusters. The "soft core" category comprises of 1521 protein clusters, whereas protein clusters having the "core" category are nearly 1269 [63].

2.2 Pan-genomics of plant pathogenic fungi

2.2.1 *Puccinia graminis*

P. graminis f. Sp. tritici (Pgt) causes wheat stem rust which is the most devastating disease of barley, wheat, and triticale [45, 46]. In cereals or grasses symptoms mainly appear on leaf and stem sheaths however sometimes also occur on glumes and leaf blades [45].

Overall 92 Mb Pgt pan-genome has been accumulated, comprising approximately 13 Mb of unique sequence. A higher proportion of this sequence is although common among numerous stem rust isolates that result in greater genomic coverage for stem rust pathogen of wheat. In dikaryotic Pgt the divergence among haploid nuclei may be the basis of variation in genomic content. A higher evolutionary variance exists between Pgt isolates [64].

Thus, it is proposed that the maximum region is not strain specific. Hence, the assembled genome has increased the sequenced genome coverage, refining the demonstration of core eukaryotic genes, and permitting the alignment of about 2000 transcripts of this region [64].

2.2.2 *Zymoseptoria tritici*

Z. tritici is responsible for one of the most detrimental diseases, *Septoria tritici* blotch on wheat [65]. These phytopathogenic fungal populations have surmounted resistance against fungicides and have overcome the genes accountable for resistance in wheat [66].

Z. tritici genome is comprised of 13 core chromosomes and almost 8 accessory chromosomes. These accessory chromosomes experience important structural changes during the process of meiosis [47]. Addition or omission of groups of transposable elements makes considerable length polymorphism in core chromosomes that are homologous to each other [67, 68].

Z. tritici pan-genome has been constructed through the clustering of protein sets. It encodes 15,749 nonredundant proteins out of which 9149 (58.1%) were coded by the core genome while 6600 (41.9%) proteins were coded by the accessory genome. The genome of *Z. tritici* possibly becomes stable at 9000 core genes. Though, the pan-genome size enlarged linearly as the accessory genes revealed by each subsidiary genome did not become stable. A conserved protein domain is encoded by 67% of core genes but this ratio is reduced to 32% for accessory and 20% for singleton genes.

The core genes were highly comprised of housekeeping genes that were significantly responsible for basic cellular functions, general metabolism, and development [69].

3 Applications of plant pathogen's pan-genomics

Several applications of pan-genome analysis of plants pathogens (Fig. 2) are discussed below in detail.

3.1 Detection and characterization of new strains

The pan-genome size and content pave the way toward a dynamic concept where genomes are repeatedly at the verge of losing genes as well as integrate foreign genetic material [70]. Pan-genome of a species may be utilized to compare and describe the genome of unidentified isolates and to attain precise typing info that proves valuable in epidemiological surveys and clinical investigations [71]. The core genome gives insight into functional potential, relations between organisms, genes necessary for distinct environmental niches, and pathogenicity; as a consequence, core genes can be used as therapeutic and environmental markers for additional characterization and in determining the likely source of diseases, or in synthetic biology.

Many methods have been formulated for the characterization of genetic variability. Whole-genome sequencing and DNA microarrays can allow several different sequence-based methods of taxonomy identification [72] and characterization of multiple pathogens and many genes in a single array assay [73]. Similarly, universally conserved genes or proteins, specific to the particular taxonomic group can serve as novel targets for species and strain identification.

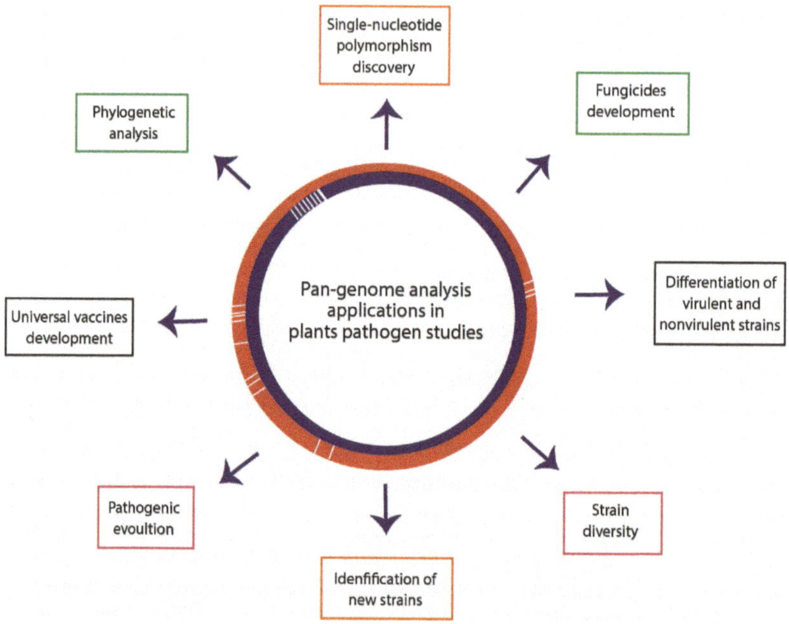

Fig. 2 Applications of pan-genome analysis in plants pathogens genome analyses.

It has become a common practice to characterize the closely resembling cluster of bacterial strains through "pan-genome" [33]. *Enterobacteriaceae* contain phytopathogens among which *Erwinia carotovora* subspecie *atroseptica* (*Eca*) was the first phytopathogenic enterobacterium to be sequenced. The Enterobacteriaceae pan-genome microarray offers a useful tool in order to determine the genetic makeup of unidentified strains of this bacterial family and can pave the way toward the investigation of phylogenetic relationships [72].

3.2 Evaluating strain diversity

The concept of pan-genome entails structural properties such as variation that may arise because of genomic recombinations. The presence or absence of gene variation causes the inefficiency to use only one organism to understand the genetic diversity. Strains with open pan-genome exhibit extreme versatility in gene content and show great potential for discovering novel genes [69]. So, the building of pan-genome is crucial to realize the degree of distinctions among genes. Several strains of some bacteria have proposed that the accessible gene pool in their pan-genome is massive and that recognition of novel genes will never stop even after multiple genomes are sequenced [74]. The genomes of several independent pathogenic isolates are necessary to realize the complications of bacterial species [33].

Comparative genomics studies have revealed intraspecific genomic variations which are believed to contribute to the ecological and phenotypic potentials pathogen requires for survival in an environment [75]. Bacterial phenotype is dependent on modifications in genomic constructs even if identical gene repertoire is available. Exchanging the site of transposable factors may place various other genes to interact with regulatory sections and trigger the expression of a gene. The phenotype of a cell is altered due to the influence of gene recombinations in prokaryotic genome [76]. Therefore, the presence of a structural pan-genome is important, as diverse genomic variants may change significant aspects, for example, pathogenicity of strain and growth rate [77].

Identification of homologous sequences sets is present in almost every comparative genomics study and is fundamental in understanding microbial diversity and evolutionary processes [78]. In addition to uncovering the genes and functions that confer distinctive features to *pathogenic* strains, the genetic variability can be explored in every gene family retrieved from its pan-genome [79].

3.3 Revealing the pathogenic evolution

The construction of pan-genome of extremely polymorphic pathogenic eukaryotes exhibited that only a single reference genome can considerably underestimate the species' gene space [35]. Relative genomic studies have revealed several ecological and metabolic variations within microbial taxa and offered raw material to understand the evolutionary processes. The core genome is the soul of phylogeny and is demonstrative of several taxonomic levels among bacterial isolates [77]. On the other hand, the huge dispensable genome offers support for adaptive evolution [80]. In eukaryotic genomes, the rearrangements in chromosomes influence the adaptive evolution by causing a disparity in gene content [81]. Maximum gene gains are the consequence of duplication, diversification, and neofunctionalization [82]. The addition and deletion of genes are important for the quick adaptation of plant pathogens to different hosts [83, 84].

In plant pathogenic fungi, intraspecific genes are particularly major factors of pathogenicity, mostly encoded by accessory genes [85]. Chromosomes having such accessory regions are rich in effector genes which are important for adaptive evolution of the pathogens [67, 86]. Moreover, several pathogenicity-related genes situated nearby repetitive sequences are assumed to accelerate evolution among pathogens [35]. Analyses of a spectrum of phytopathogens have shown that these fast-developing effectors were mostly situated in rapidly growing compartments of the genome [25]. In phytopathogens, compartmentalization of the genome and strong linkage of transposable elements and effectors in similar compartments has been labeled as the "two-speed genome" model in the evolution of pathogens [84]. Pan-genome analysis has allowed the study of genome plasticity as well as the unidentified factors of pathogenicity [35].

Size of pan-genome can effectively be determined by population size [87]. In the absence of selection, enormous populations must conserve pan-genomes by avoiding accessory genes to be disappeared from the gene pool by means of random drift. The study of the origin of the evolution of accessory genes provides an important understanding of the occurrence of extremely polymorphic pan-genomes and their contribution to adaptive evolution [88]. The degree of the accessory genome picks up many important questions with respect to the origin of the evolution of polymorphism, its part in adaptive evolution, and the route of possible redundant gene functions [35].

3.4 Development of universal vaccines

The genome sequence of a single strain can reveal many biological aspects of a species and predicts the initiation of pathogenicity among bacterial species, restricting the genome-wide screening of vaccine candidates or the antimicrobial targets for single strains [33]. Since the publishing of the first genomic sequence of bacteria, the idea of pan-genome has exhibited that the primary scheme of sequencing few genomes of a species is insufficient. Thus it is essential to sequence multiple strains in order to build up the basic knowledge of bacterial species along with the eradication of problem that comes with gene variation [35].

In the previous years, a pathogens' genome sequencing struggles have extended to take account of multiple representatives of single species and this pan-genome concept has revealed the great potential to make vaccines that were once difficult to design [89]. Reverse vaccinology is an advancement in the genomic era, which has entirely transformed the method to improve vaccines initiating from the genomic data instead of developing the causal agents [90]. Therefore, the idea of pan-genome exhibits progress in the utilization of reverse vaccinology, because it signifies the potential to look at multiple genomes of same bacterial species to surmount the difficulties characterized by gene variability and presence [90].

The idea of reverse vaccinology originates from genome sequences and, through in silico analysis, proposes certain antigens that are expected to be potential vaccine candidates [35, 75]. In addition, the core genome is highly significant for the scientific community for a number of reasons because essential genes are likely to reside inside these genomic regions [91] that may be utilized for antibiotic and vaccine targets together with universally available genes in pathogenic strains [92]. Hence, the presence of whole-genome sequences has completely altered our concept toward the development of vaccine and presents a new approach to understand the process.

3.5 Role in SNP discovery

Single-nucleotide polymorphisms (SNPs) are single base pair substitutions that arise inside and outside the genes. Those genes that contain one or more SNPs may contribute in two or more allelic arrangements of mRNAs. These variations in mRNA may keep various biological roles as an outcome of changes in the primary or upper order structures that work together with other cellular elements [93]. Single reference genome has been used in various studies to understand SNPs among several entities until now. But the understanding of SNP would be increased in numerous ways by means of a pan-genome as it is a whole-genome information of any species.

Furthermore, the pan-genome will be used for the discovery of SNP [94] which would save time and struggle to study several individuals at a time. Therefore, the results of SNPs that come from various references would not need to be joined into single SNP after investigation. Pan-genome is very important for the discovery of SNP as it distinguishes SNPs in its core and variable areas [94]. Additionally, the development of an organism and its response to the environment is also influenced by sequence variations [95]. Panseq is a software package, which helps to find the core and variable SNPs but also identifies the discriminatory loci among the core gene SNPs or variable loci [96]. Regardless of the procedure used to recognize SNPs, if pan-genome is used as a reference for read mapping, it displays presence-absence variations. Therefore, it would be easy to understand the nature of SNPs donated by each and every individual in the pan-genome. If the reference would be constructed on a single individual then it identifies the increased number of SNPs [97, 98]. There are many uses of SNP markers including crop improvement, analysis of genetic diversity, construction of high-resolution genetic maps, phylogenetic analysis, and LD-based association mapping [99].

3.6 Differentiation of virulent and nonvirulent strains

Pan-genome analysis of a pathogen may give imperative understandings into the biological features of species and helps in finding new ways for the treatment of diseases. There is limited data available regarding pan-genomes of phytopathogenic bacteria. Additionally, the analysis executed in other species has signified its importance to study multiple genomes [100] for high-resolution evolutionary analysis within the same species [101].

Pan-genome analysis of *X. arboricola* and atypical *X. arboricola* strains has determined the presence of unique phylogenetic basal lineage in these species, which is linked to a broad range of hosts. Low virulent strains present in this group appeared to initiate disease among monocotyledonous plants such as banana and barley [102, 103]. The detailed comparative study of those virulent genes helped in the determination of bacterial lineage that was slightly different from those considered as extremely virulent regarding multiple characteristics related to their pathogenesis. However, these variations are not explained based solely on the PCR typing, therefore whole-genome sequencing of strains permit an accurate inference of phylogenetic location within a species.

The genomic studies show a sequence of genes possibly associated with the pathogenicity of *X. arboricola* and has practical implication in controlling the bacterial spot disease of almonds and stone fruits, thus offering new means for its diagnosis. Eventually, to expand the information on the pathogenic capacity and diversity of bacteria will ultimately open new ways for the improvement of inventive control tactics for the diseases caused by pathogens [100]. Pan-genomic studies determined a series of novel genes for every intra-subspecific category of pathogens that might be exciting targets for the formulation of new accurate diagnostic tools [104].

3.7 Development of fungicides

Evolutionary studies have shown that the increase or decrease in gene families of pathogens is highly associated with its specific host [102, 105, 106]. The structural variations in genomes of pathogens influence their host range. Many pathogenic genomes are amenable to precise and whole-genome associations by means of long-read sequencing technologies [67, 107]. Complete genome analyses of the similar species are likely to expose segregating chromosomal polymorphism and are essential to cover the pan-genome of a species.

The difference between core and accessory genomic areas is relevant because such compartments are frequently on different evolutionary trajectories [108]. Pan-genomes assembled in the previous decade gave insight into genomic variability in bacterial species [109].

The environmental changes might evolve those pathogenic species much faster that have dispensable chromosomes. Pathogens must be observed so wisely by adding markers on their core regions and accessory genome to understand the fundamental principles of evolution and genomics. There is always a chance of the appearance of a new gene in species, which causes hindrances in the formation of pan-genomic markers. Hence, core genes are always considered preferable to discover targets of fungicides. It is not possible to make markers of the pan-genome that are undiscovered, so there is an opportunity for the unexpected appearance of formerly unidentified genes in such species. Therefore, fungicide targets should be selected within the genes that reside on the core chromosomes instead of the accessory genome.

Fungal diseases may be controlled by considering the pattern of fungal evolution and changes in lifestyle through comparative genomics. Pan-genomic analysis has revealed that *Z. tritici* rapidly developed resistance to fungicides and have overcome main resistance genes in wheat [66]. Phytopathogenic fungi that disrupt the early biotrophic phase of infection or attack the transition from biotrophic or necrotrophic growth might be promising approaches for the development of new fungicides. Fungicides may be developed against the specific target organisms by the recognition of particular genes or gene families without or less affecting the environment [110]. Multiple applications of pan-genome analysis give insight into future studies that focus on various genomic parameters

Table 2 Phytopathogen pan-genome applications

S.No	Applications	Features	References
1	Detection and characterization of new strains	• Comparison and description unidentified isolates • Characterization of closely resembling pathogenic species	[72]
2	Evaluating strain diversity	• Genomic recombinations • Phylogenetic analysis • Taxonomy identification	[69, 75]
3	Revealing the pathogenic evolution	• Evolutionary relationships and uncovering microbial diversity • Insights into gene gain and loss/polymorphism	[81, 83, 84]
4	Development of universal vaccines	• Clinical investigations • Epidemiological surveys • Reverse vaccinology	[35, 75]
5	Role in SNP discovery	• Unraveling allelic variabilities • Distinguishes conserved and variable regions in genomes • Gene mapping	[93, 99]
6	Differentiation of virulent and nonvirulent strains	• Novel drug targets • Improvement in phylogenetic lineages	[104]
7	Development of fungicides	• Pan-genomic markers • Recognition of gene families	[66, 102, 105]

of phytopathogens pertaining to conception of pan-genome. The prominent features of the aforementioned applications are presented in Table 2.

4 Analyzing pan-genomes

The field of pan-genomics is growing and therefore has not yet achieved the depth and breadth of analyses which can be easily demonstrated. However, the central idea of any pan-genome analysis is to perform the genomes-based comparisons of different strains of the same species. Three basic types of information that is retrieved being fundamental to such studies include: (1) the estimation of the size of the core genome indicating all the genes or genes families which are shared among all individuals in a species, (2) the estimation of the size of the pan-genome indicating the size of all the genes or genes families which are present within a species, and (3) the estimation of the increase in the genes or genes families with the addition of each new individual/sample in the analysis.

For this purpose whole-genome comparisons of multiple strains/haplotypes/samples are performed utilizing three different methods including proteins vs protein comparisons, nucleotide vs nucleotide comparisons, or translated proteins against genomic sequences [92]. The gene is considered conserved if the sequence alignments of a gene show 50% sequence conservation considering the 50% of a gene or protein length from any of these methods. Only the new genes are compared with the addition of each new genome comparison in such analyses. This allows assessing the increase in genes/genes families per genome comparison and in turn the open/closed status of the genomes (Fig. 1). The evolution and in-depth review of the models used for the analysis of pan-genomes are discussed in Ref. [1].

4.1 Approaches

A pan-genome analysis workflow clearly depends on the underlying research questions which could alter the sequence of steps that are needed to perform for deeper insights. However, a basic workflow based on series of steps required to conduct a pan-genome analysis based on the reviewed studies in this chapter particularly and others, in general, is graphically illustrated in Fig. 1C. This workflow may particularly be beneficial to novices in the area. The first step is acquiring the input data which can be in several different formats including existing linear reference genomes and their variants, haplotype reference panels, and raw sequencing data (coming from sequencing machines) [111]. The next logical step is to perform alignment and assembly of the genomes after the data quality control checking. The approaches used for the representation of pan-genomes can be broadly categorized into multiple sequence alignment (MSA) based approaches, *k*-mer-based approaches and graph-based approaches [111]. The MSA-based approaches utilize matrix-like data structures to store the homologous characters in the same column and are suitable for performing the analysis of shorter genomic regions and closely associated genomes [112]. Thus, the multiple whole-genomes-based alignments represent the pan-genome. Sophisticated algorithms taking into consideration the bookkeeping data structures, compressed string tree representations and co-linear block-based alignments have been used as extensions to classical alignment methods to cater whole-genome alignments for pan-genome analysis [113]. The approach is particularly suitable for high coverage sequencing samples which can be assembled using de novo approaches [114, 115].

The *k*-mer-based approaches utilize the concept of representing sequences as a collection of strings of length *k*. These *k*-mers can then be represented and visualized using a special graph structure named De Brujin Graph (DBG) which was actually designed for the task of sequence assembly. In the context of pan-genomes, the colored DBGs have been used to construct a nonredundant graph-like structures where each node represents a *k*-mer and the edges (based on overlapping characters $k-1$) between them allows for tracing of the original sequences of *k*-mers in whole genomes [114, 116]. The colors of the nodes are distinct based on the input samples. Various advantages of *k*-mer-based approaches include its efficiency, speed, and robustness.

The graph-based approaches include further extensions to pan-genome analysis using graph-based structures without necessarily using MSA-based alignments or fixed length sequence strings [111]. These approaches include cyclic and acyclic graph structure based representations where nodes and edges form the basis of the underlying coordinate system. Various successful developments using this approach have already been used for pan-genome analysis in Refs. [117–120]. Finally, the downstream analyses can be extended beyond simple full-length genomic comparisons toward more personalized analyses based on the underlying research questions.

There are several influencing factors in pan-genome analysis as highlighted by Ref. [1]. The choice of sequence alignment algorithm (such as BLAST/FASTA) and the associated parameters (minimum alignment length, percentage of similarity, and identity) for orthologous clustering is one important aspect. The orthologous gene detection is an important part of the pan-genome analysis in order to estimate the composition of pan-genome (core and dispensable genomes). It is performed to identify the gene families and functional annotation of all of the genes across individuals of the same species for their subsequent identification as core and variable genomes. The automated prediction of orthologs in databases may often return false positives and thus the filtering criteria along with the prediction methods being used holds significance in the final interpretation of the results [1]. The most common methods include BLAST like searches or OrthoMCL [121, 122].

The samples diversity is yet another influencing factor in conducting the pan-genome analysis. The selection of a sufficient number of samples and their diversity is a critical factor for calculating realistic estimates for the pan-genome content. It is because the small number of samples from a close population may not represent the complete heterogeneity of the genomes. The quality of alignment assembly and annotation are two other critical factors for pan-genome analyses. The quality of alignment assembly defined by the fragment sizes, the choices of assembly operations, gene/orthologous identification approach may affect the downstream analyses. Besides, there are several ab initio gene prediction-based and evidence-based methods for genes annotation. Most of the automated pipelines prefer to use hybrid methodologies for gene annotation in order to decrease the false positives and improve the real estimates. Other influencing factors include phylogenetic resolution, the pan-genome analysis model used to estimate its completeness status, the approaches used for variation analysis and the all-against-all level of comparison.

4.2 Overview of pan-genome analysis tools

The development of pan-genome analysis tools is progressing very fast because of the enhanced understanding of its role in efficient identification of the virulent target genes aiding the vaccine and drug developments. There is a range of tools and software which are available now for pan-genome analysis although most of them were developed for dealing with comparatively smaller genomes of prokaryotes (such as bacteria and viruses). These tools can perform a multitude of analyses which were broadly categorized into seven types including homologous genes clustering, SNPs identification, pan-genomics profiles visualization, phylogenetic analysis based on orthologous genes or gene families based information, pan-genome visualization, curation, and function-based searching in Ref. [123]. However, every tool has its own specifications and limitations making the users dependent to utilize several of them to perform the complete analysis at a single instance along with making the room of further improvements.

Most of the mature tools and software for pan-genomics analysis present today were developed initially to deal with microbial genomes such as PanOCT, PGAP, and GET_HOMOLOGUES. Since many of the identified plant pathogens (organisms that cause infectious disease) belong to bacteria, viruses, viroids, fungi, nematodes, and/or protozoa (composed of smaller genomes compared to complex eukaryotic organisms) therefore the initial set of tools designed for pan-genome analysis provided good guides for them as well. The underlying methodologies of these tools were further refined and several additional features as mentioned before were incorporated. Some of the pan-genome analysis tools that could and have been used for plants pathogens are described in detail below.

The identification of different types of homologous genes (orthologous/paralogous) allows finding the functionally equivalent genes in different strains/individuals in a species. This is specifically required for fine categorization of gene families as a part of core or variable in pan-genomes. This analysis is based on classical methods developed for identifying orthologous gene clusters. The identification of orthologous genes/genes families is particularly of interest to plants pathogens because genome rearrangements and variations occur in them frequently and the accuracy may be affected with different approaches. Several tools including PanOCT, PGAP, Roary perform orthologous gene clustering. PGAP and PanOCT use all-against-all sequence alignment through BLAST where Roary and PanOCT additionally utilize the functions like conserved gene neighborhood for clustering of orthologs of closely related strains efficiently.

Moreover, PGAP, ITEP, Harvest, and GET_HOMOLOGUES have been identified as the pan-genomics analysis tools that can perform several functions previously categorized by Ref. [123]. PGAP is a stand-alone pan-genome analysis pipeline which performs five functions including cluster analysis of functional genes, pan-genome profile analysis, genetic variation analysis of functional genes, species evolution analysis and function enrichment analysis of gene clusters [124]. It uses GeneFamily (GF) and MultiParanoid (MP) methods to identify homologous and orthologous genes, respectively. The GF method utilizes pBLAST for sequence alignments and MCL algorithm for clustering. The MP method uses Inparanoid for identification of orthologs and paralogs with the help of BLAST whereas MP performs clustering [124]. ITEP is another stand-alone Python and BASH scripts-based tool which also integrates SQLite database [125]. It can predict protein families, orthologous genes, functional domains, pan-genome (core and variable genes), and metabolic networks for related microbial species. It particularly allows the customized workflow design and caters for the unannotated/missing data. Harvest is a suite of tools that performs core genomes alignment (using multialigner), variant calls, recombination detection, and phylogenetic trees and interactive visualization of massive core-genome alignments [126]. GET_HOMLOLOGUES is a stand-alone tool built using Perl and R. It was developed for both pan-genome and comparative analysis of bacterial strains. It can perform sequence feature extraction, homologous gene identification, pan-genomic profiling, and phylogenetic analysis. It utilizes BLAST+ and HMMER for orthologous gene clustering [127].

5 Conclusions and future directions

Above and beyond the broad array of applications of pan-genome studies, some views related to the future analysis and pan-genome conception are discussed further. Above all pan-genome analysis requires comprehensively annotated genomic sequences in order to model pan-genomes. Besides that, the well-assembled huge and repetitive plant genome is expensive and challenging. Among many species of plants, gene duplications and the events of their contraction and expansion have geared evolution with respect to diversity in some genomic regions whereas other parts of the genome remain unchanged. The large repetitive regions are highly fragmented because of short length sequencing reads rendering the assemblage of those repetitive parts nearly impossible. Novel technologies such as single-molecule sequencing have the advantage of delivering longer reads but with less accuracy. Production of high-quality algorithms which promise the assemblage of long reads resulting in better quality genomes, which will enable future pan-genome analysis. Analysis of organisms with millions of genes requires highly innovative tools that can serve quick and reliable identification of orthologous genes from closely resembling organisms, phylogenetic studies, profiling of pan-genomes, and broader view to enable the exploration of pan-genome. Another challenge is the preservation and display of the pan-genomic outcomes that is why the genomic databases need to contain detailed knowledge about the pan-genome such as transposable elements, SNPs, noncoding RNAs, and indels. The incorporation of data on genomes and gene expression are also mandatory, interlinking the expression levels, core, and variable genome. Recently, a SuperGenome has been anticipated that is a display of MSA with an extra coordination system. Moreover, the addition of databases for pan-genomes will offer quick access to data. Hence, there is a need to focus on fast-track pan-genomic studies to gather the wealth of genomic data.

References

[1] G. Vernikos, D. Medini, D.R. Riley, H. Tettelin, Ten years of pan-genome analyses, Curr. Opin. Microbiol. 23 (2015) 148–154.

[2] J. Kämper, R. Kahmann, M. Bölker, L.-J. Ma, T. Brefort, B.J. Saville, et al., Insights from the genome of the biotrophic fungal plant pathogen *Ustilago maydis*, Nature 444 (2006) 97.

[3] C.R. Buell, V. Joardar, M. Lindeberg, J. Selengut, I.T. Paulsen, M.L. Gwinn, et al., The complete genome sequence of the Arabidopsis and tomato pathogen Pseudomonas syringae pv. tomato DC3000, Proc. Natl. Acad. Sci. U. S. A. 100 (2003) 10181–10186.

[4] A.R. da Silva, J.A. Ferro, F. Reinach, C. Farah, L. Furlan, R. Quaggio, et al., Comparison of the genomes of two Xanthomonas pathogens with differing host specificities, Nature 417 (2002) 459.

[5] B.M. Tyler, S. Tripathy, X. Zhang, P. Dehal, R.H. Jiang, A. Aerts, et al., Phytophthora genome sequences uncover evolutionary origins and mechanisms of pathogenesis, Science 313 (2006) 1261–1266.

[6] S. Massart, A. Olmos, H. Jijakli, T.J.V.R. Candresse, Current impact and future directions of high throughput sequencing in plant virus diagnostics, Virus Res. 188 (2014) 90–96.

[7] H. Li, P. Vikram, R.P. Singh, A. Kilian, J. Carling, J. Song, et al., A high density GBS map of bread wheat and its application for dissecting complex disease resistance traits, BMC Genomics 16 (2015) 216.

[8] J. Aylward, E.T. Steenkamp, L.L. Dreyer, F. Roets, B.D. Wingfield, M.J. Wingfield, A plant pathology perspective of fungal genome sequencing, IMA Fungus 8 (2017) 1–45.

[9] L.-J. Ma, H.C. Van Der Does, K.A. Borkovich, J.J. Coleman, M.-J. Daboussi, A. Di Pietro, et al., Comparative genomics reveals mobile pathogenicity chromosomes in Fusarium, Nature 464 (2010) 367.

[10] C. Plissonneau, J. Benevenuto, N. Mohd-Assaad, S. Fouché, F.E. Hartmann, D. Croll, Using population and comparative genomics to understand the genetic basis of effector-driven fungal pathogen evolution, Front. Plant Sci. 8 (2017) 119.

[11] M. Salanoubat, S. Genin, F. Artiguenave, J. Gouzy, S. Mangenot, M. Arlat, et al., Genome sequence of the plant pathogen Ralstonia solanacearum, Nature 415 (2002) 497.

[12] M. Van Sluys, M. De Oliveira, C. Monteiro-Vitorello, C. Miyaki, L. Furlan, L. Camargo, et al., Comparative analyses of the complete genome sequences of Pierce's disease and citrus variegated chlorosis strains of Xylella fastidiosa, J. Bacteriol. 185 (2003) 1018–1026.

[13] J. Schirawski, G. Mannhaupt, K. Münch, T. Brefort, K. Schipper, G. Doehlemann, et al., Pathogenicity determinants in smut fungi revealed by genome comparison, Science 330 (2010) 1546–1548.

[14] J.T. Greenberg, B.A. Vinatzer, Identifying type III effectors of plant pathogens and analyzing their interaction with plant cells, Curr. Opin. Microbiol. 6 (2003) 20–28.

[15] S. Huang, E.A. Van Der Vossen, H. Kuang, V.G. Vleeshouwers, N. Zhang, T.J. Borm, et al., Comparative genomics enabled the isolation of the R3a late blight resistance gene in potato, Plant J. 42 (2005) 251–261.

[16] E.H. Stukenbrock, B.A. McDonald, The origins of plant pathogens in agro-ecosystems, Annu. Rev. Phytopathol. 46 (2008) 75–100.

[17] I.P. Adams, R.H. Glover, W.A. Monger, R. Mumford, E. Jackeviciene, M. Navalinskiene, et al., Next-generation sequencing and metagenomic analysis: a universal diagnostic tool in plant virology, Mol. Plant Pathol. 10 (2009) 537–545.

[18] E.R. Mardis, The impact of next-generation sequencing technology on genetics, Trends Genet. 24 (2008) 133–141.

[19] S. Massart, M. Perazzolli, M. Höfte, I. Pertot, M.H.J.B. Jijakli, Impact of the omic technologies for understanding the modes of action of biological control agents against plant pathogens, Biocontrol 60 (2015) 725–746.

[20] C. Zong, S. Lu, A.R. Chapman, X.S.J.S. Xie, Genome-wide detection of single-nucleotide and copy-number variations of a single human cell, Science 338 (2012) 1622–1626.

[21] Deleted in Review

[22] S.R. Eichten, R.A. Swanson-Wagner, J.C. Schnable, A.J. Waters, P.J. Hermanson, S. Liu, et al., Heritable epigenetic variation among maize inbreds, PLoS Genet. 7 (2011) e1002372.

[23] R.K. Saxena, D. Edwards, R.K. Varshney, Structural variations in plant genomes, Brief. Funct. Genomics 13 (2014) 296–307.
[24] S. Kamoun, Molecular genetics of pathogenic oomycetes, Eukaryot. Cell 2 (2003) 191–199.
[25] S. Raffaele, S.J.N.R.M. Kamoun, Genome evolution in filamentous plant pathogens: why bigger can be better, Nat. Rev. Microbiol. 10 (2012) 417.
[26] R.A. Farrer, D.A. Henk, T.W. Garner, F. Balloux, D.C. Woodhams, M.C. Fisher, Chromosomal copy number variation, selection and uneven rates of recombination reveal cryptic genome diversity linked to pathogenicity, PLoS Genet. 9 (2013) e1003703.
[27] D.E. Cooke, L.M. Cano, S. Raffaele, R.A. Bain, L.R. Cooke, G.J. Etherington, et al., Genome analyses of an aggressive and invasive lineage of the Irish potato famine pathogen, PLoS Pathog. 8 (2012) e1002940.
[28] S.F. Sarkar, D.S.J.A. Guttman, E. Microbiology, Evolution of the core genome of Pseudomonas syringae, a highly clonal, endemic plant pathogen, Appl. Environ. Microbiol. 70 (2004) 1999–2012.
[29] H.J.P. Kistler, Genetic diversity in the plant-pathogenic fungus Fusarium oxysporum, Phytopathology 87 (1997) 474–479.
[30] U. Dobrindt, B. Hochhut, U. Hentschel, J. Hacker, Genomic islands in pathogenic and environmental microorganisms, Nat. Rev. Microbiol. 2 (2004) 414.
[31] D. O'Sullivan, P. Tosi, F. Creusot, B. Cooke, T.-H. Phan, M. Dron, et al., Variation in genome organization of the plant pathogenic fungus Colletotrichum lindemuthianum, Curr. Genet. 33 (1998) 291–298.
[32] J. Bishop, A. Dean, T. Mitchell-Olds, Rapid evolution in plant chitinases: molecular targets of selection in plant-pathogen coevolution, Proc. Natl. Acad. Sci. U. S. A. 97 (2000) 5322–5327.
[33] H. Tettelin, V. Masignani, M.J. Cieslewicz, C. Donati, D. Medini, N.L. Ward, et al., Genome analysis of multiple pathogenic isolates of Streptococcus agalactiae: implications for the microbial "pan-genome" Proc. Natl. Acad. Sci. U. S. A. 102 (2005) 13950–13955.
[34] L. Carlos Guimaraes, L. Benevides de Jesus, M. Vinicius Canario Viana, A. Silva, R. Thiago Juca Ramos, S. de Castro Soares, et al., Inside the pan-genome-methods and software overview, Curr. Genomics 16 (2015) 245–252.
[35] C. Plissonneau, F.E. Hartmann, D. Croll, Pangenome analyses of the wheat pathogen Zymoseptoria tritici reveal the structural basis of a highly plastic eukaryotic genome, BMC Biol. 16 (2018) 5.
[36] L. Pritchard, R.H. Glover, S. Humphris, J.G. Elphinstone, I.K. Toth, Genomics and taxonomy in diagnostics for food security: soft-rotting enterobacterial plant pathogens, Anal. Methods 8 (2016) 12–24.
[37] Q. Chen, H.S. Mason, T. Mor, A. Sutherland, G.A. Cardineau, C.J.A.C.O. Tacket, Subunit vaccines produced using plant biotechnology, in: New Generation Vaccines, fourth ed., CRC Press, 2016, pp. 664–667.
[38] J. Glasner, M. Marquez-Villavicencio, H.-S. Kim, C. Jahn, B. Ma, B. Biehl, et al., Niche-specificity and the variable fraction of the Pectobacterium pan-genome, Mol. Plant-Microbe Interact. 21 (2008) 1549–1560.
[39] S. Zoledowska, A. Motyka-Pomagruk, W. Sledz, A. Mengoni, E. Lojkowska, High genomic variability in the plant pathogenic bacterium Pectobacterium parmentieri deciphered from de novo assembled complete genomes, BMC Genomics 19 (2018) 751.
[40] T.A. Coutinho, S.N. Venter, Pantoea ananatis: an unconventional plant pathogen, Mol. Plant Pathol. 10 (2009) 325–335.
[41] R.A. Mann, T.H. Smits, A. Bühlmann, J. Blom, A. Goesmann, J.E. Frey, et al., Comparative genomics of 12 strains of Erwinia amylovora identifies a pan-genome with a large conserved core, PLoS One 8 (2013).
[42] R. Nandakumar, A. Shahjahan, X. Yuan, E. Dickstein, D. Groth, C. Clark, et al., Burkholderia glumae and B. gladioli cause bacterial panicle blight in rice in the southern United States, Plant Dis. 93 (2009) 896–905.
[43] T.G. Lessie, W. Hendrickson, B.D. Manning, R. Devereux, Genomic complexity and plasticity of Burkholderia cepacia, FEMS Microbiol. Lett. 144 (1996) 117–128.
[44] A.J.G. Simpson, F.C. Reinach, P. Arruda, F.A. Abreu, M. Acencio, R. Alvarenga, et al., The genome sequence of the plant pathogen Xylella fastidiosa, Nature 406 (2000) 151.

[45] K.J. Leonard, L.J. Szabo, Stem rust of small grains and grasses caused by Puccinia graminis, Mol. Plant Pathol. 6 (2005) 99–111.
[46] R. Park, Stem rust of wheat in Australia, Aust. J. Agric. Res. 58 (2007) 558–566.
[47] D. Croll, M. Zala, B.A. McDonald, Breakage-fusion-bridge cycles and large insertions contribute to the rapid evolution of accessory chromosomes in a fungal pathogen, PLoS Genet. 9 (2013).
[48] L. Gardan, C. Gouy, R. Christen, R. Samson, Elevation of three subspecies of Pectobacterium carotovorum to species level: Pectobacterium atrosepticum sp. nov., Pectobacterium betavasculorum sp. nov. and Pectobacterium wasabiae sp. nov, Int. J. Syst. Evol. Microbiol. 53 (2003) 381–391.
[49] V. Duarte, S. De Boer, L. Ward, A. De Oliveira, Characterization of atypical Erwinia carotovora strains causing blackleg of potato in Brazil, J. Appl. Microbiol. 96 (2004) 535–545.
[50] L.R. Triplett, Y. Zhao, G.W. Sundin, Genetic differences between blight-causing Erwinia species with differing host specificities, identified by suppression subtractive hybridization, Appl. Environ. Microbiol. 72 (2006) 7359–7364.
[51] B. Ma, M.E. Hibbing, H.-S. Kim, R.M. Reedy, I. Yedidia, J. Breuer, et al., Host range and molecular phylogenies of the soft rot enterobacterial genera Pectobacterium and Dickeya, Phytopathology 97 (2007) 1150–1163.
[52] P. De Maayer, W.Y. Chan, E. Rubagotti, S.N. Venter, I.K. Toth, P.R. Birch, et al., Analysis of the Pantoea ananatis pan-genome reveals factors underlying its ability to colonize and interact with plant, insect and vertebrate hosts, BMC Genomics 15 (2014) 1.
[53] G. Moreno-Hagelsieb, K. Latimer, Choosing BLAST options for better detection of orthologs as reciprocal best hits, Bioinformatics 24 (2007) 319–324.
[54] W. Bonn, T. van der Zwet, Distribution and economic importance of fire blight, in: J.L. Vanneste (Ed.), Fire Blight: The Disease and Its Causative Agent, Erwinia amylovora, CABI Publishing, Wallingford, UK, 2000.
[55] P.S. McManus, A.L. Jones, Genetic fingerprinting of Erwinia amylovora strains isolated from tree-fruit crops and Rubus spp, Phytopathology 85 (1995) 1547–1553.
[56] T.H. Smits, F. Rezzonico, B. Duffy, Evolutionary insights from Erwinia amylovora genomics, J. Biotechnol. 155 (2011) 34–39.
[57] J. Blom, S.P. Albaum, D. Doppmeier, A. Pühler, F.-J. Vorhölter, M. Zakrzewski, et al., EDGAR: a software framework for the comparative analysis of prokaryotic genomes, BMC Bioinf. 10 (2009) 154.
[58] J.H. Ham, R.A. Melanson, M.C. Rush, Burkholderia glumae: next major pathogen of rice? Mol. Plant Pathol. 12 (2011) 329–339.
[59] O.O. Bochkareva, E.V. Moroz, I.I. Davydov, M.S. Gelfand, Genome rearrangements and selection in multi-chromosome bacteria, *Burkholderia*, spp., BMC Genomics, 19 (1) (2018) 965.
[60] Y.-S. Seo, J.Y. Lim, J. Park, S. Kim, H.-H. Lee, H. Cheong, et al., Comparative genome analysis of rice-pathogenic Burkholderia provides insight into capacity to adapt to different environments and hosts, BMC Genomics 16 (2015) 349.
[61] V.S. da Silva, C.S. Shida, F.B. Rodrigues, D.C. Ribeiro, A.A. de Souza, H.D. Coletta-Filho, et al., Comparative genomic characterization of citrus-associated Xylella fastidiosa strains, BMC Genomics 8 (2007) 474.
[62] A.M. Varani, C.B. Monteiro-Vitorello, L.G. de Almeida, R.C. Souza, O.L. Cunha, W.C. Lima, et al., Xylella fastidiosa comparative genomic database is an information resource to explore the annotation, genomic features, and biology of different strains, Genet. Mol. Biol. 35 (2012) 149–152.
[63] A. Giampetruzzi, M. Saponari, G. Loconsole, D. Boscia, V.N. Savino, R.P. Almeida, et al., Genome-wide analysis provides evidence on the genetic relatedness of the emergent Xylella fastidiosa genotype in Italy to isolates from Central America, Phytopathology 107 (2017) 816–827.
[64] N.M. Upadhyaya, D.P. Garnica, H. Karaoglu, J. Sperschneider, A. Nemri, B. Xu, et al., Comparative genomics of Australian isolates of the wheat stem rust pathogen Puccinia graminis f. sp. tritici reveals extensive polymorphism in candidate effector genes, Front. Plant Sci. 5 (2015) 759.
[65] A. O'Driscoll, S. Kildea, F. Doohan, J. Spink, E. Mullins, The wheat–Septoria conflict: a new front opening up? Trends Plant Sci. 19 (2014) 602–610.
[66] C. Cowger, M. Hoffer, C. Mundt, Specific adaptation by Mycosphaerella graminicola to a resistant wheat cultivar, Plant Pathol. 49 (2000) 445–451.

[67] C. Plissonneau, A. Stürchler, D. Croll, The evolution of orphan regions in genomes of a fungal pathogen of wheat, MBio 7 (2016).

[68] B. McDonald, J. Martinez, Chromosome length polymorphisms in a Septoria tritici population, Curr. Genet. 19 (1991) 265–271.

[69] C. Zhong, M. Han, S. Yu, P. Yang, H. Li, K. Ning, Pan-genome analyses of 24 Shewanella strains re-emphasize the diversification of their functions yet evolutionary dynamics of metal-reducing pathway, Biotechnol. Biofuels 11 (2018) 193.

[70] P. Lapierre, J.P. Gogarten, Estimating the size of the bacterial pan-genome, Trends Genet. 25 (2009) 107–110.

[71] J.H. Chan, Y.-S. Ong, S.-B. Cho, Computational Systems-Biology and Bioinformatics: First International Conference, CSBio 2010, Bangkok, Thailand, November 3–5, 2010, Proceedings, vol. 115, Springer, 2010.

[72] O. Lukjancenko, Analysis of Pan-Genome Content and Its Application in Microbial Identification, Technical University of Denmark (DTU), 2014.

[73] A. Rasooly, K.E. Herold, Food microbial pathogen detection and analysis using DNA microarray technologies, Foodborne Pathog. Dis. 5 (2008) 531–550.

[74] Y. He, Bacterial whole-genome determination and applications, in: Molecular Medical Microbiology, second ed., Elsevier, 2015, pp. 357–368.

[75] S. Chaillou, M. Daty, F. Baraige, A.-M. Dudez, P. Anglade, R. Jones, et al., Intraspecies genomic diversity and natural population structure of the meat-borne lactic acid bacterium Lactobacillus sakei, Appl. Environ. Microbiol. 75 (2009) 970–980.

[76] V. Periwal, V. Scaria, Insights into structural variations and genome rearrangements in prokaryotic genomes, Bioinformatics 31 (2014) 1–9.

[77] A. Mira, A.B. Martín-Cuadrado, G. D'Auria, F. Rodríguez-Valera, The bacterial pan-genome: a new paradigm in microbiology, Int. Microbiol. 13 (2010) 45–57.

[78] J. Zhou, J.H. Miller, Microbial genomics—challenges and opportunities: the 9th International Conference on Microbial Genomes, J. Bacteriol. 184 (2002) 4327–4333.

[79] J. Mosquera-Rendón, A.M. Rada-Bravo, S. Cárdenas-Brito, M. Corredor, E. Restrepo-Pineda, A. Benítez-Páez, Pangenome-wide and molecular evolution analyses of the Pseudomonas aeruginosa species, BMC Genomics 17 (2016) 45.

[80] S.C. Watkinson, L. Boddy, N. Money, The Fungi, Academic Press, 2015.

[81] C. Feschotte, E.J. Pritham, DNA transposons and the evolution of eukaryotic genomes, Annu. Rev. Genet. 41 (2007) 331–368.

[82] S. Ohno, Gene duplication and the uniqueness of vertebrate genomes circa 1970–1999, in: Seminars in Cell & Developmental Biology, 1999, pp. 517–522.

[83] C.F. Olson-Manning, M.R. Wagner, T. Mitchell-Olds, Adaptive evolution: evaluating empirical support for theoretical predictions, Nat. Rev. Genet. 13 (2012) 867.

[84] S. Dong, S. Raffaele, S. Kamoun, The two-speed genomes of filamentous pathogens: waltz with plants, Curr. Opin. Genet. Dev. 35 (2015) 57–65.

[85] J.D. Jones, J.L. Dangl, The plant immune system, Nature 444 (2006) 323.

[86] D. Croll, B.A. McDonald, The accessory genome as a cradle for adaptive evolution in pathogens, PLoS Pathog. 8 (2012).

[87] J.O. McInerney, A. McNally, M.J. O'Connell, Why prokaryotes have pangenomes, Nat. Microbiol. 2 (2017) 17040.

[88] A.A. Golicz, P.E. Bayer, G.C. Barker, P.P. Edger, H. Kim, P.A. Martinez, et al., The pangenome of an agronomically important crop plant Brassica oleracea, Nat. Commun. 7 (2016) 13390.

[89] H. Tettelin, The bacterial pan-genome and reverse vaccinology, in: Microbial Pathogenomics, vol. 6, Karger Publishers, 2009, pp. 35–47.

[90] Z. Xiang, Y. He, Vaxign: a web-based vaccine target design program for reverse vaccinology, Procedia Vaccinol. 1 (2009) 23–29.

[91] R.C. Shields, L. Zeng, D.J. Culp, R.A. Burne, Genomewide identification of essential genes and fitness determinants of streptococcus mutans UA159, mSphere 3 (2018).

[92] A. Muzzi, V. Masignani, R. Rappuoli, The pan-genome: towards a knowledge-based discovery of novel targets for vaccines and antibacterials, Drug Discov. Today 12 (2007) 429–439.
[93] L.X. Shen, J.P. Basilion, V.P. Stanton, Single-nucleotide polymorphisms can cause different structural folds of mRNA, Proc. Natl. Acad. Sci. U. S. A. 96 (1999) 7871–7876.
[94] B. Hurgobin, D. Edwards, SNP discovery using a pangenome: has the single reference approach become obsolete? Biology 6 (2017) 21.
[95] T. Jehan, S. Lakhanpaul, Single Nucleotide Polymorphism (SNP)—Methods and Applications in Plant Genetics: A Review, (2006).
[96] C. Laing, C. Buchanan, E.N. Taboada, Y. Zhang, A. Kropinski, A. Villegas, et al., Pan-genome sequence analysis using Panseq: an online tool for the rapid analysis of core and accessory genomic regions, BMC Bioinf. 11 (2010) 461.
[97] D.L. Hyten, Q. Song, Y. Zhu, I.-Y. Choi, R.L. Nelson, J.M. Costa, et al., Impacts of genetic bottlenecks on soybean genome diversity, Proc. Natl. Acad. Sci. U. S. A. 103 (2006) 16666–16671.
[98] J.F. Doebley, B.S. Gaut, B.D. Smith, The molecular genetics of crop domestication, Cell 127 (2006) 1309–1321.
[99] J.A. Rafalski, Novel genetic mapping tools in plants: SNPs and LD-based approaches, Plant Sci. 162 (2002) 329–333.
[100] J. Garita-Cambronero, A. Palacio-Bielsa, M.M. López, J. Cubero, Pan-genomic analysis permits differentiation of virulent and non-virulent strains of Xanthomonas arboricola that cohabit Prunus spp. and elucidate bacterial virulence factors, Front. Microbiol. 8 (2017) 573.
[101] T. Tsuru, I. Kobayashi, Multiple genome comparison within a bacterial species reveals a unit of evolution spanning two adjacent genes in a tandem paralog cluster, Mol. Biol. Evol. 25 (2008) 2457–2473.
[102] R. Baroncelli, D.B. Amby, A. Zapparata, S. Sarrocco, G. Vannacci, G. Le Floch, et al., Gene family expansions and contractions are associated with host range in plant pathogens of the genus Colletotrichum, BMC Genomics 17 (2016) 555.
[103] A.N. Ignatov, E.I. Kyrova, S.V. Vinogradova, A.M. Kamionskaya, N.W. Schaad, D.G. Luster, Draft genome sequence of Xanthomonas arboricola strain 3004, a causal agent of bacterial disease on barley, Genome Announc. 3 (2015).
[104] J. Garita-Cambronero, A. Palacio-Bielsa, M.M. López, J. Cubero, Comparative genomic and phenotypic characterization of pathogenic and non-pathogenic strains of Xanthomonas arboricola reveals insights into the infection process of bacterial spot disease of stone fruits, PLoS One 11 (2016).
[105] R.A. Ohm, N. Feau, B. Henrissat, C.L. Schoch, B.A. Horwitz, K.W. Barry, et al., Diverse lifestyles and strategies of plant pathogenesis encoded in the genomes of eighteen Dothideomycetes fungi, PLoS Pathog. 8 (2012).
[106] P. Gladieux, J. Ropars, H. Badouin, A. Branca, G. Aguileta, D.M. De Vienne, et al., Fungal evolutionary genomics provides insight into the mechanisms of adaptive divergence in eukaryotes, Mol. Ecol. 23 (2014) 753–773.
[107] H.A. Gibriel, B.P. Thomma, M.F. Seidl, The age of effectors: genome-based discovery and applications, Phytopathology 106 (2016) 1206–1212.
[108] H. Tettelin, D. Riley, C. Cattuto, D. Medini, Comparative genomics: the bacterial pan-genome, Curr. Opin. Microbiol. 11 (2008) 472–477.
[109] O. Lukjancenko, T.M. Wassenaar, D.W. Ussery, Comparison of 61 sequenced Escherichia coli genomes, Microb. Ecol. 60 (2010) 708–720.
[110] W.A. Vargas, J.M.S. Martín, G.E. Rech, L.P. Rivera, E.P. Benito, J.M. Díaz-Mínguez, et al., Plant defense mechanisms are activated during biotrophic and necrotrophic development of Colletotricum graminicola in maize, Plant Physiol. 158 (3) (2012) 1342–1358.
[111] Computational Pan-Genomics Consortium, T. Marschall, M. Marz, T. Abeel, L. Dijkstra, B. E. Dutilh, A. Ghaffaari, P. Kersey, W.P. Kloosterman, V. Mäkinen, A.M. Novak, B. Paten, D. Porubsky, E. Rivals, C. Alkan, J.A. Baaijens, P.I.W. De Bakker, V. Boeva, R.J. P. Bonnal, F. Chiaromonte, R. Chikhi, F.D. Ciccarelli, R. Cijvat, E. Datema, C.M. Van Duijn, E. E. Eichler, C. Ernst, E. Eskin, E. Garrison, M. El-Kebir, G.W. Klau, J.O. Korbel, E.W. Lameijer, B. Langmead, M. Martin, P. Medvedev, J.C. Mu, P. Neerincx, K. Ouwens, P. Peterlongo,

N. Pisanti, S. Rahmann, B. Raphael, K. Reinert, D. de Ridder, J. de Ridder, M. Schlesner, O. Schulz-Trieglaff, A.D. Sanders, S. Sheikhizadeh, C. Shneider, S. Smit, D. Valenzuela, J. Wang, L. Wessels, Y. Zhang, V. Guryev, F. Vandin, K. Ye, A. Schönhuth, Computational pan-genomics: status, promises and challenges, Brief. Bioinform. 19 (2016) 118–135.
[112] C. Notredame, Recent evolutions of multiple sequence alignment algorithms, PLoS Comput. Biol. 3 (2007).
[113] R. Rahn, D. Weese, K. Reinert, Journaled string tree—a scalable data structure for analyzing thousands of similar genomes on your laptop, Bioinformatics 30 (2014) 3499–3505.
[114] A.A. Golicz, J. Batley, D. Edwards, Towards plant pangenomics, Plant Biotechnol. J. 14 (2016) 1099–1105.
[115] B. Kehr, K. Trappe, M. Holtgrewe, K. Reinert, Genome alignment with graph data structures: a comparison, BMC Bioinf. 15 (2014) 99.
[116] Z. Iqbal, M. Caccamo, I. Turner, P. Flicek, G. McVean, De novo assembly and genotyping of variants using colored de Bruijn graphs, Nat. Genet. 44 (2012) 226.
[117] U. Baier, T. Beller, E. Ohlebusch, Graphical pan-genome analysis with compressed suffix trees and the Burrows–Wheeler transform, Bioinformatics 32 (2015) 497–504.
[118] T. Beller, E. Ohlebusch, Efficient construction of a compressed de Bruijn graph for pan-genome analysis, in: Annual Symposium on Combinatorial Pattern Matching, 2015, , pp. 40–51.
[119] S. Marcus, H. Lee, M. Schatz, SplitMEM: graphical pan-genome analysis with suffix skips, bioRxiv 30 (24) (2014) 3476–3483.
[120] S. Sheikhizadeh, M.E. Schranz, M. Akdel, D. de Ridder, S. Smit, PanTools: representation, storage and exploration of pan-genomic data, Bioinformatics 32 (2016) i487–i493.
[121] S.F. Altschul, W. Gish, W. Miller, E.W. Myers, D.J. Lipman, Basic local alignment search tool, J. Mol. Biol. 215 (1990) 403–410.
[122] F. Chen, A.J. Mackey, C.J. Stoeckert Jr., D.S. Roos, OrthoMCL-DB: querying a comprehensive multi-species collection of ortholog groups, Nucleic Acids Res. 34 (2006) D363–D368.
[123] J. Xiao, Z. Zhang, J. Wu, J.J.G. Yu, A brief review of software tools for pangenomics, Genomics Proteomics Bioinformatics 13 (2015) 73–76.
[124] Y. Zhao, J. Wu, J. Yang, S. Sun, J. Xiao, J. Yu, PGAP: pan-genomes analysis pipeline, Bioinformatics 28 (2011) 416–418.
[125] M.N. Benedict, J.R. Henriksen, W.W. Metcalf, R.J. Whitaker, N.D. Price, ITEP: an integrated toolkit for exploration of microbial pan-genomes, BMC Genomics 15 (2014) 8.
[126] T.J. Treangen, B.D. Ondov, S. Koren, A.M. Phillippy, The harvest suite for rapid core-genome alignment and visualization of thousands of intraspecific microbial genomes, Genome Biol. 15 (2014) 524.
[127] B. Contreras-Moreira, P. Vinuesa, GET_HOMOLOGUES, a versatile software package for scalable and robust microbial pan-genome analysis, Appl. Environ. Microbiol. 79 (24) (2013) 7696–7701.

CHAPTER 7

Pan-genomics of food pathogens and its applications

César Toshio Facimoto, Luciana Balbo, Roberta Torres Chideroli, Ulisses de Pádua Pereira
State University of Londrina, Londrina, Brazil

1 Introduction

Foodborne diseases are defined as clinical manifestations as a consequence of the ingestion of food contaminated by pathogenic organisms (such as, viruses, bacteria, and parasites), or due to toxins that are usually byproducts of a microorganism such as botulinum (exotoxin produced by *Clostridium botulinum*) and toxins from *Staphylococcus aureus*. Outbreaks of foodborne diseases usually occur due to consumption of food from a specific event or place and develop with similar clinical manifestations among affected individuals [1–3]. According to the World Health Organization (WHO), 1 out of 10 people display illness by consumption of food. Beyond that, foodborne diseases yield high mortality rates in children, which is equivalent to one-third of the mortality. Half of the foodborne disease occurrences are associated with diarrhea manifestations, affecting 550 million individuals and causing 230,000 deaths. In the American continent, 95% of the diseases are reported as gastroenteritis.

Foodborne diseases are still a great concern to public health and, among the bacterial gastroenteritis manifestations worldwide, species such as *Campylobacter*, *Escherichia coli*, *Salmonella*, *Listeria monocytogenes*, *Staphylococcus aureus*, and *Clostridium botulinum* are the most important due to occurrence or severity of disease. Although these pathogens have been well studied, there is a lack of description exploring their pan-genomes. In this chapter, we will describe the most recent pan-genomics approaches applied to the above-mentioned bacteria (Table 1).

2 Pan-genomics of *E. coli*

E. coli is a commonly found species in the intestinal gut microbiota of humans [4]. Fast growth and ease in altering its genetic features make this bacterium a study model of prokaryotic microorganisms; thus, a variety of information is available for this species. *E. coli* is associated with a variety of clinical manifestations in a wide range of hosts. In humans, it is mostly associated with diarrhea caused by different

Pan-genomics: Applications, Challenges, and Future Prospects
https://doi.org/10.1016/B978-0-12-817076-2.00007-X
© 2020 Elsevier Inc.
All rights reserved.

Table 1 Key findings from foodborne diseases pathogens

Bacteria	Disease	Pan-genome characteristics	Implementation	Outcome of the analysis
Escherichia coli	Enteritis	Open pan-genome ranging from 9000 to 16,000 genes and core genome ranging from 1000 to 3000 genes	Core genome MLST, SNP analysis of core genome and proteomic analysis of core genome	Find putative protective antigens for all pathotypes; predict function based on the TUGs and/or accessory genome found in specific strains
Salmonella enterica	Salmonellosis and typhoid fever	Closed pan-genome ranging from 4000 to 25,300 genes and core genome ranging from 1500 to 3720 genes	SNP analysis of core genome	Specific markers for serovars Typhimurium, Heidelberg, Newport and Enteritis
Clostridium botulinum	Botulism	Open pan-genome with approximately 20,000 genes and core genome ranging from 1000 to 3000 genes	SNP analysis of core genome	A better refinement of the strains clustered in group I
Clostridium perfringens	Enteritis	Open pan-genome ranging from 8000 to 12,000 genes and core genome ranging 1000 to 2392 genes; Unique genes in this species represent 44% of average genome	Core genome phylogeny	Identified strains associated with food poisoning in the same clade
Listeria monocytogenes	Listeriosis	Open pan-genome[a] ranging from 3560 to 6612 genes and core genome ranging from 2014 to	Phylogeny analysis of genes present in the core and accessory genome	Suggested that strains from lineage I and III diverged from lineage II; Identify why

Table 1 Key findings from foodborne diseases pathogens—cont'd

Bacteria	Disease	Pan-genome characteristics	Implementation	Outcome of the analysis
		2647 genes; Highly stable		lineage I is more associated with humans and more related to animals
Staphylococcus aureus	Staphylococcal intoxication	Open pan-genome ranging from 2800 to 7000 genes and core genome ranging from 1000 to 2300 genes. Core genome presents approximately 56% of the average genome	Functional analysis of genes present in the core and accessory genome	Suggested that enterotoxin related genes are part of the accessory genome

[a]Lineage III is considered a closed pan-genome.

pathovars. However, some pathovars are known to cause disease outside the gastrointestinal tract (ExPEC), such as neonatal meningitis and urinary tract infections in adults [5]. Diarrhoeagenic pathovars vary in clinical presentation, host age, and virulence factors. In total, five groups are characterized through molecular biology and are described as following: Enteroaggregative (EAEC), which affects a wide range of groups; however, molecular assays were not able to identify clearly the mechanism behind the action of this pathogen. However, it is known that EAEC is capable of forming biofilm in the surface of the colon, followed by secretion of toxins and cytolytic factor [6, 7]. Enterohemorrhagic (EHEC) is responsible for causing blood diarrhea in the host through the secretion of bacterial factors capable of destroying the colon epithelium [8]. Enteropathogenic (EPEC) uses a similar mechanism of EHEC disrupting the epithelial layer, but, presenting a tropism for the small intestine [9]. Enteroinvasive (EIEC) is the only pathovar capable of invading enterocytes, thus, evading host innate immune responses [10]. Enterotoxigenic (ETEC) is capable of secreting stable and labile toxins (ST and LT) into the intestinal lumen altering the water balance resulting in watery diarrhea [11, 12].

The pan-genome of *E. coli* presents an open pan-genome, also referred to as infinite by some authors, indicating that the species is evolving by gene acquisition and diversification. A variety of studies on *E. coli* pan-genome are available and some of them also consider *Shigella* species, in which there is still a debate regarding taxonomy [13]. The pan-genome of this species generally ranges from 9000 to 16,000 genes [14–16]; however, there is a report of a pan-genome as large as 42,000 [17], which reinforces the concept of the infinite pan-genome. The wide occurrence of genetic variation that this species present explains the large range of the pan-genome repertoire. On the other hand, core genome sizes fluctuate from 1000 to 3000 genes, in this case, following the expectation that the core genome shrinks as the number of strains included in the analysis increase [15]. However, due to the open pan-genome of the species, continued sequencing can add approximately 300 novel genes per genome. Functional annotation of core genes suggests that these are likely associated to the metabolic process. Truly unique genes (TUG) indicate a particular mechanism of survival/adaptation found in only one genome when compared with others of the same species/group. In the case of *E. coli,* a high deviation of TUGs counting from 20 to 300 genes is observed which may be related to the different clinical presentation observed in the host. The majority of TUGs was not functionally predicted; thus, these may represent novel biosynthetic or pathogenic features which should be more explored [18].

Genes shared by an *E. coli* pathovar are expected to be related to the clinical presentation of the respective group. However, just a few pathovar-specific genes are found when analyzing the pan-genome of this species. Following a study model using 17 genomes, the count of pathovar-specific genes is modest. The EHEC presents a significant proportion of pathovar-specific genes, with more than 120 genes. Among the 120 genes shared by the EHEC pathovar, 43% are associated with prophage and phage elements that may carry genes related to unidentified toxins or virulence factors. Fewer pathovar-specific genes were shared with other commensal or laboratory-adapted strains belonging to ETEC, EPEC, and EAEC groups. The ExPEC genomes share a significant level of similarity, suggesting that outside the gastrointestinal tract, *E. coli* uses common molecular mechanisms of interaction with the host [18]. Furthermore, analysis of the core genome SNPs (single-nucleotide polymorphisms) has demonstrated its efficacy to strain-type bacteria together with other gene-by-gene methods such as Core Genome Multilocus Sequence Typing (cgMLST), which are handful tools to estimate the epidemiology of an outbreak when associated with year of isolation, origin (of outbreak and contamination), and disease association [6, 19].

New perspectives on *E. coli* genome plasticity emerged after a 2011 outbreak in Germany caused by an *E. coli* strain harboring EAEC and EHEC pathotypes virulence factors [20]. Since then, it is proposed that novel vaccine strategies focusing on conserved features among *E. coli* might be more effective than using pathotype features. Using the information of the core genome of *E. coli* associated with a proteomic assay to evaluate expression of these genes, it was possible to identify the YncE protein (associated with binding to single-stranded DNA) as a highly immunogenic and protective antigen for all pathotypes using murine models of bacteremia [21].

3 Pan-genomics of *Salmonella enterica*

Salmonella is a great public health concern due to its association with food poisoning and infection outbreaks. The species *Salmonella enterica* is divided into six subspecies: *enterica*, *salamae*, *arizonae*, *diarizonae*, *houtenae*, and *indica;* however, over 99% of disease cases are caused by *enterica* subspecies [22]. Furthermore, *S. enterica* subsp. *enterica* is also classified in more than 1500 serovars, in which, Typhimurium, Enteritidis, Newport, Typhi, Paratyphi A, Paratyphi C, and Choleraesuis are most related to diseases in humans and domestic animals [23, 24]. Disease caused by *S. enterica* subsp. *enterica* is often associated with consumption of poultry products, most commonly the serovars Enteritidis, Newport, and Typhimurium [25–27].

Studies on the *Salmonella* pan-genome report a high variable pan-genome size, which is most related to the amount of strains used in the study. In addition, the disbalanced count of different species, subspecies, and serovar can alter the result. Pan-genomes range from 4000 (7 strains) to 25,300 genes (4939 strains). The core genome is estimated from 1500 to 3720 genes [28, 29]. The most recent and largest study on *Salmonella* included 4939 strains and considered all the available strains in the genus. Furthermore, the *S. enterica* pan-genome displays a small increase of the pan-genome and slight shrinkage of the core genome, indicating that this genus demonstrates a closed pan-genome [28]. The pan-genome of *S. enterica* is more distributed among the genomes of this group, where 70% of the pan-genome belongs in 100 or fewer genomes; therefore, this would explain the high variability of the pan-genome size even being a closed pan-genome [29].

Considering the core genome present in at least 90% of *S. enterica* (soft core), a total of 404 genes were putatively found among *enterica* serovars, and SNP analysis demonstrated a high number of specific markers for serovars Typhimurium, Heidelberg, Newport, and Enteritidis. Although none of the markers were exclusive for a serovar, the use of a subset of markers could differentiate eight of the serovars [29], contrasting with smaller studies that were able to identify unique gene families among *Salmonella* serovars. Furthermore, typhi had the most count of serovar-specific genes while enteritidis had the least count of serovar-specific genes [28]. On the other hand, enteritidis shared the highest count of putative genes of the subspecies, which indicates that this serovar is the closest to the "core genome" among *enterica* serovars [29].

4 Pan-genomics of *Clostridium* spp.

The *Clostridium* genus is an important group of bacteria affecting both humans and animals, mainly due to *C. botulinum* and *C. perfringens* that are considered foodborne pathogens due to their ability to produce toxins. *Clostridium* spp. are Gram positive bacteria, producers of heat-resistant spores that are widely dispersed in the environment. In the absence of oxygen, the spores germinate producing toxins that allow them to contaminate food [30, 31].

Botulism is a disease caused by *C. botulinum,* mostly associated with the ingestion of neurotoxins in contaminated food, usually canned products. *C. botulinum* is currently grouped in four clusters from I to IV; however, group IV is not associated with animal or human hosts. Neurotoxins of this species are classified from A to G [32, 33]. *C. perfringens,* on the other hand, is responsible for causing disease in individuals who consume food contaminated with this bacterium, which is followed by germination in the gastrointestinal tract and production of toxins. The toxins of this group (α, β, ε, and ι) are indicators for classifying strains into toxinotypes A to E [34, 35].

Pan-genomics studies on genus level are scarce, mainly due to the diversity of species in the genus harboring nonpathogenic *Clostridium* or species of biotechnological interest. Thus, studies on foodborne *Clostridium* focus on *C. botulinum* or *C. perfringens*. The most recent pan-genome analysis of the whole genus estimated a pan-genome of 19,941 genes, with 546 genes forming a core genome, 7450 forming the accessory genome and 11,945 unique genes. *Clostridium* spp. presents high genome plasticity and an open pan-genome that allows the incorporation of unique and accessory genes such as virulence, metabolism, and information storage related, which provides the ability to colonize different niches [13, 36, 37]. Genes presented in the core are associated with information storage and processing, more specifically, to translation, ribosomal structure and biogenesis, DNA replication, recombination and repair, cofactor biosynthesis, and general metabolism [37].

Individually, the *C. botulinum* species pan-genome is estimated to have around 20,000 genes with a core genome ranging from 1000 to 3000 genes. The core genome count in this case is bigger than the core for genus; however, it is still very strict due to the high plasticity in the four groups harbored by *C. botulinum* species. On the other hand, the large pan-genome reflects the large number of genes necessary to adapt to various environments [33, 36]. The core is more adapted to the species and codes for heavy metal and antibiotic resistance, cell wall components, virulence, metabolic genes, nitrogen fixation, and bacteriocins. Functional analysis of the core and accessory and unique genes classifies the majority of them as metabolism and information storage. These clusters comprise the metabolism of carbohydrates, amino acids, nucleotides, coenzymes, lipids, inorganic ions and secondary metabolites production, transport, and secretion, production, and conversion of energy. Noteworthy, some genes are poorly characterized, which may be due to the lack of information about these specific gene functions or it may be related to a particular pathway involved in its pathogenesis [36]. A phylogenetic analysis of an SNPs matrix on *Clostridium* was able to estimate a more defined distance among strains, and demonstrated a high diversity within this genus, classifying some *C. botulinum* group III closer to *C. novyi*, *C perfringens* closer to *C. botulinum* group II, and *C. botulinum* group I closer to *C. tetani*. Further, an analysis of 25,555 core genome SNPs revealed that the

C. botulinum group I is composed of five lineages exhibiting a variety of toxin types such as A, F, or B in the same cluster. These results should be more explored in epidemiology studies of outbreaks caused by this pathogen. A total of 3817 SNPs were unique for lineage 2 in group I [38].

The *C. perfringens* pan-genome ranges from 8000 to 12,000 genes with 1000 to 2392 genes in the core. Unique genes in this species represented 44% of the genome with genes reinforcing the high diversity in this species. Phylogeny analysis using the core genome displays four main clades, where strains associated with food poisoning are clustered in the same lineage (clade 1). On the other hand, clades 2 and 3 harbor strains of a wide range of hosts/sources (human, chicken, sheep, dog, horse, and soil). In this case, core genome analysis may generate nonsufficient information to classify these genomes according to types of environment and/or hosts. In addition, it is hypothesized that genes associated with toxinotypes are present in the accessory genome, since the core phylogeny presents high similarity between different toxinotypes. At the functional level, the accessory genome presents 849 genes assigned with replication, recombination, and repair functions, comprising mainly transposases, integrases, and phage proteins. In addition, the frequency of defense mechanism related genes coding for efflux pumps, restriction enzymes, and ABC transporters were higher in the accessory genome. The core genome reinforces its major role in metabolism, presenting the double of genes associated with carbohydrate, amino acid, and lipid metabolism [30].

5 Pan-genomics of *L. monocytogenes*

L. monocytogenes is a foodborne bacterial pathogen responsible for listeriosis in humans. The main manifestations are related to a milder gastroenteritis form or severe invasive infection, which may include disease outcomes such as meningoencephalitis, sepsis, and stillbirth. The bacterium has been isolated from a range of sources including environment and foods, and it is also considered capable of adaptation to diverse ecological niches. Strains of *L. monocytogenes* can be grouped into four evolutionary lineages classified in 12 serotypes. Lineage I was found to be overrepresented among human clinical isolates and epidemic outbreaks in most studies, while lineage II is sporadically isolated from humans and animals. Lineages III and IV are rare and predominantly identified in animals [39]. Even though listeriosis incidence is low compared to that of other foodborne pathogens, mortality rates are up to 30% in positive cases [40]. *L. monocytogenes* survives under stress caused by processing and storing food such as refrigeration, high salt concentration, acidic pH, and low oxygen level [41]. However, the infective dose to cause disease from food is considered high ($>10^4$ CFU/g) [42].

Previous comparative genomic studies about the *L. monocytogenes* pan-genome indicate a range from 3560 to 6612 genes, 2014 to 2647 core genes, and 2033 to 4598 accessory genes [43–48]. These researches also reveal that the pan-genome of *L. monocytogenes* is highly stable but open, suggesting an ability to adapt to new niches and increase emerging genetic information. In contrast, other studies relying on the hybridization of lineage III strains found a closed species pan-genome [46].

Studies on the genome structure of this species shows that most of the accessory genes identified are present in the beginning of the chromosome, while core genes are located in the final quarter of the circular chromosome [45]. Also, accessory genes were located in different hotspots (localization where there are at least three nonhomologous insertions between mutually conserved genomes) and are composed mostly of mobile genetic elements, genes involved in sugar transport, cell wall components, and transcriptional regulators. For this reason, the majority of gene-scale differences are represented by the accessory genome resulted from variable hotspots, different prophages, transposons, and genomic islands [48].

Study with phylogenetic analysis comparing lineages, serotypes, and strains according to genomics and genetic content created a core-genome tree. Generally, this tree shows distances between strains based on small adaptations inside mutually conserved genes and which are clustered inside three clearly separated lineages [49]. For the *Listeria* genus, differential acquisition and loss of genes in accordance to various evolutionary offspring may be due to the relative correspondence of SNPs and the gene scale [48].

Through the use of the analyses of pan-genome, genetic localization, and sequence composition it was found that ancestral strains of lineage I and III possibly diverged from lineage II by loss of genes related to carbohydrate metabolism and gain of hypothetical and surface-associated genes [48]. Mainly, the surface-associated genes present in strains of Lineage I suggest an adaptation related to the virulence factors crucial for the pathogenesis of the disease.

In the pan-genome of *L. monocytogenes*, disparately distributed genes (DDGs) defined as genes that are highly conserved in Lineages I and II and are either absent or different in genomes of Lineage III, were also detected. The distribution and conservation of DDGs are deemed noteworthy as they possibly correlate with differences in ecological fitness and pathogenicity of different strains in the host [50]. These genes are associated with (i) metabolism and transport of carbohydrates; (ii) regulation of transcription, and (iii) gastrointestinal tract adaption. The authors reported that the predominance of strains belonging to Lineage I and II in human infections could be due to their ability to use different carbon sources. On the other hand, most Lineage III strains have been shown to possess virulence factors for intracellular replication [46].

The whole genome sequencing of five *L. monocytogenes* strains representing lineages I-III and eight strains of other *Listeria* species demonstrated that the evolution of the *L. monocytogenes* genome involved loss rather than acquisition of virulence characteristics [45].

A study evaluating the association of the speed of growth at 2 °C and *L. monocytogenes* used the accessory genome to identify 114 genes related to this ability. Some genes were already described to be involved in the cold adaptation mechanism such as genes coding for RNA helicase [51] and precursors of internalin A [52]. A total of 13% of the genes corresponded to mobile genetic elements (phage capsid family proteins or transposases) and 61% were hypothetical proteins [47].

The genomic data can be exploited with many different bioinformatics methods like SNP, cgMLST, and whole-genome multilocus sequence typing (wgMLST) [53]. It is well recognized that *L. monocytogenes* genomes are syntenic, leading to lower genomic diversity, as reflected in SNP differences, than other organisms. The importance of genome-scale analysis is already observed in some surveillance studies, detection of outbreaks, or tracing of infection sources [54, 55]. However, these methodologies must be standardized to allow an easy understanding of the evolutionary events of this pathogen worldwide.

6 Pan-genomics of *S. aureus*

S. aureus is a Gram-positive bacterium known as a commensal and opportunistic pathogen. This bacterium can grow in a wide range of temperatures (7–48 °C), pH (4.2–9.3), sodium chloride concentration (up to 15% NaCl) and it is tolerant to dry and stressful environments that allow *S. aureus* to grow on a variety of food products [56, 57]. Food poisoning is the main occurrence associated with *S. aureus*, usually due to consumption of water or contaminated foods. Outbreaks of *S. aureus* toxins contaminating food usually result in vomit, abdominal pain, and diarrhea. Further, *S. aureus* is able to produce staphylococcal enterotoxins (SEs) that are stable, resistant to heat, freezing, dry, and gastrointestinal tract conditions. More than 20 enterotoxins have been described; however, the SEA toxin is the most common staphylococcal food-poisoning cause [58–60].

S. aureus is classified as an open pan-genome, with size ranging from 2800 to 7000 genes and core genome estimated around 1000 to 2300 genes [61–63]. The average *S. aureus* genome is 2800 genes and the core genome presents approximately 56% of the whole genome in this species while in species such as *E. coli* the core genome represents 40% of the average genome [62].

A total of 90 different virulence factors are estimated to be present in the pan-genome of *S. aureus*, in which 35 are located in the core genome and involved in the synthesis of polysaccharide capsule (PC), Panton-Valentine leucocidin (PVL), gamma-hemolysin, and iron-regulated proteins (cell adherence). Other proteins that are present in the majority of the strains (more than 90%) are protein A (disrupts phagocytosis) and alpha toxins (disrupts membranes) [62]. Although enterotoxins are the main cause of food poisoning by *S. aureus*, specific enterotoxins seem to be associated with the accessory genome [59, 63]. In terms of functionality, genes classified as metabolic are mainly present in the core genome. On the other hand, the unique genome harbor 62% of genes related to mobile elements indicating that horizontal gene transfer (HGT) is an important factor on *S. aureus* acquiring resistance and virulence genes [62, 63].

7 Conclusions and future directives

The use of the pan-genomics approach in microbiology/foodborne pathogens research has the potential to provide a wide range of information to comprehend the structure and dynamics of the genome interacting with the environment and/or diseases outbreaks. However, few studies are focusing on bacteria causing foodborne diseases, specially using different bioinformatic strategies according to featured information of pan-genomic data of each group of foodborne pathogens studied (Fig. 1).

The core genome approaches to infer the phylogeny of bacteria has demonstrated the importance that these studies collaborate to understand the epidemiology of a disease and reclassification of new species/groups/types that may be relevant information if used as genetic markers related to virulence or environmental adaptation. Further, it is possible to observe that phylogeny studies are tending to switch to WGS analysis due to its high accuracy compared to the investigation of specific genes to analyze.

Besides phylogeny, pan-genomics studies can make use of the core genome to estimate proteins useful for the development of universal vaccine targets against toxins and pathogens, as well as finding better targets for drugs.

Overall, in regard to foodborne pathogens, some may show a more stable pan-genome (closed pan-genome), toxins being the major virulence factors (Fig. 1). As

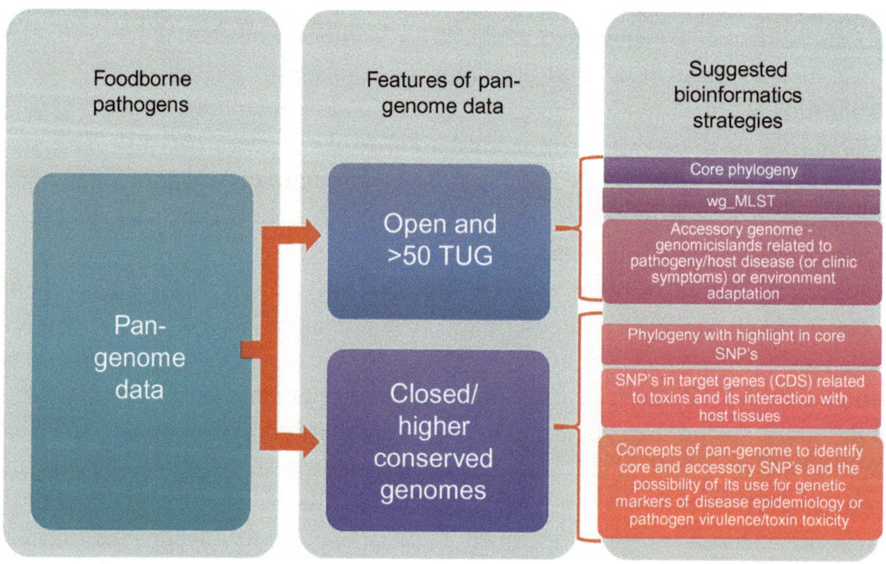

Fig. 1 Bioinformatics approaches for analysis of foodborne pathogens pangenomes.

perspectives, the "micro pan-genome analysis" or "magnifier pan-genomic approaches" evaluating the small differences (SNP's and indels) in toxin groups, would support a better understanding of pathogen/toxin/host interactions.

References

[1] M. Addis, D. Sisay, A review on major food borne bacterial illnesses, J. Trop. Dis. 3 (4) (2015) 1–7.

[2] A. Aljoudi, A. Al-Mazam, A. Choudhry, Outbreak of food borne Salmonella among guests of a wedding ceremony: the role of cultural factors, J. Fam. Community Med. 17 (1) (2010) 29.

[3] D.M. Nunes, F.J. de Paula Júnior, J.S. Melo, E.C. de Oliveira, V.C. Meneguini, F. Dias, Surto de doença transmitida por alimento em evento de massa de populações indígenas em Cuiabá, Mato Grosso, Brasil, no ano de 2013, Epidemiol. Serv. Saúde 25 (1) (2016) 1–10.

[4] E. Thursby, N. Juge, Introduction to the human gut microbiota, Biochem. J. 474 (11) (2017) 1823–1836.

[5] S.J. Salipante, D.J. Roach, J.O. Kitzman, M.W. Snyder, B. Stackhouse, S.M. Butler-Wu, et al., Large-scale genomic sequencing of extraintestinal pathogenic Escherichia coli strains, Genome Res. 25 (1) (2015) 119–128.

[6] T.J. Dallman, M.A. Chattaway, L.A. Cowley, M. Doumith, R. Tewolde, D.J. Wooldridge, et al., A. Cloeckaert (Ed.) An investigation of the diversity of strains of enteroaggregative Escherichia coli isolated from cases associated with a large multi-pathogen foodborne outbreak in the UK, PLoS One 9 (5) (2014).

[7] M.A. Croxen, B.B. Finlay, Molecular mechanisms of Escherichia coli pathogenicity, Nat. Rev. Microbiol. 8 (1) (2010) 26–38.

[8] Y. Nguyen, V. Sperandio, Enterohemorrhagic *E. coli* (EHEC) pathogenesis, Front. Cell. Infect. Microbiol. 2 (2012) 90.

[9] J.L. Thomassin, J.R. Brannon, J. Kaiser, S. Gruenheid, H. Le Moual, Enterohemorrhagic and enteropathogenic Escherichia coli evolved different strategies to resist antimicrobial peptides, Gut Microbes 3 (6) (2012) 556–561.

[10] M. Pasqua, V. Michelacci, M.L. Di Martino, R. Tozzoli, M. Grossi, B. Colonna, et al., The intriguing evolutionary journey of Enteroinvasive E. coli (EIEC) toward pathogenicity, Front. Microbiol. 8 (2017) 2390.

[11] A.A.M. Lima, M.C. Fonteles, From Escherichia coli heat-stable enterotoxin to mammalian endogenous guanylin hormones, Braz. J. Med. Biol. Res. 47 (3) (2014) 179–191.

[12] A. von Mentzer, T.R. Connor, L.H. Wieler, T. Semmler, A. Iguchi, N.R. Thomson, et al., Identification of enterotoxigenic Escherichia coli (ETEC) clades with long-term global distribution, Nat. Genet. 46 (12) (2014) 1321–1326.

[13] L. Rouli, V. Merhej, P.E. Fournier, D. Raoult, The bacterial pangenome as a new tool for analysing pathogenic bacteria, New Microbes New Infect. 7 (2015) 72–85.

[14] H. Willenbrock, P.F. Hallin, T.M. Wassenaar, D.W. Ussery, Characterization of probiotic Escherichia coli isolates with a novel pan-genome microarray, Genome Biol. 8 (12) (2007) R267.

[15] R.S. Kaas, C. Friis, D.W. Ussery, F.M. Aarestrup, Estimating variation within the genes and inferring the phylogeny of 186 sequenced diverse Escherichia coli genomes, BMC Genomics 13 (1) (2012) 577.

[16] X.Z. Ge, J. Jiang, Z. Pan, L. Hu, S. Wang, H. Wang, et al., M. Skurnik (Ed.) Comparative genomic analysis shows that avian pathogenic Escherichia coli isolate IMT5155 (O2:K1:H5; ST complex 95, ST140) shares close relationship with ST95 APEC O1:K1 and human ExPEC O18:K1 strains, PLoS One 9 (11) (2014).

[17] L. Snipen, T. Almøy, D.W. Ussery, Microbial comparative pan-genomics using binomial mixture models, BMC Genomics 10 (1) (2009) 385.

[18] D.A. Rasko, M.J. Rosovitz, G.S.A. Myers, E.F. Mongodin, W.F. Fricke, P. Gajer, et al., The pangenome structure of Escherichia coli: comparative genomic analysis of E. coli commensal and pathogenic isolates, J. Bacteriol. 190 (20) (2008) 6881–6893.

[19] A.C. Schürch, S. Arredondo-Alonso, R.J.L. Willems, R.V. Goering, Whole genome sequencing options for bacterial strain typing and epidemiologic analysis based on single nucleotide polymorphism versus gene-by-gene–based approaches, Clin. Microbiol. Infect. 24 (4) (2018) 350–354.

[20] H. Karch, E. Denamur, U. Dobrindt, B.B. Finlay, R. Hengge, L. Johannes, et al., The enemy within us: lessons from the 2011 European Escherichia coli O104:H4 outbreak, EMBO Mol. Med. 4 (9) (2012) 841–848.

[21] D.G. Moriel, L. Tan, K.G.K. Goh, M.-D. Phan, D.S. Ipe, A.W. Lo, et al., A novel protective vaccine antigen from the Core *Escherichia coli* genome, mSphere 1 (6) (2016) e00326-16.

[22] M.D. Kirk, S.M. Pires, R.E. Black, M. Caipo, J.A. Crump, B. Devleesschauwer, et al., L. von Seidlein Ed. World Health Organization Estimates of the Global and Regional Disease Burden of 22 Foodborne Bacterial, Protozoal, and Viral Diseases, 2010: a data synthesis, PLoS Med. 12 (12) (2015)

[23] K. Chan, S. Baker, C.C. Kim, C.S. Detweiler, G. Dougan, S. Falkow, Genomic comparison of *Salmonella enterica* serovars and Salmonella bongori by use of an *S. enterica* serovar typhimurium DNA microarray, J. Bacteriol. 185 (2) (2003) 553–563.

[24] S.H. Park, H.J. Kim, W.H. Cho, J.H. Kim, M.H. Oh, S.H. Kim, et al., Identification of *Salmonella enterica* subspecies I, *Salmonella enterica* serovars typhimurium, enteritidis and typhi using multiplex PCR, FEMS Microbiol. Lett. 301 (1) (2009) 137–146.

[25] Centers for Disease Control and Prevention (CDC), Surveillance for Foodborne Disease Outbreaks United States, 2016: Annual Report, U.S. Department of Health and Human Services, CDC, Atlanta, GA, 2018.

[26] A.J. Taylor, V. Lappi, W.J. Wolfgang, P. Lapierre, M.J. Palumbo, C. Medus, et al., D.J. Diekema (Ed.) Characterization of foodborne outbreaks of Salmonella enterica Serovar Enteritidis with whole-genome sequencing single nucleotide polymorphism-based analysis for surveillance and outbreak detection, J. Clin. Microbiol. 53 (10) (2015) 3334–3340.

[27] S.M. Crim, S.J. Chai, B.E. Karp, M.C. Judd, J. Reynolds, K.C. Swanson, et al., *Salmonella enterica* serotype Newport infections in the United States, 2004–2013: increased incidence investigated through four surveillance systems, Foodborne Pathog. Dis. 15 (10) (2018) 612–620.

[28] A. Jacobsen, R.S. Hendriksen, F.M. Aaresturp, D.W. Ussery, C. Friis, The *Salmonella enterica* Pan-genome, Microb. Ecol. 62 (3) (2011) 487–504.

[29] C.R. Laing, M.D. Whiteside, V.P.J. Gannon, Pan-genome analyses of the species *Salmonella enterica*, and identification of genomic markers predictive for species, subspecies, and serovar, Front. Microbiol. 8 (2017) 1345.

[30] R. Kiu, S. Caim, S. Alexander, P. Pachori, L.J. Hall, Probing genomic aspects of the multi-host pathogen *Clostridium perfringens* reveals significant pangenome diversity, and a diverse array of virulence factors, Front. Microbiol. 8 (2017) 2485.

[31] S. Fleck-Derderian, M. Shankar, A.K. Rao, K. Chatham-Stephens, S. Adjei, J. Sobel, et al., The epidemiology of foodborne botulism outbreaks: a systematic review, Clin. Infect. Dis. 66 (Suppl. 1) (2017) S73–S81.

[32] E.A. Johnson, M. Bradshaw, Clostridium botulinum and its neurotoxins: a metabolic and cellular perspective, Toxicon 39 (11) (2001) 1703–1722.

[33] H. Söderholm, K. Jaakkola, P. Somervuo, P. Laine, P. Auvinen, L. Paulin, et al., Comparison of Clostridium botulinum genomes shows the absence of cold shock protein coding genes in type E neurotoxin producing strains, Botulinum J. 2 (3/4) (2013) 189.

[34] S. Brynestad, P.E. Granum, Clostridium perfringens and foodborne infections, Int. J. Food Microbiol. 74 (3) (2002) 195–202.

[35] F.A. Uzal, J.C. Freedman, A. Shrestha, J.R. Theoret, J. Garcia, M.M. Awad, et al., Towards an understanding of the role of Clostridium perfringens toxins in human and animal disease, Future Microbiol. 9 (3) (2014) 361–377.

[36] T. Bhardwaj, P. Somvanshi, Pan-genome analysis of Clostridium botulinum reveals unique targets for drug development, Gene 623 (2017) 48–62.

[37] Z. Udaondo, E. Duque, J.L. Ramos, The pangenome of the genus Clostridium, Environ. Microbiol. 19 (7) (2017) 2588–2603.

[38] N. Gonzalez-Escalona, R. Timme, B.H. Raphael, D. Zink, S.K. Sharma, Whole genome SNP analysis for discrimination of Clostridium botulinum group I strains, Appl. Environ. Microbiol. 80 (2014) 2125–2132.
[39] F. Allerberger, M. Wagner, Listeriosis: a resurgent foodborne infection, Clin. Microbiol. Infect. 16 (2010) 16–23.
[40] E.B. Nyarko, C.W. Donnelly, Listeria monocytogenes: strain heterogeneity, methods, and challenges of subtyping, J. Food Sci. 80 (12) (2015) M2868–M2878.
[41] T. Kramarenko, M. Roasto, K. Meremäe, M. Kuningas, P. Põltsama, T. Elias, Listeria monocytogenes prevalence and serotype diversity in various foods, Food Control 30 (1) (2012) 24–29.
[42] S.T. Ooi, B. Lorber, Gastroenteritis due to listeria monocytogenes, Pediatr. Infect. Dis. J. 24 (9) (2005) 854.
[43] T. Hain, R. Ghai, A. Billion, C.T. Kuenne, C. Steinweg, B. Izar, et al., Comparative genomics and transcriptomics of lineages I, II, and III strains of listeria monocytogenes, BMC Genomics 13 (1) (2012) 144.
[44] A. Hilliard, D. Leong, A. O'Callaghan, E.P. Culligan, C.A. Morgan, N. DeLappe, et al., Genomic characterization of listeria monocytogenes isolates associated with clinical Listeriosis and the food production environment in Ireland, Genes (Basel) 9 (3) (2018) 171.
[45] H.C. den Bakker, C.A. Cummings, V. Ferreira, P. Vatta, R.H. Orsi, L. Degoricija, et al., Comparative genomics of the bacterial genus listeria: genome evolution is characterized by limited gene acquisition and limited gene loss, BMC Genomics 11 (1) (2010).
[46] X. Deng, A.M. Phillippy, Z. Li, S.L. Salzberg, W. Zhang, Probing the pan-genome of listeria monocytogenes: new insights into intraspecific niche expansion and genomic diversification, BMC Genomics 11 (1) (2010) 500.
[47] L. Fritsch, J.-F. Mariet, L. Guillier, F. Palma, M.-Y. Mistou, N. Radomski, et al., Insights from genome-wide approaches to identify variants associated to phenotypes at pan-genome scale: application to *L. monocytogenes*' ability to grow in cold conditions, Int. J. Food Microbiol. 291 (2018) 181–188.
[48] C. Kuenne, A. Billion, M.A. Mraheil, A. Strittmatter, R. Daniel, A. Goesmann, et al., Reassessment of the listeria monocytogenes pan-genome reveals dynamic integration hotspots and mobile genetic elements as major components of the accessory genome, BMC Genomics 14 (1) (2013) 47.
[49] R.H. Orsi, H.C.d. Bakker, M. Wiedmann, Listeria monocytogenes lineages: Genomics, evolution, ecology, and phenotypic characteristics, Int J. Med. Microbiol. 301 (2011) 79–96.
[50] S. Lomonaco, D. Nucera, V. Filipello, The evolution and epidemiology of listeria monocytogenes in Europe and the United States, Infect. Genet. Evol. 35 (2015) 172–183.
[51] A. Markkula, M. Mattila, M. Lindström, H. Korkeala, Genes encoding putative DEAD-box RNA helicases in listeria monocytogenes EGD-e are needed for growth and motility at 3°C, Environ. Microbiol. 14 (8) (2012) 2223–2232.
[52] J. Kovacevic, C. Arguedas-Villa, A. Wozniak, T. Tasara, K.J. Allen, Examination of food chain-derived listeria monocytogenes strains of different serotypes reveals considerable diversity in inlA genotypes, mutability, and adaptation to cold temperatures, Appl. Environ. Microbiol. 79 (6) (2013) 1915–1922.
[53] C. Henri, P. Leekitcharoenphon, H.A. Carleton, N. Radomski, R.S. Kaas, J.-F. Mariet, et al., An assessment of different genomic approaches for inferring phylogeny of listeria monocytogenes, Front. Microbiol. 8 (2017).
[54] B.R. Jackson, C. Tarr, E. Strain, K.A. Jackson, A. Conrad, H. Carleton, et al., Implementation of Nationwide real-time whole-genome sequencing to enhance Listeriosis outbreak detection and investigation, Clin. Infect. Dis. 63 (3) (2016) 380–386.
[55] Y. Chen, N. Gonzalez-Escalona, T.S. Hammack, M.W. Allard, E.A. Strain, E.W. Brown, Core genome multilocus sequence typing for identification of globally distributed clonal groups and differentiation of outbreak strains of listeria monocytogenes, Appl. Environ. Microbiol. 82 (20) (2016) 6258–6272.
[56] J. Kadariya, T.C. Smith, D. Thapaliya, Staphylococcus aureus and staphylococcal food-borne disease: an ongoing challenge in public health, Biomed. Res. Int. 2014 (2014) 1–9.
[57] P. Chaibenjawong, S.J. Foster, Desiccation tolerance in Staphylococcus aureus, Arch. Microbiol. 193 (2) (2011) 125–135.

[58] Y. Le Loir, F. Baron, M. Gautier, Staphylococcal food poisoning, Foodborne Dis. third ed. 2 (1) (2017) 367–380.
[59] M.Á. Argudín, M.C. Mendoza, M.R. Rodicio, Food poisoning and Staphylococcus aureus enterotoxins, Toxins (Basel) 2 (7) (2010) 1751–1773.
[60] J.A. Hennekinne, M.L. De Buyser, S. Dragacci, Staphylococcus aureus and its food poisoning toxins: characterization and outbreak investigation, FEMS Microbiol. Rev. 36 (4) (2012) 815–836.
[61] D. Chaves-Moreno, M.L. Wos-Oxley, R. Jáuregui, E. Medina, A.P.A. Oxley, D.H. Pieper, C. Gibas (Ed.) Application of a novel "Pan-genome"-based strategy for assigning RNAseq transcript reads to Staphylococcus aureus strains, PLoS One 10 (12) (2015).
[62] E. Bosi, J.M. Monk, R.K. Aziz, M. Fondi, V. Nizet, B.Ø. Palsson, Comparative genome-scale modelling of Staphylococcus aureus strains identifies strain-specific metabolic capabilities linked to pathogenicity, Proc. Natl. Acad. Sci. 113 (26) (2016) E3801–E3809.
[63] S. Åvall-Jääskeläinen, S. Taponen, R. Kant, L. Paulin, J. Blom, A. Palva, et al., Comparative genome analysis of 24 bovine-associated staphylococcus isolates with special focus on the putative virulence genes, PeerJ 6 (2018).

CHAPTER 8

Pan-genomics of aquatic animal pathogens and its applications

Nguyen Thanh Luan[a], Hai Ha Pham Thi[b]
[a]Department of Veterinary Medicine, Institute of Applied Science, Ho Chi Minh City University of Technology—HUTECH, Ho Chi Minh City, Vietnam
[b]Faculty of Biotechnology and Environmental Technology, Nguyen Tat Thanh University, Ho Chi Minh City, Vietnam

1 Genome study of aquaculture pathogens

1.1 The spread of aquatic pathogens and advent of next-generation sequencing

As the fastest growing food-producing sector, industrial aquaculture ensures food security and economic welfare worldwide through a sustained increase in production. Currently, different types of pathogenic bacteria such as *Yersinia ruckeri*, *Flavobacterium psychrophilum*, *Aeromonas salmonicida*, *Edwardsiella tarda*, and *Vibrio aestuarianus* cause the mass mortality of many fish species and are a serious issue in intensive aquaculture [1]. The emergence of novel pathogens and spread of infectious disease can be a consequence of evolution or global expansion of previously characterized pathogens. Host switching, a result of intensive mixed farming, is generating a new pathology caused by novel strains [2]. From a background of commensal organisms, new pathogen strains might alternatively evolve as a result of mutation or horizontal acquisition of virulence genes through the recombination of previously isolated pathogen populations [3]. Versatile adaptive strategies of pathogens are derived from a complication of multiple host-pathogen interactions. These interactions cause genetic variations such as point mutations, gene insertions or deletions, recombinations, and copy number variations. Therefore, the strains may rapidly adapt to distinct environments, and persist for prolonged periods in a broad spectrum of disease phenotypes, leading to clinical complications and difficult diagnostic interpretations. The characterization and systematic understanding of genotype-phenotype correlations in infectious diseases are major challenges in a fundamental and clinical studies as well as developing a sustainable biocontrol method, such as vaccine strategies. The next-generation sequencing (NGS) approach is a convenient and efficient tool, as it has been recently shown by the reduction in time and cost per genome sequenced, and the increase in associated metadata. This technique is used to describe the characteristics of the genome and the entire virulence gene repertoire of bacterial pathogens through computing the sum of the core and dispensable genomes, which are subsequently used to control

Pan-genomics: Applications, Challenges, and Future Prospects
https://doi.org/10.1016/B978-0-12-817076-2.00008-1
© 2020 Elsevier Inc.
All rights reserved.

diseases in farmed fish. In particular, this approach can create new opportunities to reconstruct the evolution of bacterial genomes at an unprecedented scale and level of resolution. This approach can also identify the signatures of host adaption and adaptive resistance pathways in the pathogens through exhaustive analysis, for example, functional interpretation from the core and accessory genes with regulatory elements [4].

By the NGS approach, bulk data of genes within a bacterial genome that has been harvested from large population samples reveals the evidence for genome-wide genetic change. In the multilocus sequence typing (MLST) dataset, first seven genes were used, and then data for potentially 100s, or 1000s of genes (see the MLST databases at www.pubmlst.org) was applied to construct genotypes. Previous studies [5, 6] have shown a geographic restriction in the layers of hitherto hidden subvariants within single strains when using MLST assay. Obviously, either the reconstruction of transmission pathways within disease outbreaks (both of human and animal pathogens) or tracking the source of foodborne pathogens can be exploited by the unprecedented discriminatory power of the whole genome sequencing approach [7]. Studies of evolution based on genomics [8, 9] can also infer microevolutionary changes within a single host as well as pathogen mutation rates in prolonged latent infection, over time scales of weeks to months.

In addition, genomic framework analysis can estimate patterns of pathogen transmission across the epidemiological scales. For instance, large and prolonged outbreaks within single hospital was determined to be due to the clonal spread of a specific strain that had genetically adapted to the hospital environment [10]. In fact, the potential for cross-species pathogen transmission is a documented route between human and animal host [11] in hospitals, farms [12], or across countries or even continents [13]. The exploitation of a genomic approach for aquatic pathogens is a disease management strategy in aqua farms [1]. While some recent studies dealt with genomic-derived aquatic pathogens (i.e. Refs. [14, 15]), they did not assess pan-genome analyses (Table 1). Therefore, we have discussed in this chapter two examples of aquatic pathogenic bacteria, including the genus *Edwardsiella* and *Aeromonas* that are of key importance in aquatic disease. We have focused on genotyping methods developed from pan-genome data, enabling us to deduce phylogenomic diversity and possible evolutionary trends of aquatic bacterial pathogen strains as compared to nonaquatic host pathogen strains as a hypothesis of zoonotic characteristics of strain and a possible implementation of effective disease mediation in an aqua farm.

1.2 The aquatic bacterial genome sequence and its open access data

The worldwide report of pathogenic bacteria isolated from aquatic environments has attracted the attention of the scientific society. There are two key groups of pathogenic bacteria that have been identified as posing aquatic animal diseases. The first major pathogenic Gram-negative genera that affects the aquaculture industry includes *Aeromonas, Edwardsiella, Flavobacterium, Francisella, Photobacterium, Piscirickettsia, Pseudomonas,*

Table 1 Comparative genomic studies implementing pan-genome analysis of selected aquatic bacterial pathogens

Aquatic pathogens	Ref. of WGS studies	Ref. of studies implementing pan-genome analysis and results
Aeromonas salmonicida subsp. *salmonicida*	[16, 17]	[18] The resulting binary matrix (i.e., the presence/absence) was used to map the characters on a phylogenetic tree based on the core genome. The analysis made it possible to determine which genes were acquired and which were lost during evolution and, consequently, may have played a role in the adaption of a given isolate. Given the mesophilic-to-psychrophilic gradient, we investigated the gene repertoires for the branch separating *A. salmonicida* subsp. *masoucida* from the mesophilic isolates and the branch separating *A. salmonicida* subsp. *masoucida* from the psychrophilic isolates
Aeromonas salmonicida subsp. *achromogenes*	[19]	
Aeromonas veronii		[20] They demonstrated that strain 17ISAe originating from imported diseased fish harbored various antibiotic-resistance genes (ARGs), class 1 integrons and transposon that might represent a very important source of ARG emergence and transmission in "domestic" bacteria
Aeromonas hydrophila		[21] Different strains harbor multiple virulence factors and ARGs
Aeromonas sobria		[22] A phylogenomic assessment including 2,154 softcore genes corresponding to 946,687 variable sites from 33 *Aeromonas* genomes confirms the status of *A. sobria* as a distinct species divided in two subclades, with 100% bootstrap support
A. hydrophila, A. caviae, and A. veronii		[23] Results of pan-genome analysis revealed an open pan-genome for all three species with pan-genome sizes of 9181, 7214 and 6884 genes for *A. hydrophila, A. veronii* and *A. caviae*, respectively
Edwardsiella tarda		[24] Results of heat map analysis of dispensable genes and phylogenetic tree, all *E. tarda* strains were divided into two groups. One was isolated from freshwater fish and the other was isolated from marine/migratory fish
Francisella noatunensis subsp. *orientalis*	[25]	

Continued

Table 1 Comparative genomic studies implementing pan-genome analysis of selected aquatic bacterial pathogens—cont'd

Aquatic pathogens	Ref. of WGS studies	Ref. of studies implementing pan-genome analysis and results
Flavobacterium psychrophilum	[26]	[27] The pan genome analysis showed that *F. psychrophilum* could hold at least 3373 genes, while the core genome contained 1743 genes. On average, 67 new genes were detected for every new genome added to the analysis, indicating that *F. psychrophilum* possesses an open pan genome. The putative virulence factors were equally distributed among isolates, independent of geographic location, year of isolation and source of isolates
Lactococcus garvieae	[28]	[29] Compared to the five *L. lactis* genomes, 484 genes (25%) were specific to Lg2 and were dominated by hypothetical proteins or proteins of unknown function, which may include functions to cause disease in fish or to survive in the environment
Piscirickettsia salmonis	[30]	
Renibacterium salmoninarum	[31]	[32] Approximately equal numbers of ORFs are part of the core set of similar genes (2273 *R. salmoninarum* ORFs, 2507 *Arthrobacter* sp. strain FB24 ORFs, and 2556 *A. aurescens* ORFs). The two Arthrobacter species share 740 protein ORF clusters (1917 ORFs) not found in *R. salmoninarum* that may have been lost in the course of genome reduction. Similar numbers of unique ORF clusters were identified in the three microorganisms (range, 818 to 933 clusters), suggesting that the levels of genomic divergence are similar
Streptococcus agalactiae	[33]	[34] The Chinese fish isolates GD201008-001 and ZQ0910 are phylogenetically distinct from the The Latin American fish-specific strains SA20-06 and STIR-CD-17, but are closely related to the human strain A909, in the context of the clustered regularly interspaced short palindromic repeats (CRISPRs), prophage, virulence-associated genes and phylogenetic relationships [35] The genomes of Thai ST7 strains are closely related to other fish ST7s, as the core genome is shared by 92%–95% of any individual fish ST7 genome. Among the fish ST7 genomes, we observed only small dissimilarities, based on the analysis of (CRISPRs), surface protein markers, insertions sequence elements and putative virulence genes

Table 1 Comparative genomic studies implementing pan-genome analysis of selected aquatic bacterial pathogens—cont'd

Aquatic pathogens	Ref. of WGS studies	Ref. of studies implementing pan-genome analysis and results
Streptococcus iniae	[36]	
Vibrio anguillarum	[37]	[38] In general, no big differences in the number of assigned genes to specific subsystems were observed among the 15 *V. anguillarum* strains. However, some strains (VIB93avir, 87-9-116avir, VaNT1avir, VIB15vir and 87-9-117vir) appear to have more genes classified into the subsystem "phages, prophages, transposable elements plasmids"
Vibrio harveyi	[39]	
Vibrio parahaemolyticus	[40]	
Yersinia ruckeri	[41]	[42] A complete nucleotide sequence-based pan-genome created from 58 genomes with GVIEW server and then analyzed using BRIG revealed very high conservation amongst the lineages, with a total length of 4,218,016 bp compared to the reference genome of 3,866,096 bp (Fig. 3C). Indeed, the majority of the difference between the genomes was explained by mobile genetic elements
Vibrio aestuarianus	[43]	
Vibrio anguillarum		[44] A pan genome analysis was conducted based on the 11 genomes and describe some structural features of superintegrons on chromosome 2s, and associated insertion sequence (IS) elements, including 18 new ISs (ISVa3–ISVa20), both of importance in the complement of *V. anguillarum* genomes
Aeromonas hydrophila	[45]	
Moritella viscosa		[46] Grouping all functional genes from the twelve *M. viscosa* genomes identified 5589 pan genomic gene clusters. Comparing the core genes to the pan genome cluster showed that the core genome accounts for 67% of the pan genome

References are referred from recent review of Bayliss and colleagues (2017) [1].

Tenacibaculum, Vibrio, Weissella, and *Yersinia*. The second genera includes main Gram-positive taxa of pathogens that are frequently discussed in aquaculture disease, including firmicute genera *Lactococcus* and *Streptococcus*, and *Renibacterium salmoninarum*, a member of the family Micrococcaceae [47, 48]. The causative agent of these pathogenic bacteria has been described in recent review studies (e.g. Ref. [1, 49]) and some are briefly described with additional genome sequence information (Table 2). The ability of the bacteria to

Table 2 Literature of selected aquatic bacterial pathogens adapted from study of Pridgeon and Klesius (2013) [49] and its genome sequence available

Pathogenic species	Disease	Host	Genome sequence available[a]
Gram-negative bacteria			
Aeromonas salmonicida	Furunculosis	Salmonids	8/–/8/27
Flavobacterium psychrophilum	Rainbow trout fry syndrome and bacterial cold water disease	Rainbow trout and coho salmon	11/5/3/48
Vibrio harvey	Vibrosis	Wide range of host from marine to freshwater	4/–/8/27
Vibrio parahaemolyticus	Vibrosis		23/2/143/680
Vibrio coralliilyticus	Vibrosis	Coral	4/1/1/8
Vibrio fluvialis	Vibrosis	Wide range of host from marine to freshwater such as catfish, turbot, flounder, carp, eel, tilapia, hybrid striped bass, seabream, yellowtail and sea bass	3/–/2/7
Vibrio anguillarum	Vibrosis		13/28/1/4
Edwardsiella tarda	Edwardsiellosis or putrefactive disease		4/–/2/11
Edwardsiella ictaluri	Enteric septicaemia of catfish		3/–/2/3
Edwardsiella piscicida	Edwardsiellosis or putrefactive disease		2/1/3/11
Photobacterium damselae	Pasteurellosis		3/–/5/12
Yersinia ruckeri	Yersiniosis, the etiological agent of enteric redmouth disease	finfish species, particularly salmonids,	5/–/5/53
Gram-positive bacteria			
Lactococcus garvieae	Fatal hemorrhagic septicaemia called lactococcosis	Multi-fish species such as yellowtail, trout, rockfish and mullet	3/–/3/18
Streptococcus iniae	Streptococcosis	Wide range of host from marine to freshwater such as tilapia, yellowtail, catfish, flounder and seabream	7/–/3/3
Streptococcus parauberis	Streptococcosis		7/–/3/4
Renibacterium salmoninarum	Bacterial kidney disease	Salmonids	7/–/3/5

[a]Comple/chromosome/scaffold/contig observed from NCBI database until Sep 2018.

cause disease persists in the aquatic environment independent of host, especially when the water temperature is warm. This asymptomatic colonization is part of the normal microbiome (microbial balance) and can occur in farmed species, resulting in very complicated disease monitoring and management. Furthermore, the most serious threat and challenge to health and national security are diseases caused by antibiotic-resistant bacteria. As a result, new antimicrobial compounds will be required regularly in the chemotherapeutic development pipeline. Thus far, it is necessary to create sustainable novel strategies to control bacterial infections focusing on both mitigating the spread of disease by systematic understanding and avoiding the conditions that trigger the transition from a balance lifestyle to a dysbiosis stage (exceeding serious pathogen-microbial imbalance) [50]. Alternative methods in integrated management can be targeted to give similar or enhanced protection to aquatic hosts. For example, the use of compounds/factors to inhibit virulence gene expression or to interrupt the signal transduction pathways of the pathogens will be the sustainable alternative therapies in the future [51]. Therefore, insights into the virulence and molecular mechanisms of pathogenicity are of crucial importance.

The aquatic bacterial genome sequence has revolutionized aquatic disease and continues to play an important role in controlling the spread of infectious disease as well as in developing a resistance to antimicrobial compounds produced in pathogens. The genome sequence has allowed a rapid and accurate identification of pathogens like of *A. salmonicida* subsp. *Salmonicida* [16], *E. tarda* [52], and *Vibrio anguillarum* [37] and also provided insights into evolutionary and host adaptation pathways. In addition, based on the comparative genome studies, the infra-subspecies genome diversification level of bacteria can be distinguished for isolates from different origins [53]. Therefore, the development of bioinformatics software contributes to the revolution of aquatic disease research. Conspicuously, new bioinformatics tools will greatly improve microbial identification and taxonomic classification for *Aeromonas* and *Vibrio*, a particularly taxonomically challenging genera with many aquatic pathogens (e.g., Refs. [14, 18]). Previously, in silico DNA-DNA hybridization (isDDH) and digital DNA-DNA hybridization (dDDH) were the first molecular biological techniques that allowed for the direct experimental comparison of two genomes based on digitally derived genome-to-genome distances [54]. Another technique, the average nucleotide identity (ANI), is based on pairwise genome comparison of all the shared orthologous protein coding genes (also called core genes) [55]. Compared to isDDH and dDDH, the ANI technique is a gold standard not only for identifying species but also for research of aquatic species [15, 21]. The ANI calculators are integrated into the genome analysis tools, for example, EDGAR (efficient database framework for comparative genome analyses using BLAST score Ratios) [56].

Another interesting application of NGS in aquatic bacterial genome sequence analysis is to search for phylogenetic and/or epidemiological maps of disease outbreaks [57].

In recent studies based on genome sequence of clinical samples [48, 58], authors have achieved rapid and precise identification of causative bacterial pathogens and their resistance genes. Moreover, ortholog groups in different parts of the pan-genome analysis, including variation in the core, accessory, and unique genome regions, greatly improve the understanding of the evolution of the strain, more specifically their pathogenicity/virulence. The epidemiological mapping, which reconstructs transmission pathways across epidemiological scales [59], will facilitate monitoring of disease outbreaks in real time in nearby fish farms and epidemiological studies at the global and the national level. This mapping is increasingly important as international trade expands. Therefore, tools that discriminate different ortholog groups in the pan-genome of pathogenic bacteria will become a standard tool for diagnostics and for preventing infectious diseases in aquaculture (Fig. 1).

2 Using the comparative pan-genome to analyze aquatic pathogenic bacteria

2.1 The proliferation of software packages and tools for infectious disease analysis

The main purpose of the pan-genome is to compare the genomes of different strains within a species (intraspecies) or genus (arising or occurring between species and interspecies) [60]. Pan-genome studies bring considerable insights into the understanding of bacterial evolution, niche adaptation, population structure, and host interaction. These studies can also be applied to issues such as the identification of virulence genes and vaccine and drug design [61]. A large number of genomes from different isolates of the same pathogen especially in aquatic pathogens (Table 2) has created the possibility of investigating several genomic characteristics that are intrinsic to one or more species [62]. However, the most critical barrier caused by rapid genome sequence data in routine practice is the lack of automated software that can interpret data and provide clinically meaningful information to microbiologists (rather than bioinformaticians) [63]. To make automated bioinformatic tools for bacterial pan-genome interpretation, several software packages and databases were constructed. These include Panseq, PGAT (prokaryotic-genome analysis tool), PanCGHweb, PanGP, ITEP (integrated toolkit for exploration of microbial pan-genomes), and PGAP (pan-genomes analysis pipeline). The major features and platforms of these different existing tools were compared in recent reviews [64, 65]. The following convenient and efficient pan-genome tools such as PanWeb, PGAP-X, and PanACEA (see Refs. [66, 67]), as well as BPGA and EDGAR 2.0 (see Ref. [64]) are more or less specialized to provide better data mining results and quality graphics for different purposes of presentation and publication.

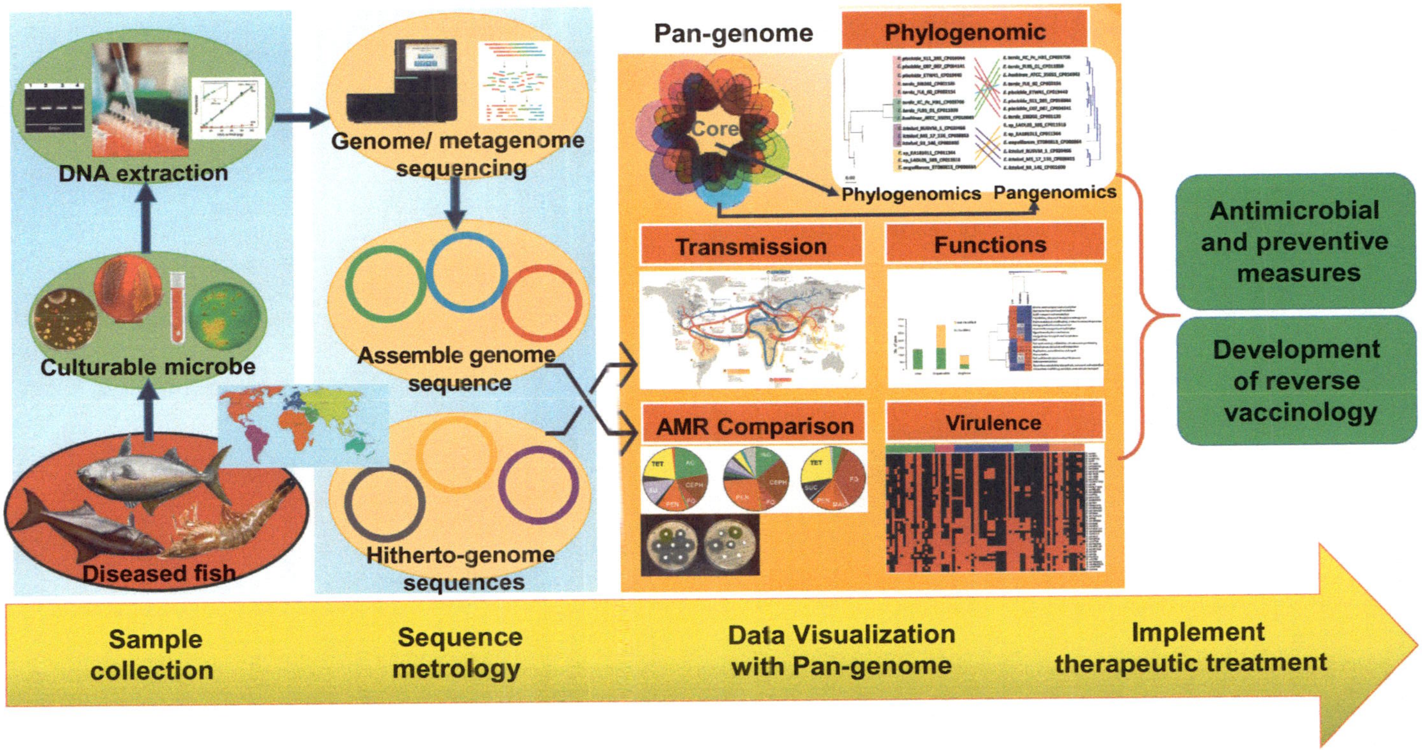

Fig. 1 The potential of integrated genome pathogen seque-cing and pan-genome analysis for the molecular epidemiology of emerging aquaculture pathogens and the development of reverse vaccinology.

2.2 Pan-genome composition of aquatic bacterial pathogens

2.2.1 Introduction

There are many well-established tools for pan-genome analysis, among the early-developed programs or databases such as PanCGHweb and Panseq, published in 2010. These programs mainly focus on grouping genes into orthologs, constructing gene-based phylogenies of related strains and isolates, and/or even determining core and noncore regions in given genomes based on MUMmer and BLASTn, as well as identifying a common type of genetic variation among the core genome [64]. While powerful and flexible toolkits integrate several useful functions, PGAP integrates analysis of functional genes and enrichment of gene clusters, pan-genome profile, and genetic variation of functional genes; PGAT can help plot the presence and absence of genes among members of a pan-genome, identify SNPs (single nucleotide polymorphisms) among orthologs and syntenic regions and compare gene orders among different strains and isolates. In addition, PGAT can identify biological pathways through the integration of several useful analysis tools, such as KEGG (Kyoto encyclopedia of genes and genomes), COG (cluster of orthologous groups of proteins), PSORT (protein subcellular localization prediction tool), SignalP (discriminating signal peptides from transmembrane regions), the TMHMM (transmembrane helices; hidden Markov model), and Pfam (protein families) pathway. Of these tools, ITEP integrates the existing bioinformatic tools with pan-genomic analysis. In addition to basic pan-genomic profiling, metabolic network integration, phylogenetic tree construction, and annotation curation, ITEP also incorporates visualization scripts that assist biologists in specific query for conserved protein domain identification.

However, there are many limitations for these online toolkits because the local database has a limited number of curated species and because of the impossible integration of new sequencing data from users, as well as a lack of intuitiveness in output files [68]. A nonstop optimization for pan-genome analysis, consisting of data interpretation and speedy data mining results, as well as functional exploration would provide better comparisons via graphical visualization. In particular, aquatic microbiologists need the intuitive and easy-to-use graphical user interface as well as great support from the host team, rather than an advanced open-code source application only usable by bioinformaticians. The integrated tools allow them to create interactive, high-quality charts based on taxonomic and functional profiling results, and even provide a popular output file extension for further analysis by others. In our experience, BPGA (bacterial pan-genome analysis tool) and EDGAR (as indicated above) are two easily accessed tools for microbial pan-genome analysis that combine all current comparative analyses. These comparative analyses include core/pan-genome calculations, singleton analysis, a phylogenetic tree-based pan-genome analysis, and ANI/AAI (average nucleotide identity/average amino acid identity) calculation (only in EDGAR). The use of BPGA requires advanced computer skills, and this efficient microbial pan-genome analysis tool provides detailed statistics, distinctive sequences. EDGAR is the most user-friendly tool and an advanced software in pan-genome studies, it is available online for the analysis of large groups of related genomes in a comparative approach. In our previous study [69] using EDGAR, the intra-species evolution of *Lactococcus* strains was computed based on the functional analysis of the core gene, pan-genome, and singleton genes. The software also supports a quick survey of evolutionary relationships and simplifies the process of obtaining binary data supporting hierarchical clustering and new biological analyses of the differential gene functions in relation to the metabolite profile of its host [70]. Our summary indicates that both BPGA and EDGAR may help obtain new biological insights into differential gene content and become useful pipeline to identify vaccine and drug targets efficiently.

2.2.2 Inside the pan-genome of aquatic pathogenic bacteria

The complexity of genotypic cluster analysis in pathogenic bacteria is intrinsically linked to their horizontal gene transfer as a consequence of ecological specialization [71]. In addition to diverse aquatic ecosystems of marine and freshwater, the tremendous diversity of fish and aquatic animal species has certainly contributed to the complication of bacterial disease. Aquatic microorganisms harbor a genome that can be efficiently energetically optimized. Previous studies [72, 73] have indicated that the horizontal acquisition of genes is highly structured by local environmental adaptation; a habitat-specific gene pool is often generated as a result of high adaptive potential to continuously change organismal interactions, such as viral predation and interference competition. By dividing a pan-genome into three parts, including all genes commonly shared by all strains of study, genes present in at least two, but not in all strains, and genes present only in a single strain, we can investigate the evolution of bacterial populations, as well as different features such as niches, adaptation, resistance, the mobilome, and global metabolism. An integrating analysis of variation in the pan-genome can improve a super-resolution view of the evolutionary events of bacterial populations. For example, our investigation of the core and pan-genome of *Edwardsiella* genera has revealed the separation of species patterns within the population (Fig. 2) which may help to reidentify the strain with high accuracy. Further analysis of the differential gene content in the accessory genome of *Edwardsiella* will significantly provide insights into the highly discriminatory molecular assays that can be routinely served for discriminating clinical strains.

Pan-genome analysis enables unparalleled resolution of the evolution of a multidrug-resistant pathogen. It also allows for a better understanding of the genetic background of pathogenicity in a variety of bacteria by comparing with virulent and avirulent strains that would remain invisible to only a core genome phylogenetic analysis [4, 38, 71]. In particular, combining a functional analysis with annotation system such as RAST (rapid annotation using subsystem technology), KEGG, WebMGA (a customizable web server

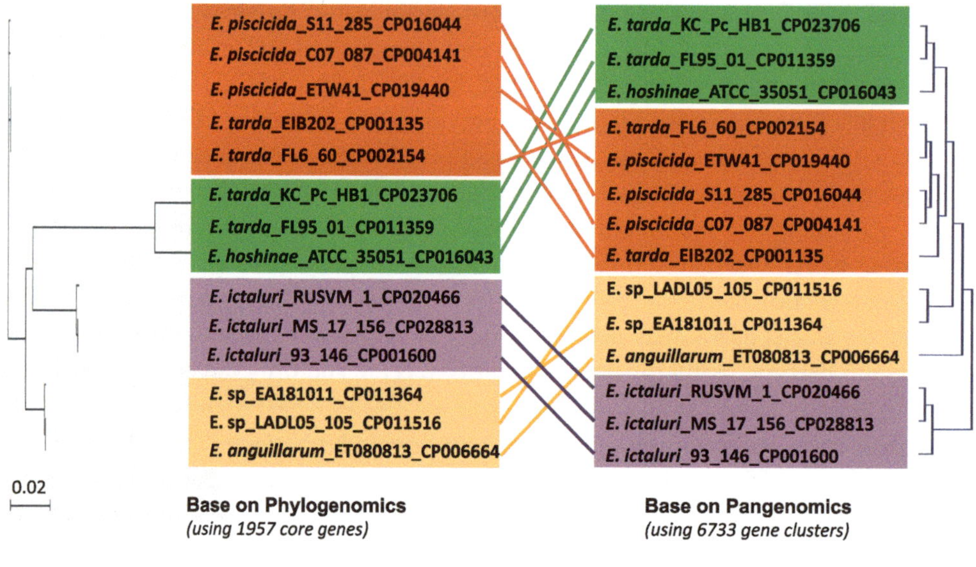

Fig. 2 Comparison of the phylogenetic tree and hierarchical clustering of *Edwardsiella* strains. Both hierarchical clustering (*right* panel), based on shared gene content, and phylogenetic tree (*left* panel), based on concatenated orthologous genes, were performed for all 14 strains. Strings connecting the same strains of both trees are used to highlight the degree of similarities between both tree methods.

for fast metagenomic sequence analysis) can help identify the key event in the emergence of a virulent strain of aquatic bacteria. Also, a broad capacity to metabolize complex sugars can be exhibited by comparing the distribution of carbohydrate-active enzymes (CAZy profiles) among strains, and CAZy signature of isolate is a selective advantage that allows them to fulfill their ecological niche [69]. Several studies have demonstrated that genes that encode the product of iron uptake systems, such as siderophores, hem, and hemoglobin, contribute to the virulence of pathogenic bacteria [74, 75]. A comprehensive comparison of virulence genes with the distribution of their horizontal genes transfer to *E. tarda* strains was reported by Nakamura and colleagues (2013) [69].

Accordingly, in contrast to an attenuated strain of *E. tarda*, which might have a loss-of-function mutation in a gene related to the type III secretion system (T3SS), fish pathogenic strains harbored type VI secretion system (T6SS) and pilus assembly genes in addition to T3SS. In particular, two pathogenicity islands of T3SS and T6SS were absent in isolates, yet existed in pathogenic *E. tarda* strains isolated from red sea bream [69]. The evolutionary analysis in previous study showed that the T3SS was able to integrate into the *E. tarda*-LEE (locus of enterocyte effacement) genome through horizontal transfer. The reason for this finding is that T3SS is homologous to the LEE in enteropathogenic and enterohemorrhagic *E. coli*. Holm and colleagues (2018) [44] have recently sequenced a genome for seven *V. anguillarum* strains, a marine bacterium causing hemorrhagic septicaemia (or vibriosis) disease in aquatic species, including fish, molluscs, and crustaceans [76]. This pathogenic species harbors clusters of highly diverse gene cassettes (VAR; *Vibrio anguillarum* repeats). These gene cassettes are mostly of unknown function, but like with *Vibrio cholera*, can be involved in substrate modification or interactions with virulence factors and DNA modification [77].

To elucidate characteristics of *Aeromonas veronii* 17ISAe, our colleagues have recently conducted a genomic comparison of their strain 17ISAe [20]. This strain was isolated from imported diseased ornamental fish, in order to isolate 44 *A. veronii* strains that included antibiotic-resistance genes (ARG). These genes were isolated using a resistance gene identifier, the database of antibiotic-resistance cassettes, and the annotation of a virulent gene using the virulence factor database (VFDB). The study showed that the strain 17ISAe is a dangerous transmission source of ARG to "domestic" bacteria. Due to various ARGs, class 1 integrons, class 1 transposons, and critical virulence genes can be integrated into the bacterial genome. Conspicuously, virulence and genotypic characteristics are not always related to phenotypic characteristics of pathogenic species like *Vibrio anguillarum* [78].

Virulence in *V. anguillarum* was defined as multi-factorial because it could not be assigned to one or a few virulence factors [38]. There was no difference in the number of gene products associated with the subsystem "virulence, disease, and defense" in RAST annotation, which were detected in different virulent or avirulent strains of the 15 *V. anguillarum*. However, some differences are still found in the comparative genome of *V. anguillarum* isolates; these differences include genes that encode for products of multidrug resistance efflux pumps (43–47 out of 65–73 genes belonged to "virulence, disease, and defense") and genes belonging to subsystem "toxins and super antigens." These genes are consistent with the ones associated with broad antibiotic resistance recorded from different isolates of *V. anguillarum* [76]. A comparative genome analysis suggested that both virulent (CNEVA NB11008 avir, VIB113 vir, and JLL143 vir) and avirulent (VIB12avir) *V. anguillarum* strains had genes involved in hem uptake and utilization, indicating that the presence of specific virulence genes could not explain the virulence in some *V. anguillarum* strains [38]. Therefore, the comparative genome is of crucial importance that unravels the presence and absence of a gene profile in given strains. This approach can also help distinguish between phenotypic and genotypic characteristics of pathogenic strains, enabling the further understanding of virulence mechanisms and expression of corresponding genes through transcriptomics, epigenetics of host-pathogen interactions, or zooming into the promoter region.

2.3 Pan-genome analysis of aquatic pathogenic species: the case of *Edwardsiella* and *Aeromonas*

2.3.1 Edwarsiella *genus*

The genus *Edwardsiella* is a member of the family Enterobacteriaceae and is known as causative agents that are present in a wide range of environments and hosts; they also cause economic losses in different commercially important fish [79]. There are three species of this genus: *Edwardsiella hoshinae*, *Edwardsiella ictaluri*, and *Edwardsiella tarda*. The pathogenic isolates are well described in association with diverse hosts, including birds and reptiles, cultured channel catfish, and cultured tilapia (see the previous review [79]). The infection of *E. tarda* has been regarded as a systemic disease associated with mass mortality in many cultured fish species, yet the species is also regarded as a versatile pathogen that can affect a wide range of other hosts, such as birds, amphibians, reptiles, marine mammals, and even humans. Moreover, its ecological niches include lakes, rivers, seawater, and intestines of healthy aquatic animals (as described in recent studies [80, 81]).

Recently, *E. piscicida* were identified as a new pathogen causing epizootics for different cultured fish species globally [80, 82]. The genus *E. anguillarum* has a discerning capacity to produce acetoin from glucose (VP positive) and to ferment arabinose from other species. This genus includes microorganisms that are potentially pathogenic to eels [80]. In fact, the taxa in genus *Edwardsiella* are difficult to distinguish from each other in 16S rDNA gene sequencing and morphological, physiological, or biochemical data. For instance, *E. piscicida* shared many phenotypic characteristics identical to *E. tarda* [80, 81, 83] and were even previously mistaken for one another [84]. In fish isolates, *E. piscicida* was demonstrated to have different genetic profiles based on molecular techniques and phylogenetic approaches. In addition, recent studies based on comparative phylogenetic approaches [81, 85] suggested that the taxon *E. tarda* presented genetically distinct groups; most fish isolates actually belonged to the species *E. piscicida*, not *E. tarda*. Therefore, pan-genome analysis approaches are needed to clarify *Edwardsiella* taxonomic position. In addition, these approaches can explore shared genes to understand their adaptation ability and species specificity.

In this chapter, we have performed a comparative genome analysis using the 14th completed genome sequence of strains belonging to the genus obtained from a bacterial genome database [National Centre for Biotechnology Information (NCBI); ftp://ftp.ncbi.nih.gov/genomes/]. This pan-genome consists of 6733 protein-encoding genes, only 29.07% of which (1957) were core genes, and the remaining 70.93% were dispensable and singleton genes within the genus of *Edwardsiella*. A pan development plot analysis shows an open pan-genome model with the value from the Heap's Law function ranging between 0 and 1 (0.302, Fig. 3A), indicating that *Edwardsiella* spp. can adapt to a variety of environments.

In addition, we confirmed that *E. tarda* EIB202 and FL6_60 belong to the group of *E. piscicida*. These findings were confirmed using phylogenomic analyses of core genes and pan-genomics (Fig. 2) or hierarchical clustering of dispensable genes (Fig. 3B). Consistent with previous studies [82, 84], our analysis shows that these *E. tarda* strains are reidentified as *E. piscicida* and four *Edwardsiella* species and were clearly distinguishable. The remaining two *Edwardsiella* sp. genomes (EA181011 and LADL05_105) showed ANIs

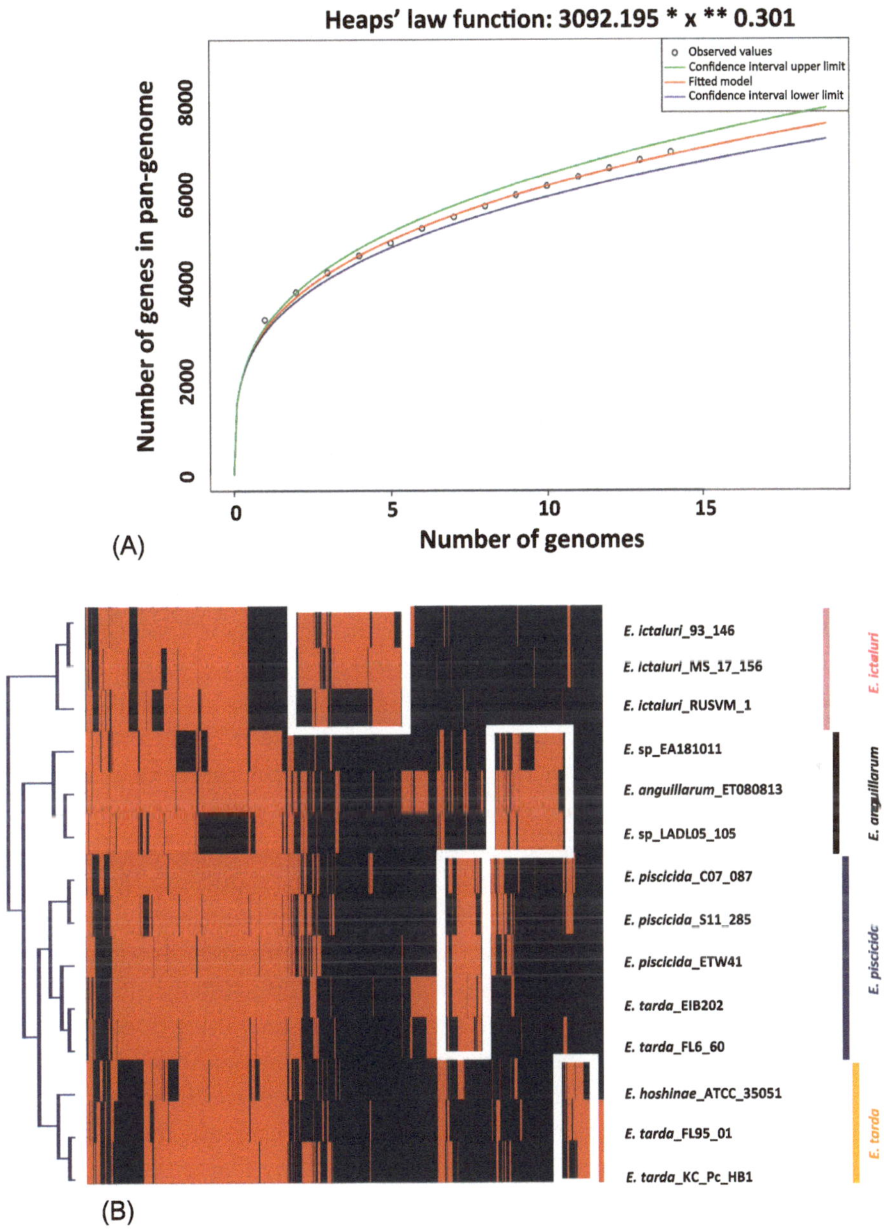

Fig. 3 Polymorphism of dispensable genes among *Edwardsiella* strains. (A) Calculated singleton gene sets to each chromosome. (B) The presence/absence of the 2578 identified genes is shown in *red/black*, respectively

value of 99.65 and 99.58, respectively (data not shown) and clustered very well with the strain of *E. anguillarum*_ET080813 (Fig. 3B). Thus, two corresponded genomes remained as "*Edwardsiella* sp. EA181011" and "*Edwardsiella* sp. LADL05_105," on the basis of the ANI and hierarchical clustering of dispensable genes that could belong to potential species *E. anguillarum*.

In particular, polymorphism of dispensable genes among *Edwardsiella* strains would provide very valuable information on species-specific control measures against Edwardsiellosis. For instance, the gene presence/absence (white box in Fig. 3B) may indicate good markers for the use of molecular techniques, an unending search to accurately identify *Edwardsiella* isolates, especially when differentiating new species from *E. tarda* [81, 82]. In further examination, the combination of phenotyping, serotyping with anti-sera, and visualization of differential gene content with their downstream analysis like KEGG/COG assignments, VFDB, ARG, and the subsystems of RAST annotation will help discover new biological insights into the evolution of pathogenesis as well as explore strain-specific drug targets against Edwardsiellosis in aquafarms. Finally, our pan-genome interpretation, pan-PCR, which is a highly discriminatory PCR assay based on highly informative identified genetic targets whose presence or absence [86], will be a routine tool in the lab that can distinguish all clinically relevant *Edwardsiella* strains.

2.3.2 Aeromonas *genus*

Aeromonasis is a considerably important bacterial disease in aquaculture reported by FAO (2017) [87] and is present in a wide range of global environments and hosts. These hosts include fresh and brackish water fish species, such as catfish, tilapia, *Puntius*, rohu, and other cyprinids. The major disease symptoms of aeromonads in fish, amphibians, and reptiles are hemorrhagic disease, ulcerative syndrome, and septicemia [88]. The *Aeromonas* species are also capable of infecting humans and other animals via food [89].

A comparative pan-genome analysis of motile aeromonads *Aeromonas hydrophila, A. veronii, Aeromonas sobria*, and *Aeromonas caviae* was performed in recent studies [21–23]. An open pan-genome was first shown in a pan-genome study using three species including *A. hydrophila, A. veronii*, and *A. caviae*. The greater genomic diversity among the given species was indicated in *A. hydrophila* [23]. Although no significant difference in virulence factors predicted among these above three species was found, the influence of homologous recombination and lateral gene transfer were identified as factors involved in the evolution of *Aeromonas* spp. isolates.

In a subsequent study, the diversity in T3SS and the conservation of type II secretion systems and T6SS, as well as various ARG from different antibiotic classes and multiple virulence factors were identified in pan-genome of *A. hydrophila* [21], supporting previous findings that *A. hydrophila* is greater hazards of pathogenesis [23]. The phylogenomic diversity of all five *A. sobria* strains were divided into two subclades with a deep dichotomy in terms of inhibitory effect against *A. salmonicida* subsp. *salmonicida*, gene contents, and codon usage [22]. This organization enabled the development of novel control strategies against pathogenic *A. salmonicida* subsp. *salmonicida* by antagonistic activities of *A. sobria* strains TM12 and TM18.

The results for ANI pairwise comparisons of 34 representative genomes of *Aeromonas* (including the species *A. hydrophila*, *A. salmonicida*, *Aeromonas rivipollensis*, *Aeromonas dhakensis*, *Aeromonas schubertii*, *A. veronii*, *A. caviae*, and *Aeromonas media*) were computed using the available ANI calculation tools in the private EDGAR project "EDGAR_Aeromonas" (Fig. 4). Previous studies have suggested that the majority of the mislabeled genomes were originally designated as *A. hydrophila* and the use of ANI analysis is recommended as the correct taxonomic affiliation [15, 21]. A misidentified *A. hydrophila* 4AK4 genome was reconfirmed in our analysis with an ANI value less than 86% observed between "*A. hydrophila* 4AK4" and other *A. hydrophila* species. The ANI between "*A. hydrophila* 4AK4" and *A. media* WS/*A. rivipollensis* KN_Mc_1 1 N1 was higher (93%), which agrees with the observation of Beaz-Hidalgo and colleagues (2015) [15].

In addition, our pan-genome analysis of 34 *Aeromonas* species resulted in total of 10,736 genes including 1573 core genes (14.6%), 4170 singleton genes (38.8%), and 4993 dispensable genes (46.5%), indicating the interstrain variation of *Aeromonas* genus. These relatively high numbers of dispensable and singleton genes could explain the impact of environmental exposure to the *Aeromonas* species [21–23]. These data are very valuable for further estimation of varying patterns and introduction of genes, which can be helpful in designing epidemiological strategies and in understanding the changing behavior of inter- and intraspecies in *Aeromonas* genus. Conspicuously, the polymorphism interpretation of dispensable genes showed that *A. hydrophila* 4AK4 clearly possesses a gene profile shared by strains *A. rivipollensis* KN_Mc_1 1 N1, *A. caviae* FDAARGOS_72, *A. caviae* 8LM, and *A. media* WS (Fig. 5). These strains were isolated from many diverse ecological environments (see the NCBI description): wild nutria (*Myocastor coypus*), South Korea in 2016 (strain KN_Mc_1 1 N1), the diarrheal stool sample of a human large intestine in the United States in 2013 (strain FDAARGOS_72), an infant male in Brazil in 2010 (strain 8LM), and water samples from East Lake, China (strain WS). The data indicate that the strains KN_Mc_1 1 N1 and 8LM have potentially zoonotic characteristics. Therefore, the calculated interstrain relationships could be considered for future analyses, especially when focusing on factors emerging potential zoonotic pathogens. For instance, T3SS has a history of virulence in humans [90]. In the United States, the importation of fish or fishery products has the potential to cause severe epidemic outbreaks in farmed catfish, the source of highly virulent *A. hydrophila*. Human activities could cause this dissemination of bacterial pathogens worldwide to either fish or humans [89]. Further studies should focus on the zoonotic invasion of *Aeromonas* strains other than *A. hydrophila*. Pan-genomic approaches are useful and powerful tools to provide more evidence of taxonomic relationships among the strains, ARG from different

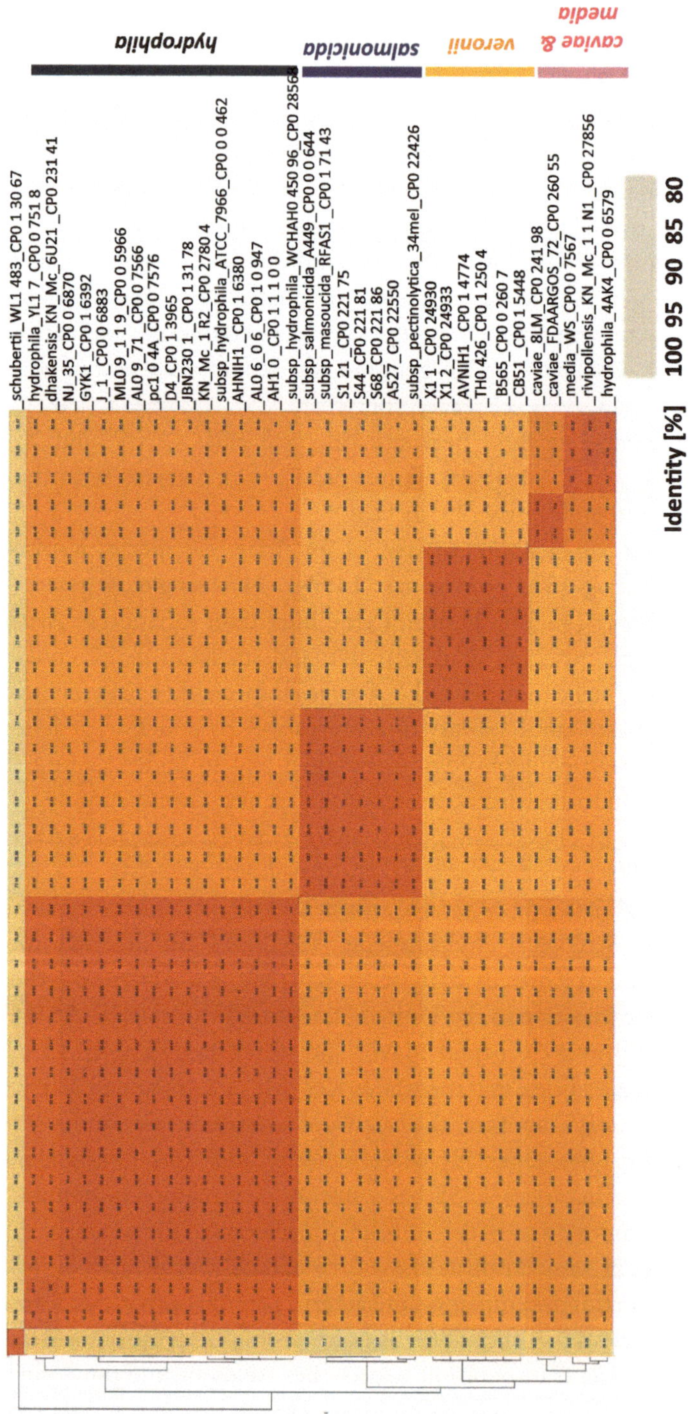

Fig. 4 Heatmap chart representing *Aeromonas* ANI inter- and intraspecies boundaries in 34 valid strains with complete genome sequences.

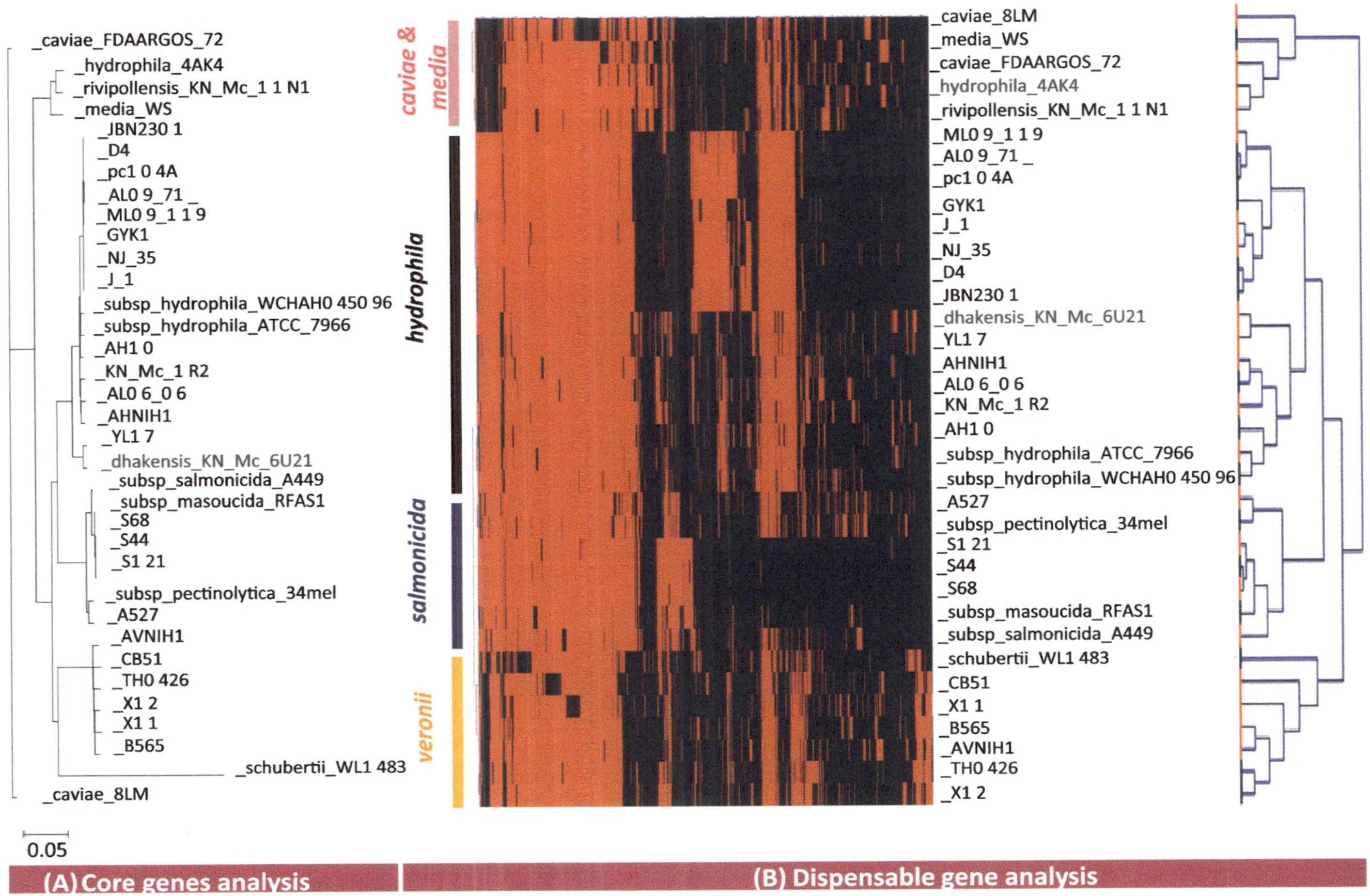

Fig. 5 Polymorphism of core and dispensable genes among *Aeromonas* strains. (A) Phylogenetic tree of 1573 core gene. (B) Map of polymorphic genes that are either present or absent among the strains. The presence/absence of 4993 genes is shown in *red/black*, respectively.

antibiotic classes, as well as the number of virulence factors for pathogenic potential of *Aeromonas* species.

3 Conclusions and the avenues of pan-genome for analyzing aquatic pathogens

The number of aquatic bacterial genome sequences deposited in the genome database of GenBank at the NCBI is exponentially growing. This data provides a huge potential for the integrated examination of etiology and epidemiology of diseases and host-pathogen interactions. Of course, pan-genome analysis is an effective tool which could possibly be extended to analysis of aquatic microorganisms and dynamic characteristics and adaptation to a broad range of their hosts and environmental niches. Integrated analyses using tools such as ANI, SNP calling, synteny block, pan-genome analysis, and phylogeny analysis are applied for the taxonomy of bacterial strains isolated from different aquatic ecosystems. For instance, *E. tarda* isolated from diseased fish is divided into freshwater group and marine/migratory group [24]. On the other hand, lateral gene transfer and homologous recombination events can be detected in phylogenomic network analysis. In terms of virulence genes, the functional annotations from VFDB, COG, and KEGG, and ARG databases can be reanalyzed based on the pan-genome categories. The description of common processes governing pathogenicity of fish pathogenic strain is of crucial importance because the virulence and pathogenicity of aquatic bacteria pathogens can be multifactorial, varies between species and strains (e.g., in the case of *Vibrio* and *Aeromonas* species [21, 38, 91]). These virulence and pathogenicity link to subcellular localization of cell proteins including chemical composition of outer-membrane proteins and capsules, surface polysaccharides, flagella, toxins, and secretion systems [91, 92]. Downstream analysis may therefore be indicative of whether an isolate is an emerging potential zoonotic pathogen or not. Furthermore, an advanced and precise interpretation of pan-genome data would not only provide deep insights into the comparison of antimicrobial resistance and virulence genes among bacterial strains but also provide further understanding of mutualistic interactions and/or host-microbe interactions. This interpretation enables the development of novel control methods against fish disease, such as antimicrobial and preventive measures, as well as a sustainable future for aquaculture [93].

In fact, a large portion of genome data has been proposed to encode a protein with unknown functions, yet in vivo function, termed as hypothetical proteins (HPs), for example, 40% of *Leptospira interrogans* proteins [92] and 60% of *Paracoccidioides lutzii* proteins [94]. The probable virulence factor proteins of HPs were predicted successfully by integrating a variety of protein classification systems, motif discovery tools as well as methods that are based on characteristic features obtained from the protein sequence. The predicted function of HPs is of diverse protein classes such as enzymes, transporters, binding proteins, regulatory proteins, and proteins involved in cellular processes or with miscellaneous functions [95]. The HPs of pathogens maybe identified as cytoplasmic and inner membrane proteins as well as surface-exposed proteins, including outer-membrane proteins, and extracellular proteins, enabling the way for drug target estimation, potential therapeutic targets and the prediction of suitable antigens as potential vaccine candidates against disease [92, 95].

Vaccines are typically formalin-killed whole cell products that provide inadequate protection against most serovars. They cannot provide cross-protection against a large number of serogroups of aquatic pathogens. For instance, serotype I (62.1%) and serotype II (36.6%) are determined in *Streptococcus parauberis* [96] while *Streptococcus iniae* was properly identified by matrix-assisted laser desorption ionization-time-of-flight mass spectrometry (MALDI-TOF MS). Isolates of that serotype were divided into cluster I (51.7%), cluster II (20.2%), and cluster III (28.1%) [97]. Since these different serotypes possess specific antigenic characteristics, a long-term and cross-protective vaccine sequence against fish pathogens urgently need to be developed.

Reverse vaccinology (RV), a revolutionary vaccine research strategy, focuses on surface-exposed proteins, including cytoplasmic, inner membrane proteins, and proteins located in the other sites of the cell. According to an RV theory, a total of 350 candidate antigens selected from the entire genome sequence of the virulent strain MC58 are applied to examine serogroup B meningococcal vaccine candidates and surface-exposed proteins conserved in sequences across a range of *Meningococcus* strains. These strains are identifiable in the sera of immunized mice (as described in Ref. [92]). Subsequently, pan-genome strategies identify potential cross-protective antigens in given genomes of the group B *Streptococcus* spp. [98]. Recently, Zeng and colleagues (2017) [92] identified 118 new candidate antigens relating to outer membrane proteins and lipoproteins through the implementation of a pan-genome analysis to screen surface-exposed proteins from 17 global *L. interrogans* strains, covering 11 epidemic serovars and 17 multilocus sequence types, enabling a future vaccine development against leptospirosis. In fact, multivalent vaccines using formalin-killed bacterins were developed for aquaculture [96], but do not include comparative pan-genome approaches. In agreement with a previous study [92], novel negative-screening strategy combined with pan-genome analysis can be further used as a standard RV method to identify numerous aquatic pathogens. Generally, RV implementing a pan-genome approach to identify candidate antigens will be a novel targeted approach toward the improvement in the cross-serotype efficacy of vaccines in farmed fish.

References

[1] S.C. Bayliss, D.W. Verner-Jeffreys, K.L. Bartie, D.M. Aanensen, S.K. Sheppard, A. Adams, E.J. Feil, The promise of whole genome pathogen sequencing for the molecular epidemiology of emerging aquaculture pathogens. Front. Microbiol. 8 (2017) 121, https://doi.org/10.3389/fmicb.2017.00121.

[2] M. Marcos-López, P. Gale, B.C. Oidtmann, E.J. Peeler, Assessing the impact of climate change on disease emergence in freshwater fish in the United Kingdom. Transbound. Emerg. Dis. 57 (2010) 293–304, https://doi.org/10.1111/j.1865-1682.2010.01150.x.

[3] P.K.M. Wijegoonawardane, N. Sittidilokratna, N. Petchampai, J.A. Cowley, N. Gudkovs, P.J. Walker, Homologous genetic recombination in the yellow head complex of nidoviruses infecting *Penaeusmonodon* shrimp. Virology 390 (2009) 79–88, https://doi.org/10.1016/j.virol.2009.04.015.

[4] A. McNally, Y. Oren, D. Kelly, B. Pascoe, S. Dunn, T. Sreecharan, et al., Combined analysis of variation in core accessory and regulatory genome regions provides a super-resolution view into the evolution of bacterial populations. PLoS Genet. 12 (2016) https://doi.org/10.1371/journal.pgen.1006280.

[5] S. Baker, W.P. Hanage, K.E. Holt, Navigating the future of bacterial molecular epidemiology. Curr. Opin. Microbiol. 13 (2010) 640–645, https://doi.org/10.1016/j.mib.2010.08.002.

[6] S.R. Harris, E.J. Feil, M.T.G. Holden, M.A. Quail, E.K. Nickerson, N. Chantratita, et al., Evolution of MRSA during hospital transmission and intercontinental spread. Science 327 (2010) 469–474, https://doi.org/10.1126/science.1182395.

[7] J. Ronholm, N. Nasheri, N. Petronella, F. Pagotto, Navigating microbiological food safety in the era of whole-genome sequencing. Clin. Microbiol. Rev. 29 (2016) 837–857, https://doi.org/10.1128/CMR.00056-16.

[8] D. Falush, Toward the use of genomics to study microevolutionary change in bacteria. PLoS Genet. 5 (2009). https://doi.org/10.1371/journal.pgen.1000627.

[9] T. Azarian, R.S. Daum, L.A. Petty, J.L. Steinbeck, Z. Yin, D. Nolan, et al., Intrahost evolution of methicillin-resistant *Staphylococcus aureus* USA300 among individuals with reoccurring skin and soft-tissue infections. J. Infect. Dis. 214 (2016) 895–905, https://doi.org/10.1093/infdis/jiw242.

[10] L. Senn, O. Clerc, G. Zanetti, P. Basset, G. Prod'hom, N.C. Gordon, et al., The stealthy superbug: the role of asymptomatic enteric carriage in maintaining a long-term hospital outbreak of ST228 methicillin-resistant *Staphylococcus aureus*. MBio 7 (2016). https://doi.org/10.1128/mBio.02039-15. e02039-15.

[11] A.E. Mather, B. Lawson, E. de Pinna, P. Wigley, J. Parkhill, N.R. Thomson, et al., Genomic analysis of *Salmonella enterica*serovar Typhimurium from wild passerines in England and Wales. Appl. Environ. Microbiol. 82 (2016) 6728–6735, https://doi.org/10.1128/AEM.01660-16.

[12] P.L. Kamath, J.T. Foster, K.P. Drees, G. Luikart, C. Quance, N.J. Anderson, et al., Genomics reveals historic and contemporary transmission dynamics of a bacterial disease among wildlife and livestock. Nat. Commun. 7 (2016). https://doi.org/10.1038/ncomms11448.

[13] D.M. Aanensen, E.J. Feil, M.T.G. Holden, J. Dordel, C.A. Yeats, A. Fedosejev, et al., Whole-genome sequencing for routine pathogen surveillance in public health: a population snapshot of invasive *Staphylococcus aureus* in Europe. MBio 7 (2016). https://doi.org/10.1128/mBio.00444-16. e00444-16.

[14] S.M. Colston, M.S. Fullmer, L. Beka, B. Lamy, J.P. Gogarten, J. Graf, Bioinformatic genome comparisons for taxonomic and phylogenetic assignments using Aeromonas as a test case. MBio 5 (2014). https://doi.org/10.1128/mBio.02136-14.

[15] R. Beaz-Hidalgo, M.J. Hossain, M.R. Liles, M.J. Figueras, Strategies to avoid wrongly labelled genomes using as example the detected wrong taxonomic affiliation for *Aeromonas* genomes in the genbank database. PLoS One 10 (2015). https://doi.org/10.1371/journal.pone.0115813.

[16] M.E. Reith, R.K. Singh, B. Curtis, J.M. Boyd, A. Bouevitch, J. Kimball, et al., The genome of *Aeromonassalmonicida* subsp. *salmonicida* A449: insights into the evolution of a fish pathogen. BMC Genomics 9 (2008) 427, https://doi.org/10.1186/1471-2164-9-427.

[17] A.T. Vincent, K.H. Tanaka, M.V. Trudel, M. Frenette, N. Derome, S.J. Charette, Draft genome sequences of two *Aeromonassalmonicida*subsp. *salmonicida*isolates harboring plasmids conferring antibiotic resistance. FEMS Microbiol. Lett. 362 (2015) 1–4, https://doi.org/10.1093/femsle/fnv002.

[18] A.T. Vincent, M.V. Trudel, L. Freschi, V. Nagar, C. Gagné-Thivierge, R.C. Levesque, et al., Increasing genomic diversity and evidence of constrained lifestyle evolution due to insertion sequences in *Aeromonassalmonicida*. BMC Genomics 17 (2016) 44, https://doi.org/10.1186/s12864-016-2381-3.

[19] J.E. Han, J.H. Kim, S.P. Shin, J.W. Jun, J.Y. Chai, S.C. Park, Draft genome sequence of *Aeromonas salmonicida* subsp. *achromogenes* AS03, an atypical strain isolated from Crucian Carp (*Carassius carassius*) in the Republic of Korea. Genome Announc. 1 (2013). https://doi.org/10.1128/genomeA.00791-13. e00791-13.

[20] H.J. Roh, B-S. Kim, A. Kim, N.E. Kim, Y. Lee Y, W.K. Chun, T.D. Ho, D.H. Kim. Whole genome analysis of multi-drug-resistant Aeromonas veronii isolated from diseased discus (*Symphysodon discus*) imported to Korea, J. Fish Dis. (2018) 1–7. https://doi.org/10.1111/jfd.12908

[21] F. Awan, Y. Dong, J. Liu, N. Wang, M.H. Mushtaq, C. Lu, Y. Liu, Comparative genome analysis provides deep insights into *Aeromonashydrophila* taxonomy and virulence-related factors. BMC Genomics 19 (1) (2018) 712, https://doi.org/10.1186/s12864-018-5100-4.

[22] J. Gauthier, A.T. Vincent, S.J. Charette, N. Derome, Strong genomic and phenotypic heterogeneity in the *Aeromonassobria* species complex. Front. Microbiol. 8 (2017) 2434, https://doi.org/10.3389/fmicb.2017.02434. eCollection 2017.

[23] S. Ghatak, J. Blom, S. Das, R. Sanjukta, K. Puro, M. Mawlong, I. Shakuntala, A. Sen, A. Goesmann, A. Kumar, S.V. Ngachan, Pan-genome analysis of *Aeromonashydrophila*, *Aeromonasveronii* and *Aeromonascaviae* indicates phylogenomic diversity and greater pathogenic potential for *Aeromonashydrophila*, Antonie Van Leeuwenhoek 109 (7) (2016) 945–956.

[24] J. Shao, Q. Guo, R. Hu, Z. Gu, Comparative genomic insights into the taxonomy of *Edwardsiellatarda* isolated from different hosts: marine, freshwater and migratory fish. Aquac. Res. 49 (2018) 197–204, https://doi.org/10.1111/are.13448.

[25] L.A. Gonçalves, S. de Castro Soares, F.L. Pereira, F.A. Dorella, A.F. de Carvalho, G.M. de Freitas Almeida, et al., Complete genome sequences of *Francisellanoatunensis* subsp. *orientalis* strains FNO12, FNO24 and FNO190: a fish pathogen with genomic clonal behavior. Stand. Genomic Sci. 11 (2016) 30, https://doi.org/10.1186/s40793-016-0151-0.

[26] A.K. Wu, A.M. Kropinski, J.S. Lumsden, B. Dixon, J.I. MacInnes, Complete genome sequence of the fish pathogen *Flavobacterium psychrophilum* ATCC 49418(T.). Stand. Genomic Sci. 10 (2015) 3, https://doi.org/10.1186/1944-3277-10-3.

[27] D. Castillo, R.H. Christiansen, I. Dalsgaard, L. Madsen, R. Espejo, M. Middelboe, Comparative genome analysis provides insights into the pathogenicity of *Flavobacterium psychrophilum*. PLoS One 11 (2016). https://doi.org/10.1371/journal.pone.0152515.

[28] G. Ricci, C. Ferrario, F. Borgo, A. Rollando, M.G. Fortina, Genome sequences of *Lactococcus garvieae* TB25, isolated from Italian cheese, and *Lactococcus garvieae* LG9, isolated from Italian rainbow trout. J. Bacteriol. 194 (2012) 1249–1250, https://doi.org/10.1128/JB.06655-11.

[29] H. Morita, H. Toh, K. Oshima, M. Yoshizaki, M. Kawanishi, K. Nakaya, et al., Complete genome sequence and comparative analysis of the fish pathogen *Lactococcus garvieae*. PLoS One 6 (2011) https://doi.org/10.1371/journal.pone.0023184.

[30] R. Pulgar, D. Travisany, A. Zuñiga, A. Maass, V. Cambiazo, Complete genome sequence of *Piscirickettsia salmonis* LF-89 (ATCC VR-1361) a major pathogen of farmed salmonid fish. J. Biotechnol. 212 (2015) 30–31, https://doi.org/10.1016/j.jbiotec.2015.07.017.

[31] O. Brynildsrud, E.J. Feil, J. Bohlin, S. Castillo-Ramirez, D. Colquhoun, U. McCarthy, et al., Microevolution of *Renibacterium salmoninarum*: evidence for intercontinental dissemination associated with fish movements. ISME J. 8 (2014) 746–756, https://doi.org/10.1038/ismej.2013.186.

[32] G.D. Wiens, D.D. Rockey, Z. Wu, J. Chang, R. Levy, S. Crane, et al., Genome sequence of the fish pathogen *Renibacterium salmoninarum* suggests reductive evolution away from an environmental *Arthrobacter* ancestor. J. Bacteriol. 190 (2008) 6970–6982, https://doi.org/10.1128/JB.00721-08.

[33] P. Pereira Ude, A. Rodrigues Dos Santos, S.S. Hassan, F.F. Aburjaile, C. Soares Sde, R.T. Ramos, et al., Complete genome sequence of *Streptococcus agalactiae* strain SA20-06, a fish pathogen associated to meningoencephalitis outbreaks. Stand. Genomic Sci. 8 (2013) 188–197, https://doi.org/10.4056/sigs.3687314.

[34] G. Liu, W. Zhang, C. Lu, Comparative genomics analysis of *Streptococcus agalactiae* reveals that isolates from cultured tilapia in China are closely related to the human strain A909. BMC Genomics 14 (2013) 775, https://doi.org/10.1186/1471-2164-14-775.

[35] P. Kayansamruaj, N. Pirarat, H. Kondo, I. Hirono, C. Rodkhum, Genomic comparison between pathogenic *Streptococcus agalactiae* isolated from Nile tilapia in Thailand and fish-derived ST7 strains. Infect. Genet. Evol. 36 (2015) 307–314, https://doi.org/10.1016/j.meegid.2015.10.009.

[36] F. El Aamri, F. Acosta, F. Real, D. Padilla, Whole-genome sequence of the fish virulent strain *Streptococcus iniae* IUSA-1, isolated from gilthead sea bream (*Sparusaurata*) and Red Porgy (*Pagruspagrus*). Genome Announc. 1 (2013) https://doi.org/10.1128/genomeA.00025-13.

[37] H. Naka, G.M. Dias, C.C. Thompson, C. Dubay, F.L. Thompson, J.H. Crosa, Complete genome sequence of the marine fish pathogen *Vibrio anguillarum* harboring the pJM1 virulence plasmid and genomic comparison with other virulent strains of *V. anguillarum* and *V. ordalii*. Infect. Immun. 79 (2011) 2889–2900, https://doi.org/10.1128/IAI.05138-11.

[38] P. Busschaert, I. Frans, S. Crauwels, B. Zhu, K. Willems, P. Bossier, C. Michiels, K. Verstrepen, B. Lievens, H. Rediers, Comparative genome sequencing to assess the genetic diversity and virulence attributes of 15 *Vibrio anguillarum* isolates. J. Fish Dis. 38 (2015) 795–807, https://doi.org/10.1111/jfd.12290.

[39] H. Kondo, P.T. Van, L.T. Dang, I. Hirono, Draft genome sequence of non-*Vibrio parahaemolyticus* acute hepatopancreatic necrosis disease strain KC13.17.5, isolated from diseased shrimp in Vietnam. Genome Announc. 3 (2015). https://doi.org/10.1128/genomeA.00978-15. e00978-15.

[40] V. Letchumanan, H.-L. Ser, K.-G. Chan, B.-H. Goh, L.-H. Lee, Genome sequence of *Vibrio parahaemolyticus* VP103 strain isolated from shrimp in Malaysia. Front. Microbiol. 7 (2016) 1496, https://doi.org/10.3389/fmicb.2016.01496.

[41] T. Liu, K.Y. Wang, J. Wang, D.F. Chen, X.L. Huang, P. Ouyang, et al., Genome sequence of the fish pathogen *Yersinia ruckeri* SC09 provides insights into niche adaptation and pathogenic mechanism. Int. J. Mol. Sci. 17 (2016) 557, https://doi.org/10.3390/ijms17040557.

[42] A.C. Barnes, J. Delamare-Deboutteville, N. Gudkovs, C. Brosnahan, R. Morrison, J. Carson, Whole genome analysis of *Yersinia ruckeri* isolated over 27 years in Australia and New Zealand reveals geographical endemism over multiple lineages and recent evolution under host selection. Microb. Genom. 2 (2016). https://doi.org/10.1099/mgen.0.000095.

[43] D. Goudenège, M.A. Travers, A. Lemire, B. Petton, P. Haffner, Y. Labreuche, et al., A single regulatory gene is sufficient to alter *Vibrio aestuarianus* pathogenicity in oysters. Environ. Microbiol. 17 (2015) 4189–4199, https://doi.org/10.1111/1462-2920.12699.

[44] K.O. Holm, C. Bækkedal, J.J. Söderberg, P. Haugen, Complete genome sequences of seven *Vibrio anguillarum* strains as derived from PacBio sequencing. Genome Biol. Evol. 10 (4) (2018) 1127–1131, https://doi.org/10.1093/gbe/evy074.

[45] C.R. Rasmussen-Ivey, M.J. Hossain, S.E. Odom, J.S. Terhune, W.G. Hemstreet, C.A. Shoemaker, et al., Classification of a hypervirulent *Aeromonas hydrophila* pathotype responsible for epidemic outbreaks in warm-water fishes. Front. Microbiol. 7 (2016). https://doi.org/10.3389/fmicb.2016.01615.

[46] C. Karlsen, E. Hjerde, T. Klemetsen, N.P. Willassen, Pan-genome and CRISPR analyses of the bacterial fish pathogen *Moritellaviscosa*. BMC Genomics 18 (1) (2017) 313, https://doi.org/10.1186/s12864-017-3693-7.

[47] B. Austin, D.A. Austin, Bacterial Fish Pathogens—Disease of Farmed and Wild Fish, fifth ed., Springer, Dordrecht, 2012.

[48] H. Hasman, D. Saputra, T. Sicheritz-Ponten, O. Lund, C.A. Svendsen, N. Frimodt-Møller, F.M. Aarestrup, Rapid whole-genome sequencing for detection and characterization of microorganisms directly from clinical samples, J. Clin. Microbiol. 52 (2014) 139–146.

[49] J.W. Pridgeon, P.H. Klesius, Major bacterial diseases in aquaculture and their vaccine development, Anim. Sci. Rev. 141 (2013).

[50] S. Carding, K. Verbeke, D.T. Vipond, B.M. Corfe, L.J. Owen, Dysbiosis of the gut microbiota in disease. Microb. Ecol. Health Dis. 26 (2015) https://doi.org/10.3402/mehd.v26.26191.

[51] T. Defoirdt, P. Sorgeloos, P. Bossier, Alternatives to antibiotics for the control of bacterial disease in aquaculture, Curr. Opin. Microbiol. 14 (2011) 251–258.

[52] Q. Wang, M. Yang, J. Xiao, H. Wu, X. Wang, Y. Lv, et al., Genome sequence of the versatile fish pathogen *Edwardsiella tarda* provides insights into its adaptation to broad host ranges and intracellular niches, PLoS One 4 (2009).

[53] V. Chaudhry, B.P. Prabhu, Genomic investigation reveals evolution and lifestyle adaptation of endophytic *Staphylococcus epidermidis*. Sci. Rep. 6 (2016) https://doi.org/10.1038/srep19263.
[54] A.F. Auch, M. von Jan, H.P. Klenk, M. Göker, Digital DNA-DNA hybridization for microbial species delineation by means of genome-to-genome sequence comparison. Stand. Genomic Sci. 2 (2010) 117–134, https://doi.org/10.4056/sigs.531120.
[55] K.T. Konstantinidis, J.M. Tiedje, Genomic insights that advance the species definition for prokaryotes, Proc. Natl. Acad. Sci. U.S.A. 102 (2005) 2567–2572.
[56] J. Blom, J. Kreis, S. Spänig, T. Juhre, C. Bertelli, C. Ernst, et al., EDGAR 2.0: an enhanced software platform for comparative gene content analyses. Nucleic Acids Res. 44 (2016) W22–W28, https://doi.org/10.1093/nar/gkw255.
[57] D.W. Eyre, T. Golubchik, N.C. Gordon, R. Bowden, P. Piazza, E.M. Batty, C.L. Ip, D.J. Wilson, X. Didelot, L. O'Connor, et al., A pilot study of rapid benchtop sequencing of *Staphylococcus aureus* and *Clostridium difficile* for outbreak detection and surveillance, BMJ Open 2 (2012).
[58] A. Kim, T.L. Nguyen, D.H. Kim, Complete genome sequence of the virulent *Aeromonas salmonicida* subsp. *masoucida* strain RFAS1. Genome Announc. 6 (2018). https://doi.org/10.1128/genomeA.00470-18. e00470-18.
[59] C.U. Köser, M.T.G. Holden, M.J. Ellington, et al., Rapid whole-genome sequencing for investigation of a neonatal MRSA outbreak, N. Engl. J. Med. 366 (2012) 2267–2275.
[60] L. Snipen, T. Almøy, D.W. Ussery, Microbial comparative pan-genomics using binomial mixture models. BMC Genomics 10 (2009) 385, https://doi.org/10.1186/1471-2164-10-385.
[61] A.V. Chaplin, B.A. Efimov, V.V. Smeianov, L.I. Kafarskaia, A.P. Pikina, A.N. Shkoporov, Intraspecies genomic diversity and long-term persistence of *Bifidobacterium longum*. PLoS One 10 (2015). https://doi.org/10.1371/journal.pone.0135658.
[62] H. Tettelin, V. Masignani, M.J. Cieslewicz, C. Donati, D. Medini, et al., Genome analysis of multiple pathogenic isolates of *Streptococcus agalactiae*: implications for the microbial "pan-genome" Proc. Natl. Acad. Sci. U. S. A. 102 (2005) 13950–13955.
[63] M.E. Torok, S.J. Peacock, Rapid whole-genome sequencing of bacterial pathogens in the clinical microbiology laboratory—pipe dream or reality? J. Antimicrob. Chemother. 67 (2012) 2307–2308.
[64] J. Xiao, Z. Zhang, J. Wu, J. Yu, A brief review of software tools for pangenomics. Genomics Proteomics Bioinformatics 13 (2015) 73–76, https://doi.org/10.1016/j.gpb.2015.01.007.
[65] T. Zekic, G. Holley, J. Stoye, Pan-genome storage and analysis techniques. Methods Mol. Biol. 1704 (2018) 29–53, https://doi.org/10.1007/978-1-4939-7463-4_2.
[66] Y. Zhao, C. Sun, D. Zhao, Y. Zhang, Y. You, X. Jia, et al., PGAP-X: extension on pan-genome analysis pipeline. BMC Genomics 19 (Suppl 1) (2018) 36, https://doi.org/10.1186/s12864-017-4337-7.
[67] T.H. Clarke, L.M. Brinkac, J.M. Inman, G. Sutton, D.E. Fouts, PanACEA: a bioinformatics tool for the exploration and visualization of bacterial pan-chromosomes, BMC Bioinform. 19 (2018) 246.
[68] M.C.F. Thomsen, J. Ahrenfeldt, J.L.B. Cisneros, V. Jurtz, M.V. Larsen, H. Hasman, et al., A bacterial analysis platform: an integrated system for analysing bacterial whole genome sequencing data for clinical diagnostics and surveillance. PLoS One 11 (2016). https://doi.org/10.1371/journal.pone.0157718.
[69] T.L. Nguyen, D.H. Kim, Genome-wide comparison reveals a probiotic strain *Lactococcus lactis* WFLU12 isolated from the gastrointestinal tract of olive flounder (*Paralichthys olivaceus*) harboring genes supporting probiotic action. Mar. Drugs 16 (2018). https://doi.org/10.3390/md16050140.
[70] T.L. Nguyen, W.-K. Chun, A. Kim, N. Kim, H.J. Roh, Y. Lee, M. Yi, S. Kim, C.-I. Park, D.-H. Kim, Dietary probiotic effect of *Lactococcus lactis* WFLU12 on low-molecular-weight metabolites and growth of olive flounder (*Paralichythys olivaceus*). Front. Microbiol. 9 (2018) 2059, https://doi.org/10.3389/fmicb.2018.02059.
[71] Y. Nakamura, T. Takano, M. Yasuike, T. Sakai, T. Matsuyama, M. Sano, Comparative genomics reveals that a fish pathogenic bacterium *Edwardsiella tarda* has acquired the locus of enterocyte effacement (LEE) through horizontal gene transfer. BMC Genomics 14 (2013) 642, https://doi.org/10.1186/1471-2164-14-642.
[72] C.S. Smillie, M.B. Smith, J. Friedman, O.X. Cordero, L.A. David, E.J. Alm, Ecology drives a global network of gene exchange connecting the human microbiome, Nature 480 (2011) 241–244.
[73] R.I. Aminov, Horizontal gene exchange in environmental microbiota, Front. Microbiol. 2 (2011) 158.

[74] C. Wandersman, I. Stojiljkovic, Bacterial heme sources: the role of heme, hemoprotein receptors and hemophores, Curr. Opin. Microbiol. 3 (2000) 215–220.

[75] M. Balado, M.A. Lages, J.C. Fuentes-Monteverde, D. Martínez-Matamoros, J. Rodríguez, C. Jiménez, M.L. Lemos, The siderophore piscibactin is a relevant virulence factor for *Vibrio anguillarum* favored at low temperatures. Front. Microbiol. 9 (2018) 1766, https://doi.org/10.3389/fmicb.2018.01766.

[76] I. Frans, C. Michiels, P. Bossier, K.A. Willems, B. Lievens, H. Rediers, *Vibrio anguillarum* as a fish pathogen: virulence factors, diagnosis and prevention, J. Fish Dis. 34 (2011) 643–661.

[77] D.A. Rowe-Magnus, A.M. Guerout, L. Biskri, P. Bouige, D. Mazel, Comparative analysis of superintegrons: engineering extensive genetic diversity in the Vibrionaceae, Genome Res. 13 (2003) 428–442.

[78] I. Frans, K. Dierckens, S. Crauwels, A. Van Assche, J. Leisner, M.H. Larsen, C.W. Michiels, K.A. Willems, B. Lievens, P. Bossier, H. Rediers, Does virulence assessment of *Vibrio anguillarum* using sea bass (*Dicentrarchus labrax*) larvae correspond with genotypic and phenotypic characterization? PLoS One 8 (2013).

[79] M.J. Griffin, T.E. Greenway, D.J. Wise, *Edwardsiella* spp, in: P.T.K. Woo, R.C. Cipriano (Eds.), Fish Viruses and Bacteria: Pathobiology and Protection, CAB International, Boston, 2017, pp. 190–210.

[80] S. Shafiei, S. Viljamaa-Dirks, K. Sundell, S. Heinikainen, T. Abayneh, T. Wiklund, Recovery of *Edwardsiella piscicida* from farmed white-fish, *Coregonus lavaretus* (L.), in Finland, Aquaculture 454 (2016) 19–26.

[81] N. Buján, H. Mohammed, S. Balboa, J.L. Romalde, A.E. Toranzo, C.R. Arias, B. Magariños, Genetic studies to re-affiliate *Edwardsiella tarda* fish isolates to *Edwardsiella piscicida* and *Edwardsiella anguillarum* species. Syst. Appl. Microbiol. 41 (2018) 30–37, https://doi.org/10.1016/j.syapm.2017.09.004.

[82] S.B. Fogelson, B.D. Petty, S.R. Reichley, C. Ware, P.R. Bowser, M.J. Crim, R.G. Getchell, K.L. Sams, H. Marquis, M.J. Griffin, Histologic and molecular characterization of *Edwardsiella piscicida* infection in large-mouth bass (*Micropterus salmoides*), J. Vet. Diagn. Investig. 28 (2016) 338–344.

[83] S. Shao, Q. Lai, Q. Liu, H. Wu, J. Xiao, Z. Shao, Q. Wang, Y. Zhang, Phylogenomics characterization of a highly virulent Edwardsiella strain ET080813(T) encoding two distinct T3SS and three T6SS gene clusters: propose a novel species as *Edwardsiella anguillarum* sp. nov, Syst. Appl. Microbiol. 38 (2015) 36–47.

[84] N. Castro, A.E. Toranzo, A. Bastardo, J.L. Barja, B. Magariños, Intraspecific genetic variability of *Edwardsiella tarda* strains from cultured turbot, Dis. Aquat. Org. 95 (2011) 253–258.

[85] T. Abayneh, D.J. Colquhoun, H. Sørum, Multi-locus sequence analysis (MLSA) of *Edwardsiella tarda* isolates from fish, Vet. Microbiol. 158 (2012) 367–375.

[86] J.Y. Yang, S. Brooks, J.A. Meyer, R.R. Blakesley, A.M. Zelazny, J.A. Segre, E.S. Snitkin, Pan-PCR, a computational method for designing bacterium-typing assays based on whole-genome sequence data, J. Clin. Microbiol. 51 (2013) 752–758.

[87] FAO (2017). Major Bacterial Diseases Affecting Aquaculture. Available at: http://www.fao.org/fi/static-media/MeetingDocuments/WorkshopAMR/presentations/07_Haenen.pdf.

[88] Y. Yano, K. Hamano, I. Tsutsui, D. Aue-Umneoy, M. Ban, M. Satomi, Occurrence, molecular characterization, and antimicrobial susceptibility of *Aeromonas* spp. in marine species of shrimps cultured at inland low salinity ponds. Food Microbiol. 47 (2015) 21–27, https://doi.org/10.1016/j.fm.2014.11.003.

[89] I.H. Igbinosa, E.U. Igumbor, F. Aghdasi, M. Tom, A.I. Okoh, Emerging *Aeromonas* species infections and their significance in public health. Sci. World J. 2012 (2012). https://doi.org/10.1100/2012/625023.

[90] B. Coburn, I. Sekirov, B.B. Finlay, Type III secretion systems and disease, Clin. Microbiol. Rev. 20 (4) (2007) 535–549.

[91] J.M. Tomás, The main *Aeromonas* pathogenic factors, ISRN Microbiol. 2012 (2012) 256261.

[92] L. Zeng, D. Wang, N. Hu, Q. Zhu, K. Chen, K. Dong, Y. Zhang, Y. Yao, X. Guo, Y.F. Chang, Y. Zhu, A novel pan-genome reverse vaccinology approach employing a negative-selection strategy for screening surface-exposed antigens against leptospirosis, Front. Microbiol. 8 (2017) 396.

[93] A. Kim, T.L. Nguyen, D.H. Kim, Modern methods of diagnosis, in: B. Austin, A. Newaj-Fyzul (Eds.), Diagnosis and Control of Diseases of Fish and Shellfish, John Wiley & Sons Ltd, Hoboken, 2017, pp. 109–145.
[94] C.A. Desjardins, M.D. Champion, J.W. Holder, A. Muszewska, et al., Comparative genomic analysis of human fungal pathogens causing paracoccidioidomycosis, PLoS Genet. 7 (2011).
[95] A.A. Naqvi, F. Anjum, F.I. Khan, A. Islam, F. Ahmad, M.I. Hassan, Sequence analysis of hypothetical proteins from *Helicobacter pylori* 26695 to identify potential virulence factors, Genomics Inform. 14 (3) (2016) 125–135.
[96] S.B. Park, S.W. Nho, H.B. Jang, I.S. Cha, M.S. Kim, W.J. Lee, T.S. Jung, Development of three-valent vaccine against streptococcal infections in live flounder, *Paralichthys olivaceus*, Aquaculture 461 (2016) 25–31.
[97] S.W. Kim, S.W. Nho, S.P. Im, J.S. Lee, J.W. Jung, J.M. Lazarte, et al., Rapid MALDI biotyper-based identification and cluster analysis of *Streptococcus iniae*. J. Microbiol. 55 (2017) 260–266, https://doi.org/10.1007/s12275-017-6472-x.
[98] D. Maione, I. Margarit, C.D. Rinaudo, V. Masignani, M. Mora, M. Scarselli, et al., Identification of a universal Group B streptococcus vaccine by multiple genome screen. Science 309 (2005) 148–150, https://doi.org/10.1126/science.1109869.

CHAPTER 9

Pan-genomics of model bacteria and their outcomes

Kanwal Naz, Nimat Ullah, Tahreem Zaheer, Muhammad Shehroz, Anam Naz, Amjad Ali
Department of Plant Biotechnology, Atta-ur-Rahman School of Applied Biosciences (ASAB), National University of Sciences and Technology (NUST), Islamabad, Pakistan

1 Introduction

The development and improvements in next-generation sequencing technologies have greatly contributed toward whole genome sequences and public databases [1].

After the first completely sequenced genome of *Haemophilus influenza* in 1995, more than 85,000 genomes including 53,000 bacterial genomes have been sequenced and are available at NCBI [2]. Another statistics obtained from GOLD (Genomes Online Database) in 2018 showed that 120,617 projects are only from the bacterial domain which is more than 50% of projects from other domains including archaea, viruses, and eukaryotes.

This increasing interest in bacterial genome sequencing projects revolutionized the study of human pathogens. Comparative genomics studies performed on this overwhelming number of microbial genomic data has developed our understanding of inter- and intraspecies diversity [3, 4]. It has been observed that several bacterial strains often acquire new genes from a large genetic reservoir [5, 6]. This rapid accumulation of genes in the gene pool of a species revealed that a single reference genome is not enough to fully understand species-level diversity.

Thus, to fully describe the genomic diversity in a bacterial species Tettelin et al. introduced the concept of pan-genome in 2005, which comprises complete gene repertoire among all the strains of a species [7]. Pan-genome is further divided into three categories, they are core genome, dispensable genome, and unique genome. Core genome consists of genes that are common among all the strains, dispensable genome consists of genes that are common only in few strains while unique genome contains strain-specific genes. Indeed, after the pioneering work of Tettelin, several other pan-genome of bacterial species were performed to describe the genetic diversity and pathogenicity of the strains [8].

Pan-genomics: Applications, Challenges, and Future Prospects
https://doi.org/10.1016/B978-0-12-817076-2.00009-3
© 2020 Elsevier Inc.
All rights reserved.

2 Technical approaches and their outcomes

Pan-genome projects have followed different strategies/parameters including number of genomes analyzed, the phylogenetic resolution (super kingdom, phylum, class, genus, and species level), the alignment search algorithm (FASTA and/or BLAST) and parameters associated with alignment algorithms (percent identity and percent aligned sequence length), threshold to define the similarity of orthology (paralogs, orthologs, and xenologs), the mathematical models (for the estimation of new genes in genomes under study), and the sequence annotation type and quality (i.e., CDSs, ORFs, genes).

In several pan-genome studies, different alignment threshold values are used for analysis. For example, Tettelin employed 50/50 rule to identify conserved genes/proteins in different genomes of a species with a minimum of 50% sequence identity over 50% of the genes/proteins length [7], while Hiller et al. [9] applied a comparatively strict threshold of 70% sequence identity/sequence length. Meric et al. [10] used a 70% identity threshold over 50% of the gene/protein sequence lengths. However, Rasko et al. [11] adopted a more strict similarity threshold of more than 80% sequence identity, whereas Bentley et al. [12] used 30% similarity threshold over 80% of gene/protein sequence lengths.

Annotation method is another parameter that needs proper consideration as the orthology characterization is dependent on the type and quality of annotation at the pre-implementation level [13]. However, when the phylogenetic resolution parameter is taken into account, broad-range taxonomic classification (phylum or kingdom level) through simple sequence similarity may produce ambiguous orthology. Thus, high-resolution algorithms like phyletic profile approach (presence and absence of genes) or PSI-BLAST (position-specific iterative basic local alignment searching tool) may be implemented to accurately classify true orthologs [14].

A sufficient number of genomes are required to fully describe the pan-genome of a species/genus to know whether it is still open or close. However, the selection of criterion with their respective thresholds can also significantly impact orthologous clustering, the core, and accessory genome size and nature of the pan-genome [15]. The estimation of pan-genome infers that either the species has an open pan-genome, when the number of genes in pan-genome increases with the addition of further genomes or a closed pan-genome, when the additional sequenced genomes do not add new genes into the existing pan-genome. It has been observed that species colonizing multiple environments can easily exchange genetic material tend to have an open pan-genome, for example, *Escherichia coli, Meningococci, Streptococci, Salmonellae, Helicobacter pylori*, etc. On the other hand, species which live in an isolated habitat with less possibility to exchange genetic material usually have closed pan-genome, for example, *Mycobacterium tuberculosis, Bacillus anthracis*, and *Chlamydia trachomatis* [8]. Hence, pan-genome analysis serve as a framework to determine and understand genomic diversity. In this chapter, we have discussed bacterial pan-genome research performed to date by using examples of some model organisms, employed technical implementations, and their outcomes (Table 1).

Table 1 History of pan-genome analysis of model organisms, technical implementations employed, and their outcomes

Organism	Technical implementations employed	No. of genome	Pan-genome	Core genome	Year/Ref.
S. agalactiae	50/50 rule	8	2667	1806	2005 [7]
	50/50 rule	15	4730	1202	2013 [16]
N. meningitidis	50/50 rule	6	3290	1337	2008 [17]
	50/50 rule	20	–	1630	2011 [18]
S. aureus	OrthoMCL algorithm	17	3155	2266	2011 [19]
	Homologous gene clustering	32	6471	2115	2018 [20]
	pairwise best bidirectional	64	7457	1441	2016 [21]
E. coli	50/50 rule and alignment	7	3470	2865	2006 [22]
	50/50 rule and alignment	32	9433	2241	2007 [23]
	BSR	17	13,000	2200	2008 [11]
	linkage clustering method	20	17,838	1976	2009 [24]
	Binomial mixture models	22	42,640	2446	2009 [25]
	BLAST matrix, BLAST Atlas	53	13,296	1472	2010 [26]
S. pyogenes	OrthoMCL algorithm	29	14,986	1957	2011 [27]
	Homologous gene clustering	186	16,373	3051	2012 [28]
	OrthoMCL	11	2500	1400	2007 [29]
	PGAP Multi-paranoid	11	2743	1366	2011 [30]
	PGAP Gene-family	11	2889	1332	2011 [30]
	BPGA USEARCH	28	2790	913	2016 [31]
	BPGA CD-HIT	28	2743	914	2016 [31]
	BPGA OrthoMCL	28	2762	855	2016 [31]
	Pan X	50	2856	970	2017 [32]
H. influenzae	Single linkage algorithm	13	2786	1461	2007 [33]
	Reverse best-hit algorithm	97	2852	935	2014 [34]
	Roray v.3.8.0	88	3424	1308	2019 [35]
S. pneumoniae	Orthologous gene clustering	17	2870	1454	2007 [9]
	Alignment	44	3221	1666	2010 [36]
	COGnitor & COGtriangles	616	5442	1194	2013 [37]
	PanX	33	3361	1188	2017 [32]

3 Pan-genomics of model bacteria

3.1 *Streptococcus agalactiae*

Streptococcus agalactiae (*S. agalactiae*) is a facultative anaerobe and major cause of illness and deaths among infants [38]. In 25% healthy women, the bacteria are a part of normal vaginal flora [39]. Moreover, they also cause septicemia, mastitis, and urogenital tract infection in cats and dogs [40] and is also a well-known fish pathogen, which represents a zoonotic hazard and compromises food safety and security [16].

S. agalactiae is the first organism studied for its pan-genome by Tettelin in 2005. In this study only eight genomes were analyzed, using 50/50 rule (as discussed above). A total of 2667 genes were identified as pan-genome, among which 67.7% of pan-genome (1806 genes) were identified as core genome. Later this criterion was adopted by various other studies. For genome comparison, all-against-all alignment search was applied where each genome is compared with all other genomes. The results of the study revealed that only eight strains are not sufficient to fully describe the pan-genome of this species. The regression analysis of the study interpreted that *S. agalactiae* pan-genome is open because whenever a new strain is sequenced, new genes are added to the gene pool of the species [7].

In 2013, another study was conducted on 15 genomes of *S. agalactiae* to understand the evolutionary relationships, genetic basis associated with the host, and to predict the virulence determinants. The same strategy of 50/50 rule and the all-against-all BLASTp search was used for pan-genome identification. The results showed that pan-genome comprises 4730 genes which include 1202 core genes, 1388 dispensable genes, and 2040 unique genes [16].

S. agalactiae is also studied under genus (*Streptococcus*) level in 2007, where 26 genomes of genus *Streptococcus* were analyzed belonging to six different species, including *S. agalactiae, Streptococcus pneumoniae, Streptococcus mutans, Streptococcus pyogenes, Streptococcus thermophilus*, and *Streptococcus suis* [29]. The results show that *S. agalactiae* exhibited little recombination in its core genome and has a large pan-genome.

By December 2018, 103 completely sequenced genomes of *Streptococcus agalactiae* were freely available at NCBI. Pan-genome analysis on such a large number of genomic data would increase our understanding of the diversity and variability within this species.

3.2 *Neisseria meningitidis*

Neisseria meningitidis (*N. meningitidis*) is a diplococcal, Gram-negative, and human commensal bacterium of the upper respiratory tract. The pathogen can invade the mucosa and gain access to the bloodstream, resulting meningitis, severe sepsis, or localized infections in joints and heart [41, 42]. Invasive meningococcal disease (IMD) can rapidly progress in healthy young adults and adolescents, and the global mortality rate is around 10%, even though effective vaccines and antibiotics are available [43, 44]. The isolates of *Neisseria* species have been sequenced extensively from the isolates MC58 (serogroup B) and

Z2491 (serogroup A) since 2000 [45, 46]. Whole-genome sequence (WGS) of *N. meningitidis* strain MC58 opened new avenues for researchers in both basic and applied research, for instance, provided the starting point for the identification of vaccine candidates in pathogen genome and developed serogroup B vaccine using reverse vaccinology [47, 48]. WGS data from *N. meningitidis* species that outran other *Neisseria* species with 91 complete genomes is available at NCBI, currently.

The first pan-genome of *N. meningitidis* was determined by Schoen et al. to understand the pathogenicity and evolution of virulence traits in sic strains (sequenced till that time) based on a 50/50 rule [7, 17]. It was estimated that the number of genes in *N. meningitidis* pan-genome is about 3290, whereas their core genome contains at least 1337 genes. The number of new genes contributed to the *N. meningitidis* pan-genome with the addition of each new genome was predicted to be at least 43 [17]. To study the population structure of *N. meningitidis*, genome comparisons of 20 strains were performed and pan-genome was estimated based on the 50/50 rule of Tettelin et al. with little modifications. It was reported that the pan-genome is growing at a very slow rate and approximately 1630 genes were present in the meningococcal core genome and each meningococcal genome is composed of approximately 79% core, 21% dispensable, and <0.1% unique genes [18].

This indicates that *N. meningitidis* pan-genome is still open and new genes continue to be added to the meningococcal pan-genome any time a new strain is sequenced. Therefore, more genomes might be needed to accurately calculate the total size of *N. meningitidis* pan-genome that capture the complete gene repertoire of meningococcal species.

3.3 *Staphylococcus aureus*

Staphylococcus aureus (*S. aureus*) is a Gram-positive opportunistic pathogen, which can be a part of normal microflora commonly colonizing skin surfaces. This organism causes a wide variety of infections in human ranging from mild skin infections to severe bacteremia [49]. The infections caused by *S. aureus* now seem to be very difficult to treat due to the acquired antimicrobial resistance in these strains, such as the acquisition of *mecA* gene, which confers methicillin resistance and designated as methicillin-resistant *Staphylococcus aureus* (MRSA). MRSA has become a significant burden on health-care systems [50]. However, the high-resolution accessible through WGS has the power to improve our understanding of the infection and management of MRSA. The first *S. aureus* genome published in 2001 provided crucial information on genome architecture and gene content. Since then, hundreds of *S. aureus* whole genomes from different sources have been sequenced [51].

This exponential increase in WGSs allowed the scientific community to estimate pan-genome for determining genetic diversity within species and to predict how many additional genomes would be required to describe the complete gene repertoire. Therefore, first pan-genome of 17 *S. aureus* genomes was performed using the percentage of shared distributed genes. This method yielded orthologous clusters based on 70% sequence identity over 70% of the shorter sequence. The clustering resulted in 3155 orthologous genes, of which 134 were unique genes, 755 were distributed genes, and 2266 were core genes [19]. In another study, 32 strains of *S. aureus* pan-genome was computed based on whole-genome alignment method as described by Herbig et al. [52]. The pan-genome was found to consist of 6471 genes, of which 2115 genes were shared by 31 strains and only 1 of the analyzed strains harbored 2032 genes [20]. Bosi et al., analyzed pan-genome for 64 *S. aureus* strains using the genome module of the Ductape suite based on pairwise best bidirectional hit approach. It yielded a pan-genome size of 7457 genes, of which the unique genome was composed of 3145 genes, the accessory genome was composed of 2871 genes, and core genome was composed of 1441 genes [21].

The above literature revealed that *S. aureus* pan-genome is still open and its genome is naturally competent for the uptake of new genes from different sources and is able to acquire a variety of DNA sequences through horizontal gene transfer (HGT). Currently, there are 352 complete genomes of *S. aureus* publicly available at NCBI. Pan-genome analysis on such a large number of genomic data will provide insights into *S. aureus* evolution and designing a universal vaccine.

3.4 *Escherichia coli*

E. coli was first isolated in 1885 from the feces of a healthy individual while the first complete genomic sequence of K-12 strain was published in September 1997 [53, 54]. *E. coli* mainly resides in the mammalian colon and is one of the best-studied organisms in biotechnology and molecular biology due to easy manipulation of its genome. Most of the strains of *E. coli* are non-pathogenic in nature, but the virulent strains may cause mild to severe diseases like hemorrhagic colitis, Crohn diseases, urinary tract infections (UTI), gastroenteritis, and neonatal meningitis [55]. Commensal *E. coli* has diverse genomic content, due to environmental and host factors that shaped the genetic structures of *E. coli*. The acquired virulent genes are involved in pathogenicity, and are also responsible for diverse diseases. The pan-genome analysis is a great way to investigate genes acquired or lost during the evolution of a species or strain [56].

The first pan-genome of *E. coli* was estimated using seven genomes (available at that time) to explore the pathogenesis of species and strain-specific adaptation. It was a comparative genomics approach in which 50/50 alignment rule was applied in order to predict the core genome and genes involved in positive selection. A total of 3470 orthologous clusters in pan-genome and 2470 were identified in *E. coli* in this study [22]. Meanwhile in another study, high-density microarray technique was used to calculate pan-genome of 32 *E. coli* genomes using 50/50 rule. The identified pan-genome consisted of 9433 genes including 2241 core genes. The study also aided in the exploration of phylogenetic association of unknown or newly sequenced *E. coli* strains through best-fitted extrapolation model and as a result 1563 core genes were estimated for an infinite number of genomes [23].

In 2008, pan-genome of more than 300 laboratory-adapted isolates of 17 *E. coli* strains was calculated by using BSR (BLAST score ratio) analysis. Pan-genome identified in this study consisted of 13,000 genes and the core genome consisted of only 2200 gene, however, each new genome added approximately 300 genes in pan-genome. They concluded that the newly added or uncharacterized genes in pan-genome may have a role in virulence and pathogenicity [11].

In 2009, pan-genome was estimated to comprehend the degree of variation within 20 *E. coli* genomes. A total of 17,838 genes were found in pan-genome using a single linkage clustering algorithm while the core genome consisted of only 1976 genes. The estimated core genome was only 20% of the pan-genome which represents diverse genomic contents of *E. coli.* [24]. Meanwhile, in the same year, another study employing a mixture model or regression method on 22 *E. coli* genomes estimated a pan-genome of 42,640 genes while only 2446 genes in the core genome [25].

In 2010, genomic diversity of *E. coli* was explored through single-linkage clustering algorithm using 53 *E. coli* genomes. The identified pan genome consisted of 13,296 genes having only 1472 core genes while 90% genes accounted for accessory genome illustrating that *E. coli* have a very high propensity of acquiring new genes [26].

Vieira et al. (2011), analyzed the diversification of *E. coli* genome at a metabolic level using 21 pathogenic and 8 commensal strains. The pan-genome has calculated a total of 14,986 genes containing 1957 core genes using the orthoMCL algorithm [27].

A more comprehensive study was conducted on 186 *E. coli* strains using homologous gene clustering method. The pan-genome identified 16,373 clusters, while core gene clusters were identified by using two criteria. The soft criteria predicted 3051 clusters, however, when the criterion was made stringent, the core gene clusters were reduced to 1702. Although the pan-genome remained the same, a significant variation in a number of shared genes was noted. This study also concluded that the pan-genome of *E. coli* is still open and the newly sequenced genome contribute to acquired genes in the pan gene pool [28].

Currently, according to NCBI genome statistics, 13,725 genome assemblies have been annotated and 667 complete genomes are currently available. The variation in complete genome, if exploited, can give a better understanding of this model bacteria.

3.5 *Streptococcus pyogenes*

Streptococcus pyogenes (*S. pyogenes*) is an anaerobic Gram-positive coccus, belonging to one of the most diverse genera. A number of pathological conditions are reported to be caused by *S. pyogenes*, like post-streptococcal glomerulonephritis, cellulitis, endocarditis, meningitis, and septic joint inflammation, making it one of the top 10 deadly pathogenic species worldwide. The availability of next-generation sequencing data and pan-genome studies have helped the researcher understand the genome statistics and diversity [57].

In 2007, the first pan-genome study of *S. pyogenes* was conducted in order to determine the acquisition of genes and recombination events at the genus level. The pan-genome of 11 strains of *S. pyogenes* obtained using alignment and MCL clustering algorithm was 2500 genes, while the core genes were 1400. This study revealed that recombinant genes determined by pairwise homoplasy index in core genome of *S. pyogenes* and Streptococcus genus were about 20% [29].

In 2011, 11 strains were used by Chen et al. to validate their newly developed Linux-based pipeline, PGAP (pan-genomes analysis pipeline). The two inbuilt methods of PGAP, that is, multi-paranoid (MP) and gene-family (GF) were employed individually at default parameters. The pan-genome identified using MP method was 2743 while core genome was found to be 1366, however, size of pan-genome estimated by using GF method was 2889 with a core genome of 1332 genes. The results of MP methods were relatively close to the study conducted in 2007 [30].

In 2016, pan-genome analyses of 28 genomic sequences were analyzed using BPGA (bacterial pan genome analysis tool) to estimate the size of *S. pyogenes*. Three algorithms OrthoMCL, CD-HIT, and USEARCH integrated into BPGA with 50% threshold were used and the size of pan-genome was calculated as 2762, 2743, and 2790 while the core genome was calculated as 855, 914, and 913, respectively [31].

In another study, pan-genome of 50 genomes of *S. pyogenes* was determined using newly designed pipeline PanX. The estimated pan-genome of *S. pyogenes* consisted of 2856 genes while only 970 genes were present in the core genome [32].

As per genome statistics of NCBI, a total of 458 genome assemblies of *S. pyogenes* have been annotated so far, including 119 complete genomes and 339 partial genomes. Since the pan-genome of *S. pyogenes* is still open, this extensive number of complete genomes can help in understanding the genomic diversity.

3.6 *Haemophilus influenzae*

In 1995, Craig Venter and his team published a full genome sequence of *H. influenzae* (*H. influenzae*); it was the first free-living organism that got its genome sequenced. *H. influenzae* is an important human pathogen implicated in a variety of invasive diseases. In 1990, a conjugate vaccine was proposed against *H. influenzae* b-serotype, but it led to the emergence of non-typeable *H. influenzae* (NTHi) [2, 58]. According to genome statistics of *H. influenzae* in NCBI, 700 annotation and genome assembly reports have been documented so far, among which 59 comprise complete genomes while rest of them concern partial genomes or contigs. The strains of *H. influenzae* exhibited enormous genetic diversity due to transformation-mediated homologous recombination where strains can acquire genes from pan-genome, transduction (bacteriophage-induced DNA delivery), and conjugation. The analysis of pan-genome can be helpful in understanding genomic diversity, drug resistance, and characterize genomic basis of disease severity [8, 59].

In order to understand the genomic diversity of *H. influenzae* strains, the first pan-genome analysis was conducted on 13 strains of NTHi. Gene clustering using single linkage algorithm (70% threshold) identified 2786 genes out of which 1461 were part of core genome. The study also predicted that supra-genome of NTHi contains around 4425–6052 genes [33]. After 7 years, pan-genome analysis of 97 NTHi strains was conducted in which 6 clades were identified through a reverse best-hit algorithm and orthology-calling approach. The study revealed a pan-genome of 2852 genes of which 935 genes were core genes [34]. Recently, 88 novel NTHi strains were isolated from Portugal to determine the genomic diversity. Pan-genome analysis of these novel Portugal isolated strains was carried out using Roary v.3.8.0 with 70% cutoff value and without splitting paralogues. A total of 3424 genes comprised pan-genome among which 1308 genes were the part of the core genome. The study concluded that the pan-genome of *H. influenzae* is still open and expanding every day due to the acquisition of genes and accessory genome can reach up to 6000 genes [35]. Pan-genome analysis of *H. influenzae* can help researchers and clinicians to understand its invasiveness and pathogenesis and by using the core genome drug/vaccine against all *H. influenza*e could be predicted.

3.7 *Streptococcus pneumoniae*

Streptococcus pneumoniae (S. pneumoniae) is a Gram-negative lactic acid bacterium that is reported to cause more than 1.1 million deaths annually, even though appropriate antibiotics and vaccine are available. *S. pneumoniae* has a diverse genome and is further classified into 85 antigenic strains on the basis of capsular polysaccharide [60]. The pathogen has 8418 genome assemblies in NCBI while only 53 complete genomes are available to date.

In 2007, the genomic diversity of 17 strains of *S. pneumoniae* was analyzed through pan-genome analysis. Pan-genome was estimated via a clustering algorithm (70% threshold) and was found to be 3170, among them 1454 core genes comprising 46% of pan-genome were identified [9].

Since an updated analysis was required, accordingly in 2010 pan-genome approach was used for the comparison of pathogenic and nonpathogenic strains to track the virulence genes. Pan-genome consistingof 3221 genes of which core genome consisted of 1666 genes was reported. The results validated the existing experimental data that *S. pneumoniae* acquire genes for their survival in the extreme environment. The study also highlighted that the core genome contains genes associated with virulence and are required for the survival of the pathogen. Hence, the core genome should be exploited for drug and vaccine development [36]. In another study, the pan-genome analysis was conducted on 616 genomes of *S. pneumoniae* to understand the recombination events and clonal expansion after therapeutic intervention. Algorithms generated by COGnitor and COGtriangles identified 5442 genes in pan-genome and 1194 genes in the core genome [37]. One such study estimated only core genome in which 1206 core genes were found in 616 annotated sequences of *S. pneumoniae* using Bayesian Decision Model [61]. Recently, PanX was used to calculate pan-genome. PanX calculated pan-genome size that was found to contain 3361 genes while the core genome consisted of 1188 genes through its own clustering strategy in 33 genomes of *S. pneumoniae* [32].

All the studies reported till date concluded that pan-genome of *S. pneumoniae* is likely to increase by the addition of newly sequenced genomes and is still open. The species still manages to maintain its genomic stability; however, new genes are also acquired for survival in the host.

4 Conclusion

The availability of thousands of genomic sequence from each species is a primary resource to better understand the diversity of bacterial species. The pan-genome studies are considered indispensable for bacterial genome comparisons. Pan-genome categories (core, accessory, and strain-specific) may provide better insight into the genome of a species, not only its evolution and diversity but also about genetic factors involved in pathogenicity and genes associated with antimicrobial resistant and also those that can be targeted for therapeutic interventions. For example, core genome encodes for basic cellular functions, accessory genome involves functions that are related to certain niches such as colonization, virulence, and antibiotic resistance. Ever since the first pan-genome study was conducted, a frequent rise in pan-genome analysis has been observed in many bacterial species. Advancement in genome sequencing technologies enabled the continuous expansion of genomic data in public repositories. Therefore, pan-genome of several species has been studied through time and with a record of increasing total number of available genomes of that species. At the same time, improvements in pan-genome analysis protocols have been seen employing different strategies leading to a numerous diverse outcome. It has also been observed from pan-genome analyses of model bacteria that several species including human pathogens exhibit an open pan-genome and a number of genomes would be required to define the complete gene repertoire of those species.

References

[1] D. Medini, M. Stella, J. Wassil, MATS: global coverage estimates for 4CMenB, a novel multicomponent meningococcal B vaccine, Vaccine 33 (23) (2015) 2629–2636.

[2] R.D. Fleischmann, et al., Whole-genome random sequencing and assembly of Haemophilus influenzae Rd, Science 269 (5223) (1995) 496–512.

[3] M.J. Pallen, B.W. Wren, Bacterial pathogenomics, Nature 449 (7164) (2007) 835.
[4] D.E. Fouts, et al., What makes a bacterial species pathogenic?: comparative genomic analysis of the genus Leptospira, PLoS Negl. Trop. Dis. 10 (2) (2016).
[5] J.O. McInerney, A. McNally, M.J. O'Connell, Why prokaryotes have pangenomes, Nat. Microbiol. 2 (4) (2017) 17040.
[6] N.A. Andreani, E. Hesse, M. Vos, Prokaryote genome fluidity is dependent on effective population size, ISME J. 11 (7) (2017) 1719.
[7] H. Tettelin, et al., Genome analysis of multiple pathogenic isolates of Streptococcus agalactiae: implications for the microbial "pan-genome", Proc. Natl. Acad. Sci. U. S. A. 102 (39) (2005) 13950–13955.
[8] D. Medini, et al., The microbial pan-genome, Curr. Opin. Genet. Dev. 15 (6) (2005) 589–594.
[9] N.L. Hiller, et al., Comparative genomic analyses of seventeen Streptococcus pneumoniae strains: insights into the pneumococcal supragenome, J. Bacteriol. 189 (22) (2007) 8186–8195.
[10] G. Méric, et al., A reference pan-genome approach to comparative bacterial genomics: identification of novel epidemiological markers in pathogenic Campylobacter, PLoS One 9 (3) (2014).
[11] D.A. Rasko, et al., The pangenome structure of Escherichia coli: comparative genomic analysis of E. coli commensal and pathogenic isolates, J. Bacteriol. 190 (20) (2008) 6881–6893.
[12] S.D. Bentley, et al., Meningococcal genetic variation mechanisms viewed through comparative analysis of serogroup C strain FAM18, PLoS Genet. 3 (2) (2007).
[13] S. Bentley, Sequencing the Species Pan-Genome, Nature Publishing Group, 2009.
[14] P. Lapierre, J.P. Gogarten, Estimating the size of the bacterial pan-genome, Trends Genet. 25 (3) (2009) 107–110.
[15] G. Vernikos, et al., Ten years of pan-genome analyses, Curr. Opin. Microbiol. 23 (2015) 148–154.
[16] G. Assis, et al., Natural coinfection by Streptococcus agalactiae and Francisella noatunensissubsp. orientalis in farmed Nile tilapia (Oreochromis niloticus L.), J. Fish Dis. 40 (1) (2017) 51–63.
[17] C. Schoen, et al., Whole-genome comparison of disease and carriage strains provides insights into virulence evolution in Neisseria meningitidis, Proc. Natl. Acad. Sci. 105 (9) (2008) 3473–3478.
[18] S. Budroni, et al., Neisseria meningitidis is structured in clades associated with restriction modification systems that modulate homologous recombination, Proc. Natl. Acad. Sci. (2011). https://doi.org/10.1073/pnas.1019751108.
[19] R. Boissy, et al., Comparative supragenomic analyses among the pathogens Staphylococcus aureus, Streptococcus pneumoniae, and Haemophilus influenzae using a modification of the finite supragenome model, BMC Genomics 12 (1) (2011) 187.
[20] S. Fuchs, et al., AureoWiki—the repository of the Staphylococcus aureus research and annotation community, Int. J. Med. Microbiol. 308 (6) (2018) 558–568.
[21] E. Bosi, et al., Comparative genome-scale modelling of Staphylococcus aureus strains identifies strain-specific metabolic capabilities linked to pathogenicity, Proc. Natl. Acad. Sci. U. S. A. 113 (26) (2016) E3801–E3809.
[22] S.L. Chen, et al., Identification of genes subject to positive selection in uropathogenic strains of Escherichia coli: a comparative genomics approach, Proc. Natl. Acad. Sci. 103 (15) (2006) 5977–5982.
[23] H. Willenbrock, et al., Characterization of probiotic Escherichia coli isolates with a novel pan-genome microarray, Genome Biol. 8 (12) (2007) R267.
[24] M. Touchon, et al., Organised genome dynamics in the Escherichia coli species results in highly diverse adaptive paths, PLoS Genet. 5 (1) (2009) e1000344.
[25] L. Snipen, T. Almøy, D.W. Ussery, Microbial comparative pan-genomics using binomial mixture models, BMC Genomics 10 (1) (2009) 385.
[26] O. Lukjancenko, T.M. Wassenaar, D.W. Ussery, Comparison of 61 sequenced Escherichia coli genomes, Microb. Ecol. 60 (4) (2010) 708–720.
[27] G. Vieira, et al., The core and pan-metabolism in the Escherichia coli species, J. Bacteriol. 193 (2011) 1461–1472.
[28] R.S. Kaas, et al., Estimating variation within the genes and inferring the phylogeny of 186 sequenced diverse Escherichia coli genomes, BMC Genomics 13 (1) (2012) 577.
[29] T. Lefébure, M.J. Stanhope, Evolution of the core and pan-genome of Streptococcus: positive selection, recombination, and genome composition, Genome Biol. 8 (5) (2007) R71.

[30] L. Chen, et al., VFDB 2012 update: toward the genetic diversity and molecular evolution of bacterial virulence factors, Nucleic Acids Res. 40 (D1) (2011) D641–D645.
[31] N.M. Chaudhari, V.K. Gupta, C. Dutta, BPGA—an ultra-fast pan-genome analysis pipeline, Sci. Rep. 6 (2016) 24373.
[32] W. Ding, F. Baumdicker, R.A. Neher, panX: pan-genome analysis and exploration, Nucleic Acids Res. 46 (1) (2017) e5.
[33] J.S. Hogg, et al., Characterization and modeling of the Haemophilus influenzae core and supragenomes based on the complete genomic sequences of Rd and 12 clinical nontypeable strains, Genome Biol. 8 (6) (2007) R103.
[34] M. De Chiara, et al., Genome sequencing of disease and carriage isolates of nontypeable Haemophilus influenzae identifies discrete population structure, Proc. Natl. Acad. Sci. (2014). https://doi.org/10.1073/pnas.1403353111.
[35] M. Pinto, et al., Insights into the population structure and pan-genome of Haemophilus influenzae, Infect. Genet. Evol. 67 (2019) 126–135.
[36] C. Donati, et al., Structure and dynamics of the pan-genome of Streptococcus pneumoniae and closely related species, Genome Biol. 11 (10) (2010) R107.
[37] N.J. Croucher, et al., Population genomics of post-vaccine changes in pneumococcal epidemiology, Nat. Genet. 45 (6) (2013) 656.
[38] A. Schuchat, J.D. Wenger, Epidemiology of group B streptococcal disease: risk factors, prevention strategies, and vaccine development, Epidemiol. Rev. 16 (2) (1994) 374–402.
[39] J.R. Campbell, et al., Group B streptococcal colonization and serotype-specific immunity in pregnant women at delivery, Obstet. Gynecol. 96 (4) (2000) 498–503.
[40] D.E. Low, Nonpneumococcal streptococcal infections, rheumatic fever, in: Goldman's Cecil Medicine, Twenty Fourth ed., Elsevier, 2012, pp. 1823–1829.
[41] L. Eriksson, et al., Whole-genome sequencing of emerging invasive Neisseria meningitidis serogroup W in Sweden, J. Clin. Microbiol. 56 (4) (2018).
[42] S.G.B. Heckenberg, et al., Clinical features, outcome, and meningococcal genotype in 258 adults with meningococcal meningitis—a prospective cohort study, Medicine 87 (4) (2008) 185–192.
[43] A.C. Cohn, L.H. Harrison, Meningococcal vaccines: current issues and future strategies, Drugs 73 (11) (2013) 1147–1155.
[44] K. Thorburn, et al., Mortality in severe meningococcal disease, Arch. Dis. Child. 85 (5) (2001) 382–385.
[45] J. Parkhill, et al., Complete DNA sequence of a serogroup A strain of Neisseria meningitidis Z2491, Nature 404 (6777) (2000) 502–506.
[46] H. Tettelin, et al., Complete genome sequence of Neisseria meningitidis serogroup B strain MC58, Science 287 (5459) (2000) 1809–1815.
[47] M. Pizza, et al., Identification of vaccine candidates against serogroup B meningococcus by whole-genome sequencing, Science 287 (5459) (2000) 1816–1820.
[48] C. Schoen, et al., Genome flexibility in Neisseria meningitidis, Vaccine 27 (2009) B103–B111.
[49] S. Lim, et al., Comparative genomic analysis of Staphylococcus aureus FORC_001 and S. aureus MRSA252 reveals the characteristics of antibiotic resistance and virulence factors for human infection, J. Microbiol. Biotechnol. 25 (1) (2015) 98–108.
[50] A.S. Lee, et al., Methicillin-resistant Staphylococcus aureus, Nat. Rev. Dis. Primers 4 (2018) 18033.
[51] J.R. Fitzgerald, M.T. Holden, Genomics of natural populations of Staphylococcus aureus, Annu. Rev. Microbiol. 70 (2016) 459–478.
[52] A. Herbig, et al., GenomeRing: alignment visualization based on SuperGenome coordinates, Bioinformatics 28 (12) (2012) i7–i15.
[53] H. Jeong, et al., Genome sequences of Escherichia coli B strains REL606 and BL21 (DE3), J. Mol. Biol. 394 (4) (2009) 644–652.
[54] F.R. Blattner, et al., The complete genome sequence of Escherichia coli K-12, Science 277 (5331) (1997) 1453–1462.
[55] F.C. Neidhardt, Escherichia coli and Salmonella typhimurium: Cellular and Molecular Biology, Vol. 1, American Society for Microbiology, 1987.

[56] O. Tenaillon, et al., The population genetics of commensal Escherichia coli, Nat. Rev. Microbiol. 8 (3) (2010) 207.
[57] J.J. Ferretti, D.L. Stevens, V.A. Fischetti, The streptococcal proteome, in: Streptococcus pyogenes: Basic Biology to Clinical Manifestations, University of Oklahoma Health Sciences Center, 2016.
[58] E.P. Price, et al., Haemophilus influenzae: using comparative genomics to accurately identify a highly recombinogenic human pathogen, BMC Genomics 16 (1) (2015) 641.
[59] E.R. Mardis, The impact of next-generation sequencing technology on genetics, Trends Genet. 24 (3) (2008) 133–141.
[60] M. Kilian, et al., Evolution of Streptococcus pneumoniae and its close commensal relatives, PLoS One 3 (7) (2008).
[61] A.J. van Tonder, et al., Defining the estimated core genome of bacterial populations using a Bayesian decision model, PLoS Comput. Biol. 10 (8) (2014).

CHAPTER 10

Pan-genomics of multidrug-resistant human pathogenic bacteria and their resistome

Mauricio Corredor, Amalia Muñoz-Gómez
GEBIOMIC Group, FCEN, University of Antioquia, Medellin, Colombia

1 Introduction

The development of bacterial genome sequencing is feasible and accessible at present, since the multiple strain genomes of the same species can be aligned in high through-put scale using DNA, RNA, and proteins. Additionally, aside from the progress of data mining, some genome databases were growing gradually in NCBI (https://www.ncbi.nlm.nih.gov/genome/microbes/) [1] (Fig. 1), EMBL (https://www.ebi.ac.uk/genomes/bacteria.html) [2], KEEG (https://www.genome.jp/kegg/genome.html) [3], PATRIC (https://www.patricbrc.org/) [4] (Fig. 1), MBGD (http://mbgd.genome.ad.jp/) [5], ENSEMBL (https://bacteria.ensembl.org/index.html) [6], JGI-IMG/M (https://img.jgi.doe.gov/) [7], and, among others (Fig. 1). These databases provide complete downloadable genomic information, which can be analyzed for intraspecies diversity to determine the species' core genome. The pan-genome analysis, which consists of characterizing the size of the gene repertoire accessible to a given species and estimating the number of whole genome sequences required for correct analysis, is increasingly used 10 years after Tettelin et al. [8] quotation. Different current pan-genomic models for the analysis and their accuracy and applicability depend on the case at hand [9].

Contrary to classical genomics, pan-genomic technique requires multiple genomes or at least two genomes of pathogenic bacteria (PB) and can provide the broadest resolution of genetic variation determinants (synteny, pathogenicity islands, virulence genes distribution, mobile elements, plasticity, evolution, etc.). Consequently, pan-genomics has promoted advances in many fields, like computing, genomics, and bioinformatics, associated with population genetics, evolution, and molecular epidemiology. One benefit of pan-genomics is that experimental data have shown for some species that new genes are being discovered even after sequencing of several strain genomes [10]. Given that, the number of unique genes is vast, the pan-genome of a bacterial species might be orders of magnitude larger than any single genome, as predicted by Medini et al. [10], more than 10 years ago.

Pan-genomics: Applications, Challenges, and Future Prospects
https://doi.org/10.1016/B978-0-12-817076-2.00010-X
© 2020 Elsevier Inc.
All rights reserved.

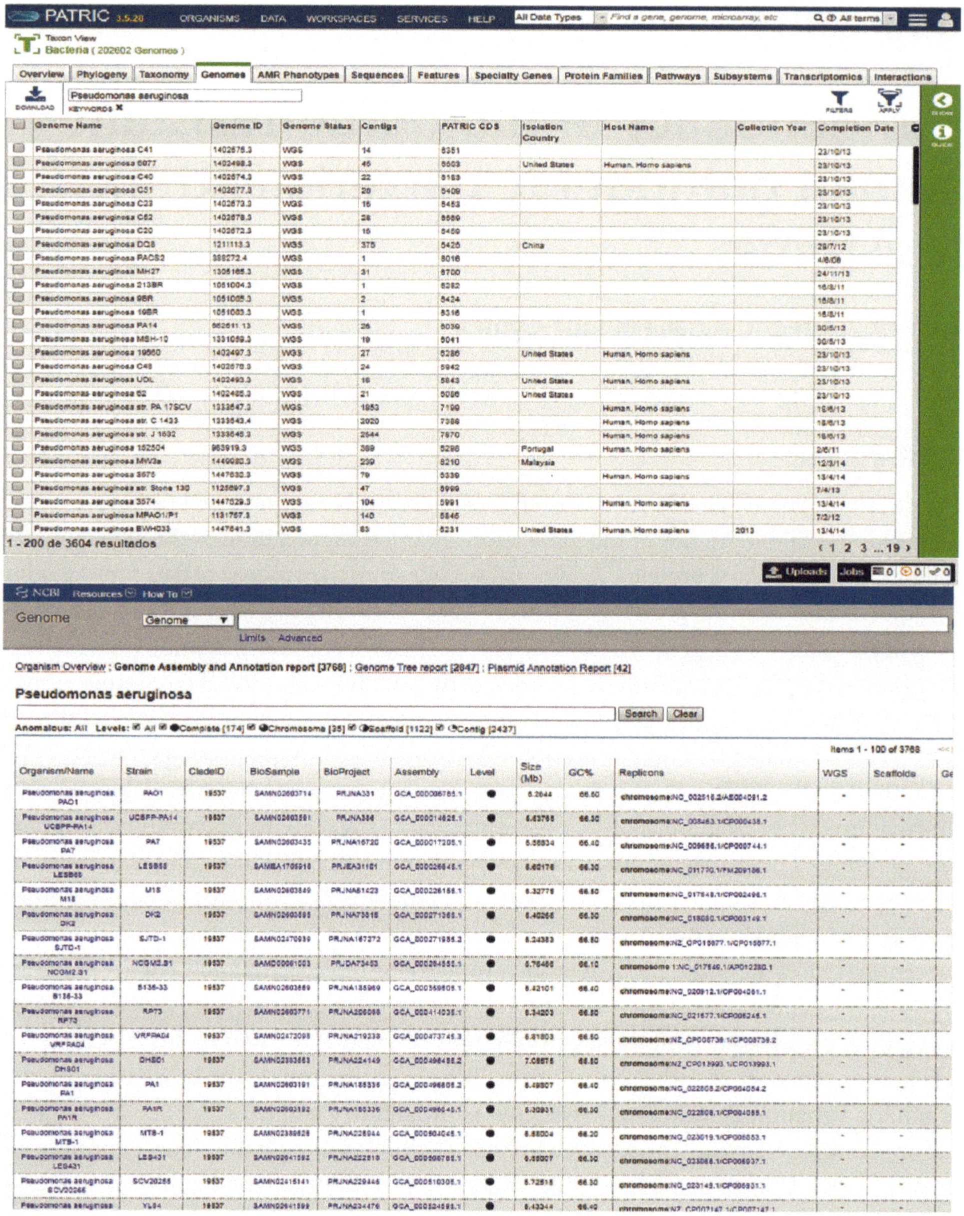

Fig. 1 Some databases, where thousands of bacterial genomes are currently available to build up pan-genomes: for instance, PATRIC (up) and NCBI (down) databases.

In pan-genome examination, the amount of the gene catalog accessible to any given species is characterized by an estimate of the number of complete genome sequences required for convenient analysis, and this technology is increasingly being employed, due to next-generation sequencing development [11]. Unfortunately, the sequence of a single genome does not reflect how genetic variability drives pathogenesis within a bacterial species, besides, it also limits genome-wide screens for vaccine candidates or for antimicrobial targets [11].

The Pan-genomics is now a cut-edge of the computational genomics field, and it is a subarea of computational biology [12]. The concept of computational pan-genomics, therefore, intentionally passes through many other bioinformatics-related disciplines.

At present, there is a great effort to understand the biological inheritance and evolution of antibiotic resistance in bacteria. An adequate approach is to perform a pan-genomic analysis to solve major questions about the resistome inside the bacterial genome [13]. Updated data will close us to the metadata available for establishing what part of resistome traits belongs to both the core genome and the accessory genome in some bacterial species, as well as a broader perspective of antibiotic resistance in bacteria. For instance, the well-known mechanism of beta-lactamase ampicillin resistance conferred by plasmid transformation (accessory-genome) in *Escherichia coli* (horizontal gene transfer), contrasts with constitutive or inducible genetic information (core genome of "vertical" gene transfer) of rRNA methyltransferases in the aminoglycoside resistance development. Undoubtedly, these two examples are related to multidrug-resistant (MDR) in the same species. However, it is important to have a previous notion to conceive the standard concepts and the pan genomic state of the art in the pathogenic antibiotic-resistant bacteria (ARB).

Core genome is the genome shared by all strains without exception and variable- or flexible- or dispensable- or accessory-genome is what genes shared among some strains inside the species, along together with unique genes (in a single strain, singletons as well). The perfect meaning description of pan-genomics was defined by Rouli et al. [14] or posted on websites such as http://www.metagenomics.wiki/pdf/definition/pan-genome [15] (Table 1), where terms as pan-metabolome and mobilome are introduced or comprised. Pan-genome and resistome terms, coined by Wright [17], were introduced almost simultaneously (2005 and 2007, respectively); moreover, currently, resistome and pan-genome are extensively being used, analyzed, and applied jointly from 2012 [13, 18–25]. Finally, the pan-resistome concept has not been quite introduced in previous studies (less in serial journals) as proposed in this review; nevertheless, it has also been proposed by a Brazilian research project. (https://bv.fapesp.br/en/auxilios/94866/pan-resistome-of-beta-lactamase-kpc-2-ctx-m-8-ctx-m-15-producing-klebsiella-pneumoniae-and-esche/, [26]). Other authors use in its project Pan-genome-resistome (http://grantome.com/grant/NIH/R43-AI129187-01, [27]) in the United States.

Table 1 Description of Pan-genome terminologies previously reported by Rouli et al. [14][a], http://www.metagenomics.wiki/pdf/definition/pan-genome[b], Paralanov et al. [16][c] or by us[d]

Accessory genome[a]	– Not unique but not in the core genome. Common to the studied strains
The variable or accessory genome (also: flexible, dispensable genome)[b]	– Refers to genes not present in all strains of a species. This includes genes present in two or more strains or even genes unique to a single strain only, for example, genes for specific strain adaptation such as antibiotic resistance
Core genome[a]	– The pool of genes common to all the studied genomes of a given species
The core genome[b]	– Represents the genes present in all strains of a species. It typically includes housekeeping genes for cell envelope or regulatory functions
Unique-genome[d]	– They are genes without homolog copy among the genomes
Singletons[c]	– Singletons are genes found only in one of the genomes
Mobilome[a]	– All mobile genetic elements of a genome
Pan-genome[a]	– The repertoire of genes for a group of genomes
Pan-genome[b]	– It is the entire gene set of all strains of a species. It includes genes present in all strains (core genome) and genes present only in some strains of a species (variable or accessory genome)
Open pan-genome[a]	– A pan-genome increasing when a new genome is added to the pan-genome
Open pan-genome[b]	– Number of genes of the pan-genome increases with the number of additionally sequenced strains
Pan-metabolome[a]	– The repertoire of metabolic reactions for a group of genomes
Pan-regulon[a]	– The group of genes co-regulated observed by transcriptomics analysis
Closed pan-genome[a]	– Finished pan-genome in which there is no change when new genomes are added
Closed pan-genome[b]	– After some sequenced strains, additional strains do not provide new genes to the species pan-genome
Resistome[d]	– Comprises all the genes and their products that contribute to resist whatever environment, substance, or some extreme grow factor
Antibiotic Resistome[a]	– The antibiotic resistome comprises all the genes and their products that contribute to antibiotic resistance
Pan-resistome[d]	– It is the entire gene set of all strains of a species that contribute to resist whatever environment, substance, or some extreme grow factor

2 The pan-genomics of human pathogens

The difficulties in the last century for characterizing PB were overcome somehow by whole-genome sequencing, since, the big-data genome information allowed new strategies for integrating the species, genera or clades in phylogenetic evolution studies of PB. However, few strains are not enough representing the bacterial species pan-genome; on the other hand, there is plenty of clinical isolates for some species such as *E. coli*, *Streptococcus pneumoniae*, *Staphylococcus aureus*, *Mycobacterium tuberculosis*, *Salmonella enterica*, *Listeria monocytogens*, *Bacillus cereus*, and *Pseudomonas aeruginosa*, which allow processing hundreds of genomes to find out the variability of each species genome. Big-data from dozen to hundreds of genomes, set up the requirement for comparisons; undoubtedly, pan-genome is and will be the straightforward and appropriate approach to solve genetic questions regarding evolution, taxonomy, biology, biotechnology, and definitely, health care, pathology, drug research, and medicine.

The pan-genome or pangenome (term currently accepted word) was introduced and coined by Tettelin et al. [11], who precisely worked with the PB-nonantibiotic resistant *S. agalactiae* (serotypes of group B streptococci, GBS). However, the term was also introduced by Sigaux [28], working with tumor cells. The first pan-genome of PB was also determined by the same authors as the first pan-genome for a particular organism. These analyses were carried out in 2005 with eight strains of *S. agalactiae* [11] and *S. saprophyticus* [29]. Table 2 summarizes a timeline for pan-genomes of the most important PB.

The most studied pan-genomes are likely for *H. pylori* [14, 30, 48–51] using 56, 6, 10, 39, 29, and 376 genomes, respectively, and *E. coli* [30, 43–46] using 19, 25, 29, 22, and 347 genomes, respectively. Other interesting species with enriched pan-genome information are: *M. tuberculosis* [14, 55–57] using 5, 5, 70, and 20 genomes; *P. aeruginosa* [59, 60, 61, 62] using 5, 181, 100, and 1311 genomes, and *S. aureus* [10, 30, 65, 66] using 2, 14, 40, and 516 genomes. *P. aeruginosa* pan-genome has perhaps the most representative samples in this type of analysis; in fact, a recent research carried out by Freschi et al. [62] compared more than 1300 genomes; however, as we said before, it was reported 376, 347, and 516 genomes for comparisons with *H. pylori*, *E. coli*, and *S. aureus* respectively, which shows the increasing interest to build pan-genomes using representative number of strains of each species. Nevertheless, there are hampers and limitations that will not ease the construction of enormous pan-genomes for some species, in terms of strains quantity. The difficulty to classify their genomes in some species will be overcome, making the gold standard use of rDNA as a taxonomic tool, an inaccurate maker nowadays. Precisely, integrating the pan-genomics tool for the species identification will provide more robust findings and correlations, provided that it determines the core genome conservation rate. Other species with complete genome data are well supported by referenced researches or depicted online. Next, some examples of online tools using some species selected from Table 2 will be introduced.

Table 2 Pan-genome in pathogenic bacteria worldwide by species, with or without pan-genome article or reported on internet database

Species	Pagenome	Authors/web	Genomes
Acinetobacter baumannii	Yes	Snipen et al. [30]/Rouli et al. [14]/http://pan-genome.de/Acinetobacter_baumannii	6/11/116
Bacillus anthracis	Yes	Zwick et al. [31]/Wang et al. [32]	45
Bacillus cereus	yes	Snipen et al. [30]/Rouli et al. [14]/http://pan-genome.de/Bacillus_cereus	8/12/48
Bordetella pertussis	No published	http://pan-genome.de/Bordetella_pertussis	357
Borrelia burgdorferi	Yes	Mongodin et al. [33]	23
Brucella melitensis	Yes	Yang et al. [34]/http://bamics2.cmbi.ru.nl/websoftware/pancgh/pancgh_start.php	42/12
Burkholderia pseudomallei	No published	http://pan-genome.de/Burkholderia_pseudomallei	83
Campylobacter jejuni	Yes	Wilson et al. [35]/Snipen et al. [30]/Rouli et al. [14]	63/5/14
Chlamydia pneumoniae	Yes	Collingro et al. [36]	6
Chlamydia trachomatis	Yes	Collingro et al. [36]/http://pan-genome.de/Chlamydia_trachomatis	4/85
Clostridium botulinum	Yes	Snipen et al. [30]/Rouli et al. [14]/Udaondo et al. [37]/Bhardwaj and Somvanshi [38]/http://pan-genome.de/Clostridium_botulinum	8/14/1/13/46
Clostridium perfringens	Yes	Trost et al. [39]/Udaondo et al. [37]/http://bamics2.cmbi.ru.nl/websoftware/pancgh/pancgh_start.php	3/1/4
Clostridium tetani	Yes	Udaondo et al. [37]/http://bamics2.cmbi.ru.nl/websoftware/pancgh/pancgh_start.php	1/4
Clostridium difficile	Yes	Rouli et al. [14] http://pan-genome.de/Clostridioides_difficile	18/50

Table 2 Pan-genome in pathogenic bacteria worldwide by species, with or without pan-genome article or reported on internet database—cont'd

Species	Pagenome	Authors/web	Genomes
Corynebacterium diphtheriae	Yes	Trost et al. [40]	13
Corynebacterium pseudotuberculosis	Yes	Soares et al. [41]	15
Enterococcus faecalis	Yes	Qin et al. [42]/http://bamics2.cmbi.ru.nl/websoftware/pancgh/pancgh_start.php	2/6
Enterococcus faecium	Yes	Qin et al. [42]	22
Escherichia coli	Yes	Rasko et al. [43]/Gordienko et al. [44]/Vieira et al. [45]/Snipen et al. [30]/Snipen and Ussery [46]	19/25/29/22-347
Francisella tularensis	Yes	Snipen et al. [30]/http://pan-genome.de/Francisella_tularensis	7/34
Haemophilus influenzae	Yes	Hogg et al. [47]/Rouli et al. [14]/http://pan-genome.de/Haemophilus_influenzae	13/9/55
Helicobacter pylori	Yes	Gressmann et al. [48]/Snipen et al. [30]/Rouli et al. [14]/Ali et al. [49]/Uchiyama et al. [50]/van Vliet [51]	56/6/10/39/29/376
Legionella pneumophila	Yes	D'Auria et al. [52]/Rouli et al. [14]	5/11
Leptospira interrogans	No published	http://bamics2.cmbi.ru.nl/websoftware/pancgh/pancgh_start.php	6
Listeria monocytogenes	Yes	Deng et al. [53]/Kuenne et al. [54]/Rouli et al. [14]	31/19/20
Mycobacterium leprae	Yes	Zakham et al. [55]	1/5
Mycobacterium tuberculosis	Yes	Zakham et al. [55]/Supply et al. [56]/Periwal et al. [57]/Rouli et al. [14]/http://pan-genome.de/Mycobacterium_tuberculosis	5/5/70/20/168
Mycoplasma pneumoniae	Yes	http://pan-genome.de/Mycoplasma_pneumoniae	48

Continued

Table 2 Pan-genome in pathogenic bacteria worldwide by species, with or without pan-genome article or reported on internet database—cont'd

Species	Pagenome	Authors/web	Genomes
Mycoplasma genitalium	Yes	Medini et al. [10]/pan-genome elaborated in this review, Fig. 2	2/5
Neisseria gonorrhoeae	Yes	Ezewudo et al. [58]/http://bamics2.cmbi.ru.nl/websoftware/pancgh/pancgh_start.php	76/3
Neisseria meningitidis	No published	http://bamics2.cmbi.ru.nl/websoftware/pancgh/pancgh_start.php	14
Pseudomonas aeruginosa	Yes	Sharma et al. [59]/Fisher et al. [60]/Mosquera-Rendón et al. [61]/Freschi et al. [62]	5/100/181/1311
Rickettsia rickettsii	Yes	Rouli et al. [14]/Wu et al. [63]/http://bamics2.cmbi.ru.nl/websoftware/pancgh/pancgh_start.php	1/8/8
Salmonella enterica	Yes	Snipen et al. [30]/Jacobsen et el. [64]/Gordienko et al. [44]/Rouli et al. [14]/http://pan-genome.de/Salmonella_enterica/http://bamics2.cmbi.ru.nl/websoftware/pancgh/pancgh_start.php	20/45/16/20/555/104
Shigella flexneri	Yes	Gordienko et al. [44]/http://bamics2.cmbi.ru.nl/websoftware/pancgh/pancgh_start.php	3/4
Staphylococcus aureus	Yes	Medini et al. [10]/Snipen et al. [30]/Gerrish et al. [65]/Jamrozy et al. [66]	2/14/40/516
Staphylococcus epidermidis	Yes	Conlan et al. [67]	35
Streptococcus agalactiae	Yes	Medini et al. [10]/Puymège et al. [68]/Rouli et al. [14]	8/303/5
Streptococcus pyogenes	Yes	Snipen et al. [30]	13

Table 2 Pan-genome in pathogenic bacteria worldwide by species, with or without pan-genome article or reported on internet database—cont'd

Species	Pagenome	Authors/web	Genomes
Streptococcus pneumoniae	Yes	Snipen et al. [30]/Donati et al. [69]/Muzzi et al. [70]/Rouli et al. [14]/ http://pan-genome.de/ Streptococcus_ pneumoniae	10/44/44/52/10
Treponema pallidum	No published	Tong et al. [71]/http:// bamics2.cmbi.ru.nl/ websoftware/pancgh/ pancgh_start.php	7/11
Ureaplasma urealyticum	Yes	Paralanov et al. [16]	14
Ureaplasma parvum	Yes	Paralanov et al. [16]	5
Vibrio cholerae	No published	Vezzulli et al. [72]/http:// pan-genome.de/Vibrio_ cholerae	2/44
Yersinia pestis	Yes	Snipen et al. [30]/Eppinger et al. [73]/Rouli et al. [14]/Yang and Cui [74]	7/9/12/42

The last column specifies the number of genomes. For example, *E. coli* pan-genomes were built and reported by Rasko et al. [43] (19 genomes), Gordienko et al. [44] (25 genomes), Vieira et al. [45] (29 genomes), Snipen et al. [30] (22 genomes), and Snipen and Ussery [46] (347 genomes).

Regarding the information revised in Table 2, some important PB species are not represented (not in table) or poorly or underrepresented, since either they have quite a few available strains or have not even been reported for a pan-genome analysis. In Table 2, we only reported underrepresented as *Chlamydia pneumoniae* with six genomes, *Clostridium perfringens* with four genomes, *C. tetani* with four genomes, *Enterococcus faecalis* with six genomes, *Leptospira interrogans* with six genomes, *M. leprae* with five genomes, *Mycoplasma genitalium* with five genomes (see Fig. 2), *Rickettsia rickettsii* with eight genomes, *Shigella flexneri* with four genomes, and *Ureaplasma parvum* with four genomes. Some of those pan-genomes can be solved using the software tools described below.

In addition to the publishing on pan-genome software tools, currently developed to perform via personal server (avoiding time-consuming for programming the suites or packages and selecting ram and cores) [9], there are currently online resources that allow to quickly building own pan-genome analysis such as PGAweb [75]: http://pgaweb.vlcc.cn/analyze; PGAT [76]: http://nwrce.org/pgat/; Panseq [77]: https://lfz.corefacility.ca/panseq/page/pan.html; Spine [78]: http://vfsmspineagent.fsm.northwestern.edu/cgi-bin/spine.cgi; PanCGHweb [79]: http://bamics2.cmbi.ru.nl/websoftware/pancgh/pancgh_start.php.

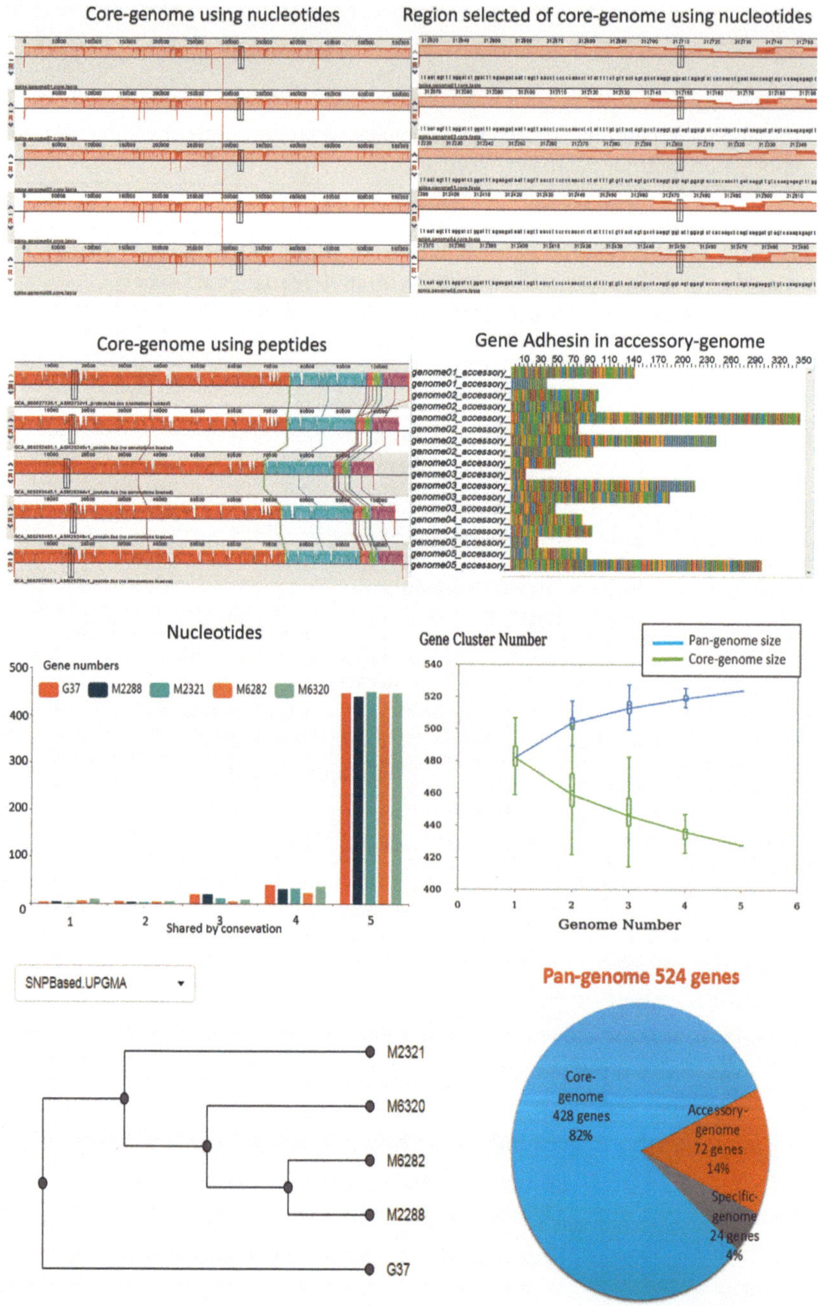

Fig. 2 Pan-genome from only five available genomes of *Mycoplasma genitalium* as a potential-antibiotic-resistant bacteria with likely multiple-antibiotic-resistance. Two first figures on top represent the core genomes aligned from five strains, displaying some selected region of five genomes aligned. Left figure marks between 300,000 and 350,000 bp and right figure shows the detail of the region selected in base pairs (312,070–312,750 bp approx.), which the alignment emphasizes the differences in red underlying. Second level figures show the core genomes aligned using peptides (left figure) with their synteny regions in *red*, *blue*, *rose*, *fluorescent green*, and *purple*, and right figure shows the selected adhesin gene from the accessory genome. Third and fourth levels figures exhibit the next pan-genome results: left plot (Nucleotides: gene numbers × shared by conservation) among genomes, right plot shows the classical trend of pan-genome size and core genes among the five strains. The last two figures are the SNP-based UPGMA tree, where G37 is the most divergent genome (left figure) and cheese-cake (right figure) shows the percentage of core genome (428 genes, 82%), accessory genome (72 genes, 14%), and specific genome or exclusive genes (24 genes, 4%).

or with the genomes published or stored on the web to build someone with panX [80]: http://pan-genome.de; Spine [78]: http://vfsmspineagent.fsm.northwestern.edu/cgi-bin/spine.cgi; Panseq [77]: https://lfz.corefacility.ca/panseq/page/pan.html; PanCGH-web [79]: http://bamics2.cmbi.ru.nl/websoftware/pancgh/pancgh_start.php.

To prove the reliability of those mentioned tools, we performed an *M. genitalium* (MG) pan-genome using PGAweb, Panseq, and Spine online tools. The results are depicted in Fig. 2, which led us to conclude that web pan-genome utilities are feasible, straightforward, and moreover easy to run.

An overall examination of the PB listed in Table 2, showed that there is an important growing interest to study its particular pan-genome. This decade disclosed an extraordinary steeping increase of pan-genome analysis from antibiotic-resistant pathogenic bacteria (ARPB) and multiple-antibiotic-resistant pathogenic bacteria (MARPB) in comparison to past decades. Therefore, this concern turns into a public health care problem and nonexclusive scientific matter. The indiscriminate use of antibiotic therapies in other fields, not necessarily hospitals; for instance, in livestock farms outbreaks, which increases the antibiotic resistance phenotype not only for PB, as well as the case for nonpathogenic gut microbiome [81]. Thus, the problematic situations have gone out of hands and we are not able to tackle these issues using old analysis strategies. The advance on pan-genomics to study and determine resistome is taking the right path towards overcoming this drawback.

3 The pan-genome of resistant bacteria

The World Health Organization (WHO) summaries antimicrobial resistance (AMR) as the resistance of a microorganism to an antimicrobial drug that was originally effective for the treatment of infections caused by themselves. Resistant microorganisms (including bacteria, fungi, viruses, and parasites) are able to withstand exposure by antimicrobial drugs, such as antibacterial molecules (e.g., antibiotics), antifungals, antivirals, and antimalarials/parasites, leading to ineffective treatment and in consequence, persisting infections, which increase the hazard associated with outspread to other microorganisms [82]. The same WHO report gives a particular definition for antibiotic resistance, which specifically refers to the resistance to antibiotics that occurs in common PB. Antimicrobial resistance is a broader term, encompassing resistance to drugs for treating infections caused by other microbes as well, such as parasites (e.g., malaria), viruses (e.g., HIV), and fungi (e.g., Candida).

The attention paid to PB brought about the development of genomics and later pan-genomics, perhaps reinforced by antibiotics resistance concern will help to find the best way in health care. Almost the most important pathogenic bacterium has a pan-genome to date (Tables 2 and 3), at least the most relevant such as *S. pneumonia* or *M. tuberculosis* [56] among others, from the former first pan-genome report [11] to the most recent publication [103] about *Haemophilus influenzae.*

Table 3 Data about the development of vaccine and multidrug resistance phenotype (MDR) with pan-genome study

Bacteria	Antibiotic resistance	Vaccine	MDR phenotype	Pan-genome	ARDB-genes (resistome)
Gram positive					
Clostridium difficile	Fluoroquinolones, cephalosporins, carbapenems, and clindamycin [83–85]	In progress	Yes	Yes	118 genes
Enterococcus faecalis	Penicillin, vancomycin, linezolid [86]	In progress	Yes	Yes	322 genes
Enterococcus faecium	Vancomycin-resistant [87, 88]	No	Yes	Yes	318 genes
Mycobacterium tuberculosis	Rifampicin-resistant [89]	In progress	Yes	Yes	53 genes
Mycoplasma genitalium	Azithromycin, moxifloxacin [90]	No	No	In this review	1 gene
Staphylococcus aureus	Methicillin-resistant [91]	In progress	Yes	Yes	516 genes
Streptococcus pyogenes	Clindamycin-resistant [92]	In progress	No	Yes	216 genes
Helicobacter pylori	Clarithromycin-resistant [93]	In progress	No	Yes	2 genes
Haemophilus influenzae	Ampicillin-resistant [94]	Yes	No	Yes	34 genes
Gram negative					
Acinetobacter baumannii	Carbapenem-resistant [95]	No	Yes	Yes	No data
Campylobacter jejuni	Fluoroquinolone-resistant [96]	In progress	Yes	Yes	19 genes
Neisseria gonorrhoeae	Ceftriaxone-resistant [97]	In progress	Yes	Yes	95 genes
Enterobacter faecium	Carbapenem-resistant, daptomycin β-Lactam resistant [98]	No	Yes	Yes	301 genes

Table 3 Data about the development of vaccine and multidrug resistance phenotype (MDR) with pan-genome study—cont'd

Bacteria	Antibiotic resistance	Vaccine	MDR phenotype	Pan-genome	ARDB-genes (resistome)
Klebsiella pneumoniae	Carbapenems, imipenem [99]	In progress	Yes	Yes	547 genes
Salmonella enterica	Fluoroquinolone-resistant [100]	In progress	Yes	Yes	840 genes
Escherichia. coli	β-Lactam-resistant [101]	In progress	Yes	**Yes**	1805 genes
Pseudomonas aeruginosa	Carbapenem-resistant [102]	No	Yes	Yes	538 genes

Exceptionally the *M. genitalium* pan-genome was fully made for this review.

Bacterial antibiotic resistance is a particular type of antimicrobial drug resistance induced by a gene battery. In the last decades, the ARB or AR [104] and currently the antibiotic resistance genes (ARGs) have been reviewed [105–107].

On the other hand, the emerging evolution of MDR bacteria will mean that these can thrive under environmental conditions in the presence of various antibiotics; leading their activation or synthesis to many ARGs for readily antibiotic metabolization likewise hamper or prevent antibiotic access to the binding site. Antibiotic resistance has become a serious global concern. The outbreak of MDR bacterial strains is expected to dramatically hinder treatment effectiveness [108].

The emergence of AMR in microorganisms is a natural phenomenon, since AMR selection has been driven by gene transfer among ARB; this development is associated with the health care and livestock practices as well as environmental driven variation. The pan-genome study of ARB and MARPB will help to elucidate the stance of health care and environmental adaptation caused by the indiscriminate use of antibiotics.

One important subject is finding new antibiotics; however, another current strategy being implemented is vaccine design, which endows our immune system, despite the difficulties in developing these against bacteria. Table 3 summarizes the current state of some MARPB, Gram-positive or negative associated with pan-genome studies and vaccines development, as another approach for tackling the drawbacks in antibiotics treatments. Vaccines for *H. influenzae,* and ampicillin-resistant bacterium, have been developed [94], despite this is not an MDR bacterium. Unfortunately, the attempts toward vaccines design for most MDR bacteria had been unsuccessful, which discouraged the efforts for conventional treatments.

It is coherent to wonder about why pan-genome was first detailed for PB, especially for ARB and MARPB, perhaps as a consequence of genomes availability. Since the annotation of diverse genomes is accessible, it is a straightforward process to research the dispersal of virulence genes, the occurrence of ARGs or the status of the whole resistome, for each strain. The variation of genome size in each strain gives the first simple question of how variable could the genome be in terms of size (Table 4). A second question is to wonder if each strain carries the same number of genes. Evidently, to answer such questions without a pan-genome analysis would be not an easy task; however, it is necessary to construct the pan-genome and establish how core and accessory genome are structured. We rapidly built the pan-genomes using PanX for some selected bacteria from Table 4, leaving out *M. genitalium* (Fig. 4A). Observing Table 4 and Fig. 4A, it is clear that the core and accessory genomes for each PB are different, which leads to assume that the quantity and location of virulence genes and ARGs are randomly distributed. The differences encountered after comparing *M. genitalium* and *Clostridium botulinum* are enormous since the core genome is 85% and 5% respectively. That means that *C. botulinum* carries 95% accessory genome and in terms of evolution, it is reasonable to think that *M. genitalium* safeguards almost all of its genomic repertoire (500 genes approx.). Nevertheless, it is not reasonable to assume that *C. botulinum* genome structure changes constantly to maintain fitness. *C. botulinum* is MARPB whilst *M. genitalium* is not properly an ARB.

MARPB *E. coli* is one of the most ubiquitous bacteria in the world. The accessory genome is 92%, calculated using PanX. This microorganism is probably the most studied bacterium at the genetic level worldwide; especially for horizontally transferred genes (HTGs), even more for ARGs. The size of resistome has not been established yet, not even for the successfully spread *E. coli* ST131 [109]. Probably we could assume that the resistome or the whole of ARGs comes from HTGs, but that is not the case, it is just a hypothesis. We will analyze this hypothesis using additional information of *P. aeruginosa* and *Yersinia pestis* genomes and comparing the three γ-proteobacteria MARPB altogether.

4 The pan-genome of emergent resistant bacteria

In February 2017, an urgent report was released by the WHO (https://www.who.int/news-room/detail/27-02-2017-who-publishes-list-of-bacteria-for-which-new-antibiotics-are-urgently-needed). This web page post was entitled: WHO publishes the list of bacteria for which new antibiotics are urgently needed [110]. "Priority 1: CRITICAL *Acinetobacter baumannii*, carbapenem-resistant *P. aeruginosa*, carbapenem-resistant Enterobacteriaceae*, carbapenem-resistant, third-generation cephalosporin-resistant, which are a large family of Gram-negative bacteria".

Table 4 The pan-genome of some selected antibiotic-resistant pathogen bacteria (ARPB) and two strain examples by species with their genome size and number of genes

	Pan-genome			Genomes of two strains		
Species	**Core genome**	**Accessory genome**	**Number of genomes**	**Strains**	**Sizes of genome**	**# Genes**
Clostridium botulinum	5	95	46	ATCC 3502/ A str. Hall	3.767.000/ 3.760.560	3767/3499
Escherichia coli	8.5	91.5	633	O157:H7/ K-12	5,528,445/ 4.646.332	5349/4331
Bacillus cereus	18.5	81.5	48	ATCC10987/ ZK	5.224.283/ 5.300.915	5603/5134
Pseudomonas aeruginosa	18.5	81.5	153	RW109/ PAO	7.049.347/ 6.264.404	6829/5697
Acinetobacter baumannii	20	80	116	AB030/ADP1	4.335.793/ 3.598.621	4296/3325
Streptococcus pneumoniae	31.5	68.5	52	TIGR4/R6	2.160.842/ 2.038.615	2125/2043
Staphylococcus aureus	32.5	67.5	344	MRSA252/ RF122	2.902.619/ 2.742.531	2744/2589
Bacillus anthracis	50	50	50	Ames/Sterne	5.227.293/ 5.228.663	5632/5263
Yersinia pestis	56.5	43.5	36	Antiqua/ Mediaevalis	4.702.289/ 4.595.065	4167/3895
Mycoplasma pneumoniae	63.5	36.5	48	M129/M29	816.394/ 857.799	1061/834
Serratia marcescens	68.5	31.5	32	U36365/ Db11	5.540.160/ 5.113.802	5341/4848
Mycobacterium tuberculosis	78.5	21.5	168	CDC1551/ H37Rv	4.403.837/ 4.411.532	4189/3999
Mycoplasma genitalium	82	18	5	G37/M2288	580,076/ 579558	566/567
Average (1731)	41.0	59.0	1731			

The value of the core genome and accessory genome adds 100%. Third column shows the number of genomes used to establish pan-genome. The last three columns on the right are the example of the genome size and respective genes of the two selected strains.

In this section, we compared the pan-genome of the three current emergent resistant bacteria (ERB), selected from Table 4, due to four fundamental reasons: (I) the three species are antibiotic emergent; (II) there is a pronounced difference for number of strains available to obtain the pan-genome; (III) in the three species, core genome is similar only for *A. baumannii* and *P. aeruginosa* (20:80/core: accessory genome), while MG is absolutely the opposite (80:20/core: accessory genome); (IV) it is quite important to notice that the three bacteria have different genome size: the smallest accounted for *M. genitalium* with ~0.88 Mb (~580 genes), medium for *A. baumannii* with ~3.8 Mb (~3600 genes), and the biggest for *P. aeruginosa* with ~6.5 Mb (~6100 genes); see Table 4. In the future, more in-deep studies on these three species pan-genome will reach a huge spectrum for identification of the whole antibiotics-resistome genes (ARGs) and certainly for other important ARB and MARPB.

The first pan-genome reported for *A. baumannii* used 6 genomes [30] and the second one was assembled 6 years after using 11 [14], likely a limited number of genomes; however, the last decade in clinical scenarios, *A. baumannii* infection raised concern due to the acquisition of multi-antibiotic resistance traits toward aminoglycosides, aminocyclitols, tetracycline, and chloramphenicol [111] or Carbapenem-resistant [95]. Nevertheless, it is possible to decrypt a pan-genome using PanX tool (http://pan-genome.de/Acinetobacter_baumannii in December 2018), which we found 116 *A. baumannii* genomes with a 20% core and 80% accessory genome ratio in 2018. Currently, there are 140 genomes in NCBI and surprisingly PATRIC only reported 17 genomes (April 2019), this is the reason why this species was not included in Table 5.

The *P. aeruginosa* pan-genome certainly came out late compared to other widely studied γ-proteobacteria. When this delay on readily available information acquisition was occurring, an emergence of antibiotic-resistant strains of *P. aeruginosa* rose. The first pan-genome for *P aeruginosa* was reported by Fisher et al. [60], using 100 genomes, and the same year Sharma et al. [59] reported a pan-genome using five genomes. Later, Mosquera-Rendón et al. [61] constructed a pan-genome with 181 genomes, and last year Freschi et al. [62] brought about the largest pan-genome published ever, analyzing 1311 genomes. Probably, the most sequenced and studied pan-genome is *P. aeruginosa*. Recently, PATRIC database reported 4681 genomes (April 2019) and 3604 genomes were reported in December 2018, over 1000 genomes released in 4 months (see Table 5).

MG belongs to the *Mollicutes* class and lacks the cell wall. This bacterium is an emerging sexually transmissible infection (STI) that causes the etiologic agent of nongonococcal urethritis (NGU) and it is persistent or recurrent urethritis. In women, available evidence suggests that MG infection is significantly associated with an increased risk of cervicitis, pelvic inflammatory disease (PID), preterm birth and spontaneous miscarriage, and the risk of infertility is also associated. There are only six MG genomes published to date in the world, and the number of genomes is minimal compared to *M. pneumonie*, *E. coli*, or *P. aeruginosa*. In this review, we released the first pan-genome using those available strains (five genomes), retrieved from PATRIC database (Fig. 2), and these data were also run using PGAweb. Azithromycin is the recommended antibiotic for MG treatment; however, resistance acquisition and coupling with moxifloxacin, meaning a high cost of medication, which has the potential to induce hepatotoxicity and unfortunately there are report cases for antibiotic resistance [112].

Table 5 Chronology of pan-genome from antibiotic-resistant pathogenic bacteria (ARPB) and multiple-antibiotic-resistant pathogenic bacteria (MARPB), from Medini et al. [10] to the assembly of NCBI and PATRIC genome databases

Species with sequenced genome(s) with pan-genome (*)	Number of genomes sequenced per species September 2005, Medini et al. [10]	Number of genomes sequenced per species in PATRIC December 2018, http://www.patricbrc.org	Number of genomes sequenced per species in NCBI December 2018, http://www.patricbrc.org	Antibiotic Resistance Genes Database December 2018 from ARDB https://ardb.cbcb.umd.edu/
*Streptococcus agalactiae**	8	1089	1098	210
*Bacillus anthracis**	8	362	233	396
Burkholderia mallei	8	81	80	205
*Burkholderia pseudomallei**	7	1924	1532	205
*Staphylococcus aureus**	6	12112	10204	516
*Streptococcus pyogenes**	6	481	463	216
*Salmonella entérica**	5	18214	13628	850
*Escherichia coli**	5	17112	14820	1806
*Bacillus cereus**	5	2859	1013	396
*Haemophilus influenzae**	5	756	700	34
*Listeria monocytogenes**	5	3987	3762	13
*Xylella fastidiosa** (plant)	5	96	44	0
Buchnera aphidicola	3	63	30	4
Burkholderia cenocepacia	3	269	299	205
*Legionella pneumophila**	3	640	615	0
Pseudomonas syringae (plant)	3	845	369	538
Streptococcus thermophilus	3	70	53	210
*Yersinia pestis**	3	497	376	180
*Streptococcus pneumoniae**	2	8779	8440	576
*Mycobacterium tuberculosis**	2	11219	5948	53

Continued

Table 5 Chronology of pan-genome from antibiotic-resistant pathogenic bacteria (ARPB) and multiple-antibiotic-resistant pathogenic bacteria (MARPB), from Medini et al. to the assembly of NCBI and PATRIC genome databases—cont'd

Species with sequenced genome(s) with pan-genome (*)	Number of genomes sequenced per species September 2005, Medini et al.	Number of genomes sequenced per species in PATRIC December 2018, http://www.patricbrc.org	Number of genomes sequenced per species in NCBI December 2018, http://www.patricbrc.org	Antibiotic Resistance Genes Database December 2018 from ARDB https://ardb.cbcb.umd.edu/
*Neisseria meningitidis**	2	1780	1930	95
*Campylobacter jejuni**	2	2585	1610	19
*Helicobacter pylori**	2	1349	1200	2
Leptospira interrogans	2	312	297	0
Mycoplasma genitalium	2	6	5	1
*Pseudomonas aeruginosa**	2	3604	3768	538
*Shigella flexneri**	2	1037	540	202
*Staphylococcus epidermidis**	2	635	504	516
Xanthomonas campestris (plant)	2	139	94	3
Total	113	92902	73655	
Various species with 1 genome	211			
Total pathogen genomes selected	113	92902	73655	
Other species	211	109700	141343	
Total genomes	324 (254 species)	202602	214998	

The last column corresponds to ARDB-Antibiotic Resistance Genes Database (https://ardb.cbcb.umd.edu/cgi/search).

5 PATRIC and other databases

At present, to develop useful resources for decrypting PB associated with antibiotic resistance and pan-genomic analyses, we suggest the currently more used web resources. There are some outstanding web services from PATRIC and ARDB web pages, which we will describe below in a summarized way.

The web page of Pathosystems Resource Integration Center, better known as PATRIC is the bacterial Bioinformatics Resource Center (https://www.patricbrc.org; Ref. [113]), a bioinformatics guideline endeavored by National Institute of Health (NIH). PATRIC currently includes more than 200 thousand genomes species with a wide set of utilities, which provide users with a platform enabling the visualization of single genomes to even a wide analysis using thousands of genomes. We compared the number of genomes used in the first pan-genome report [10] with the number of genomes deposited in PATRIC database (Table 5 and Fig. 3). It displays five tabs to choose a genome or species: organisms, data, workspace, services, and help. The organism tab allows picking out bacteria, archaea, and viruses from environmental niches or even for a given eukaryotic host. Data resources provide wide possibilities as Antibiotic Resistance, Genomes, Genomic Features, Pathways, Protein Families, Specialty Genes, Transcriptomics, and downloadable data linking FTP server; it is also likely to find PATRIC Driving Biology Projects (DBPs), besides the NIH: National Institute of Allergy and Infectious Diseases links to Clinical Proteomics, Genome Sequencing, Structural Genomics, Systems Biology, and Functional Genomics. The Services section is composed of five tabs: (1) Genomics with wide set of options: Assembly, Annotation, Comprehensive Genome Analysis, BLAST, Similar Genome Finder, Variation Analysis, Tn-Seq Analysis, Phylogenetic Tree, Metagenome Binning, and Genome Alignment,

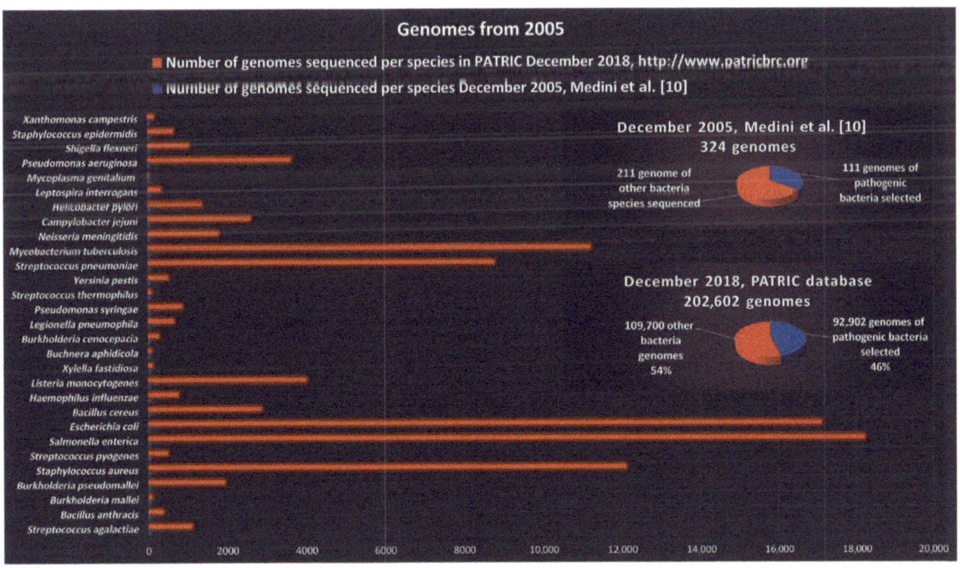

Fig. 3 Timeline of construction of pan-genomes from the year 2005 to December 2018. The plot displays the number of genomes registered in PATRIC database to December 2018 (*orange*, 202.602 genomes) and the number of genomes registered in December 2005, published by Medini et al. [10] (*blue*, 324 genomes). See detail of cheeses at the right.

(2) Transcriptomics (Expression Import, RNA-Seq Analysis), (3) Protein Tools (Protein Family Sorter and Proteome Comparison), (4) Metabolomics (Comparative Pathway and Model Reconstruction), and (5) Data with ID Mapper that associates metadata. Wattam et al. [113] updatings to PATRIC, all-bacterial bioinformatics database and analysis resource center published in *Nucleic acids research* journal, *45*(D1), D535–D542.

The workspace interface of PATRIC is an excellent user experience, allowing researchers direct access to data and tools, creating genome groups with detailed summaries and a variety of visual aids for data imaging, such as plots, graphs, and drawings. Users have the possibility to analyze their private data and compare them to available public data in PATRIC web. To perform a pan-genome, it is clear that it is impossible to find out the core or accessory genome. Nonetheless, the other services will be useful for constructing and analyzing bacterial-pathogens pan-genomes mainly those related to resistome.

Another interesting web resource is Antibiotic Resistance Genes Database or ARDB [114] (https://ardb.cbcb.umd.edu/index.html), an online service supported by the Center for Bioinformatics and Computational Biology University of Maryland College Park, MD 20742. This database has a large collection of genes concerning antibiotic resistome from pathogen bacteria (see Tables 3 and 5 and Fig. 4B). This database intends to study the resistome in PB, which promotes specific searches for particular ARGs found in the isolates of interest. For instance, this database was first used for pan-genomic analysis in *M. tuberculosis* [14].

Today, the STRING (https://string-db.org/cgi/network.pl) or KEGG networks do not consider proteins, microRNA or metabolites linked to the resistome conception, taking caution measures, since there is not well-supported information. The GEAR database (Genomic Elements Associated with Drug Resistance) compiles only human genes associated with bacterial resistance; this also aims to provide comprehensive information about genomic patterns bearing resistance for human drugs database, but not specifically to bacterial species.

Gillings [13], citing other authors [8, 10, 115] states: "The phenotypes exhibited by the global microbiome are encoded by the microbial pan-genome, that is, the set of genes present in all the genomes of all the prokaryotes in the biosphere." Antibiotics that battle PB are inadequate, inefficient or precarious when bacteria have become resistant. They are included in the whole microbiome and the global resistome shall be unveiled. Pan-genomics will help to prevent this sort of restriction since global resistome can be recognized by the pan-genome study.

6 Core and accessory genomes of antibiotic-resistant bacteria

Resistome genes comprise essential and nonessential genes, which are part of accessory or core genome, likely being gained by horizontal transfer. Evidently, the full set of genes is not drug resistance-related, since it includes genes for cold or heat, salt concentration,

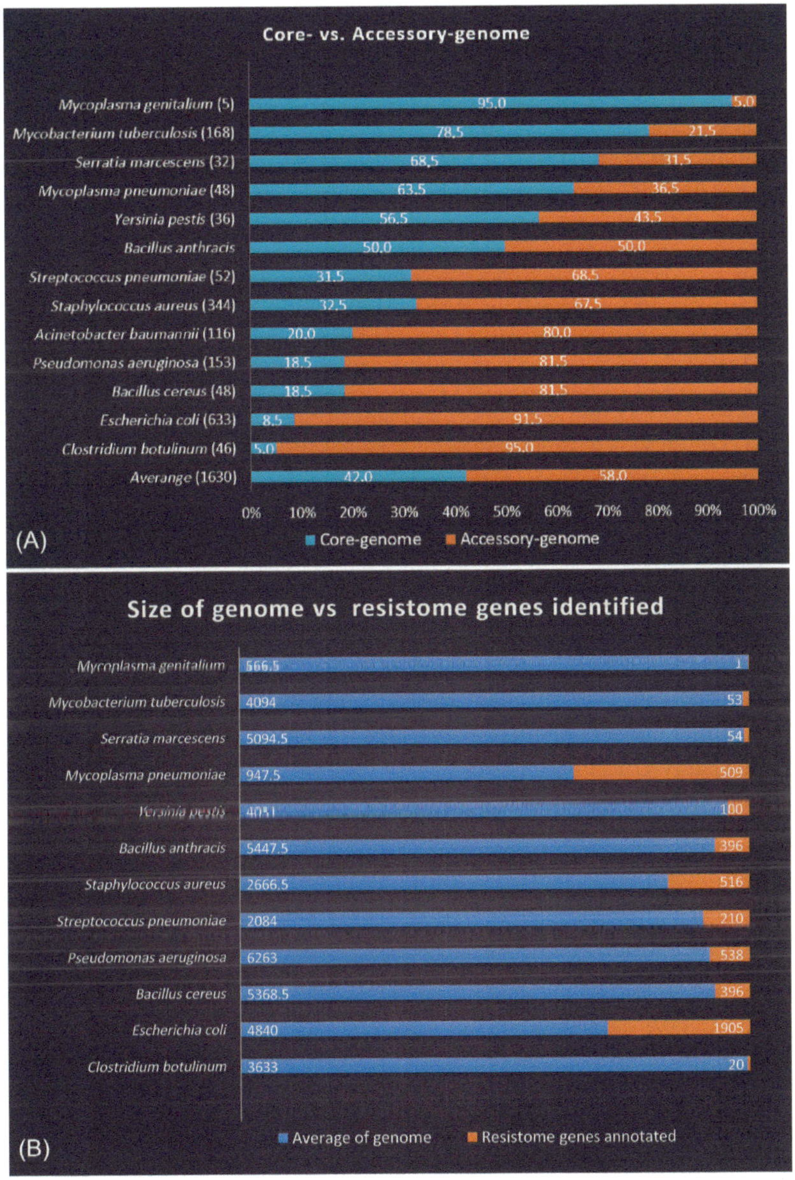

Fig. 4 (A) The ratio of the core and accessory genome carried out with panX: Pan-genome Analysis & Exploration (http://pan-genome.de/). However, *M. genitalium* was performed for this manuscript. In parenthesis are the numbers of genomes used. (B) Shows the relationship of an average of the genome size (average of two genomes in Table 4) on blue bar and the resistome genes currently identified in each species from ARDB-Antibiotic Resistance Genes Database (https://ardb.cbcb.umd.edu/cgi/search), in orange bar.

pH, osmosis, dryness, and toxics. There are also genes associated to natural antibiotics or for synthesis of related metabolites.

Phan et al. [109] identified 315 essential genes in *E. coli* EC958, 231 (73%), which were also essential for the well-known *E. coli* K-12. They encountered the indispensable genes set required for *in vitro* growth and the serum resistome (specifically essential genes for resistance to human serum and not for antibiotics), finding that the serum resistome comprised 56 genes, the majority encoding membrane proteins or factors involved in lipopolysaccharide (LPS) biosynthesis [109].

Fig. 4A summarizes the core and accessory genome percentage (100%) of some selected ARB. As mentioned before, the pan-genome of *A. baumannii* and *P. aeruginosa* shows 20:80/core:accessory genome, while MG pan-genome exhibits 80:20/core:accessory genome; however, other bacteria as *C. botulinum* shows 5:95/core:accessory genome, *M. tuberculosis* exhibits 21.5:78.5/core:accessory genome, *Yersinia pestis* shows 56.5:43.5/core:accessory genome, *Bacillus anthracis* exhibits 50:50/core:accessory genome (Fig. 4A). These data corroborate the great variability of core and accessory genomes ratios for ARBs.

Both the accessory genome and the core genome are variable in PB. Currently, it is too imperative to assert that there is a relationship between pan-genome and pathogenicity-virulence or pan-genome and antibiotics resistance. However, the most essential aspect regarding pan-genome analysis in PB and ARB or MRAB is that the resistome size could be associated with the pan-genome features that bear the bacterium. At the moment, it is possible to speculate that for a bacterium with a large core genome, its resistome apparently could be located in this genomics clustering. Otherwise, its resistome genes would be located in the accessory genome. Looking close to Fig. 4A and B, it could be said that there is no connection between the variation of core and accessory genome; nevertheless, it is clear that depending on the bacterial species, it could be stated that some of these genes coding antibiotic resistance traits are located in core genome and others in accessory genome. Next section describes pan-genome and resistome studies for other bacterial species.

7 Pan-genome and resistome

The antibiotic resistome comprises all ARGs. It includes resistance elements found not only in PB but also in antibiotic-producing bacteria, bearing cryptic resistance genes (not necessarily expressed), which are present in bacterial chromosomes or as a result of horizontal gene transfer from plasmids, or by chemical mechanisms for co-optation. However, the resistome is highly redundant and interlocked [116].

The close connection between pan-genome and resistome is given by ARGs, which are highly mobile and have been classified as determinants of resistance risk increase. However, the transfer network of the mobile resistome and the forces, which drive mobile ARGs are largely unknown [117].

As we mentioned before, an excellent database for studying the resistome in PB is ARDB [114], which can be used to browse specifically genes associated with resistance for isolates of interest. For instance, this database was used for *M. tuberculosis* analysis. The data shown in Fig. 4B summarizes the amount of antibiotic resistome genes from selected ARB.

The resistome also includes epigenetic mechanisms as methylation, acetylation, and other DNA or RNA taggings. Obviously, resistance genes are present in the microbial pan-genome, regardless human selection [22]. The same authors emphasized that "our current concept of the antibiotic resistome includes the obvious and the obscure; both genes that cause phenotypic resistance in the clinic and those that lie hidden and silent in the environmental pan-genome have the potential to cause treatment failure."

The pioneering work in pan-genome and resistome was published by Bhardwaj and Somvanshi [38]. They carried out a *C. botulinum* pan-genome analysis using 13 strains to understand the symptoms linked to *C. botulinum*, in terms of the broad host's spectrum. The successive calculation and characterization of the core and pan-genome subset revealed more specific targets for drug design and vaccine development. In Fig. 4A, the small size of core genomes (only 5%) of *C. botulinum*, indicates that 95% of the accessory genome is likely mobile. At the moment, the number of genes associated with antibiotic resistome is low, according to ARDB database, which comprise just 20 genes by now. ResFinder is a web service or software for easy identification of ARGs in bacteria; however, it compiles a short number of genes and species. Bhardwaj and Somvanshi [38] used this tool for analyzing *C. botulinum*. This research was carried out by comparing query genomes to individual level against a database containing 1862 horizontally acquired resistance genes related to 12 antimicrobial classes at a minimum threshold value of 95%, and a minimum length of 50% [118]. They concluded that "an open genome endowed by Clostridial lineage indicates the possibility of the addition of new gene sets with every genome introduction along with novel strain-specific genes (Singletons)."

Other recent research with *C. difficile* addresses the identification of pan-genome and resistome pipeline developed by Knight et al. [119]. They found 11 CRISPR elements, 233 prophages, 7 AMR genes, and also 2 transposons sequences. They used RT014, defined by a large "open" pan-genome (7587 genes) comprising a core genome of 2296 genes (30.3% of the total gene repertoire) and an accessory genome of 5291 (69.7%) genes. Following the research with *C. botulinum* (Fig. 4A), we ran the *C. difficile* data in PanX web tool, although the core and accessory genome were 40% and 60% respectively; these results are slightly different from those obtained in this study (Table 2, 50 strains). It would be ideal to combine these 50 strains from PanX with those 44 used by Knight et al. [119] and reconstruct a new and broader pan-genome. Further, Knight et al. [119] found AMR genotypes and phenotypes, which vary across host populations for tetracycline [*tetM, tetA(P), tetB(P),* and *tetW*], clindamycin/erythromycin (*ermB*), and aminoglycosides (*aph3-III-Sat4A-ant6-Ia*). The resistance was mediated by mobile genetic elements, such as Tn*6194* (harboring *ermB*) and a novel variant of Tn*5397* (harboring *tetM*).

8 New challenges of pan-genome strategy

Virtually some bacterium in the world can be resistant to antibiotics, and pan-genome and resistome will evidence this fact. And so, resistome from all strains in the world will be a pan-resistome. This pan-resistome would be a part of the core and accessory genome. In other words, there will be a core genome from the pan-resistome and an accessory genome as well.

The pan-resistome, pan-RNAome, pan-mobilome, pan-metabolome, pan-proteome, and pan-epigenome will be the new challenges in the MDR-bacteria pan-genomics setting. Surely, those Pan-omics will share abundant genes, proteins, mobile elements, prophages, microRNA, and methylations and, therefore, structures and functions, which will be involved in a pan-regulon. Probably a pan-regulon will be more plastic than a pan-genome, since the relationship between core and accessory genome is based on structure and not necessarily on function.

Pan-resistome probably integrates a pan-genome with a pan-RNAome, a pan-proteome (coding or not), and a pan-epigenome. The contrast of open and close pan-genomes will be the challenge for decrypt resistome in each species, for instance *C. dificil, M. genitalium, E. coli*, or *M. pneumoniae*, and comparing, for example, their pan-genomes integrated with other pan-omics.

We hypothesized the size of resistome (pan-resistome) in *E. coli, P. aeruginosa, and Y. pestis*, arranging the size of resistome compared with the size of the core and accessory genome (Fig. 5). The resistome is a transparent circle, which intersects the core or accessory genome, but it takes the external area to other genomes (mobile or transferred external elements). Taking into account Fig. 4B that contains the genes of resistome, the closest to resistome will be for *E. coli* (1905 genes) and *M. pneumoniae* (509 genes), but other species still have a small number of annotated genes.

In the future, the use of pan-genomics will be the best tool for the classification of one strain within one species, for example, a new unknown bacterial pathogen. In the past, we sequenced a small fragment of rDNA and loaded it to NCBI-BLAST, and the tool searched the most identical sequence (*E* value); therefore, today we already have multiple online tools where it is possible to load a genome and determine its core genome. This tool will be much more powerful for the strain identification and species classification.

Note that each nuclear and accessory genome has different sizes, resembling the different environmental and evolutionary circumstances that each species must overcome. Therefore, there are no exact and intentional pan-genomes among the strains of each species. Perhaps the case of MG is peculiar in nature, which shows a low number of strains and a great core genome conservation. It also shows that there are no rules in nature for the arrangement between bacterial pan-genome and their resistome.

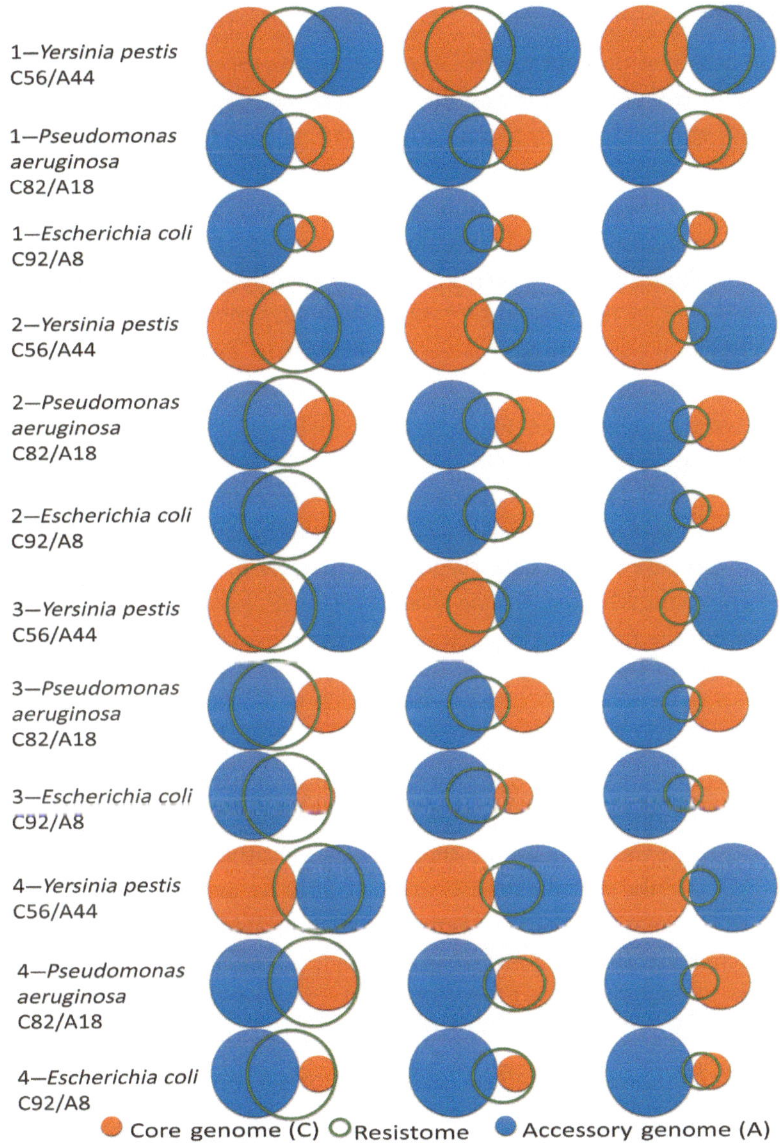

Fig. 5 The different hypothesis of the size of resistome (pan-resistome) in relationship to the core (C) and accessory (A) genome using three species of antibiotic-resistant bacteria (C56+A44=100). In step 1 the resistome (transparent circle) has the same hypothetical size as the accessory genome but those genes have three possibilities: they are in the same proportion both in core as in the accessory genome or they are whether mostly in core genome or vice versa. In step 2 the resistome does not matter its size, therefore, there is always a proportionality of sharing the genes in core or accessory genomes. In step 3 as in step 2 does not matter the size of resistome (being variable) and most of genes from resistome are part of the core genome, while in step 4 the size of resistome is not important, and most of genes are in the accessory genome.

9 Conclusion

Pan-genome was first thought for PB, not just as a methodological issue rather than a necessity. The best way to face the antibiotic resistance acquiescence concern is to construct the pan-genome posterior to the genome sequencing and annotation. However, it is hard to establish the relationship between pan-genome-resistome while there is still uncertainty whether resistome fits inside the pan-genome. What is the real size of resistome inside pan-genome? And, what are the correct genes set for resistome? Anyhow, we can raise much questions; however, we consider that pan-genomics has tackled some important concerns, which would be impossible to solve using classical molecular biology or descriptive genomics: it is very important to define the core and accessory genome for establishing the plasticity of resistome, since *E. coli* and *C. botulinum* have a large accessory genome, allowing us to show that resistome is probably enriched by genes contained in accessory genome, and it was the case for *E. coli*, which let proving that horizontal or lateral gene transfer works, studied by β-lactamase and ampicillin resistance. Evidently, it was our first successful tool for classification in molecular biology, but today, the pan-genome of both bacteria depicts a draft for the evolutive landscape of antibiotic resistome to the MG and *Serratia marcescens* species, since the core genome for both species empowers the resistome by enclosing the invading microorganism genes or carefully selecting them; in this way both species close the variability by losing antibiotic resistance phenotype.

To conceptually define resistome would be an excellent challenge, especially during the current careless antibiotic consumption and its dispersion in the environment. Thousands of unknown bacteria and microorganisms are exposed to manufactured antibiotics, leading us to assume that there are no means to prevent this catastrophe. In opposition, many scientists, just like us, believe that pan-genomics is a powerful approach to prevent such a disaster. We must move towards the sequencing of known and unknown species, classify them, and establish their antibiotic resistance status, their pan-genome, and come out with new alternatives for reducing antibiotic consumption nowadays. Recently in Korea, a successful trial as a new alternative for the treatment of *P. aeruginosa* is again considering the use of low concentration of aminoglycosides [120].

The major challenge today will be establishing the resistome inside of pan-genome, as it is happening for the core or accessory genome. The resistome genome not only provides information for resistance to antibiotics, but also provides one perspective since resistome size will be considered an integrative part of core and accessory genome. Of course, there is a difference between inheritance of core genome and those HTGs, HRGs, in the accessory genome. This statement remains undetermined since we do not know whether HRGs are also part of the core genome, mainly because bacterial conjugation probably carries new or constitutive genes (repeats). We know for certain that plasmid and virus are part of accessory genomes, just when the core genome does not share this information in each strain, however, how to explain the transposons, CRISPR-cas 9 or microRNAs in pan-genome. Precisely, since transposons, CRISPR-cas 9, and microRNAs do take part in resistome and reflect the genomic plasticity of resistome, which varies with the range of core and accessory genome within the the evolution of species. Finally, it will be important to consider the size of resistome as a pan-resistome and why not a pan-RNAome (pan-transcriptome), all coupled by pan-genome analysis.

Acknowledgment

We thank the University of Antioquia for financial support, project CODI-2017-15753 and special thanks to Dr. Debmalya Barh, and Professor Olga Gil.

Conflict of interest

The authors declare that they have no conflict of interest.

References

[1] NCBI. n.d. https://www.ncbi.nlm.nih.gov/genome/microbes/ (Accessed December 2018).
[2] EMBL. n.d. https://www.ebi.ac.uk/genomes/bacteria.html (Accessed December 2018).
[3] KEEG Genome. n.d. https://www.genome.jp/kegg/genome.html (Accessed December 2018).
[4] PATRIC. n.d. https://www.patricbrc.org/ (Accessed December 2018).
[5] MBGD. n.d. http://mbgd.genome.ad.jp/ (Accessed December 2018).
[6] ENSEMBL. n.d. https://bacteria.ensembl.org/index.html (Accessed December 2018).
[7] JGI-IMG/M. n.d. https://img.jgi.doe.gov/ (Accessed December 2018).
[8] H. Tettelin, D. Riley, C. Cattuto, D. Medini, Comparative genomics: the bacterial pan-genome, Curr. Opin. Microbiol. 12 (2008) 472–477.
[9] J. Xiao, Z. Zhang, J. Wu, J. Yu, A brief review of software tools for pangenomics, Genomics Proteomics Bioinformatics 13 (2015) 73–76.
[10] D. Medini, C. Donati, H. Tettelin, V. Masignani, R. Rappuoli, The microbial pan-genome, Curr. Opin. Genet. Dev. 15 (2005) 589–594.
[11] H. Tettelin, V. Masignani, M.J. Cieslewicz, C. Donati, D. Medini, N.L. Ward, et al., Genome analysis of multiple pathogenic isolates of *Streptococcus agalactiae*: implications for the microbial "pan-genome". Proc. Natl. Acad. Sci. U. S. A. 102 (39) (2005) 13950–13955.
[12] T. Marschall, M. Marz, T. Abeel, L. Dijkstra, B.E. Dutilh, A. Ghaffaari, et al., Computational pan-genomics: status, promises and challenges, Brief. Bioinform. 19 (2018) 118–135.
[13] M.R. Gillings, Evolutionary consequences of antibiotic use for the resistome, mobilome and microbial pangenome, Front. Microbiol. 4 (4) (2013) 1–10.
[14] L. Rouli, V. Merhej, P.E. Fournier, D. Raoult, The bacterial pangenome as a new tool for analysing pathogenic bacteria, New Microbes New Infect. 7 (2015) 72–85.
[15] Metagenomics. n.d. http://www.metagenomics.wiki/pdf/definition/pangenome (Accessed December 2018).
[16] V. Paralanov, J. Lu, L.B. Duffy, D.M. Crabb, S. Shrivastava, B.A. Methé, J. Inman, S. Yooseph, L. Xiao, G.H. Cassell, K.B. Waites, J.I. Glass, Comparative genome analysis of 19 *Ureaplasma urealyticum* and *Ureaplasma parvum* strains, BMC Microbiol. 12 (2012) 88.
[17] G.D. Wright, The antibiotic resistome: the nexus of chemical and genetic diversity, Nat. Rev. Microbiol. 5 (3) (2007) 175–186.

[18] C. Bertelli, G. Greub, Rapid bacterial genome sequencing: methods and applications in clinical microbiology, Clin. Microbiol. Infect. 19 (2013) 803–813.

[19] B.M. Forde, N.L.B. Zakour, M. Stanton-Cook, M.D. Phan, M. Totsika, K.M. Peters, K.G. Chan, M.A. Schembri, M. Upton, S.A. Beatson, The complete genome sequence of Escherichia coli EC958: a high quality reference sequence for the globally disseminated multidrug resistant *E. coli* O25b: H4-ST131 clone, PLoS One 9 (8) (2014).

[20] K. Lafevers, Horizontal gene transfer spreads antibiotic resistance among human gut microbiota, Microrev. Cell Mol. Biol. 4 (1) (2018).

[21] Moreno Switt, A., 2013. Genomic Characterization of *Salmonella* Free-Living Phages, Plasmids and Chromosomally Inserted Mobile Elements. Cornell Theses and Dissertations.

[22] J.A. Perry, E.L. Westman, G.D. Wright, The antibiotic resistome: what's new? Curr. Opin. Microbiol. 21 (2014) 45–50.

[23] M. Reguero, V. Flores, L.P. Uribe, E.B. González, J.R. Mantilla, E.M. Valenzuela de Silva, L. Falquet, E. Barreto-Hernández, Genomic analysis of the resistome of the strain of *Acinetobacter baumannii* ABIBUN 107m multi-resistant and persistent in colombian hospitals, Rev. Colomb. Biotecnol. 16 (2) (2014) 104–113.

[24] J. Rusakovica, J. Hallinan, A. Wipat, P. Zuliani, Probabilistic latent semantic analysis applied to whole bacterial genomes identifies common genomic features, J. Integr. Bioinform. 11 (2) (2014) 243.

[25] G.D. Wright, H. Poinar, Antibiotic resistance is ancient: implications for drug discovery, Trends Microbiol. 20 (4) (2012) 157–159.

[26] FAPESP. https://bv.fapesp.br/en/auxilios/94866/pan-resistome-of-beta-lactamase-kpc-2-ctx-m-8-ctx-m-15-producing-klebsiella-pneumoniae-and-esche/, 2016-2018 (Accessed December 2018).

[27] Grantome. http://grantome.com/grant/NIH/R43-AI129187-01, 2017 (Accessed December 2018).

[28] F. Sigaux, Cancer genome or the development of molecular portraits of tumors [in French], Bull. Acad. Natl Med. 184 (7) (2000) 1441–1447; discussion 1448-1449.

[29] M. Kuroda, A. Yamashita, H. Hirakawa, M. Kumano, K. Morikawa, M. Higashide, A. Maruyama, Y. Inose, K. Matoba, H. Toh, S. Kuhara, M. Hattori, T. Ohta, Whole genome sequence of *Staphylococcus saprophyticus* reveals the pathogenesis of uncomplicated urinary tract infection, Proc. Natl. Acad. Sci. U. S. A. 102 (37) (2005) 13272–13277.

[30] L. Snipen, T. Almøy, D.W. Ussery, Microbial comparative pan-genomics using binomial mixture models, BMC Genomics 10 (2009) 385.

[31] M.E. Zwick, S.J. Joseph, X. Didelot, P.E. Chen, K.A. Bishop-Lilly, A.C. Stewart, et al., Genomic characterization of the *Bacillus cereus* sensu lato species: backdrop to the evolution of *Bacillus anthracis*, Genome Res. 22 (2012) 1512–1524.

[32] D.B. Wang, B. Tian, Z.P. Zhang, J.Y. Deng, Z.Q. Cui, R.F. Yang, et al., Rapid detection of *Bacillus anthracis* spores using a super-paramagnetic lateral-flow immunological detection system, Biosens. Bioelectron. 42 (2013) 661–667.

[33] E.F. Mongodin, S.R. Casjens, J.F. Bruno, Y. Xu, E.F. Drabek, D.R. Riley, B.L. Cantarel, P.E. Pagan, Y.A. Hernandez, L.C. Vargas, J.J. Dunn, S.E. Schutzer, C.M. Fraser, W.-G. Qiu, B.J. Luft, Inter-and intra-specific pan-genomes of *Borrelia burgdorferi* sensu lato: genome stability and adaptive radiation, BMC Genomics 14 (2013) 693.

[34] X. Yang, Y. Li, J. Zang, Y. Li, P. Bie, Y. Lu, Q. Wu, Analysis of pan-genome to identify the core genes and essential genes of *Brucella spp*, Mol. Gen. Genomics. 291 (2) (2016) 905–912.

[35] M.K. Wilson, A.B. Lane, B.F. Law, W.G. Miller, L.A. Joens, M.E. Konkel, B.A. White, Analysis of the pan genome of *Campylobacter jejuni* isolates recovered from poultry by pulsed-field gel electrophoresis, multilocus sequence typing (MLST), and repetitive sequence polymerase chain reaction (rep-PCR) reveals different discriminatory capabilities, Microb. Ecol. 58 (4) (2009) 843–855.

[36] A. Collingro, P. Tischler, T. Weinmaier, T. Penz, E. Heinz, R.C. Brunham, T.D. Read, P.M. Bavoil, K. Sachse, S. Kahane, Unity in variety—the pan-genome of the *Chlamydiae*, Mol. Biol. Evol. 28 (12) (2011) 3253–3270.

[37] Z. Udaondo, E. Duque, J.L. Ramos, The pangenome of the genus *Clostridium*, Environ. Microbiol. 19 (7) (2017) 2588–2603.

[38] T. Bhardwaj, P. Somvanshi, Pan-genome analysis of *Clostridium botulinum* reveals unique targets for drug development, Gene 623 (2017) 48–62.

[39] B. Trost, M. Haakensen, V. Pittet, B. Ziola, A. Kusalik, Analysis and comparison of the pan-genomic properties of sixteen well-characterized bacterial genera, Environ. Microbiol. 10 (2010) 258.

[40] E. Trost, J. Blom, S. de Castro Soares, I.H. Huang, A. Al-Dilaimi, J. Schroder, et al., Pangenomic study of *Corynebacterium diphtheriae* that provides insights into the genomic diversity of pathogenic isolates from cases of classical diphtheria, endocarditis, and pneumonia, J. Bacteriol. 194 (12) (2012) 3199–3215.

[41] S.C. Soares, A. Silva, E. Trost, J. Blom, R. Ramos, A. Carneiro, A. Ali, A.R. Santos, A.C. Pinto, C. Diniz, E.G. Barbosa, F.A. Dorella, F. Aburjaile, F.S. Rocha, K.K.F. Nascimento, L.C. Guimara, S. Almeida, S.S. Hassan, S.M. Bakhtiar, U.P. Pereira, V.A.C. Abreu, M.P.C. Schneider, A. Miyoshi, A. Tauch, V. Azevedo, The pan-genome of the animal pathogen *Corynebacterium pseudotuberculosis* reveals differences in genome plasticity between the biovar ovis and equi strains, PLoS One 8 (1) (2013).

[42] X. Qin, J.R. Galloway-Peña, J. Sillanpaa, J.H. Roh, S.R. Nallapareddy, S. Chowdhury, A. Bourgogne, T. Choudhury, D.M. Muzny, C.J. Buhay, Y. Ding, S. Dugan-Rocha, W. Liu, C. Kovar, E. Sodergren, S. Highlander, J.F. Petrosino, K.C. Worley, R.A. Gibbs, G.M. Weinstock, B.E. Murray, Complete genome sequence of Enterococcus faecium strain TX16 and comparative genomic analysis of Enterococcus faecium genomes, *BMC Microbiol.* 12 (2012 Jul 7) 135.

[43] D.A. Rasko, M.J. Rosovitz, G.S. Myers, E.F. Mongodin, W.F. Fricke, P. Gajer, J. Crabtree, M. Sebaihia, N.R. Thomson, R. Chaudhuri, I.R. Henderson, V. Sperandio, J. Ravel, The pangenome structure of *Escherichia coli*: comparative genomic analysis of *E. coli* commensal and pathogenic isolates, J. Bacteriol. 190 (20) (2008) 6881–6893.

[44] E.N. Gordienko, M.D. Kazanov, M.S. Gelfand, Evolution of pan-genomes of *Escherichia coli, Shigella* spp., and *Salmonella enterica*, J. Bacteriol. 195 (12) (2013) 2786–2792.

[45] G. Vieira, V. Sabarly, P.Y. Bourguignon, M. Durot, F. Le Fèvre, D. Mornico, et al., The core and pan-metabolism in the *Escherichia coli* species, J. Bacteriol. 193 (6) (2011) 1461–1472.

[46] L.G. Snipen, D.W. Ussery, A domain sequence approach to pangenomics: applications to *Escherichia coli*, F1000Res. 1 (2013) 19.

[47] J.S. Hogg, F.Z. Hu, B. Janto, R. Boissy, J. Hayes, R. Keefe, J.C. Post, G.D. Ehrlich, Characterization and modeling of the *Haemophilus influenzae* core and supragenomes based on the complete genomic sequences of Rd and 12 clinical nontypeable strains, Genome Biol. 8 (2007) R103.

[48] H. Gressmann, B. Linz, R. Ghai, K.-P. Pleissner, R. Schlapbach, Y. Yamaoka, C. Kraft, S. Suerbaum, T.F. Meyer, M. Achtman, Gain and loss of multiple genes during the evolution of *Helicobacter pylori*, PLoS Genet. 1 (4) (2005) e43, 0419-0428.

[49] A. Ali, A. Naz, S.C. Soares, M. Bakhtiar, S. Tiwari, S.S. Hassan, F. Hanan, R. Ramos, U. Pereira, D. Barh, H.C. Pereira, H.C.P. Pereira Figueiredo, D.W. Ussery, A. Miyoshi, A. Silva, V. Azevedo, Pan-genome analysis of human gastric pathogen *H. pylori:* comparative genomics and pathogenomics approaches to identify regions associated with pathogenicity and prediction of potential core therapeutic targets, Biomed. Res. Int. 2015 (2015) 139580.

[50] I. Uchiyama, J. Albritton, M. Fukuyo, K.K. Kojima, K. Yahara, I. Kobayashi, A novel approach to *Helicobacter pylori* pan-genome analysis for identification of genomic islands, PLoS One 11 (8) (2016).

[51] A.H. van Vliet, Use of pan-genome analysis for the identification of lineage-specific genes of *Helicobacter pylori*, FEMS Microbiol. Lett. 364 (2017), fnw296.

[52] G. D'Auria, N. Jiménez-Hernández, F. Peris-Bondia, A. Moya, A. Latorre, Legionella pneumophila pangenome reveals strain-specific virulence factors, BMC Genomics 11 (2010) 181.

[53] X. Deng, A.M. Phillippy, Z. Li, S.L. Salzberg, W. Zhang, Probing the pan-genome of Listeria monocytogenes: new insights into intraspecific niche expansion and genomic diversification, BMC Genomics 11 (2010) 500.

[54] C. Kuenne, A. Billion, M.A. Mraheil, A. Strittmatter, R. Daniel, A. Goesmann, S. Barbuddhe, T. Hain, T. Chakraborty, Reassessment of the *Listeria monocytogenes* pan-genome reveals dynamic integration hotspots and mobile genetic elements as major components of the accessory genome, BMC Genomics 14 (2013) 47.

[55] F. Zakham, L. Belayachi, D. Ussery, M. Akrim, A. Benjouad, R. El Aouad, M. Ennaji, Mycobacterial species as case-study of comparative genome analysis, Cell. Mol. Biol. 57 (2011) OL1462–OL1469.

[56] P. Supply, M. Marceau, S. Mangenot, D. Roche, C. Rouanet, V. Khanna, et al., Genomic analysis of smooth tubercle bacilli provides insights into ancestry and pathoadaptation of *Mycobacterium tuberculosis*, Nat. Genet. 45 (2013) 172–179.

[57] V. Periwal, A. Patowary, S.K. Vellarikkal, A. Gupta, M. Singh, A. Mittal, S. Jeyapaul, R.K. Chauhan, A.V. Singh, P.K. Singh, P. Garg, V.M. Katoch, K. Katoch, D.S. Chauhan, S. Sivasubbu, V. Scaria, Comparative whole-genome analysis of clinical isolates reveals characteristic architecture of *Mycobacterium tuberculosis* pangenome, PLoS One 10 (4) (2015).

[58] M.N. Ezewudo, S.J. Joseph, S. Castillo-Ramirez, D. Dean, C. Del Rio, X. Didelot, J.-A. Dillon, R.F. Selden, W.M. Shafer, R.S. Turingan, M. Unemo, T.D. Read, Population structure of *Neisseria gonorrhoeae* based on whole genome data and its relationship with antibiotic resistance, PeerJ. 3 (2015).

[59] A. Sharma, N. Sangwan, V. Negi, P. Kohli, J.P. Khurana, D.L.N. Rao, R. Lal, Pan-genome dynamics of *Pseudomonas* gene complements enriched across hexachlorocyclohexane dumpsite, BMC Genomics 16 (2015) 313.

[60] S. Fischer, J. Klockgether, P. Morán Losada, P. Chouvarine, N. Cramer, C.F. Davenport, S. Dethlefsen, M. Dorda, A. Goesmann, R. Hilker, S. Mielke, Intraclonal genome diversity of the major *Pseudomonas aeruginosa* clones C and PA 14, Environ. Microbiol. Rep. 8 (2) (2016) 227–234.

[61] J. Mosquera-Rendón, A.M. Rada-Bravo, S. Cárdenas-Brito, M. Corredor, E. Restrepo-Pineda, A. Benítez-Páez, Pangenome-wide and molecular evolution analyses of the *Pseudomonas aeruginosa* species, BMC Genomics 17 (2016) 45.

[62] L. Freschi, A.T. Vincent, J. Jeukens, J.G. Emond-Rheault, I. Kukavica-Ibrulj, M.J. Dupont, S.J. Charette, B. Boyle, R.C. Levesque, The *Pseudomonas aeruginosa* pan-genome provides new insights on its population structure, horizontal gene transfer and pathogenicity, Genome Biol. Evol. 11 (1) (2018) 109–120.

[63] J. Wu, T. Yu, Q. Bao, F. Zhao, Evidence of extensive homologous recombination in the core genome of *Rickettsia*, Comp. Funct. Genomics 2009 (2009) 510270.

[64] A. Jacobsen, R.S. Hendriksen, F.M. Aaresturp, D.W. Ussery, C. Friis, The *Salmonella enterica* pan-genome, Microb. Ecol. 62 (2011) 487–504.

[65] R.S. Gerrish, A.L. Gill, V.G. Fowler, S.R. Gill, Development of pooled suppression subtractive hybridization to analyze the pangenome of *Staphylococcus aureus*, J. Microbiol. Methods 81 (1) (2010) 56–60.

[66] D.M. Jamrozy, S.R. Harris, N. Mohamed, S.J. Peacock, C.Y. Tan, J. Parkhill, A.S. Anderson, M.T. Holden, Pan-genomic perspective on the evolution of the *Staphylococcus aureus* USA300 epidemic, Microb. Genom. 2 (5) (2016).

[67] S. Conlan, L.A. Mijares, NISC Comparative Sequencing Program, J. Becker, R.W. Blakesley, G.G. Bouffard, S. Brooks, H. Coleman, J. Gupta, N. Gurson, M. Park, B. Schmidt, P.J. Thomas, M. Otto, H.H. Kong, P.R. Murray, J.A. Segre, *Staphylococcus epidermidis* pan-genome sequence analysis reveals diversity of skin commensal and hospital infection-associated isolates, Genome Biol. 13 (7) (2012) R64.

[68] A. Puymège, S. Bertin, G. Guédon, S. Payot, Analysis of *Streptococcus agalactiae* pan-genome for prevalence, diversity and functionality of integrative and conjugative or mobilizable elements integrated in the tRNA Lys CTT gene, Mol. Gen. Genomics. 290 (2015) 1727–1740.

[69] C. Donati, N.L. Hiller, H. Tettelin, A. Muzzi, N.J. Croucher, S.V. Angiuoli, M. Oggioni, J.C. Dunning Hotopp, F.Z. Hu, D.R. Riley, A. Covacci, T.J. Mitchell, S.D. Mitchell, M. Kilian, G.D. Ehrlich, R. Rappuoli, E.R. Moxon, V. Masignani, Structure and dynamics of the pan-genome of *Streptococcus pneumoniae* and closely related species, Genome Biol. 11 (2010) R107.

[70] A. Muzzi, C. Donati, Population genetics and evolution of the pan-genome of *Streptococcus pneumoniae*, Int. J. Med. Microbiol. 301 (2011) 619–622.

[71] M.L. Tong, Q. Zhao, L.L. Liu, X.Z. Zhu, K. Gao, H.L. Zhang, R. Lin, J.-J. Niu, Z.-L. Ji, T.C. Yang, Whole genome sequence of the *Treponema pallidum* subsp. *pallidum* strain Amoy: an Asian isolate highly similar to SS14, PLoS One 12 (8) (2017).

[72] L. Vezzulli, C. Grande, G. Tassistro, I. Brettar, M.G. Höfle, R.P.A. Pereira, D. Mushi, A. Pallavicini, P. Vassallo, C. Pruzzo, Whole-genome enrichment provides deep insights into *Vibrio cholerae* metagenome from an African river, Microb. Ecol. 73 (3) (2017) 734–738.

[73] M. Eppinger, P.L. Worsham, M.P. Nikolich, D.R. Riley, Y. Sebastian, S. Mou, M. Achtman, L.E. Lindler, J. Ravel, Genome sequence of the deep-rooted *Yersinia pestis* strain Angola reveals new insights into the evolution and pangenome of the plague bacterium, J. Bacteriol. 192 (6) (2010) 1685–1699.
[74] C. Yang, Y. Cui, Genome-wide variation analysis of *Yersinia pestis*, in: *Yersinia pestis* Protocols, Springer, Singapore, 2018, pp. 61–66.
[75] X. Chen, Y. Zhang, Z. Zhang, Y. Zhao, C. Sun, M. Yang, J. Wang, Q. Liu, B. Zhang, M. Chen, J. Yu, J. Wu, Z. Jin, J. Xiao, PGAweb: a web server for bacterial pan-genome analysis, Front. Microbiol. 9 (2018) 1910.
[76] M.J. Brittnacher, C. Fong, H.S. HaydeN, M.A. Jacobs, M. Radey, L. Rohmer, PGAT: a multistrain analysis resource for microbial genomes, Bioinformatics 27 (17) (2011) 2429–2430.
[77] C. Laing, C. Buchanan, E.N. Taboada, Y. Zhang, A. Kropinski, A. Villegas, J.E. Thomas, V.P.J. Gannon, Pan-genome sequence analysis using Panseq: an online tool for the rapid analysis of core and accessory genomic regions, BMC Bioinform. 11 (2010) 461.
[78] E.A. Ozer, J.P. Allen, A.R. Hauser, Characterization of the core and accessory genomes of *Pseudomonas aeruginosa* using bioinformatic tools Spine and AGEnt, BMC Genomics 15 (2014) 737.
[79] J.R. Bayjanov, R.J. Siezen, S.A.F.T. van Hijum, PanCGHweb: a web tool for genotype calling in pangenome CGH data, Bioinformatics 26 (9) (2010) 1256–1257.
[80] Ding, W., 2017. Pan-Genome Analysis, Visualization and Exploration. Dissertation. Universität Tübingen.
[81] W. van Schaik, The human gut resistome, Philos. Trans. R. Soc. B 370 (2015).
[82] WHO (World Health Organization), 2014. Antimicrobial Resistance Fact Sheet N°194" Archived From the Original on March 10, 2015. Retrieved December 15, 2018.
[83] R. Baxter, G.T. Ray, B.H. Fireman, Case-control study of antibiotic use and subsequent *Clostridium difficile*-associated diarrhea in hospitalized patients, Infect. Control Hosp. Epidemiol. 29 (1) (2008) 44–50.
[84] A.H. Gifford, K.B. Kirkland, Risk factors for *Clostridium difficile*-associated diarrhea on an adult hematology-oncology ward, Eur. J. Clin. Microbiol. Infect. Dis. 25 (12) (2006) 751–755.
[85] T.N. Palmore, S. Sohn, S.F. Malak, J. Eagan, K.A. Sepkowitz, Risk factors for acquisition of *Clostridium difficile*-associated diarrhea among outpatients at a cancer hospital, Infect. Control Hosp. Epidemiol. 26 (8) (2005) 680–684.
[86] C.J. Kristich, L.B. Rice, C.A. Arias, Enterococcal infection—treatment and antibiotic resistance, in: Enterococci: From Commensals to Leading Causes of Drug Resistant Infection [Internet], Massachusetts Eye and Ear Infirmary, 2014 Feb 6.
[87] R.D. Gonzales, P.C. Schreckenberger, M.B. Graham, S. Kelkar, K. DenBesten, J.P. Quinn, Infections due to vancomycin-resistant *Enterococcus faecium* resistant to linezolid, Lancet 357 (9263) (2001) 1179.
[88] M.B. Edmond, J.F. Ober, D.L. Weinbaum, M.A. Pfaller, T. Hwang, M.D. Sanford, R.P. Wenzel, Vancomycin-resistant *Enterococcus faecium* bacteremia: risk factors for infection, Clin. Infect. Dis. 20 (5) (1995) 1126–1133.
[89] S.E. Dorman, S.G. Schumacher, D. Alland, P. Nabeta, D.T. Armstrong, B. King, S.L. Hall, S. Chakravorty, D.M. Cirillo, N. Tukvadze, N. Bablishvili, W. Stevens, L. Scott, C. Rodrigues, M.I. Kazi, M. Joloba, L. Nakiyingi, M.P. Nicol, Y. Ghebrekristos, I. Anyango, W. Murithi, R. Dietze, R.L. Peres, A. Skrahina, V. Auchynka, K.K. Chopra, M. Hanif, X. Liu, X. Yuan, C.C. Boehme, J.J. Ellner, C.M. Denkinger, Xpert MTB/RIF Ultra for detection of *Mycobacterium tuberculosis* and rifampicin resistance: a prospective multicentre diagnostic accuracy study, Lancet Infect. Dis. 18 (2018) 76–84.
[90] J. Gratrix, S. Plitt, L. Turnbull, P. Smyczek, J. Brandley, R. Scarrott, P. Naidu, P. Parker, B. Blore, A. Bull, S. Shokoples, Prevalence and antibiotic resistance of *Mycoplasma genitalium* among STI clinic attendees in Western Canada: a cross-sectional analysis, BMJ Open 7 (7) (2017) e016300.
[91] M.E. Velázquez-Meza, *Staphylococcus aureus* methicillin-resistant: emergence and dissemination, Salud Publica Mex. 47 (5) (2005) 381–387.

[92] G. Cornaglia, M. Ligozzi, A. Mazzariol, M. Valentini, G. Orefici, R. Fontana, Rapid increase of resistance to erythromycin and clindamycin in Streptococcus pyogenes in Italy, 1993-1995. The Italian Surveillance Group for Antimicrobial Resistance, Emerg. Infect. Dis. 2 (4) (1996) 339–342.
[93] M. Li, T. Oshima, T. Horikawa, K. Tozawa, T. Tomita, H. Fukui, J. Watari, H. Miwa, Systematic review with meta-analysis: Vonoprazan, a potent acid blocker, is superior to proton-pump inhibitors for eradication of clarithromycin-resistant strains of *Helicobacter pylori*, Helicobacter 23 (4) (2018).
[94] K. Misawa, N. Tarumoto, S. Tamura, M. Osa, T. Hamamoto, A. Yuki, Y. Kouzaki, K. Imai, R.L. Ronald, T. Yamaguchi, T. Murakami, S. Maesaki, Y. Suzuki, A. Kawana, T. Maeda, Single nucleotide polymorphisms in genes encoding penicillin-binding proteins in β-lactamase-negative ampicillin-resistant Haemophilus influenzae in Japan, BMC Res. Notes 11 (2018) 53.
[95] C.J. Norsigian, E. Kavvas, Y. Seif, B.O. Palsson, J.M. Monk, iCN718, an updated and improved genome-scale metabolic network reconstruction of Acinetobacter baumannii AYE, Front. Genet. 9 (2018) 121.
[96] L. Collado, N. Muñoz, L. Porte, S. Ochoa, C. Varela, I. Muñoz, Genetic diversity and clonal characteristics of ciprofloxacin-resistant *Campylobacter jejuni* isolated from Chilean patients with gastroenteritis, Infect. Genet. Evol. 58 (2018) 290–293.
[97] K. Town, H. Bolt, S. Croxford, M. Cole, S. Harris, N. Field, G. Hughes, *Neisseria gonorrhoeae* molecular typing for understanding sexual networks and antimicrobial resistance transmission: a systematic review, J. Infect. 76 (2018) 507–514.
[98] V. Menon, R. Davis, N. Shackel, B.A. Espedido, A.G. Beukers, S.O. Jensen, S.J. van Hal, Failure of daptomycin β-Lactam combination therapy to prevent resistance emergence in *Enterococcus faecium*, Diagn. Microbiol. Infect. Dis. 90 (2) (2018) 120–122.
[99] T. Pillonel, P. Nordmann, C. Bertelli, G. Prod'hom, L. Poirel, G. Greub, Resistome analysis of a carbapenemase (OXA-48)-producing and colistin-resistant *Klebsiella pneumoniae* strain, Antimicrob. Agents Chemother. 62 (2018), e00076-18.
[100] R.A. Kingsley, C.L. Msefula, N.R. Thomson, S. Kariuki, K.E. Holt, M.A. Gordon, D. Harris, L. Clarke, S. Whitehead, V. Sangal, K. Marsh, M. Achtman, M.E. Molyneux, M. Cormican, J. Parkhill, C.A. MacLennan, R.S. Heyderman, G. Dougan, Epidemic multiple drug resistant *Salmonella typhimurium* causing invasive disease in sub-Saharan Africa have a distinct genotype, Genome Res. 19 (2009) 2279–2287.
[101] C.F. Flach, M. Genheden, J. Fick, D.G. Joakim Larsson, A comprehensive screening of *Escherichia coli* isolates from Scandinavia's largest sewage treatment plant indicates no selection forantibiotic resistance, Environ. Sci. Technol. 52 (19) (2018) 11419–11428.
[102] D. Shortridge, M.A. Pfaller, M. Castanheira, R.K. Flamm, Antimicrobial activity of ceftolozane-tazobactam tested against *Enterobacteriaceae* and *Pseudomonas aeruginosa* with various resistance patterns isolated in US hospitals (2013–2016) as part of the surveillance program: program to assess ceftolozane-tazobactam susceptibility, Microb. Drug Resist. 24 (5) (2018) 563–578.
[103] M. Pinto, A. González-Díaz, M.P. Machado, S. Duarte, L. Vieira, J.A. Carriço, S. Martib, M.P. Bajanca-Lavadog, J.P. Gomes, Insights into the population structure and pan-genome of *Haemophilus influenzae*, Infect. Genet. Evol. 67 (2019) 126–135.
[104] S.B. Levy, B. Marshall, Antibacterial resistance worldwide: causes, challenges and responses, Nat. Med. 10 (12S) (2004) S122–S129.
[105] C. Baker-Austin, M.S. Wright, R. Stepanauskas, J. McArthur, Co-selection of antibiotic and metal resistance, Trends Microbiol. 14 (4) (2006) 176–182.
[106] I. Bueno, J. Williams-Nguyen, H. Hwang, J.M. Sargeant, A.J. Nault, R.S. Singer, Systematic Review: impact of point sources on antibiotic-resistant bacteria in the natural environment, Zoonoses Public Health 65 (1) (2018) e162–e184.
[107] R.W. Meek, H. Vyas, L.J.V. Piddock, Nonmedical uses of antibiotics: time to restrict their use? PLoS Biol. 13 (10) (2015).
[108] A.H. Holmes, L.S. Moore, A. Sundsfjord, M. Steinbakk, S. Regmi, A. Karkey, P.J. Guerin, L.J. Piddock, Understanding the mechanisms and drivers of antimicrobial resistance, Lancet 387 (10014) (2016) 176–187.
[109] M.D. Phan, K.M. Peters, S. Sarkar, S.W. Lukowski, L.P. Allsopp, D. Gomes Moriel, M.E.S. Achard, M. Totsika, V.M. Marshall, M. Upton, S.A. Beatson, M.A. Schembri, The serum resistome of a globally disseminated multidrug resistant uropathogenic *Escherichia coli* clone, PLoS Genet. 9 (10) (2013).

[110] WHO (World Health Organization), 2017. Global Priority List of Antibiotic-Resistant Bacteria to Guide Research, Discovery, and Development of New Antibiotics. http://www.who.int/medicines/publications.WHO-PPL-Short_Summary_25Feb-ET_NM_WHO.pdf.

[111] R. Xie, X.D. Zhang, Q. Zhao, B. Peng, J. Zheng, Analysis of global prevalence of antibiotic resistance in *Acinetobacter baumannii* infections disclosed a faster increase in OECD countries, Emerg. Microbes Infect. 7 (2018) 31.

[112] M. Bissessor, S.N. Tabrizi, Y. Twin, H. Abdo, K.C. Fairley, M.Y. Chen, L.A. Vodstrcil, J.S. Jensen, J.S. Hocking, S.M. Garland, C.S. Bradshaw, Macrolide resistance and azithromycin failure in a *Mycoplasma genitalium*-infected cohort and response of azithromycin failures to alternative antibiotic regimens, Clin. Infect. Dis. 60 (8) (2014) 1228–1236.

[113] A.R. Wattam, J.J. Davis, R. Assaf, S. Boisvert, T. Brettin, C. Bun, et al., Improvements to PATRIC, the all-bacterial bioinformatics database and analysis resource center, Nucleic Acids Res. 45 (2016) D535–D542.

[114] B. Liu, M. Pop, ARDB—antibiotic resistance genes database, Nucleic Acids Res. 37 (Suppl 1) (2008) D443–D447. Published online 2 October 2008.

[115] P. Lapierre, J.P. Gogarten, Estimating the size of the bacterial pan-genome, Trends Genet. 25 (3) (2009) 107–110.

[116] I. Sultan, S. Rahman, A.T. Jan, M.T. Siddiqui, A.H. Mondal, O.M.R. Haq, Antibiotics, resistome and resistance mechanisms: a bacterial perspective, Front. Microbiol. 9 (2018) 2066.

[117] Y. Hu, X. Yang, J. Li, N. Lv, F. Liu, J. Wu, I.Y.C. Lin, N. Wu, B.C. Weimer, G.F. Gao, Y. Liu, B. Zhua, The bacterial mobile resistome transfer network connecting the animal and human microbiomes, Appl. Environ. Microbiol. 82 (2016) 6672–6681.

[118] E. Zankari, H. Hasman, S. Cosentino, M. Vestergaard, S. Rasmussen, O. Lund, F.M. Aarestrup, M.V. Larsen, Identification of acquired antimicrobial resistance genes, J. Antimicrob. Chemother. 67 (2012) 2640–2644.

[119] D.R. Knight, M.M. Squire, D.A. Collins, T.V. Riley, Genome analysis of *Clostridium difficile* PCR ribotype 014 lineage in Australian pigs and humans reveals a diverse genetic repertoire and signatures of long-range interspecies transmission, Front. Microbiol. 7 (2138) (2016) 2138. Published online 2017 Jan 11.

[120] Y.A. Kim, Y.S. Park, T. Youk, H. Lee, K. Lee, Correlation of aminoglycoside consumption and amikacin- or gentamicin-resistant *Pseudomonas aeruginosa* in long-term nationwide analysis: is antibiotic cycling an effective policy for reducing antimicrobial resistance? Ann. Lab. Med. 38 (2018) 176–178.

Further reading

[121] Y. Zhang, T. Luo, C. Yang, X. Yue, R. Guo, X. Wang, M. Buren, Y. Cui, … X. Dai, Phenotypic and molecular genetic characteristics of *Yersinia pestis* at an emerging natural plague focus, Junggar Basin, China, Am. J. Trop. Med. Hyg. 98 (1) (2018) 231–237.

[122] A. Kandavelmani, S. Piramanayagam, Comparative genomics of Mycoplasma: insights on genome reduction and identification of potential antibacterial targets, Biomed. Biotechnol. Res. J. 3 (1) (2019) 9.

[123] M.C. Fookes, J. Hadfield, S. Harris, S. Parmar, M. Unemo, J.S. Jensen, N.R. Thomson, *Mycoplasma genitalium*: whole genome sequence analysis, recombination and population structure, BMC Genomics 18 (1) (2017) 993.

[124] J.I. Glass, N. Assad-Garcia, N. Alperovich, S. Yooseph, M.R. Lewis, M. Maruf, C.A. Hutchison, H. O. Smith, J.C. Venter, Essential genes of a minimal bacterium, Proc. Natl. Acad. Sci. 103 (2) (2006) 425–430.

[125] C.M. Fraser, J.D. Gocayne, O. White, M.D. Adams, R.A. Clayton, R.D. Fleischmann, C.J. Bult, A. R. Kerlavage, G. Sutton, J.M. Kelley, J.L. Fritchman, The minimal gene complement of *Mycoplasma genitalium*, Science 270 (5235) (1995) 397–404.

CHAPTER 11

Pan-genomics of virus and its applications

Marta Giovanetti[a,b], Alvaro Salgado[b], Vagner de Souza Fonseca[a,b,c], Fraga de Oliveira Tosta[b], Joilson Xavier[a,d], Jaqueline Goes de Jesus[a,d], Felipe Campos Melo Iani[b,e], Talita Emile Ribeiro Adelino[b,e], Fernanda Khouri Barreto[d,f], Nuno Rodrigues Faria[g], Tulio de Oliveira[c], Luiz Carlos Junior Alcantara[a,b]

[a]Laboratório de Flavivírus, IOC, Fundação Oswaldo Cruz, Rio de Janeiro, Brazil
[b]Laboratório de Genética Celular e Molecular, ICB, Universidade Federal de Minas Gerais, Belo Horizonte, Brazil
[c]KwaZulu-Natal Research Innovation and Sequencing Platform (KRISP), College of Health Sciences, University of KwaZulu-Natal, Durban, South Africa
[d]Laboratório de Patologia Experimental, Instituto Gonçalo Moniz, Fiocruz Bahia, Salvador, Brazil
[e]Fundação Ezequeil Dias (Funed), Belo Horizonte, Brazil
[f]Instituto Multidisciplinar em Saúde—IMS, Universidade Federal da Bahia (UFBA), Salvador, Brazil
[g]Department of Zoology, University of Oxford, Oxford, United Kingdom

1 Next-generation sequencing strategies

Exploring the genetic information of viruses has been made possible due to the technology of DNA sequencing. Currently, there are 4958 described virus species, according to the International Committee on Taxonomy of Viruses (ICTV, 2018). Although many species still do not have their genomes sequenced, viral species with medical, biotechnological, and environmental relevance usually have more than one complete or partial publicly deposited genomes [1].

This large and diverse viral genetic information available in public databases allows us to begin addressing the genetic complexity of viruses, which originates from several molecular mechanisms, including insertion/deletion events, different rates of nucleotide substitution, as well as intra- and inter-genotype recombination and reassortment events [2]. These mechanisms directly affect the genetic repertoire of viral populations of different hosts and habitats, leading to important implications in molecular diagnosis, pathogenesis, and viral epidemiology [3].

Partial viral genomes sequencing has been used to: (i) detect drug resistance in both DNA and RNA viruses [4, 5] and (ii) perform phylogenetic analyses for the assignment of genotypes [6]. Despite the broad array of discoveries and advancements brought by that approach, whole genome sequencing (WGS) consistently provides more information than the sequencing of a reduced number of genes. Therefore, WGS allows for: (i) the detection of all known drug-resistant variants and the identification of new

Pan-genomics: Applications, Challenges, and Future Prospects
https://doi.org/10.1016/B978-0-12-817076-2.00011-1
© 2020 Elsevier Inc.
All rights reserved.

ones; (ii) the identification of mutations associated with disease transmission or severity; (iii) better phylogenetic resolution; and (iv) genomic surveillance [7–10].

Regarding the different methods of DNA sequencing, the first strategy applied was that of chain termination with dideoxynucleotides (ddNTPs)—(Sanger sequencing) [11]. Commonly used for confirmation in some cases, given its high accuracy, it has an extremely low throughput, as well as being laborious and time consuming. The scenario started to change in 2004, with the emergence of the first DNA sequencing technologies known as NGS (next-generation sequencing), allowing for a new approach of large-scale sequencing (HTS—high-throughput sequencing) [12].

In the following years, several *Second-Generation Sequencing* platforms were developed, based mainly on the following technologies: (i) sequencing by ligation (SOLiD), (ii) ion sensing synthesis technology (IonTorrent), and (iii) sequencing by synthesis (Illumina) [13]. Second-generation platforms allowed for a more in-depth characterization of the genomic variability of viruses, while providing large amounts (millions of reads) of data for each individual sequence for the same clone or amplicon. Despite this, its major limitation is the size of each individual sequenced, not being possible to obtain complete viral genomes in a single sequencing reaction.

Recently, the development of single-molecule third-generation sequencing approaches is now providing the first promising results in the sequencing of complete viral genomes [14,15]. Two platforms are currently available: Pacific Biosciences (PacBio) RS and RS II systems, and the Oxford Nanopore Technologies (ONT) systems (MinION, GridION, and PromethION). PacBio uses Single-Molecule, Real Time Technology Sequencing (SMRT) (PacBio, http://www.pacb.com/).

Each SMRT cell of the PacBio RS II system has a typical throughput of 0.5–1 GB, with an average read length of 10 kb. Despite this, PacBio reads still present a significantly higher error rate when compared to second- and first-generation sequencing technologies (>10%–15%) [16].

In 2014, the MinION from ONT was released to early access users [17], heralding the potential for highly portable "lab-in-a-suitcase" sequencing, which is capable of sequencing DNA or RNA in a real-time scale, with ultra-long-reads. The MinION is pocket sized and is controlled and powered through a laptop USB connection.

In this technology, the DNA or RNA strands passes through various nanopores, which connect the two sides of a semiconductive layer, anchored by specialized proteins. A voltage is applied between the surfaces of the layer and, as the DNA or RNA strands move through the nanopores, each of its nucleotides creates a characteristic disruption in the electrical current flowing through the pore. This nanopore signal, which is different for each type of nucleotide, is used to determine the sequence of bases on the DNA or RNA strand [18].

Ongoing improvements to the launched barcoding kits in the nanopore sequencing technology had the potential to increase the number of generated genomes per sequencing run from 12 to 96, which could also increase the number of genome sequences available from affected regions and allow more detailed investigations of the association between pathogens mutations and environmental context with less costs.

However, nanopore technology also has a lower accuracy when compared to older technologies, with an error profile <10% insertion-deletion mutations (indels) rate [19].

For data analysis, most bioinformatics tools take FASTA or FASTQ files as input, where base calling has already been done during the sequencing process or offline with the sequencer. For new platforms in their early stages, however, original raw data files may be useful for some applications. Currently, the MinION outputs one FAST5 file per read. Much like the h5 file format adopted by PacBio, the FAST5 file format is based on the hierarchical data format 5 (HDF5) standard (https://www.hdfgroup.org). FAST5 files have a hierarchical structure, meaning that they can store both the metadata associated with a read, along with the events (such as aggregated bulk current measurements) preprocessed by the sequencing device [19]. Despite this, nanopore long reads simplify assembly and sequencing of repetitive regions and speed up the identification of new species and metagenomic experiments. For those reasons, the MinION sequencer is getting much attention from the genomic community, mainly for genomic viral surveillance and genomic epidemiology areas, as they can benefit from the real-time nature of this sequencing platform. Importantly, the MinION has been used in field situations, including in diagnostic tent laboratories during the Ebola epidemic [20,21] and in a roving bus-based mobile laboratory in Brazil as part of the ZIBRA project (http://www.zibraproject.org) [22]. Others have taken the MinION to more extreme environments where even the smallest traditional benchtop sequencer could not go, including the Arctic [23] and Antarctic [24], a deep mine [25], and zero gravity aboard the reduced gravity aircraft [26], and the International Space Station [27].

The shortage of complete genomic sequences represents a limiting point for the study of viral genetic divergence as well as of population dynamics (genotypes and subgenotypes), pathogenesis and vectors associated with virus transmission among human populations [9, 28]. In this context, sequencing of viral genomes plays an important role in the fight against emerging and reemerging epidemics, as well as in the early detection and/or identification of new potential emerging pathogens through metagenomics approaches. Metagenomics, in this sense, can be used as a tool to monitor, at an early stage, the introduction of new pathogens in specific regions. It may have important applications for the epidemiological surveillance, outbreak investigation, and diagnosis of infectious diseases [29] of both known and unknown pathogens. NGS-based metagenomics, therefore, can be used as a complementary tool to monitor the emergence and spread of new human pathogens, a central concern in public health in tropical regions.

Therefore, as sequencing chemistry and technologies progress, such techniques are likely to become key tools for the construction of viral pan-genomes. We expect that computational pan-genomics will allow increased power and accuracy, for example, by allowing the pan-genome structure of a viral population to be directly compared with that of a susceptible host population. Portable genome sequencing technology and digital epidemiology platforms form the foundation for both real-time pathogen and disease surveillance systems and outbreak response efforts, all of which exist within the One Health context, in which surveillance, outbreak detection, and response span the human, animal and environmental health domains.

2 Genomic surveillance

Infectious diseases continue to be one of the leading causes of death worldwide [30] and pathogens such as viruses can be considered notorious mutation machines. They can evolve and spread rapidly, leading to the emergence of newly mutated human pathogens, more virulent strains, as well as antibiotic- and drug-resistant organisms [31,32]. In this context, genomic surveillance aims are: (i) to perform global surveillance of pathogens using WGS; (ii) to understand drug resistance, emergence, and spread of viral pathogens; and (iii) to provide actionable data.

Several approaches have been developed and are widely used for the quick detection and identification of viral pathogens (i.e., diagnostics). Some of them are based on different serological and molecular strategies including, for example, assays based on real-time polymerase chain reaction [33]. Even though these kinds of approaches present high sensitivity and specificity for their purpose, they are more suitable for diagnostics only and cannot provide detailed genomic information [34].

Bearing these limitations in mind, the main point of developing new genomic surveillance tools is to answer the following inquiry: what sort of questions are important for genomic surveillance that cannot be addressed by conventional RT-qPCR or serology? (i) RT-qPCR assays do not allow genotype classification, neither does it help identify particular and/or characteristic transmission routes; (ii) RT-qPCR assays also do not allow to determine how fast a viral pathogen is being transmitted and in what direction it is spreading; (iii) serological and molecular assays also cannot help identify epidemiologically linked individuals, neither predict future outbreaks; and (iv) finally, serological and some molecular approaches cannot help to identify novel pathogenic agents and are, therefore, unsuitable for pathogen discovery [34].

NGS technologies produce significantly more raw data than other molecular diagnostic assays, including Sanger sequencing, and are also capable of informing not just pathogen diagnostics but also epidemiology [35]. This is why WGS of viral genomes by using new technologies plays an important role in the fight against emerging and reemerging epidemics [36,37]. The availability of high-throughput sequencing has also provided immense insights into the ecology of health-care-associated pathogens [38]. Therefore, real-time sequencing of entire pathogen genomes has become a standard and indispensable research tool for the critical role of genomic surveillance in the prevention and control of emerging infectious diseases [39], which justifies why NGS can be considered a powerful strategy that also allows the discovery of novel potential viral pathogens [34,40].

Considering pathogen surveillance in mind, bioinformatics tools and the combination of genomic and epidemiological data from viral infections can give essential information for understanding the past and the future of an epidemic, because genomic data generated by real-time sequencing can provide important information on how and when viruses were introduced in a particular site, their pattern, and determinants of dissemination in neighboring locations and the extent of genetic diversity, that is, its dynamics, making it possible to establish an effective surveillance framework on tracking the spread of infections to other geographic regions [28,40]. In this context, recently established international networks for real-time, portable genomic sequencing, genomic surveillance, and data analysis made it possible to monitor the evolution of viral genomes, to understand the origins of outbreaks and epidemics, to predict future outbreaks and to assist in the maintenance of updated diagnostic methods [40–43]. In addition, genomic surveillance framework allows to determine, through genome sequencing, the real-time molecular epidemiology of viruses circulating and cocirculating in different regions in a specific area, and also to detect and characterize the early emergence of new pathogens in large urban centers, generating data that can inform outbreak control responses [28,43]. Generated data regarding the molecular, epidemiological, phylogenetic, and geographical aspects of circulating viral pathogens in a specific setting contribute to a better understanding of those viral infections in a national and international context, assuming an important role in solving issues relevant to Public Health [44]. As a result, studies involving more in-depth molecular and dispersion analysis of circulating pathogens may help the World Health Organization appropriately adopt measures to control epidemics and to monitor the dynamics and spreading of new viral strains. However, even though NGS has advantages over diagnostics routine, all of the different strategies and technologies, developed by Illumina, Thermo Scientific, Oxford Nanopore, and others, are not yet considered a panacea. Remaining challenges include dealing with high data throughput, which requires sophisticated computational processing as well as the annotation of large amounts of sequencing data, high DNA or RNA input sample requirements (in some cases hundreds of nanograms), which often raises the need for previous PCR-based amplification approaches. On top of all this, there are relatively few researchers in the area with sufficient bioinformatics expertise and who are able to engage in near-patient or disease surveillance activities [44].

3 Genomic epidemiology

Genomic epidemiology has been applied to many outbreaks in the past few years and is becoming a widely accepted method to investigate outbreaks [45]. The use of WGS to understand infectious disease transmission and epidemiology is crucial to understanding the direction of an outbreak both in national and international contexts. The characterization of the evolutionary history and the geographic and temporal dissemination of viral pathogens could allow the identification of strains associated with a greater epidemic potential, suggesting targets for the development of more effective therapeutic interventions, and then allowing the establishment of an effective surveillance framework in the tracking of the spread of these strains to other geographic regions [46]. The goal of this kind of approach is to use the population structure of the pathogen to understand the overall dynamics of the epidemic [47]. Moreover, with improvements in sequencing technology and continuing optimization and standardization of bioinformatics algorithms, genomic epidemiology investigations can now be conducted during the course of an ongoing outbreak to provide real-time guidance for infection control interventions [47].

In addition, the rapid development of sequencing technologies has led to an explosion of pathogen sequencing data, which are increasingly collected as part of routine surveillance or clinical diagnostics. While sequencing has become cheaper, the analysis of sequence data has become a critical bottleneck.

Molecular epidemiological techniques can reconstruct the temporal and spatial spread of an outbreak. Similarly, by linking samples that originate from different geographic locations, phylogeographic methods can reconstruct the geographic spread and can differentiate distinct introductions. In this context, the use of powerful bioinformatics tools in the field of phylodynamics, defined as the study of how epidemiological, immunological, and evolutionary processes act and potentially interact to shape viral phylogenies, can support genomic surveillance and epidemiology. Phylodynamic models may aid in dating epidemic and pandemic origins and viral spread by mapping the geographic movement of a particular pathogen population in a specific area. Phylodynamic approaches have also been used to better understand viral transmission dynamics and spread within infected hosts. Such approaches can also be useful in ascertaining the effectiveness of viral control efforts, particularly for diseases with low reporting rates [48].

The potential exists to move from pathogen genomics, providing static "snapshots" of epidemics, often months after the cases occurred, to a situation where data are produced in real time, providing a detailed picture of the epidemic that is only a few days old. Such rapid results are crucial if the intention is to intervene in an outbreak rather than simply document it in retrospect.

4 Bioinformatic tools

NGS techniques have transformed genomic studies from the analysis of single or few genomes to an ever-increasing amount of genomic data, bringing with it the need to develop novel techniques to efficiently treat, novel tools to assemble, analyze, and derive useful information from overwhelmingly large datasets.

One of the ways to derive meaningful and useful information from a large genomic dataset is through pan-genomics. According to Vernikos et al. [49], a pan-genome defines the whole genetic repertoire of a phylogenetic clade and describes the set of all sequence entities (ORFs, genes, etc.) belonging to the genomes of interest. The union, intersection, and subsetting of units in the pan-genome can be classified as core genes, dispensable genes, and strain-specific genes.

The analysis of pan-genomes can uncover significant information regarding the genomes of interest. According to Carlos Guimaraes et al. [50], pan-genomic studies can help understand pathogen evolution, niche adaptation, population structure, and host interaction. Furthermore, it can help in vaccine and drug design, as well as in the identification of virulence genes.

In the context of virus investigations, pan-genomics, and bioinformatics in general face great challenges. Rapid extraction of genomic features with an evolutionary signal will facilitate evolutionary analyses ranging from the reconstruction of species phylogenies to tracing epidemic outbreaks.

4.1 Bioinformatic tools used in pan-genomic studies

According to Xiao et al. [51], Panseq [52], and PGAP (pan-genomes analysis pipeline) [53] were ranked as the two top most popular packages based on cumulative citations of peer-reviewed scientific publications at the end of 2014. Other tools applicable to virus pan-genomics include: EDGAR, ITEP, GET_HOMOLOGUES, CASTOR, and Genome Detective. Most pan-genome bioinformatic tools are based on orthologous and paralogous gene identification [50], and the functions of these software packages and tools usually include categorizing orthologous genes, calculating pan-genomic profiles, integrating gene annotations, and constructing phylogenies [51].

4.1.1 Panseq—Pan-genome sequence analysis program

As mentioned by Carlos Guimaraes et al. [50], Panseq is a freely available web-tool written in BioPerl, which is available at http://76.70.11.198/panseq. Panseq defines the core and accessory genome based on the sequence identity and segmentation length. The NRF (novel region finder) module first splits the genome sequence into fragments with predefined sizes, then the MUMmer alignment program [54] identifies the sequences and contiguous regions that are present or absent in the database. Next, the CAGF module (Core and Accessory Genome Finder) compares each individual fragment sequence to all sequences, adding single sequences that fit in with predefined parameters to the pan-genome. Each newly added to fragment sequence is used for subsequent comparisons, continuing this loop until all of the fragment sequences have been tested [53]. Panseq, according to Laing et al. [52], is able to determine core and accessory regions of genome assemblies and identify SNPs among the core genomic regions. In addition, it can select the most discriminatory loci among the accessory loci or core gene SNPs. Panseq, however, is not able to provide pan-genomic profile and functional enrichment analysis that is important for discriminating the functional relevance of the pan-genomic elements.

4.1.2 PGAP—Pan-genome analysis pipeline

PGAP is a stand-alone tool available at http://pgap.sf.net developed by Laing et al. [52] to perform pan-genome analysis, genetic variation, evolution, and function analysis of gene clusters [50]. The software uses two methods to calculate all of the analyses: (i) the GF method to detect homologous genes, and (ii) the MP method to detect orthologous genes.

The GF method is based on the protein BLAST and MCL (Markov clustering) algorithms. All of the protein sequences are brought together, and protein BLAST is performed; the results are filtered and clustered using the MCL algorithm [55,56]. The MP method is based on two algorithms: (i) Inparanoid to search orthologous and paralogous genes using BLAST. Then, the pairwise ortholog clusters are moved to (ii) MultiParanoid, which was specifically developed to search for gene clusters among multiple strains [50,55,57–59].

4.1.3 EDGAR (efficient database framework for comparative genome analyses using BLAST score ratios)

EDGAR is a web-tool available at https://edgar.computational.bio.uni-giessen.de/ [50]. It is designed to automatically perform genome comparisons in a high-throughput approach. It provides novel analysis features and significantly simplifies the comparative analysis of related genomes. The software supports a quick survey of evolutionary relationships and simplifies the process of obtaining new biological insights into the differential gene content of kindred genomes. Visualization features, like synteny plots or Venn diagrams, are offered to the scientific community through a web-based and therefore platform-independent user interface, where the precomputed data sets can be browsed [60]. According to Carlos Guimaraes et al. [50], this software performs homology analyses based on a specific cutoff that is automatically adjusted to the query data. The orthology analysis to calculate pan-genome, core-genome, and singletons is performed using BLAST score ratio values.

4.1.4 ITEP—Integrated toolkit for the exploration of microbial pan-genomes

ITEP is a stand-alone toolkit that is available for download at https://price.systemsbiology.net/itep [50]. It was developed to predict protein families, orthologous genes, functional domains, pan-genome, and metabolic networks for related microbial species [61]. Its workflow consists of a three-step process: data input, database building (startup scripts), and database analysis [50]. ITEP receives three different types of data: GenBank file format, organism file format, and groups file format, and all of the inputs require preprocessing before running the ITPEP toolkit (for more details, see the ITEP documentation). In database building, scripts are run to predict the gene locations, BLAST results, and clustering results. Finally, the package can perform core and variable genes analyses, phylogenies, metabolic reconstructions, and gene gain and loss patterns [50].

4.1.5 GET_HOMOLOGUES

GET_HOMOLOGUES is a stand-alone and open-source toolkit that was written in Perl and R that can be installed on personal machines. It was developed to perform pan-genome and comparative-genomic analysis [50,62].

PanFunPro: PAN-genome analysis based on FUNctionalPROfiles

PanFunPro is a stand-alone tool for pan-genome analysis using functional domains from HMM (hidden Markov models) to group homologous proteins into families based on their functional domain content [50,63,64]. In addition to pan-genome analyses, the software performs homology detection and genome annotation using HMM, genome and proteome estimation as well as gene ontology (GO) information [65].

4.1.6 CASTOR

The classification and annotation of virus genomes constitute important assets in the discovery of genomic variability, taxonomic characteristics, and disease mechanisms. Existing classification methods are often designed for specific well-studied families of viruses. Thus, the viral comparative genomic studies could benefit from more generic, fast, and accurate tools for classifying and typing newly sequenced strains of diverse virus families [66].

According to Rose et al. [67], CASTOR is a virus classification platform based on machine learning methods, inspired by a well-known technique in molecular biology: restriction fragment length polymorphism. It simulates, in silico, the restriction digestion of genomic material by different enzymes into fragments. It uses two metrics to construct feature vectors for machine learning algorithms in the classification step. The performance of CASTOR, its genericity, and robustness could permit performing novel and accurate large-scale virus studies. The CASTOR web platform provides an open access, collaborative, and reproducible machine learning classifiers. CASTOR can be accessed at http://castor.bioinfo.uqam.ca.

4.1.7 Genome Detective

According to Vilsker et al. [68], the analysis of viral genomes is especially challenging because of their high variability and deviation from reference genomes. This is aggravated by the increasing speed of identification, the continuous emergence of new viruses, and the relative rareness of viral fragments in metagenomic analyses.

Genome Detective (http://www.genomedetective.com/app/typing tool/virus/) was developed to address this problem [69]. It is an easy to use web-based software application that assembles the genomes of viruses quickly and accurately, designed to generate and analyze whole or partial viral genomes directly from NGS reads within minutes. The application gains accuracy by using a novel alignment method that uses a combination of amino acids and nucleotide scores to construct genomes by the reference-based linking of de novo contigs. Speed and accuracy were also gained by using DIAMOND with a UniProt90 reference dataset to sort viral taxonomy units. The use of DIAMOND and UniRef90 allowed Genome Detective to identify viral short reads at least 1000 times faster than if we used Blastn and the viral nt database of NCBI [70]. The software was optimized using synthetic datasets to represent the great diversity of virus genomes. The application was then validated with NGS data of hundreds of viruses. User time is minimal, and it is limited to the time required to upload the data [69]. According to the authors [69], Genome Detective accepts unprocessed paired-end or single reads generated by NGS platforms in FASTQ format and/or processed FASTA sequences. Candidate viral reads are identified using the protein-based alignment method, DIAMOND [70]. It uses the viral subset of the Swiss-Prot UniRef90 protein database, which contains representative clusters of proteins linked to taxonomy IDs, to improve sensitivity and speed, which was also improved by first sorting short reads into groups, or buckets. The objective is to run a separate metagenomic de novo assembly in each bucket; so, all reads of one virus species have to be assigned to the same bucket. Each bucket is then identified using the taxonomy ID of the lowest common ancestor of the hits identified by DIAMOND.

Once all of the reads have been sorted in buckets; each bucket is then de novo assembled separately using SPAdes [71] for single-ended reads or metaSPAdes [71] for paired-end reads. Blastx and Blastn are used to search for candidate reference sequences against the NCBI RefSeq virus database. Genome Detective combines the results for every detected contig at the amino acid and nucleotide (nt) level by calculating a total score that is a sum of the total nt score plus total amino acid score. It then chooses the five best scoring references for each contig to be used during the alignment. The contigs for each individual species are joined using Advanced Genome Aligner (AGA) [72]. AGA is designed to compute the optimal global alignment considering simultaneously the alignment of all annotated coding sequences of a reference genome. This makes alignments using Genome Detective more sensitive and accurate as both nt and protein scores are taken into account in order to produce a consensus sequence from the de novo contigs. A report is generated, referring to the final contigs and consensus sequences, available in

FASTA format. The report also contains detailed information on filtering, assembly, and consensus sequence. Web-based graphics are also available. In addition, the user can produce a bam file with BWA [73] using the reference or de novo consensus sequence by selecting the detailed report and access viral phylogenetic identification tools [74] directly from the interface. The authors found that, for large NGS and metagenomic datasets, Genome Detective substantially reduces computational cost without compromising the quality of the result. However, the construction of de novo whole genomes from metagenomic samples depends on the number of reads, the virus genome size, and read length. Genome Detective is linked to popular virus-specific typing tools [74], which allow phylogenetic classification below species level.

4.2 Future improvements

According to Xiao et al. [51], additional annotation information, such as that of epigenetics, noncoding RNAs, insertion elements, conserved structural elements, and pseudogenes remains to be implemented into the relevant software packages. The authors highlight the transition from the representation of reference genomes as strings to representations as graphs as a prominent example for a computational paradigm shift. In addition, improvements on genome assembly using machine learning techniques are proposed by Padovani de Souza et al. [75]. Finally, in order to help better use all the information acquired by high-throughput real-time sequencing and its analysis, text mining and knowledge discovery techniques, integrated with medical and scientific literature and gene family and metabolic pathway databases, could help generate new insights and speed up discoveries.

5 Conclusions

High-throughput real-time NGS projects have transformed the field of bioinformatics from single-genome studies to pan-genome analyses. The limiting factor now is no longer data rarity, but immense data availability and dimensionality. In this new context, bottom-up analysis stemming from big data provides great challenges but also great rewards.

References

[1] ICTV Master Species List 2018a v1, International Committee on Taxonomy of Viruses (ICTV), Available from https://talk.ictvonline.org/files/master-species-lists/m/msl/7992%3e, 2018.
[2] S. Duffy, L.A. Shackelton, E.C. Holmes, Rates of evolutionary change in viruses: patterns and determinants, Nat. Rev. Genet. 9 (2008) 267–276.
[3] E. Domingo, Mechanisms of viral emergence, Vet. Res. 41 (2010) 38–312.
[4] C.J. Houldcroft, J.M. Bryant, D.P. Depledge, B.K. Margetts, J. Simmonds, S. Nicolaou, H.J. Tutill, R. Williams, A.J.J. Worth, S.D. Marks, P. Veys, E. Whittaker, J. Breuer, Detection of low frequency

multi-drug resistance and novel putative maribavir resistance in immunocompromised pediatric patients with cytomegalovirus, Front. Microbiol. 7 (2016) 13–17.
[5] H. Zaraket, R. Saito, Y. Suzuki, T. Baranovich, C. Dapat, I. Caperig-Dapat, H. Suzuki, Genetic makeup of amantadine-resistant and oseltamivir-resistant human influenza A/H1N1 viruses, J. Clin. Microbiol. 48 (2010) 1085–1092.
[6] T. Gräf, H. Machado Fritsch, R.M. de Medeiros, D. Maletich Junqueira, S. Esteves de Matos Almeida, A.R. Pinto, Comprehensive characterization of HIV-1 molecular epidemiology and demographic history in the Brazilian region most heavily affected by AIDS, J. Virol. 90 (2016) 8160–8168.
[7] S. Ramirez, L.S. Mikkelsen, J.M. Gottwein, J. Bukh, Robust HCV genotype 3a infectious cell culture system permits identification of escape variants with resistance to sofosbuvir, Gastroenterology 151 (2) (2016) 973–985.
[8] L. Yuan, X.-Y. Huang, Z.-Y. Liu, F. Zhang, X.-L. Zhu, J.-Y. Yu, X. Ji, Y.-P. Xu, G. Li, C. Li, H.-J. Wang, Y.-Q. Deng, M. Wu, M.-L. Cheng, Q. Ye, D.-Y. Xie, X.-F. Li, X. Wang, W. Shi, B. Hu, P.-Y. Shi, Z. Xu, C.-F. Qin, A single mutation in the prM protein of Zika virus contributes to fetal microcephaly, Science 358 (2017) 933–936.
[9] N.R. Faria, J. Quick, I.M. Claro, J. Theze, J.G. de Jesus, M. Giovanetti, et al., Establishment and cryptic transmission of Zika virus in Brazil and the Americas, Nature 546 (2017) 406–410.
[10] J.L. Gardy, N.J. Loman, Towards a genomics-informed, real-time, global pathogen surveillance system, Nat. Rev. Genet. 19 (1) (2018) 9–20.
[11] F. Sanger, S. Nicklen, A.R. Coulson, DNA sequencing with chain-terminating inhibitors, Proc. Natl. Acad. Sci. U. S. A. 74 (12) (1977) 5463–5467.
[12] T. Jarvie, Next generation sequencing technologies, Drug Discov. Today Technol. 2 (3) (2005) 255–260.
[13] S. Goodwin, J.D. McPherson, W.R. McCombie, Coming of age: ten years of next-generation sequencing technologies, Nat. Rev. Genet. 17 (2016) 333–351.
[14] J. Wang, N.E. Moore, Y.M. Deng, D.A. Eccles, R.J. Hall, MinION nanopore sequencing of an influenza genome, Front. Microbiol. 6 (2015) 766.
[15] N. Beerenwinkel, H.F. Günthard, V. Roth, K.J. Metzner, Challenges and opportunities in estimating viral genetic diversity from next-generation sequencing data, Front. Microbiol. 3 (2012) 329.
[16] N. Nagarajan, M. Pop, Sequence assembly demystified, Nat. Rev. Genet. 14 (2013) 157–167.
[17] M. Jain, H.E. Olsen, B. Paten, M. Akeson, The Oxford Nanopore MinION: delivery of nanopore sequencing to the genomics community, Genome Biol. 17 (2016) 239.
[18] C.L. Ip, M. Loose, J.R. Tyson, M. de Cesare, B.L. Brown, M. Jain, et al., MinION analysis and reference consortium: phase 1 data release and analysis, F1000Res 4 (2015) 1075.
[19] T. Laver, J. Harrison, P.A. O'Neill, K. Moore, A. Farbos, K. Paszkiewicz, et al., Assessing the performance of the Oxford Nanopore technologies MinION, Biomol. Detect. Quantif. 3 (2015) 1–8.
[20] J. Quick, N.J. Loman, S. Duraffour, J.T. Simpson, E. Severi, L. Cowley, Real-time, portable genome sequencing for Ebola surveillance, Nature 530 (2016) 228.
[21] T. Hoenen, A. Groseth, K. Rosenke, R.J. Fischer, A. Hoenen, S.D. Judson, Nanopore sequencing as a rapidly deployable Ebola outbreak tool, Emerg. Infect. Dis. 22 (2016) 331.
[22] N.R. Faria, E.C. Sabino, M.R. Nunes, L.C.J. Alcantara, N.J. Loman, O.G. Pybus, Mobile real-time surveillance of Zika virus in Brazil, Genome Med. 8 (2016) 97.
[23] A. Edwards, A.R. Debbonaire, B. Sattler, L.A. Mur, A.J. Hodson, Extreme Metagenomics Using Nanopore DNA Sequencing: A Field Report From Svalbard, 78N, 2016.
[24] S.S. Johnson, E. Zaikova, D.S. Goerlitz, Y. Bai, S.W. Tighe, Real-time DNA sequencing in the Antarctic dry valleys using the Oxford Nanopore sequencer, J. Biomol. Tech. 28 (2017) 2–7.
[25] A. Edwards, A. Soares, S. Rassner, P. Green, J. Felix, A. Mitchell, Deep sequencing: intra-terrestrial metagenomics illustrates the potential of off-grid Nanopore DNA sequencing, bioRxiv (2017).
[26] A.B. McIntyre, L. Rizzardi, M.Y. Angela, N. Alexander, G.L. Rosen, D.J. Botkin, Nanopore sequencing in microgravity, NPJ Microgravity 2 (2016) 16035.
[27] S.L. Castro-Wallace, C.Y. Chiu, K.K. John, S.E. Stahl, K.H. Rubins, A.B. McIntyre, Nanopore DNA sequencing and genome assembly on the International Space Station, Sci. Rep. 7 (2017) 18022.

[28] N.R. Faria, M.U. Kraemer, S. Hill, J.G. de Jesus, R.S. de Aguiar, F.C. Iani, et al., Genomic and epidemiological monitoring of yellow fever virus transmission potential, Science (2018) https://doi.org/10.1126/science.aat7115.

[29] S. Sardi, S. Somasekar, S.N. Naccache, A.C. Bandeira, L.B. Tauro, G.S. Campos, et al., Co-infections from Zika and chikungunya virus in Bahia, Brazil identified by metagenomic next-generation sequencing, J. Clin. Microbiol. 54 (9) (2016) 2348–2353.

[30] D.M. Morens, G.K. Folkers, A.S. Fauci, The challenge of emerging and re-emerging infectious diseases, Nature 430 (2004) 242–249.

[31] P. Daszak, A.A. Cunningham, A.D. Hyatt, Emerging infectious diseases of wildlife—threats to biodiversity and human health, Science 287 (2000) 443–449.

[32] S.S. Morse, Factors in the emergence of infectious diseases, Emerg. Infect. Dis. 1 (1995) 7–15.

[33] J. Versalovic, J.R. Lupski, Molecular detection and genotyping of pathogens: more accurate and rapid answers, Trends Microbiol. 10 (2002) 15–21.

[34] A.J. Sabat, A. Budimir, D. Nashev, R. Sá-Leão, J.M. van Dijl, F. Laurent, et al., Overview of molecular typing methods for outbreak detection and epidemiological surveillance, Euro Surveill. 18 (2013) 20380.

[35] J. Shendure, H. Ji, Next-generation DNA sequencing, Nat. Biotechnol. 26 (2008) 1135–1145.

[36] B.L. Haagmans, A.C. Andeweg, A.D.M.E. Osterhaus, The application of genomics to emerging zoonotic viral diseases, PLoS Pathog. 5 (2009).

[37] A.C. McHardy, B. Adams, The role of genomics in tracking the evolution of influenza A virus, PLoS Pathog. 5 (2009).

[38] P. Tang, J.L. Gardy, Stopping outbreaks with real-time genomic epidemiology, Genome Med. 6 (2014) 104.

[39] E.C. Holmes, Viral evolution in the genomic age, PLoS Biol. 5 (2007).

[40] J. Gardy, N.J. Loman, A. Rambaut, Real-time digital pathogen surveillance—the time is now, Genome Biol. 16 (2015) 155.

[41] J. Quick, N.D. Grubaugh, S.T. Pullan, I.M. Claro, A.D. Smith, K. Gangavarapu, Multiplex PCR method for MinION and Illumina sequencing of Zika and other virus genomes directly from clinical samples, Nat. Protoc. 12 (2017) 1261.

[42] N.D. Grubaugh, J.T. Ladner, M.U. Kraemer, G. Dudas, A.L. Tan, K. Gangavarapu, Genomic epidemiology reveals multiple introductions of Zika virus into the United States, Nature 546 (2017) 401.

[43] J. Thézé, T. Li, L. du Plessis, J. Bouquet, M.U. Kraemer, S. Somasekar, Genomic epidemiology reconstructs the introduction and spread of Zika virus in Central America and Mexico, Cell Host Microbe 23 (2018) 855–864.

[44] N.J. Loman, C. Constantinidou, J.Z.M. Chan, M. Halachev, M. Sergeant, C.W. Penn, et al., High-throughput bacterial genome sequencing: an embarrassment of choice, a world of opportunity, Nat. Rev. Microbiol. 10 (2012) 599–606.

[45] K.J. Popovich, E.S. Snitkin, Whole genome sequencing—implications for infection prevention and outbreak investigations, Curr. Infect. Dis. Rep. 19 (2017) 15.

[46] S. Reuter, M.J. Ellington, E.J.P. Cartwright, C.U. Köser, M.E. Török, T. Gouliouris, et al., Rapid bacterial whole-genome sequencing to enhance diagnostic and public health microbiology, JAMA Intern. Med. 173 (2013) 1397–1404.

[47] M.R. Halachev, J.Z.-M. Chan, C.I. Constantinidou, N. Cumley, C. Bradley, M. Smith-Banks, et al., Genomic epidemiology of a protracted hospital outbreak caused by multidrug-resistant Acinetobacter baumannii in Birmingham, England, Genome Med. 6 (2014) 70.

[48] A.J. Drummond, O.G. Pybus, A. Rambaut, R. Forsberg, A.G. Rodrigo, Measurably evolving populations, Trends Ecol. Evol. 18 (2003) 481–488.

[49] G. Vernikos, D. Medini, D.R. Riley, H. Tettelin, Ten years of pan-genome analyses, Curr. Opin. Microbiol. 23 (2015) 148–154.

[50] L. Carlos Guimaraes, et al., Inside the pan-genome—methods and software overview, Curr. Genomics 16 (4) (2015) 245–252.

[51] J. Xiao, Z. Zhang, J. Wu, J. Yu, A brief review of software tools for pangenomics, Genomics Proteomics Bioinformatics 13 (1) (2015) 73–76.

[52] C. Laing, et al., Pan-genome sequence analysis using Panseq: an online tool for the rapid analysis of core and accessory genomic regions, BMC Bioinformatics 11 (1) (2010) 461.

[53] Y. Zhao, J. Wu, J. Yang, S. Sun, J. Xiao, J. Yu, PGAP: pan-genomes analysis pipeline, Bioinformatics 28 (3) (2012) 416–418.

[54] S. Kurtz, et al., Versatile and open software for comparing large genomes, Genome Biol. 5 (2) (2004) 34–87.

[55] S.F. Altschul, W. Gish, W. Miller, E.W. Myers, D.J. Lipman, Basic local alignment search tool, J. Mol. Biol. 215 (3) (1990) 403–410.

[56] A.J. Enright, S. Van Dongen, C.A. Ouzounis, An efficient algorithm for large-scale detection of protein families, Nucleic Acids Res. 30 (7) (2002) 1575–1584.

[57] A. Alexeyenko, I. Tamas, G. Liu, E.L.L. Sonnhammer, Automatic clustering of orthologs and inparalogs shared by multiple proteomes, Bioinformatics 22 (14) (2006) 9–15.

[58] G. Ostlund, et al., InParanoid 7: new algorithms and tools for eukaryotic orthology analysis, Nucleic Acids Res. 38 (2010) 196–203.

[59] M. Remm, C.E. Storm, E.L. Sonnhammer, Automatic clustering of orthologs and in-paralogs from pairwise species comparisons, J. Mol. Biol. 314 (5) (2001) 1041–1052.

[60] J. Blom, et al., EDGAR: a software framework for the comparative analysis of prokaryotic genomes, BMC Bioinformatics 10 (2009) 154.

[61] M.N. Benedict, J.R. Henriksen, W.W. Metcalf, R.J. Whitaker, N.D. Price, ITEP: an integrated toolkit for exploration of microbial pan-genomes, BMC Genomics 15 (2014) 8.

[62] B. Contreras-Moreira, P. Vinuesa, GET_HOMOLOGUES, a versatile software package for scalable and robust microbial pangenome analysis, Appl. Environ. Microbiol. 79 (24) (2013) 7696–7701.

[63] S.R. Eddy, Hidden Markov models, Curr. Opin. Struct. Biol. 6 (3) (1996) 361–365.

[64] O. Lukjancenko, M.C. Thomsen, M. Voldby Larsen, D.W. Ussery, PanFunPro: PAN-genome analysis based on FUNctionalPROfiles, F1000Res. 2 (2013) 15.

[65] The Gene Ontology Consortium, The gene ontology project in 2008, Nucleic Acids Res. 36 (2008) 440–444.

[66] M.A. Remita, A. Halioui, A.A. Malick Diouara, B. Daigle, G. Kiani, A.B. Diallo, A machine learning approach for viral genome classification, BMC Bioinformatics 18 (1) (2017) 208–239.

[67] R. Rose, B. Constantinides, A. Tapinos, D.L. Robertson, M. Prosperi, Challenges in the analysis of viral metagenomes, Virus Evol. 2 (2) (2016) 207–323.

[68] M. Vilsker, Y. Moosa, S. Nooij, V. Fonseca, Y. Ghysens, K. Dumon, R. Pauwels, L.C. Alcantara, E. VandenEynden, A.M. Vandamme, K. Deforche, T. de Oliveira, Genome Detective: an automated system for virus identification from high-throughput sequencing data, Bioinformatics 4 (2018) 32–103.

[69] B. Buchfink, C. Xie, D.H. Huson, Fast and sensitive protein alignment using DIAMOND, Nat. Methods 12 (2014) 59.

[70] A. Bankevich, et al., SPAdes: a new genome assembly algorithm and its applications to single-cell sequencing, J. Comput. Biol. 19 (5) (2012) 455–477.

[71] K. Deforche, An alignment method for nucleic acid sequences against annotated genomes, bioRxiv (2017).

[72] H. Li, R. Durbin, Fast and accurate short read alignment with Burrows-Wheeler transform, Bioinformatics 25 (14) (2009) 1754–1760.

[73] T. de Oliveira, K. Deforche, S. Cassol, M. Salminen, D. Paraskevis, C. Seebregts, J. Snoeck, E.J. van Rensburg, A.M. Wensing, D.A. van de Vijver, C.A. Boucher, R. Camacho, A.M. Vandamme, An automated genotyping system for analysis of HIV-1 and other microbial sequences, Bioinformatics 21 (19) (2005) 3797–3800.

[74] Computational pan-genomics: status, promises and challenges, Brief. Bioinform. 19 (1) (2018) 118–135.

[75] K. Padovani de Souza, J.C. Setubal, A.C. Ponce de Leon F. de Carvalho, G. Oliveira, A. Chateau, R. Alves, Machine learning meets genome assembly, Brief. Bioinform. 18 (1) (2018) 533.

CHAPTER 12

Pan-genomics of fungi and its applications

Rodrigo Bentes Kato[a,*], Arun Kumar Jaiswal[b,*], Sandeep Tiwari[b], Debmalya Barh[c], Vasco Azevedo[b], Aristóteles Góes-Neto[a]
[a]Molecular and Computational Biology of Fungi Laboratory, Department of Microbiology, Institute of Biological Sciences (ICB), Federal University of Minas Gerais (UFMG), Belo Horizonte, Brazil
[b]PG Program in Bioinformatics, Institute of Biological Sciences, Federal University of Minas Gerais (UFMG), Belo Horizonte, Brazil
[c]Centre for Genomics and Applied Gene Technology, Institute of Integrative Omics and Applied Biotechnology (IIOAB), Purba Medinipur, India

1 Introduction

Fungi are an evolutionary lineage within Opisthokonta, comprising one of the largest and most diverse groups of Eukarya on planet Earth [1]. Their multicellular non-motile bodies (mycelia) are constructed of apically growing, walled, and tubular cells (hyphae) or can be unicellular in which each adult individual is a single cell. These two distinct morphological groups are the so-called mycelial (or filamentous) fungi and yeasts, respectively [2]. The fungi are eukaryotic and chemoheterotrophic organisms, exhibiting osmotrophic nutrition with partial external digestion [3].

Fungi play a key role in the global carbon cycle especially in terrestrial biomes [4], and survive by using three basic trophic modes: (i) as saprotrophs, breaking down dead organic matter, (ii) as parasitic (and pathogens), or (iii) mutualistic symbionts with other living organisms [5]. Fungi and their by-products have a great economic importance. Fungi can be used to produce fermented food (e.g., beers, wines, breads, and cheeses) [6], primary and secondary metabolites (e.g., ethanol, hydrolases, and oxidoreductases enzymes, organic acids, and many vitamins, hypocholesterolemics, antineoplastics) [7], and inoculants and biocides (for mycorrhization and as biological control agents) [8]. Furthermore, fungi can be used for bioremediation of solid residues, effluents, and gaseous emissions [9], as well as to produce new biomaterials, such as mycocomposites [10] and nanomycomaterials [11].

Fungi constitute one of the major clades of organisms with approximately 145,000 species already described [12], however, there must be many more species since estimates suggest the existence of 5.1 million species, despite only about 10% were described [13]. Although there is still no consensus in the hierarchical taxonomical classification of the

* These authors contributed equally to this work.

Pan-genomics: Applications, Challenges, and Future Prospects
https://doi.org/10.1016/B978-0-12-817076-2.00012-3
© 2020 Elsevier Inc.
All rights reserved.

main groups inside of Kingdom Fungi since many recent classification schemes proposed different number of phyla [14–16], the great majority of the species are in the well-established and consensual phyla Ascomycota and Basidiomycota, which form the sub-kingdom Dikarya [17].

With the advancement and the low cost of high-throughput sequencing technology, these days leads in excess of 250,000 genome projects registered at the Genome Online Database (GOLD) (https://gold.jgi.doe.gov/measurements) [18]. These endeavors have made big change in the study of fungal genes and genomic association. The fungal genomic datasets can be exploited for adaptive and environmental behavior study. Substantial genomics and transcriptomics dataset of fungi have empowered the utilization of novel strategies and molecular evolution studies of fungi [19,20]. The combinations of experimental and computational methods have a great potential for point-by-point investigation of the fungal evolution and its biology [20].

The pan-genomics is a comparative genomics-based methodology that identifies the core and the dispensable genomes. The dispensable genome is composed of genes that present in some but not in all the strains studied, as well as the strain-specific genes. The dispensable genome helps to its fundamental way of life but rather present particular points of interest including antifungal resistance, niche adaptation, and the capacity to colonize new hosts [21]. As gene content and genome copy number can vary in distinct populations of a single species, the inventory of the variation at genomic level in different isolates is crucial to characterize the complete set of genes (core and accessory) that exists in a fungal species [20]. In this chapter, our work aimed to perform an extensive literature review and meta-analysis of this customized database in order to depict the state of the art of fungal pan-genomics.

2 Application of pan-genomics of fungi based on meta-analysis

The metadata related to genomics, comparative genomics and pan-genomics analyses on fungi, were mined through the literature. The NCBI (National Center for Biotechnology Information) genome and JGI (Joint Genome Institute) genome databases were also used for the search of data related to these areas. Thereafter, abstract and full-text level manual curations were performed. For the description of the data, histograms, and pie charts were constructed using the R version 3.5.0 software [22] and the ggplot2 version 2.2.1 package [23]. The localization on global map was done using the GPS Visualizer tool [24]. As a result, among the obtained 159 articles, only the most cited articles were considered for manual text curation. From the metadata from the databases, we found 97 species and 16.5% of these species has more than 16 distinct isolates fully sequenced. Then, we used this threshold (~16%) to our analyses. The obtained metadata related to fungal pan-genomics from published articles were divided in four groups.

(1) *Technological*: This group contains fungal isolates that are used to produce several products related to pharmaceuticals, food, dairy products, and alcohol industries. In our search, we found many papers using fungi to produce foods as fermented products and alcohol beverages.

(2) *Environmental*: It is a collection of published research results on the study of fungal diversity in the natural environment. We found some works that use fungi in biological process of biodegradation and biological treatment of lagoons or rivers. Industry and biodegradation areas have interesting in fungi that act in enzyme activities. These fungi are very important in agricultural and ecological contexts because they maintain the balance of the environment decomposing plant debris, degrade toxic substances, help plants to grow, and protect themselves against enemies.

(3) *Host pathogen*: In this group, we discussed some studies on the fungal diversity in the host-pathogen interactions. Some of these fungi contribute to pharmacy industry to cure some disease.

(4) *Laboratory*: some fungi were comparable with model, laboratory strains, and then we created this group.

Fig. 1 shows the frequency of number of works in each group. We found that most of the works were related to the technological importance fungal group (~61%). The reason behind the several studies in this area may be the presence of many industries, which invest large amount of money to its research and development. On the other hand, we found that host-pathogen related fungal research were the second bigger group with around 30%, mainly supported by agribusiness.

This metadata analysis was done to show the impact of pan-genomics in comparative fungal genomics. Furthermore, we found 1567 species related to 12 genera of fungi, and

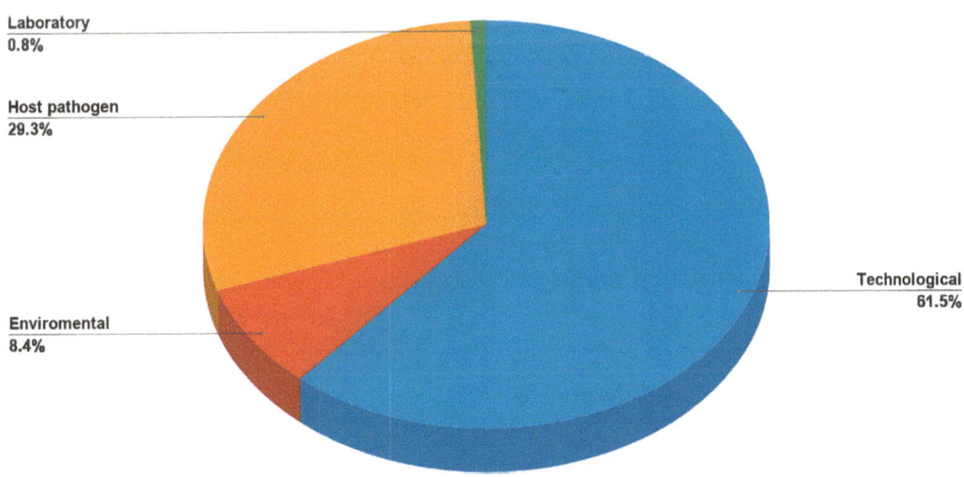

Fig. 1 The pie chart demonstrates the frequency of the studied fungi from each group.

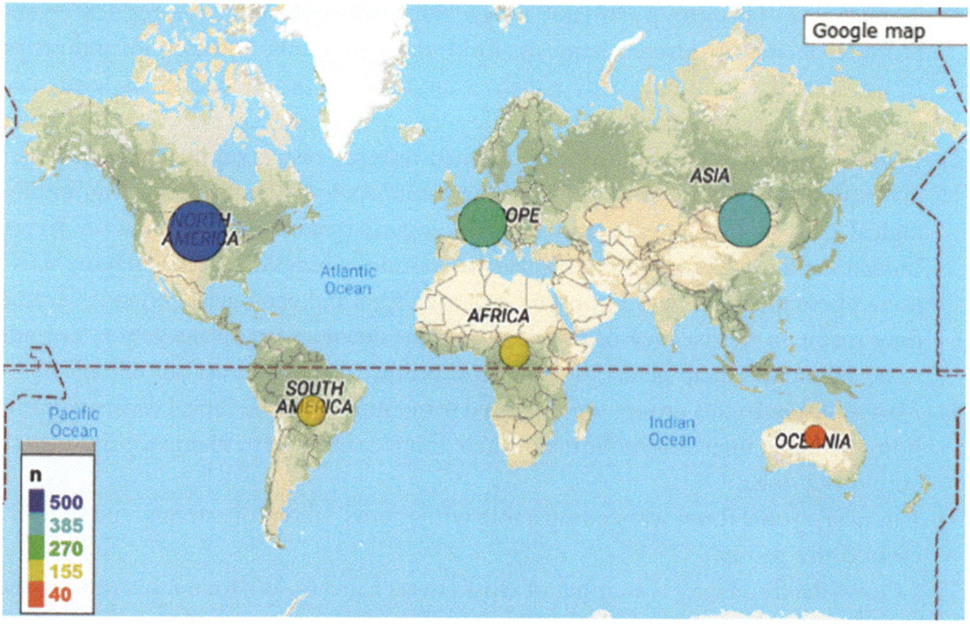

Fig. 2 The image shows the geographical a distribution of 1567 species of 12 different genera of fungal isolates.

they were distributed globally, where *n* means number of isolates for each country (Fig. 2).

Among these 12 genera we found that *Saccharomyces cerevisiae* is the most frequently (Fig. 3) studied genus. Around 83% of researches have been related to *S. cerevisiae* in the last 2 years of published research article [25].

2.1 Application of pan-genomics on advantageous fungus

The set of complete genes in all the strains of a specific species is known as pan-genome [20]. The genus *Saccharomyces* is among the most important and broadly studied model eukaryotic organisms. The fermented beverages production commonly used *S. cerevisiae* yeasts, dates at least as back 7000 BC, in china [26]. In order to comprehend the significance of selection during domestication and understand the levels of genetic diversity among wine yeasts, a number of pan-genome analyses have been done using commercial wine yeasts and industrial yeasts. A set of 83 strains of *S. cerevisiae* was used for the pan-genome analysis to identify the copy number variations in this yeast distributed in different industrial environments [26].

Another comparative work of 43 strains of *S. cerevisiae* isolated from fermenting grape was used to analyze genome renewal, and they propose that natural wine yeast strains can undergo such modifications and, thereby, change a multiple heterozygote into

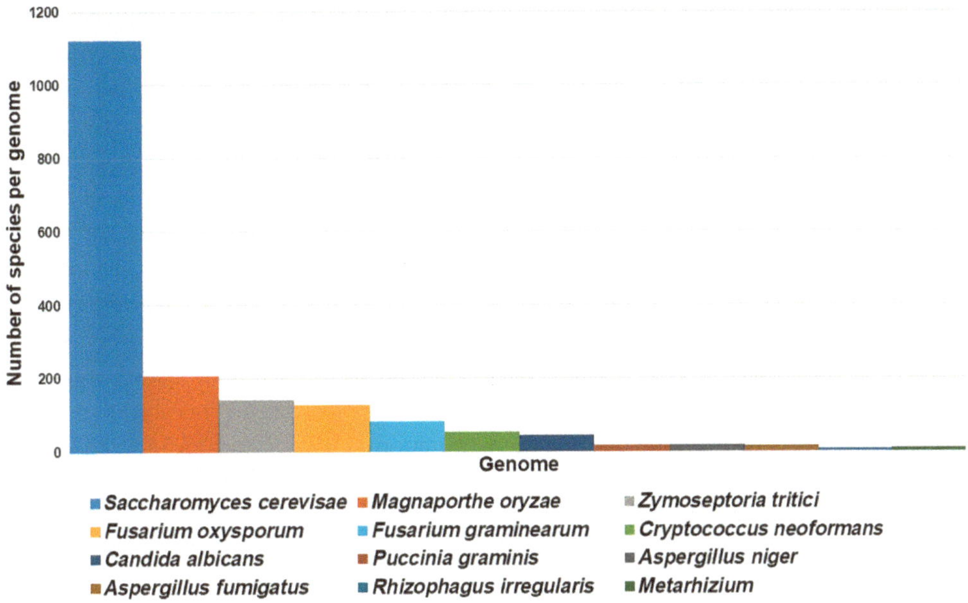

Fig. 3 The figure represents the most frequently studied fungi among the 12 genera.

completely homozygous diploids, some of which may replace the original heterozygous diploid [27]. The pan-genome studies of very polymorphic eukaryotic pathogens utilizing the accessory genome gives a better understanding for adaptive evolution. The genomics study of this yeast has enhanced our understanding of the evolutionary dynamics of natural populations when comparing with the domesticated strains, during infections, and during laboratory experiments [28]. Apart from *S. cerevisiae* [29], population genomic studies have also characterized the metabolic, genetic, and biogeographic diversity of *Saccharomyces paradoxus* [30], *Saccharomyces kudriavzevii* [31], and *Saccharomyces uvarum* [32]. As well as all the organisms, yeast genome sequences largely describe their genetic makeups; however, the comparative genomic studies have given better shape to the historical and genetic processes in their evolution [28].

2.2 Application of pan-genomics in disadvantageous fungus

Effect of chromosomal rearrangements on genes can lead to functional variation between individuals and influenced the expression of phenotypic attributes [33]. The inter and intraspecific structural variation among genomes of fungi has already been reported [34]. This structural variation among the pathogens can affect their host range. For instance, in the fungus *Melanopsichium pennsylvanicum*, gene loss are responsible for the hosts jump from dicotyledonous to monocotyledon plant hosts [33,35]. Unexpected number of fungal and fungal-like diseases have been recently afflicted animals and plants, and some of them are the most severe die-offs and extinctions ever witnessed in wild species, and are a peril for food security [20]. Emerging infectious diseases (EIDs) brought about by fungi are progressively perceived as exhibiting a danger to food security around the world [36]. Until date, several fungal genomes are responsible to accurate and complete genome assemblies using long-read sequencing technologies [37,38]. Various symbiotic interactions have been recorded between insects and fungi. Although the genomics has already been elucidated in many fungi that expanded our knowledge on this group, there is still much to explore the genomic features of the insect-commensal relationships [39]. *Zymoseptoria tritici* is a pathogen of wheat causing Septoria tritici blotch, and a recently published work using pan-genome analysis of this pathogen identified that host specialization has evolved by gene deletions and chromosomal rearrangements. In this aforementioned study, the authors used five isolates for the pan-genome analysis and 15,749, 9,149, and 6600 nonredundant proteins were identified as pan-genome, core, and accessory genome, respectively [33] (Table 1).

Table 1 Pan-genomics studies on different fungi

Fungi	Importance	Comparative Genomics	Strains/ isolates	Reference
Saccharomyces cerevisiae	Industrial important	**1.** Pangenome Analysis of *Saccharomyces cerevisiae*	25	[40]
		2. Report of the whole-genome sequencing and phenotyping of 1011 *Saccharomyces cerevisiae isolates*	1011	[25]
Rhizophagus irregularis	Plant pathogen	Genome assembly and gene annotation of the model strain *Rhizophagus irregularis* DAOM197198, and gene comparison with five different isolates of *Rhizophagus irregularis*	6	[41]
Puccinia graminis f. sp. *tritici*	Plant pathogen	Comparative genomics of Australian isolates of the wheat stem rust pathogen *Puccinia graminis f.* sp. *tritici* and draft genome was built for a founder Australian *Pgt* isolate	16	[42]

Table 1 Pan-genomics studies on different fungi—cont'd

Fungi	Importance	Comparative Genomics	Strains/ isolates	Reference
Zymoseptoria tritici	Plant pathogen	Pangenome analysis of *Zymoseptoria tritici*	123	[33, 38]
Metarhizium spp.	Insect pathogen	Pangenome analysis for Metarhizium spp.	7	[43]
Coccidioides posadasii, Coccidioides immitis and other fungus of order Onygenales	Human fungal pathogen	Genome sequencing and comparison of the primary human pathogens *C. immitis* and *C. posadasii*	17	[44]
Fusarium graminearum	Cereal pathogen	Sequencing of genomes of 60 diverse *F. graminearum* isolates from North America, and also the assembly of the first pan-genome for *F. graminearum* to clarify population-level differences in gene content potentially contributing to pathogen diversity.	70	[45]
Fusarium meridionale/ Fusarium Asiaticum/Fusarium graminearum	Plant/Cereal pathogen	Genomic comparison and gene content analysis of six newly isolates from the species complex, including the first available genomes of *F. asiaticum* and *F. meridionale*, with four other genomes	10	[46]

3 Conclusions and future prospective

Although people frequently demonstrate a nonmycophilic or even mycophobic relation with fungi, these group of organisms are vital on many aspects of human life, including medicine, food, and farming, and also play key roles in nature, such as in the carbon bio-geochemical cycle. The comparative genomics approach based on sequence similarity with statistical analysis helps in identifying the essential genomic content common among all fungal isolates of a same species as well as the subset of genes encoding novel functions as variable genome. Biotechnology consists of the use of organisms for the development of processes and products of economic or social interest. It is recognized as one of the technologies for the 21st century with higher potential impact on global problems (diseases, nutrition, and environmental pollution) and sustainable industrial development (use of renewable resources, "green technology," and reduction of global warming). Based on the search and discovery of industrially exploitable biological resources, the scientific and technological advances achieved by fungal pan-genomics studies in recent years have revolutionized traditional approaches to the exploitation of biological resources for biotechnology.

References

[1] F. Badotti, F.S. de Oliveira, C.F. Garcia, A.B. Vaz, P.L. Fonseca, L.A. Nahum, et al., Effectiveness of ITS and sub-regions as DNA barcode markers for the identification of Basidiomycota (Fungi), BMC Microbiol. 17 (1) (2017) 42.

[2] D. Moore, 21st Century Guidebook to Fungi, Q. Rev. Biol. 87 (4) (2012) 396.

[3] Sarah C. Watkinson NPM, Lynne Boddy. The Fungi. San Diego, Elsevier Science Publishing Co Inc.

[4] G.M. Gadd, Fungi in Biogeochemical Cycles, CAB International, 2006.

[5] N.H. Nguyen, Z. Song, S.T. Bates, S. Branco, L. Tedersoo, J. Menke, et al., FUNGuild: An open annotation tool for parsing fungal community datasets by ecological guild, Fungal Ecol. 20 (2016) 241–248.

[6] M. Hofrichter, The Mycota: A Comprehensive Treatise on Fungi as Experimental Systems for Basic and Applied Research, second ed., Springer, 2010.

[7] Fungal Biomolecules: Sources, Applications and Recent Developments, Wiley-Blackwell, 2015.

[8] T.M. Butt, C. Jackson, N. Magan (Eds.), Fungi as Biocontrol Agents: Progress Problems and Potential, CABI, 2001.

[9] H. Singh, Mycoremediation, Wiley, 2006.

[10] C. Girometta, A. Picco, R. Baiguera, D. Dondi, S. Babbini, M. Cartabia, et al., Physico-mechanical and thermodynamic properties of mycelium-based biocomposites: A review, Sustainability 11 (1) (2019).

[11] R. Prasad, Fungal Nanotechnology, Springer, 2017.

[12] R. F. Species 2000 & ITIS Catalogue of life, 20th February 2019. 2019.

[13] M. Blackwell, The fungi: 1, 2, 3 … 5.1 million species? Am. J. Bot. 98 (3) (2011) 426–438.

[14] J.W. Spatafora, M.C. Aime, I.V. Grigoriev, F. Martin, J.E. Stajich, M. Blackwell, The fungal tree of life: from molecular systematics to genome-scale phylogenies, Microbiol. Spectr. 5 (5) (2017).

[15] L. Tedersoo, S. Sánchez-Ramírez, U. Kõljalg, M. Bahram, M. Döring, D. Schigel, et al., High-level classification of the fungi and a tool for evolutionary ecological analyses, Fungal Divers. 90 (1) (2018) 135–159.

[16] J. Choi, S.-H. Kim, A genome tree of life for the fungi kingdom, Proc. Natl. Acad. Sci. 114 (35) (2017) 9391–9396.

[17] D.S. Hibbett, M. Blackwell, T.Y. James, J.W. Spatafora, J.W. Taylor, R. Vilgalys, Phylogenetic taxon definitions for Fungi, Dikarya, Ascomycota and Basidiomycota, IMA Fungus 9 (2018) 291–298.

[18] Genome Online Database (GOLD) n.d. [Internet]. [cited March 2019]. Available from: https://gold.jgi.doe.gov/measurements.

[19] D.S. Hibbett, J.E. Stajich, J.W. Spatafora, Toward genome-enabled mycology, Mycologia 105 (6) (2013) 1339–1349.

[20] J.E. Stajich, Fungal genomes and insights into the evolution of the kingdom, Microbiol. Spectr. 5 (4) (2017).

[21] H. Tettelin, D. Riley, C. Cattuto, D. Medini, Comparative genomics: the bacterial pan-genome, Curr. Opin. Microbiol. 11 (5) (2008) 472–477.

[22] R Core Team R: A language and environment for statistical computing. R Foundation for Statistical Computing, Vienna, Austria. Available from: https://www.r-project.org/.
[23] Wickham H. ggplot2 2009.
[24] Schneider, A GPS visualizer. Available from: http://www.gpsvisualizer.com/map_input?form=data.
[25] J. Peter, M. De Chiara, A. Friedrich, J.-X. Yue, D. Pflieger, A. Bergström, et al., Genome evolution across 1,011 *Saccharomyces cerevisiae* isolates, Nature 556 (7701) (2018) 339–344.
[26] B. Dunn, C. Richter, D.J. Kvitek, T. Pugh, G. Sherlock, Analysis of the *Saccharomyces cerevisiae* pan-genome reveals a pool of copy number variants distributed in diverse yeast strains from differing industrial environments, Genome Res. 22 (5) (2012) 908–924.
[27] R.K. Mortimer, P. Romano, G. Suzzi, M. Polsinelli, Genome renewal: a new phenomenon revealed from a genetic study of 43 strains of *Saccharomyces cerevisiae* derived from natural fermentation of grape musts, Yeast 10 (12) (1994) 1543–1552.
[28] C.T. Hittinger, A. Rokas, F.Y. Bai, T. Boekhout, P. Goncalves, T.W. Jeffries, et al., Genomics and the making of yeast biodiversity, Curr. Opin. Genet. Dev. 35 (2015) 100–109.
[29] P.K. Strope, D.A. Skelly, S.G. Kozmin, G. Mahadevan, E.A. Stone, P.M. Magwene, et al., The 100-genomes strains, an S. *cerevisiae* resource that illuminates its natural phenotypic and genotypic variation and emergence as an opportunistic pathogen, Genome Res. 25 (5) (2015) 762–774.
[30] G. Liti, D.M. Carter, A.M. Moses, J. Warringer, L. Parts, S.A. James, et al., Population genomics of domestic and wild yeasts, Nature 458 (7236) (2009) 337–341.
[31] C.T. Hittinger, P. Gonçalves, J.P. Sampaio, J. Dover, M. Johnston, A. Rokas, Remarkably ancient balanced polymorphisms in a multi-locus gene network, Nature 464 (7285) (2010) 54–58.
[32] P. Almeida, C. Gonçalves, S. Teixeira, D. Libkind, M. Bontrager, I. Masneuf-Pomarède, et al., A Gondwanan imprint on global diversity and domestication of wine and cider yeast *Saccharomyces uvarum*, Nat. Commun. 5 (1) (2014).
[33] C. Plissonneau, F.E. Hartmann, D. Croll, Pangenome analyses of the wheat pathogen *Zymoseptoria tritici* reveal the structural basis of a highly plastic eukaryotic genome, BMC Biol. 16 (1) (2018).
[34] M.E. Zolan, Chromosome-length polymorphism in fungi, Microbiol. Rev. 59 (4) (1995) 686–698.
[35] R. Sharma, B. Mishra, F. Runge, M. Thines, Gene loss rather than gene gain is associated with a host jump from monocots to dicots in the smut fungus *Melanopsichium pennsylvanicum*, Genome Biol. Evol. 6 (8) (2014) 2034–2049.
[36] M.C. Fisher, D.A. Henk, C.J. Briggs, J.S. Brownstein, L.C. Madoff, S.L. McCraw, et al., Emerging fungal threats to animal, plant and ecosystem health, Nature 484 (7393) (2012) 186–194.
[37] L. Faino, M.F. Seidl, E. Datema, G.C.M. van den Berg, A. Janssen, A.H.J. Wittenberg, et al., Single-molecule real-time sequencing combined with optical mapping yields completely finished fungal genome, MBio 6 (4) (2015).
[38] C. Plissonneau, A. Stürchler, D. Croll, The evolution of orphan regions in genomes of a fungal pathogen of wheat, MBio 7 (5) (2016).
[39] Y. Wang, M. Stata, W. Wang, J.E. Stajich, M.M. White, J.-M. Moncalvo, et al., Comparative genomics reveals the core gene toolbox for the fungus-insect symbiosis, MBio 9 (3) (2018).
[40] J. Schacherer, G. Song, B.J.A. Dickins, J. Demeter, S. Engel, B. Dunn, et al., AGAPE (automated genome analysis pipeline) for pan-genome analysis of *Saccharomyces cerevisiae*, PLoS One 10 (3) (2015).
[41] E.C.H. Chen, E. Morin, D. Beaudet, J. Noel, G. Yildirir, S. Ndikumana, et al., High intraspecific genome diversity in the model arbuscular mycorrhizal symbiont *Rhizophagus irregularis*, New Phytol. 220 (4) (2018) 1161–1171.
[42] N.M. Upadhyaya, D.P. Garnica, H. Karaoglu, J. Sperschneider, A. Nemri, B. Xu, et al., Comparative genomics of Australian isolates of the wheat stem rust pathogen *Puccinia graminis f.* sp. *tritici* reveals extensive polymorphism in candidate effector genes, Front. Plant Sci. 5 (2015).
[43] X. Hu, G. Xiao, P. Zheng, Y. Shang, Y. Su, X. Zhang, et al., Trajectory and genomic determinants of fungal-pathogen speciation and host adaptation, Proc. Natl. Acad. Sci. 111 (47) (2014) 16796–16801.
[44] T.J. Sharpton, J.E. Stajich, S.D. Rounsley, M.J. Gardner, J.R. Wortman, V.S. Jordar, et al., Comparative genomic analyses of the human fungal pathogens Coccidioides and their relatives, Genome Res. 19 (10) (2009) 1722–1731.

[45] A.C. Kelly, T.J. Ward, Population genomics of *Fusarium graminearum* reveals signatures of divergent evolution within a major cereal pathogen, PLoS One 13 (3) (2018).

[46] S. Walkowiak, O. Rowland, N. Rodrigue, R. Subramaniam, Whole genome sequencing and comparative genomics of closely related Fusarium Head Blight fungi: *Fusarium graminearum*, *F. meridionale* and *F. asiaticum*, BMC Genomics 17 (1) (2016).

CHAPTER 13

Genomics of algae: Its challenges and applications

Anupriya Minhas, Bineypreet Kaur, Jaspreet Kaur
University Institute of engineering and Technology (UIET), Department of Biotechnology, Panjab University, Chandigarh, India

1 Diversity in algae and their evolutionary insights

Algae comprise a divergent and notable group of living organisms due to their morphological diverse features, large biomass, abundant metabolites synthesis, and vast economic future [1]. Algae and plants are the most dominant and evolutionarily diverse primary producers on our planet. The "algae" are the predominantly aquatic photosynthetic eukaryotes that could vary from unicells prokaryotes (few microns in diameter) to the complex multicellular eukaryotic forms (giant kelps of more than 30 m in length). Our prokaryotes photosynthetic ancestors (cyanobacteria) evolved some 3.6 billion years ago. Most of the eukaryotes co-opted the photosynthetic trait some 1.8 billion years ago from prokaryotes by the process of primary endosymbiosis of engulfing and stably integrating a photoautotrophic prokaryotic cyanobacteria [2]. After these billion years of coevolution, both the eukaryotic host and the endosymbiont have achieved an excellent amalgamation indicated by generating a diverse group of primary producers nurturing life on land as well as in water. This process eventually resulted in a cell holding a photosynthetic plastid with lessened cyanobacterial genome [2, 3]. The event of primary endosymbiosis finally gave rise to the Archaeplastida, a major group of autotrophic eukaryotes, comprising the red algae (Rhodophyta), the green algae, land plants (Chloroplastida), and small group of freshwater unicellular algae called glaucophytes.

Various ultrastructural, biochemical, and genetic evidence support the additional secondary endosymbiotic events occurring multiple times, engulfing not a cyanobacterium but a green or red photosynthetic eukaryote. This phenomenon gave rise to cryptophytes, haptophytes, heterokonts, dinoflagellates, and other photosynthetic eukaryotes [4, 5]. However, both the primary and secondary endosymbiosis events have resulted in a massive loss of genes from the engulfed genome, hence contributed widely to the evolution of algae. In the very beginning of kingdom classification, the unicellular algae along with other unicellular life forms were designated a broad kingdom defined by Ernst Haeckel as "Protista" [6]. However, later molecular phylogenetic studies by Rothschild (1989) [7] revealed that the protists are deeply rooted within diverse groups and found Haeckel's view of Protista as "a kingdom of primitive life forms" still accurate. Therefore,

Pan-genomics: Applications, Challenges, and Future Prospects
https://doi.org/10.1016/B978-0-12-817076-2.00013-5
© 2020 Elsevier Inc.
All rights reserved.

analysis of algal genomes can indeed shed light on the early origins of photosynthetic eukaryotes.

Taxonomic groups ranging from eukaryotic micro- and macroalgae to prokaryotic cyanobacteria capable of performing oxygenic photosynthesis have been grouped as Marine photosynthetic organisms (MPOs) [8]. Microalgae (called as phytoplankton) include diatoms and dinoflagellates, accounting for about half of global primary productivity. Macroalgae (called seaweeds) include brown algae, green algae, and red algae, carry out the bulk of photosynthesis in coastal regions, safeguard juvenile fish and invertebrates, restrict seabed shifting to support many coastal ecosystems and are of ecological and economical significance. Prokaryote MPOs, Cyanobacteria (blue-green algae) are the simplest form of algae contributing to 20%–40% of chlorophyll biomass and carbon fixation in the oceans. Although being a true relative of bacteria, their photosynthetic lamellae contain chlorophyll *a* and several accessory pigments like phycoerythrin (PE) and phycocyanin analogs of the eukaryotic thylakoid membranes. Since cyanobacteria represent one of the few cultured groups of marine microorganisms, the study of their physiology, biochemistry, and complete genomes sequencing can accelerate our understanding of the metabolism of these organisms and can provide novel insights into the genetic adaptations of these important marine microbes to their environment. Since they colonize a relatively simple environment, they can, therefore, provide an excellent model system to identify the molecular mechanisms priming their success in water columns with an enormous vertical gradation of light and nutrients. Heterocyst-forming species of cyanobacteria can "fix" atmospheric nitrogen in rice field while "Spirulina" is used as a staple food in parts of Africa and Mexico (as a source of all essential amino acids). Worldwide blue-green algae are commercially used as a source of vitamins, drug compounds, and growth factors. They are also exploited for the production of hydrogen gas and fertilizers [9]. Major free-living marine cyanobacterial species are currently known to belong to three N_2-fixers genera: *Trichodesmium*, *Crocosphaera*, *Nodularia spumigena*, and two nondiazotrophs genera *Synechococcus* and *Prochlorococcus* [10–12].

2 Advancements in genomics and its importance to ecologists

Whole-genome sequencing has acted as a bloom for algal research. It throws light on our understanding of various questions related to the evolution of different lineages of algae, the molecular processes of algae playing a role in their adaptation to a stressful environment and climatic change. Still, there are few red algal genomes that have been analyzed and comprehended. The recent advancement of "Next-Generation Sequence" technologies has tremendously increased the sequencing output by several orders of magnitude with substantial cost cutting per sequenced base compared to traditional methods. This development allowed the application of these sequencing procedures not only to the broad range of species including non-model organisms but has also initiated new fields of genomics such as metagenomics and metatranscriptomics, helping to access the uncultivated organisms and complex biological communities of worth [13]. Consequently, ecologists have started employing these genomic approaches to answer queries of ecological importance extending from the adaptation of organisms in changing environments to the evolution of complex phenotypes. Whereas, molecular biologists are maneuvering the expression analysis and functional assays to study intra and interspecies variability. Therefore, environmental or ecological genomics target to understand the relationships between an organism with its abiotic and biotic environment by inspecting the structure, function, and evolution of the genome [13].

Last decade has witnessed a major advancement toward the complete analysis of red algal genomes with the advent of new high-end molecular biology techniques and efficient, affordable, and reliable DNA sequencing. Comparative analysis of the existing genetic information from the available database is yet a challenge that biologists need to direct toward. To improve the accessibility to the online analysis of red algae genome and transcriptome data, realDB (realDB.algaegenome.org) provides an exciting platform for ecologists and researchers [14]. For morphological analysis, databases such as AlgaeBase (www.algaebase.org) and *Porphyra* database (www.porphyra.org) are also available.

3 Ecological and economic importance of algae

The Red algae or Rhodophytes are a rich and diverse group of algae, comprising of seven classes with enormous species number. These inhabit in a range of aquatic environments, mostly marine. Other niches include freshwater habitats and hot water springs. As a terrestrial niche, they have been reported to grow in tropical rainforests as well [15]. Fossil evidence indicates rhodophytes as one of the founding multicellular lineages and ancient groups of photoautotrophic eukaryotes (Archaeplastida). It is now a proven fact that genes with nuclear as well as plastid origin from the ancestral red algae have contributed significantly to enhance eukaryotic evolution and diversity [16, 17]. The size of red algae ranges from single cellular forms to multicellular forms to several feet in length. Rhodophytes are broadly classified into mesophiles, extremophiles, and economically important seaweeds.

Rhodophytes are prominently red in color due to the presence of water-soluble pigments phycobiliprotein and PE. Other colors are imparted by phycocyanin (blue) and blue-greenish allophycocyanin [18]. The pigment responsible for photosynthesis is chlorophyll a, and it is associated by accessory phycobiliproteins, forming a light-harvesting complex known as phycobilisome, present on the surface of thylakoids. There are a number of common structural and biochemical characteristics that these Rhodophytes share with other lineages of algae. But, definitely, they have a certain unique set of cellular features which are confined to this class only, for example, presence of light-harvesting complex, phycobilisomes, unstacked thylakoids, pit connections between adjacent cells and are marked by the absence of parenchyma. They play a major role as carbon sync along with the synthesis of many fatty acids and metabolites [19]. The biogeochemical influence of rhodophytes on the ecological aquatic foodwebs and global climate is indispensable. Red algae have a valuable role in producing oxygen in the seawater. Various species of red algae are a source of food for many aquatic organisms like fishes, worms, etc. Not only this, certain algal species are responsible for the formation of tropical reefs [20]. In fact, red algae have far more contributed to the reef structure than any other organism.

The commercial value of representatives of rhodophytes is undeniable. They are used as an important economic resource for humans. Red seaweeds are used as a food source for human consumption and several industries are based on food processing of nori (*Porphyra spp*) and many other algae. The high and rich content of vitamins, carotenoids, proteins, and antioxidants of red algae-derived foods has made them an attractive and popular choice in the health food industry for more than a thousand years. Other industries are based on phycocolloid production, which harvests a range of red seaweeds like carrageenophytes and agarophytes for the production of agar and carrageen gels [15]. Aquaculture based industries use algae as additives to feed. Phycobiliproteins have also been used as a fluorescent tag for the localization and quantification of antigens [21].

Brown algae or Phaeophyceae is characteristically defined by golden-brownish color due to the presence of fucoxanthin, a carotenoid pigment and in few species, tannins [22]. Taxonomically, this group belongs to multicellular species group, Stramenopiles. Brown algae are mostly found in marine segments and dominant seaweeds in rocky coastal ecosystems. Less than 1% of the approximately 2000 brown algal species occur in freshwater ecosystems. These include various seaweeds, which play an important role in marine ecosystems as they are a source of food to so many diverse organisms and facilitate habitats for others. The size of brown algae ranges from small filamentous forms to large and complex marcroscopic kelps, which could range from 1 m to more than 100 m. Brown algae hold valuable economical aspect in food and pharmaceutical industries. Seaweeds are commercially important for the production of alginates, fucoids, etc. Despite their ecological, evolutionary, and economical value, very few brown algal genomes have been sequenced so far. The useful whole-genome sequencing is need of the hour and more brown algal genomes are required to be sequenced in order to explore their versatility and sustainability toward ecological conditions.

4 Genomics of microalgae

4.1 Picoplanktonic marine cyanobacterial species

Trichodesmium genus is known to form dense, widespread blooms most frequent in the northern Arabian Sea, the western Indian Ocean, and the south-eastern Pacific. So far, only *T. erythraeum* IMS101, has been sequenced, however, no formal description of this genome is available to date. *Crocosphaera watsonii* (a single species ocean-wide) is usually found in warm (>27°C) oligotrophic subsurface waters [23, 24]. Interestingly, phylogenetic studies of natural populations and strains of *Crocosphaera* isolated from diverse areas revealed a very low level of genetic divergence, despite a significant variability at the phenotypic level [25]. Comparative genomics of two strains (WH8501 and WH0003) with distinct cell sizes, growth temperature range, and N_2 fixation rate, isolated respectively from the south Atlantic and the north Pacific oceans, confirmed a remarkably high similarity at the nucleotide level, genome-wide (over 80% of each genome was >98% identical to the other strain), despite a large number of genome rearrangements, insertions, or deletions [26].

Of these two major nondiazotrophs genera, *Synechococcus* is ubiquitous but usually less adapted to very oligotrophic environments. This genus represents a great deal of genetic diversity with strains well adapted to specific ecological niches [13, 27, 28] sequenced 2.4-megabase genome of *Synechococcus* sp. strain WH8102. The comparative genomic study identified 1314 open reading frames (ORFs) common between *Synechococcus* sp. strain WH8102 and two *Prochlorococcus* strains. WH8102 had 736 specific ORFs indicating a specific ecological adaptation of WH8102. Of these ORFs, 23% are common with freshwater cyanobacterium PCC6803, with a BLAST cut-off *e*-value of e^{-10}. These common ORFs might be partly responsible for functional phycobilisome for light harvesting and a functional nitrate reductase with molybdenum cofactors (for utilizing nitrate as the nitrogen source) in PCC6803 and WH8102 strains, unlike *Prochlorococcus* strains. Unlike PCC6803, WH8102 can use organic nitrogen and phosphorus sources and have acquired more sodium-dependent transporters indicating nutritionally more adaptable. The presence of more phage integrases in *Synechococcus* indicates more frequent horizontal gene transfer (HGT) as compared to *Prochlorococcus*. The transferred genes are associated with cell surface modification and provide swimming motility of *Synechococcus*. However, all *Synechococcus* are not versatile in their transport abilities and require partial or complete genome information of additional marine cyanobacteria. Furthermore, to conserve limited iron stores, *Synechococcus* has adapted to use nickel and cobalt in some enzymes. *Synechococcus* has reduced regulatory machinery accordant with the opinion that open sea constitutes a far more constant and buffered environment than freshwater.

The unicellular cyanobacterium *Prochlorococcus* comprises dominating phytoplankton in the tropical and subtropical oceans contributing significantly to global photosynthesis [29, 30]. For sequencing *Prochlorococcus* MED4 genome, whole-genome shotgun libraries have been constructed by cloning 2–3 kb g-DNA fragments into pUC18. Cloned plasmids were sequenced using PE BigDye Terminator chemistry and sequences have been resolved using PE 377 Automated DNA Sequencers. The whole-genome sequence of *Prochlorococcus* MED4 is assembled from 27,065 end sequences using PHRAP (P. Green) and primer walking is used for gap filling. *Prochlorococcus* MED4 (oxygen-evolving marine autotroph) is a high-light-adapted ecotype with smallest genome (1,657,990 base pairs) encoding 1716 genes. The genome of the low-light-adapted counterpart (MIT9313) is markedly larger (2,410,873 base pairs) encoding 2275 number of genes. Continuously changing architectures of these two strains reveal dynamic genomes constantly changing under diverse selection pressures. Both the strains share common 1350 number of genes and 65% of the genome can be assigned a functional category. Besides sharing a common ancestor, a significant number of genes are unshared between two and may have been acquired through duplication or lateral transfer. These differential genes might play roles in relative fitness of the ecotypes in diverse environmental conditions, and hence regulating their abundance and dispersal in the oceans [31]. Low-light-adapted ecotypes are most abundant in surface waters whereas high-light-adapted ecotypes are plentiful in deep waters. The presence of transfer RNAs flanking point of breakage in rearrangement loci in the orthologous gene clusters, suggests rearrangement by internal homologous recombination or phage integration events [31].

Coexisting *Prochlorococcus* cells differing in their ribosomal DNA sequence by <3% are reported to have different requirements of optimal light intensities for growth, light-harvesting efficiencies, pigment contents, nitrogen usage abilities, cyanophage specificities, and sensitivities to trace metals, suggesting considerable niche differentiation [30, 32–36].

The genome of another strain, *P. marinus* SS120 has been sequenced by constructing two shotgun genomic libraries by cloning with inserts of size 7 and 10 kb in a low-copy plasmid (pCNS). End sequencing of purified plasmid DNA is carried out using dye-primer and dye-terminator chemistries (50/50) on Licor 4200L and ABI3700 sequencers and data is assembled using PHRAP (www.phrap.org). Glimmer, GeneMarks, and Critica have been used to identify ORFs in the genome. *P. marinus* SS120 genome is found to be composed of a single circular chromosome of 1,751,080 bp with an average GC content of 36.4%. Cluster of Orthologous Groups (COG) database (www.ncbi.nlm.nih.gov/COG) and the National Center for Biotechnology Information (NCBI) protein database (www.ncbi.nlm.nih.gov) are used for genome annotation by using BLAST and PSI-BLAST with manual verification. The genome contains 1884 predicted ORFs with an average size of 825 bp. Few of photosynthetic genes, genes involved in DNA repair, solute uptake, intermediary metabolism, and many other systems of signal transduction and environmental stress response show a particularly drastic reduction in *P. marinus* SS120 compared to *Synechocystis* sp. PCC 6803 and *Anabaena* (*Nostoc*) sp.PCC 7120 genome [37–39]. This represents *P. marinus* SS120 as an oxyphototrophic organism with nearly minimal gene complement, consistent with the fact that the oligotrophic marine environment where it preferentially thrives is much more stable than freshwaters. Phylogenies based on 16S rRNA genes show SS120 at an intermediate position between the "high-light clade," represented by MED4 and another "low-light clade" containing MIT9313, that is located near the base of the radiation [30, 40, 41]. The MED4 strain has an even more compact genome than SS120 (1.66 vs 1.75 Mbp, respectively), whereas that of MIT9313 is larger (2.41 Mbp) [31]. The lower diversity within the high-light clade suggests that it has appeared more recently than the more highly divergent low-light clades [20, 35]. Thus, evolution in the genus *Prochlorococcus* would have tended toward genome reduction [41]. However, this phenomenon would certainly not be enough to account for the large differences in genome sizes and complexity between marine *P. marinus* SS120 and the freshwater cyanobacteria strains and therefore still awaits phylogenetic analysis of large gene regions, from more complete cyanobacterial genomes [41]. Strain represents an extreme within the *Prochlorococcus* genus because of its ability to grow at very low light levels. They contain unique divinyl derivatives of chlorophyll *a* and *b* (Chl *a2* and *b2*) as their major pigments [42]. Unlike typical cyanobacteria, *Prochlorococcus* lacks phycobilisomes, large extrinsic multi-subunit light-harvesting complexes, instead have Chl *a2_b2*-binding proteins known as Pcbs. Pcbs are analogous in a function but are structurally and phylogenetically distinct from the light-harvesting complexes of higher plants [43].

4.2 Eukaryotic phytoplankton

To understand the ocean ecology of phytoplankton, genomic studies have been focused on the bacterial component of the plankton, the majority belonging to *Prochlorococcus* and *Synechococcus*, with very less focus on eukaryotic phytoplankton such as *Thalassiosira pseudonana* (diatom), and *Ostreococcus tauri* [31, 44, 45]. *O. tauri* OTH95, smallest free-living unicellular marine eukaryote green alga known so far, belongs to Prasinophyceae class of unicellular green algae, one of the lineages giving rise to the present form of terrestrial green plants (the green lineage) [46, 47]. Its small size, naked, nonflagellated cell harboring single mitochondrion and chloroplast, and ease in culturing render *O. tauri* an excellent model organism [48]. *Ostreococcus* is widely distributed from coastal to oligotrophic waters [49–52]. Like *Prochlorococcus, Ostreococcus* strains isolated from surface deep waters show distinct genetical and physiological ecotypes, with different light-regulated growth optima compared to those from deep chlorophyll maximum [31, 34, 53]. Genome sequencing of *Ostreococcus* was carried out by creating shotgun libraries and cloning DNA fragments from 1 to 5 kb into pBluescript II KS (Stratagene). Using universal forward and reverse M13 primers and the ET DYEnamic terminator kit around 60,000 clones were sequenced with MegaBace 1000 automated sequencers (GE Healthcare). Phred-Phrap and Consed software packages are used for data analysis and contigs assembly. *O. tauri* strain OTH95 has 12.56 Mb genome size similar in size to yeasts *Saccharomyces cerevisiae* and *Schizosaccharomyces pombe* distributed in 20 chromosomes. Genome size is smaller than any other oxyphototrophic eukaryote known, including the red alga *Cyanidioschyzon merolae* [45, 54].

Out of 6265 genes identified in this genome, 46% shows homology to plant orthologs [45]. A second remarkable feature of *O. tauri* genome is the intense degree of genome compaction, due to shortened intergenic regions (average 196 bp), shorter than that of other eukaryotes with a similar genome size. Like other photosynthetic eukaryotic systems, *O. tauri* possesses multigene families involved in pigment biosynthesis, photosynthesis, and carbon fixation, but at lower copy number [45, 55]. *O. tauri* lacks the typical genes encoding the major light-harvesting complex proteins associated with photosystem II (LHCII), instead, paralogs encoding prasinophyte-specific chlorophyll-binding proteins are present, as observed in *Mantoniella squamata*, a green algae [55]. However, the presence of a small set of five lhcA genes, encoding a LHCI antenna supports the hypothesis that the LHCI antenna type is more ancestral than LHCI [55]. Regarding carbon assimilation machinery, only one unique carbonic anhydrase (CA), similar to bacterial CA, was identified instead of carbon concentrating mechanism (CCM) genes similar to *C. reinhardtii (*a single-cell green alga) or the organisms actively or passively enhancing inorganic carbon influx were found [45, 56]. Putative genes encoding all the enzymes of C4 photosynthetic pathway were identified in *O. tauri* genome unlike unicellular organisms [44, 57]. Interestingly, C4 photosynthesis has been observed in only one member of the Chlorophyta, the macroalga *Udotea*. Adapting the costly C4 photosynthetic pathway, *O. tauri* could constitute a critical ecological advantage in the CO_2-limiting conditions of phytoplankton blooms when competitors have lower CCM efficiencies. *O. tauri* has two NADP-ME orthologs (one targeted to chloroplast) most similar to *Hydrilla verticillata* [45, 58]. Unlike other eukaryotic algae, *O. tauri* seems to have developed survival strategies in the competitive marine environment. *O. tauri* is known to grow on a variety of substrates such as nitrate, ammonium, and urea [45, 50]. Four genes encoding ammonium transporters (two being green lineage-related and the other two of prokaryote-lineage), eight genes related to nitrate acquisition and assimilation (on chromosome 10), four genes related to urea assimilation (on chromosome 15) have been identified in genome of *O. tauri* reminding of gene organization in prokaryotic cyanobacteria [31, 45]. Overall *O. tauri* with the capacity of C4 photosynthesis, tiny phytoplankter with relatively large surface area to volume ratio, with variety of ways of optimizing nitrogen assimilation render it major competitive edge over other unicellular phytoplankton, especially in intense bloom condition with limited resources [45].

Instead of light-independent protochlorophyllide reductase, presence of two copies of light-dependent protochlorophyllide oxidoreductase gene similar to land plants, *O. tauri* synthesizes chlorophyll only during the day time. Like *Arabidopsis* and *Chlamydomonas*, a large number of kinase-encoding and calcium-binding domains are part of phosphorelay-based calcium-dependent signal transduction systems in *O. tauri* [45].

5 Genomics of macroalgae

5.1 Green algae

Single-celled green algae, also known as green microalgae, play an important role in the world's ecosystems and in habitat oceans to lands, including deserts. Microalgae are promising alternative to terrestrial crops as renewable fuel feedstocks due to their high production potential and unique cultivation capacity on nonarable lands [59] The oleaginous chlorophyte microalgal *Chlorella vulgaris* represents a promising green microalgae model synthesizing and accumulating large quantities of fuel intermediates in the form of stored lipids [60–64]. Recently, the genome of *Chlorella* has been sequenced using Illumina HiSeq 2000 technology with 108 cycles. Final 168,611,711 reads of which 165,874,962 remained as pairs have been assembled using a DeBruijn method. 113 scaffolds at $\geq$1000 × depth of coverage are generated, 24 of which are longer than 100 kb and 566 are of 20–100 kb, generating a total genome assembly size of 37.34 Mb (with 61.5% GC content). Omics analyses of *C. vulgaris* UTEX 395 revealed complete gene sets encoding fatty acid and triacylglyceride biosynthetic pathways and nitrogen assimilation inventory includes genes for nitrate/nitrite transporters and reductases. This highlights the genetic pathways for potential marker less strain-engineering strategies targeting lipid accumulation in the absence of stress induction, for cost-competitive biofuel production [65].

Green algae are primary biomass producers and are viable sources of commercial compounds in the food, fuel, and pharmaceutical industries. Analyzing genomics and related phenotyping studies are necessary to better understand the ecology and physiology of microscopic algae (microphytes), both at the local and global scales, and to optimize their cultivation and yield of bioproducts for industrial applications [66]. Currently, species with superior growth characteristics for large scale cultivation remain understudied, underdeveloped, and underexploited [67].

Green alga *Chloroidium sp.* UTEX 3007, is a lipid-producing alga, capable of surviving in the desert. This species accumulates 78.2±7.7% fatty acids, and ~41.8% of total fatty acids consisted of palmitic acid at the time of harvest. Accumulating palmitic acid instead of other longer and more unsaturated fatty acids is expected to increase fitness at higher temperatures because palmitic acid is more thermostable. *Chloroidium sp. UTEX 3007* has been sequenced using Illumina HiSeq 2500 (Illumina, San Diego, USA). Genome is assembled with CLC Genomics Workbench assembler (v8.5, CLC Genomics, Qiagen, Aarhus, Denmark) with a kmer length of 45 (word size of 22) and reads are mapped to contigs. The average size of insert for paired genomic reads (2 × 100 bp) is ~720 bp with a maximum of 1200 bp. The reads assembled into 710 scaffolds with a total length of 52.5 Mbp. Final of 323,780,836 (98.6%) reads have been matched to the final assembly. *Chloroidium sp.* has 52.5 Mbp genome with 16 nuclear chromosomes compared to *Coccomyxa subellipsoidea* C-169 (20 chromosomes) and *Chlorella variabilis* NC64A (12 chromosomes) [68].

Unlike cyanobacteria, acidophilic eukaryotic phototrophs are capable of photosynthesis in acidic environments (pH <4.0) such as acid mine drainage (AMD) and geothermal hot springs [69, 70]. The low pH facilitates metal solubility in water; therefore, acidic waters tend to have high concentrations of metals. Thus, acidophilic eukaryotic algae usually possess the ability to cope with toxic heavy metals in addition to low pH, both of which are lethal to most eukaryotes [69]. The extremely low pH of these waters is due to the dissolution and oxidation of sulfur that is exposed to water and oxygen and produces sulfuric acid [71]. Acidophilic algae are distributed throughout different branches of the eukaryotes, such as in red and green algae, stramenopiles, and euglenids. Evolving from neutrophilic relatives. Three thermo-acidophilic red algae, *Cyanidioschyzon merolae, Galdieria sulphuraria, and Galdieria phlegrea*, have been sequenced [54, 72, 73]. Acidophilic green alga *Chlamydomonas eustigma* NIES-2499 showed that phytochelatin synthase genes of bacterial HGT origin played an important role in the tolerance to cadmium (Olsson et al.). The shotgun and paired-end libraries of size 8 kb of *C. eustigma* are sequenced using Roche 454 GS FLX+ Titanium (Roche Diagnostics) and paired-end (400-bp) library are sequenced using HiSeq 2500 with 100 base-paired end format with the Tru-Seq SBS kit v3 (Illumina, Inc.). The paired-end (800-bp) and mate-pair libraries of size 3, 5, and 8 kb have been sequenced using MiSeq (Illumina, Inc.) with the MiSeq reagent kit version 3 (600 cycles; Illumina). The MiSeq reads were filtered using ShortReadManager, based on a 17-mer frequency. Genomic analysis revealed upregulation of genes encoding heat-shock proteins (HSPs) and plasma membrane H+-ATPase (PMA), loss of fermentative genes that produce organic acids and thus reduce cytosolic pH, the acquisition of an energy shuttle and buffering system, and the acquisition and multiplication of genes involved in arsenic biotransformation and detoxification have contributed to the adaptation of *C. eustigma* to acidic conditions [74].

After successful sequencing and analyzing from more than 1000 species of green plants representing most of the known diversity within Viridiplantae under 1 KP (1000 Plants), the focus is on sequencing complete genomes from more than 10,000 plants and protists 10 KP [75–78]. The project launched at the 19th International Botanical Congress 2017, Shenzhen, China is focused on addressing fundamental questions in plant evolution and diversity, providing data on more than 10,000 species representing every major clade of embryophytes (land plants), green algae (chlorophytes and streptophytes), and protists (photosynthetic and heterotrophic) [79].

5.2 Red algae

Tracing back the history, the first complete red algal genome that was sequenced and elucidated belongs to ultrasmall unicellular red algae *Cyanidioschyzon merolae* which inhabits sulfate rich hot water springs. *C. merolae* is a model organism for the genomic studies as it has very basic eukaryotic cell structure without a rigid cell wall and minimal membrane-bound organelles like a golgi apparatus with only two cisternae, a single endoplasmic reticulum and a few lysozymes like organelles. It provided information on not only essential genes of *C. merolae* but also on the origin and evolution of eukaryotic cells. The genome size of this alga is the smallest of all the photosynthetic eukaryotic organisms. The genome comprises 16,520,305 base pairs on 20 chromosomes with 5331 genes. The expressed genes account for almost 86.3% of the total genome. Along with nuclear sequencing, its mitochondria and plastid have also been sequenced which accounts for 32,211 bp and 149.987 bp respectively. In *C. merolae*, two genes for the mitochondrial division were found, *Dnm1* and *Dnm2*. Five proteins related to kinesin family were also encoded by the genome of *C. merolae*. Genes related to photosynthesis and photosystems PS1 and PSII were also found. The mosaic region of enzymes involved in Calvin cycle was found to be conserved and indicated a strong supporting evidence for primary plastid endosymbiosis [54, 80]. *C. merolae* genome was sequenced by the whole-genome random sequencing method. About 335,000 insert ends were sequenced, which covered the genome 11 times. BAC libraries with two subsets were constructed and a large-scale full-length cDNA library prepared from cells cultured under various growth conditions. The sequences were assembled using Phrap, further examined by referring to another assembly using ARACHNE, and edited using CONSED. The scaffolds were built within the hybridization groups using read-pair information from the BAC, shotgun, and cDNA clones. The gaps between the contigs were closed by primer walking PCR, and mate-pair clone and BAC clone sequencings.

The next unicellular algal group targeted for identification of essential genes and novel gene functions also included Cyanidiophyceae group. Comparative analysis for *G. sulphuraria* and *C. merolae* genome sequence became available in 2005 [81]. An extraordinary adaptation of unicellular *G. sulphuraria* is metabolic versatility to grow in extreme acidic thermophilic environments with unusual physiological traits and heterotrophic and mixotrophic mode of nutrition with multiple substrates. The genome size of the species has been reported to be in the range of 10 to 16 Mbp in a similar context with *C. merolae*. Based on comparative analysis, more than 30% of *G. sulphuraria* genes did not match the genes of *C. merolae*. The comparative analysis revealed that there are more introns in *G. sulphuraria* than in *C. merolae,* which contains introns only in its 26 genes [54]. Although, *C. merolae* is an obligate photoautotroph whereas *G. sulphuraria* adopts a heterotrophic mode of nutrition and is the only member of Cyanidiales, the putative enzymes responsible for sugar and polyol metabolism exhibit a lot of similarities. The genome of *G. sulphuraria* encodes a number of kinases which exhibit similarity to prokaryotic enzymes like fructokinases, glucokinases, galactokinases, ribokinases, glycerokinases, and xylulokinases [82]. One of the contrasting characteristics of both these species is the presence of carbohydrate transporters in *G. sulphuraria* which is one of the significant reasons responsible for its mixotrophic and heterotrophic growth. Another striking difference is in the metabolism of glycerol. *C. merolae* lacks aquaporin type of glycerol permeases whereas *G. sulphuraria* contains at least 4 genes encoding these putative glycerol permeases which would allow *G. sulphuraria* to take up glycerol from the environment, although glycerol kinase is encoded by genomes of both the species [83–85].

Colonies were randomly picked using a GeneMachines Mantis Colony and Plaque Picker (GeneMachines, San Carlos, CA), and plasmid DNA were prepared from overnight cultures using a Qiagen 3000 robot (Qiagen USA, Valencia, CA). DNA sequences were determined by cycle sequencing and sequence analysis using an ABI PRISM 3700 DNA Analyzer (Applied Biosystems, Foster City, CA). All sequence data and chromatograms were stored on a Geospiza Finch server (Geospiza, Seattle).

HGT has an important role to play in the evolution of ancestral lineages of red algae. In a similar context, the genome of mesophilic red algae, *Porphyridium purpureum* was drafted. The genome of the algae is of the size of 19.7 Mbp with tightly packed coding regions and is intron poor with 8355 genes. Genomic analysis reveals evidence for sexual reproduction. The presence of 8 out of 9 meiosis related genes is consistent with the maintenance of sexual reproduction in the organism. The genes required for the methylerythritol phosphate pathway for isoprenoid biosynthesis were also identified. Phylogenomic analysis indicated HGT between prokaryotes and photosynthetic eukaryotes with red algae as mediators. As discussed previously, red algae have phycobilisomes associated with light-harvesting complex (LHC), *P. purpureum* was the first organism with its phycobilisomes isolated along with identification of seven LHC (light-harvesting complex) proteins. Genomic analysis also revealed that there exist two subunits along with several linker proteins (L_{CM}, L_{RC}, L_C, and $4\gamma^{PE}$) for PE that is, alpha and beta. Comparative analysis with *C. merolae* reveals the cell wall polysaccharide complexity of *P. purpureum*. The genome of *P. purpureum* encodes a total of 114 of glycoside hydrolases and glycosyltransferases whereas that of *C. merolae* encodes only 83. There are 33% more CAZy families in *P. purpureum* (14 glycoside hydrolases and 34 glycosyltransferases) than in *C. merolae* (9 glycoside hydrolases and 27 glycosyltransferases). Out of 8355 predicted genes, almost 3.4% genes account for solute transporters, pumps, and channels, along with a sodium-potassium ATPase pump. These findings suggest that the pump regulates the sodium and potassium homeostasis when the organism is exposed to high salt levels in the environment. Not only this, the genome of the red algae also codes for several other putative transporters like sodium-bicarbonate transporter, sodium-dependent phosphate transporter, and a sodium-glucose transporter [86].

A total of 7.4 Gbp of *P. purpureum* CCMP 1328 paired-end (150_150 bp) genome data generated using two flow cell lanes in the Illumina GAIIx were assembled with the CLC Genomics Workbench tools (http://www.clcbio.com/products/clc-genomics-workbench/) into 4770 contigs with a N50 of 20,296 bp. As with other genome studies, these estimates are subjected to further validation with a better assembly of more sequence data. Thereafter, 4.1 Gbp of Illumina mRNA-seq data (150×150 bp reads) were used to train the ab initio gene predictors resulting in a set of 8355 weighted consensus gene structures that were used for downstream analyses.

From microalgae, the focus of the genomic biologists shifted toward macroalgae with considerable economic importance that is, Susabi-nori or *Pyropia Yezoensis*. It is an edible marine red algae (seafood) and the genome of the species was drafted by next-generation sequencing techniques. The genome size of the algae was found to be 43 Mbp. Almost 60% of the genes lacked introns. A new enzyme methionine synthase was identified from the genome which might explain the symbiotic interaction of *P. yezoensis* with its environment which was reported for the first time. Phycobilisome related genes were also analyzed from the genome of this red alga with the possibility of their use as DNA markers for the improvement of the cultivar. The second homolog of *NblA* gene was hypothesized to be involved in the characteristic color of *P. yezoensis*. Out of 10,327 genes, the function of 35% of genes remains unknown and 2069 genes show the similarity of those of *C. merolae* [87].

From the protoplast DNA sample of *P. yezoensis*, whole-genome shotgun libraries were prepared for two platforms, a Roche 454 GS-FLX/FLX+ (Roche Diagnostics, Branford, CT) and an Illumina Genome Analyzer IIx (Illumina, Inc. San Diego, CA), respectively. The 454-pyrosequencing library for single-end reads was constructed from the sheared DNA by GS Titanium RapidLibrary Preparation Kit (Roche Diagnostics). For the IlluminaGenome Analyzer IIx, a 75-bp paired-end shotgun library (insert sizes of 500 bp) was prepared according to the manufacturer's protocols. The read data obtained have been deposited in DDBJ/EMBL/GenBank under accession number SRA061934. Both reads were assembled using CLC Assembly CellTM version4.06 beta (CLC bio, Aarhus N, Denmark). For the preliminary *de novo* assembly, the contigs obtained still contained the sequences of organelles (mitochondrion and chloroplast) and an unknown bacterium of the genus Agarivorans. To remove such nonnuclear sequences, reference sequences were prepared. For organelles, the sequences of the chloroplast (accession: NC_007932) and mitochondrion (NC_017837) genomes of *P. yezoensis* were downloaded from the GenBank. In addition, *Agarivorans albus* strain MKT 106 [24] was purchased from the National Institute of Technology and Evaluation, Japan (NBRC) and the genomic sequences were read with 454 GS FLX+ (Text S1). The sequences of the organelles and bacterium were then used to clean the assembly of *P. yezoensis* nuclear DNA sequences [87].

Another genome of red algal seaweed belonging to the Bangiophyceae lineage, *Porphyra umbilicalis* has been recently drafted by whole-genome shotgun sequencing methods. Its unique features include growth in harsh and stressful environments along with the undeniable economic value. The genome size is 87.7 Mbp with around 13,125 genes [88]. Photoprotection is one mechanism by which it copes up with the ecological stress with the help of "Redcap" genes, high-light-induced or one-helix proteins (OHPs), which have a role to play for photoacclimation and cell viability. *P. umbilicalis* has genes coding for high- affinity iron transport complex for iron uptake processes to obtain nutrients during stressful high tides [89]. The accumulation of enormous amount of data with the advent of efficient sequencing techniques has raised many questions. The foremost is the role of repeats and transposable elements in the variation of genome size of red algae. Secondly, if there is any history of gene duplication, expansion of reduction in rhodophytes. Thirdly, if there is any evidence regarding the suppression of these transposable elements in rhodophytes. In an effort to answer these questions, Lee and coworkers [90] sequenced the genome of one of the most economically important red seaweed, *Gracilariopsis chorda*, which belongs to florideophyceae class. The size of the genome is around 92.1 Mb and 10.806 protein-coding genes. Almost 61.2% of the genome is comprised of transposable elements. Comparative analysis of unicellular red algae like *G. sulphuraria, C. merolae*, and *P. purpureum*, revealed that transposable elements had a significant impact on the size expansion of the genome of multicellular red algae. These TEs might have led to an increase in the genome size of red algae through duplication or transposition as in land plants [91, 92]. The study also indicated the absence of polyploidization in red algae. This gives a valid and appropriate explanation for non-expansion of gene inventory in red seaweeds, despite growth in genome size. The researchers hypothesized the role of epigenetic mechanisms such as DNA methylation in regulating the sexual reproduction in red algae [90] (Table 1 and Fig. 1).

5.3 Brown algae

Ectocarpus siliculosus is the most prominent filamentous brown alga, is a model organism and is closely related to the kelps [93]. The genomic analysis of *E. siliculosus* has provided

Table 1 Comparative genome statistics of different red algae

	Genomic characteristics				
Algae	**Size of genome (Mbp)**	**Protein coding genes**	**G+C content (%)**	**Repeats (%)**	**References**
Cyanidioschyzon merolae	16	4803	55	20	Matsuzaki et al. [54]
Galdieria sulphuraria	13	7174	–	16	Barbier et al. [91]
Porphyridium purpureum	19.7	8355	–	3.9	Bhattacharya et al. [85]
Pyropia yezoensis	43	10,327	63.6	1.4	Nakamura et al. [86]
Porphyra umbilicalis	87.7	13,125	65.8	43.9	Brawley et al. [87]
Gracilariopsis chorda	92.1	10,806	49.26	61.2	Lee et al. [89]

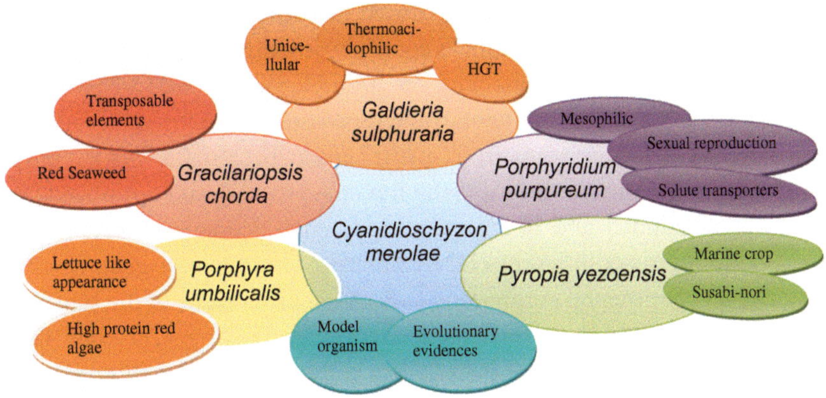

Fig. 1 Highlighting features of representative *red* algae.

detailed information, reflecting the evolutionary basics of the group. The genome size of *E. siliculosus* is 214 Mbp and its genome is rich in introns, with a frequency of seven per gene. The number of protein-coding genes is 16,256. About 22.7% of the *E. siliculosus* comprises repeated sequences including DNA transposons and retrotransposons. *Ectocarpus* inhabits harsh conditions of shoreline, where it is exposed to different kinds of biotic and abiotic stress. Several such gene families have been identified which have played a possible role in the adaptation of *Ectocarpus* to such an environment. One such indication is the presence of a large number of LHC genes. LHC genes comprises a cluster of genes belonging to L1818 light-stress related family and other photosynthetic genes such as a light-independent protochlorophyllide reductase (DPOR) to adapt to harshful light conditions. Sequencing has also revealed the presence of phenolic compounds in order to protect the organism from ultraviolet (UV) rays, a kind of abiotic stress. Homologues of genes responsible for flavonoid biosynthesis present in land plants were also present in *Ectocarpus*. To combat osmotic and light stress, enzymes required for the metabolism of reactive oxygen species were also coded by the genome of *Ectocarpus*. Apart from this, the genes for new biosynthetic pathways such as halide metabolism also support the adaptation of the organism to the variable intertidal environment. Although in comparison to kelps like *Laminaria digitata* where large families of haloperoxidases are found, the genome of *Ectocarpus* encoded only one vanadium-dependent bromoperoxidase (vBPO). Brown algae cell walls contain alginates and fucans, which were not found in the *Ectocarpus* genome. However, it contains a large number of polysaccharide modifying enzymes like mannuronan C5 epimerases, sulfotransferases, and sulfatases. The carbon storage system is unique by virtue of production of mannitol and the β-1, 3-glucan laminarin. The genomic analysis has also revealed the presence of signal transduction genes, like a family of receptor kinases, ensuring the evolution of multicellularity in this lineage. Sequence analysis also identified 26 microRNAs, which after comparison with other eukaryotic groups indicates their presence during early evolutionary stages [94].

Genome and cDNA sequencing were carried out using *Ectocarpus siliculosus* strain Ec 32, which is a meiotic offspring of a field sporophyte collected in 1988 in San Juan de Marcona, Peru. The genome sequence was assembled using 2,233,253 and 903,939 paired, end-sequences from plasmid libraries with 3 and 10 kbp inserts respectively, plus 58,155 paired, end-sequence reads from a small-insert bacterial artificial chromosome library. Annotation was carried out using the EuGene program and optimized by manual correction of gene models and functional assignments. Sequencing of 91,041 cDNA reads, corresponding to six different cDNA libraries, and a whole-genome tiling array analysis provided experimental confirmation of a large proportion of the transcribed part of the genome. Small RNAs were characterized by generating 7,114,682 sequencing reads from two small RNA libraries on a Solexa Genome Analyzer (Illumina). Analyses of the methylation state of genomic DNA and of specific transposon families were carried out using HPLC analysis of nucleotide methylation and McrBC digestion, respectively. Full information about the *Saccharina japonica* is the first sequenced reference genome from kelps and belongs to the Laminariaceae family and the second genome from brown algae. It is economically most important seaweed in the northwest coast of the Pacific Ocean. Alginates produced by kelps are cultivated in North America and Europe which are widely used in food, pharmaceutical, and other industrial processes. The genome size of the brown alga is 537 Mbp with 18,733 protein-coding regions. *S. japonica* is unique in possessing complex differentiation, large blades, a higher polysaccharide content and significant iodine accumulating abilities, when compared with *E. siliculosus*. Comparative genome analysis with *E. siliculosus* showed that genomes of both these species share 4309 gene families. It accounts to 17,379 genes in *S. japonica* and 14,136 genes in *E. siliculosus,* hypothesizing gene expansion in *S. japonica*. The genome of this microalga has shown significant gene expansion in about 58 families. Genomic analysis of *S. japonica* revealed important genes associated with cell wall synthesis and various stress-related mechanisms like halogen concentrations. Brown macroalgae are known as effective iodine accumulators among all living organisms. In *S. japonica* genome, a phylogenetic analysis of vBPOs and iodoperoxidases (vIPOs) showed the evolution of red and brown algae from a common ancestor with the genes Vanadium-dependent chloroperoxidase (vCPOs) genes of fungi. Comparative genomics of *S. japonica* and *E. siliculosus* showed similarity in the carbohydrate metabolic pathways. The formation of cell walls has been related to the expansion of cellulose synthase, mannuronan C-5-epimerase, and alpha-(1, 6)-fucosyltransferase gene families. It depicts the independent evolution of brown algae from higher plants and animals [95] (Table 2).

Cladosiphon okamuranus is one of the most important edible seaweed, which is cultivated for fucoidan in Japan and it belongs to chordariaceae family. Fucoidan has various significant physiological and biological roles. The genome size of *C. okamuranus* is approximately 140 Mbp and contains 13,640 protein-coding genes. Genes have been identified in the genome of *C. okamuranus* responsible for enzymes related to

Table 2 Comparative genomic statistics of *E. siliculosus*, *S. japonica*, and *C. okamuranus*

Genomic Characteristics	Algae		
	Ectocarpus siliculosus (Cock et al. [94])	*Saccharina japonica* (Ye et al. [95])	*Cladosiphon okamuranus* (Nishitsuji et al. [96])
Size of the sequenced genome (Mbp)	214	545	140
Protein coding regions	16,256	18,733	13,640
Genomic G+C content (%)	54	50	54
Average gene length (bp)	7696	6859	9587
Number of introns	113,619	–	–
Number of exons	129,875	–	–

phlorotannin biosynthesis pathway. Comparable to *E. siliculosus* genome, the genome of *C. okamuranus* is also intron rich with the frequency of 8.26 per gene. Comparative analysis of genomes of *C. okamuranus* and *E. siliculosus* has revealed the genes required for the biosynthesis of fucans and alginates. Comparative genomics disclosed several genes associated with transcription factors and signaling molecules in the biosynthesis of polysaccharides and phlorotannin. There are 214 genes coding for transcription factors in *C. okamuranus* as compared to 260 in *E. siliculosus*. The most abundant transcription factor was found to be from Myb domain family with 58 genes. The various other domains that were identified include HSF, Myb, bZIP, Zinc Finger, bHLH, CCAAT-binding, homeobox, AP2-EREBP, Nin-like, TAF, E2F-DP, CBF/NF-Y/archaeal, and Sigma-70 r2/r3/r4. The genome of this brown algae also contains genes for specific receptor kinases as cell-cell signaling molecules. Genes related to alginate biosynthesis pathway were predicted such as Mannose-6-phosphate isomerase (MPI), Phosphomannomutase (PMM), GDP-mannose 6-dehydrogenase (GMD), Mannuronan synthase (MS), and Mannuronate C5-epimerase (MC5E). Genomics also revealed the conservation of key enzyme GMD in brown algae. The alga is also reported to contain enzymes for fucoidan biosynthesis pathway. Four genes encoding fucosyltransferase, six genes for sulfotransferase, two genes for GDP-mannose 4, 6-dehydratase and one for GDP-L-fucose synthase were found in both *C. okamuranus* and *E. siliculosus*. Presence of these enzymes in *C. okamuranus* for sulfated fucan biosynthetic pathway indicates that it is a rich source of fucoidans. Similarly, the genes encoding for enzymes responsible for alginate biosynthesis pathway were also reported such as GMD and MC5E in *C. okamuranus* like *E. siliculosus*. Phloratannin are structured analogs of condensed tannins, with multiple ecological roles [97]. Two candidate genes PKS 1 and PKS 2 (Polyketide synthase) responsible for phlorotannin biosynthesis were found in *C. okamuranus* [96].

6 Conclusions

Last decade witnessed a huge genomic data generation by accomplishing sequencing projects in a range of species. However, the path toward algal genomics has yet not touched its horizons, required for pan-genomics. Keeping in view the enormous number of algal species belonging to different classes, sequencing few genomes per phyla is not sufficient to develop a "core" genome, representing ~80% of the genome of any single algal strain. Phylogenetic and BLAST analysis have been used as potential tools for comparative genomic studies evaluating the representative algal species diversity. However, these approaches can only explain the presence, absence, and variability among known genetic loci. However, they do not throw light on strain-specific novel genes which are absent in the reference genome. Multiple species need to be extensively sequenced to comprehend our understanding of the basics of algal species. Future gene reservoir for algal pan-genomics is speculated to be tremendously vast and novel genes will continue to emerge [98].

References

[1] Ohno, M., Critchley, A.. Seaweed Resources of the world. Available at: https://scholar.google.com/scholar_lookup?title=Seaweed%20Resources%20of%20the%20World.%20Kanagawa%20International%20Fisheries%20Training%20Centre&author=AT.%20Critchley&author=M.%20Ohno&publication_year=1998, 1998 (Accessed 26 March 2019).

[2] S.B. Gould, R.F. Waller, G.I. McFadden, Plastid evolution, Annu. Rev. Plant Biol. 59 (2008) 491–517, https://doi.org/10.1146/annurev.arplant.59.032607.092915.

[3] M.S. Parker, T. Mock, E.V. Armbrust, Genomic insights into marine microalgae, Annu. Rev. Genet. 42 (2008) 619–645, https://doi.org/10.1146/annurev.genet.42.110807.091417.

[4] U. Kutschera, K.J. Niklas, Macroevolution via secondary endosymbiosis: a Neo-Goldschmidtian view of unicellular hopeful monsters and Darwin's primordial intermediate form, Theory Biosci. 127 (2008) 277–289, https://doi.org/10.1007/s12064-008-0046-8.

[5] C.F. Delwiche, J.D. Palmer, Rampant horizontal transfer and duplication of rubisco genes in eubacteria and plastids, Mol. Biol. Evol. 13 (1996) 873–882, https://doi.org/10.1093/oxfordjournals.molbev.a025647.

[6] E. Haeckel, Generelle Morphologie Der Organismen. Allgemeine Grundzüge Der Organischen Formen-Wissenschaft, Mechanisch Begründet Durch Die Von Charles Darwin Reformirte Descendenztheorie, G. Reimer, Berlin, 1866, https://doi.org/10.5962/bhl.title.3953.

[7] L.J. Rothschild, Protozoa, Protista, Protoctista: what's in a name?, J. Hist. Biol. 22 (1989) 277–305. Available at: http://www.ncbi.nlm.nih.gov/pubmed/11542176.

[8] P.G. Falkowski, T. Fenchel, E.F. Delong, The microbial engines that drive Earth's biogeochemical cycles, Science 320 (2008) 1034–1039, https://doi.org/10.1126/science.1153213.

[9] J. Snijder, R.J. Burnley, A. Wiegard, A.S.J. Melquiond, A.M.J.J. Bonvin, I.M. Axmann, et al., Insight into cyanobacterial circadian timing from structural details of the KaiB-KaiC interaction, Proc. Natl. Acad. Sci. 111 (2014) 1379–1384, https://doi.org/10.1073/pnas.1314326111.

[10] R.J. Langlois, D. Hummer, J. LaRoche, Abundances and distributions of the dominant nifH phylotypes in the Northern Atlantic Ocean, Appl. Environ. Microbiol. 74 (2008) 1922–1931, https://doi.org/10.1128/AEM.01720-07.

[11] N.L. Goebel, K.A. Turk, K.M. Achilles, R. Paerl, I. Hewson, A.E. Morrison, et al., Abundance and distribution of major groups of diazotrophic cyanobacteria and their potential contribution to N2

fixation in the tropical Atlantic Ocean, Environ. Microbiol. 12 (2010) 3272–3289, https://doi.org/10.1111/j.1462-2920.2010.02303.x.
[12] P.H. Moisander, R.A. Beinart, M. Voss, J.P. Zehr, Diversity and abundance of diazotrophic microorganisms in the South China Sea during intermonsoon, ISME J. 2 (2008) 954–967, https://doi.org/10.1038/ismej.2008.51.
[13] S.M. Coelho, N. Simon, S. Ahmed, J.M. Cock, F. Partensky, Ecological and evolutionary genomics of marine photosynthetic organisms, Mol. Ecol. 22 (2013) 867–907, https://doi.org/10.1111/mec.12000.
[14] F. Chen, J. Zhang, J. Chen, X. Li, W. Dong, J. Hu, et al., realDB: a genome and transcriptome resource for the red algae (phylum Rhodophyta), Database (2018) https://doi.org/10.1093/database/bay072.
[15] Gurgel, C.F.D. and Lopez-Bautista, J. (2007). Red algae, In: *Encyclopedia of Life Sciences*. Wiley, Chichester, UK.
[16] J.W. Stiller, et al., The evolution of photosynthesis in chromist algae through serial endosymbioses, Nat. Commun. 5 (2014) 5764.
[17] H. Qiu, D.C. Price, E.C. Yang, H.S. Yoon, D. Bhattacharya, Evidence of ancient genome reduction in red algae (Rhodophyta), J. Phycol. 51 (2015) 624–636.
[18] A.R. Grossman, M.R. Schaefer, G.G. Chiang, J.L. Collier, The phycobilisome, a light-harvesting complex responsive to environmental conditions, Microbiol. Rev. 57 (1993) 725–749.
[19] L.D. Graham, L.W. Wilcox, Algae, Prentice-Hall, USA, 2000.
[20] I.A. Abbott, Marine Red Algae of the Hawaiian Islands, Bishop Muse, 1999.
[21] A.R. Grossman, Paths towards algal genomics, Plant Physiol. 137 (2005) 410–417.
[22] V.G. Hoek, D.G. Mann, H.M. Jahns, Algae—An Introduction to Phycology, Cambridge, Cambridge University Press, Cambridge, 1995, p. 623.
[23] S.L. Mazard, N.J. Fuller, K.M. Orcutt, O. Bridle, D.J. Scanlan, PCR analysis of the distribution of unicellular cyanobacterial diazotrophs in the Arabian Sea, Appl. Environ. Microbiol. 70 (2004) 7355–7364, https://doi.org/10.1128/AEM.70.12.7355-7364.2004.
[24] L. Campbell, et al., Picoplankton community structure within and outside a Trichodesmium bloom in the southwestern Pacific Ocean, Vie et Mileu 55 (2005) 185–195.
[25] E.A. Webb, I.M. Ehrenreich, S.L. Brown, F.W. Valois, J.B. Waterbury, Phenotypic and genotypic characterization of multiple strains of the diazotrophic cyanobacterium, *Crocosphaera watsonii*, isolated from the open ocean, Environ. Microbiol. 11 (2009) 338–348, https://doi.org/10.1111/j.1462-2920.2008.01771.x.ench.
[26] S.R. Bench, I.N. Ilikchyan, H.J. Tripp, J.P. Zehr, Two strains of crocosphaera watsonii with highly conserved genomes are distinguished by strain-specific features, Front. Microbiol. 2 (2011) 261, https://doi.org/10.3389/fmicb.2011.00261.
[27] D.J. Scanlan, N.J. West, Molecular ecology of the marine cyanobacterial genera Prochlorococcus and Synechococcus, FEMS Microbiol. Ecol. 40 (2002) 1–12, https://doi.org/10.1111/j.1574-6941.2002.tb00930.x.
[28] B. Palenik, B. Brahamsha, F.W. Larimer, M. Land, L. Hauser, P. Chain, et al., The genome of a motile marine Synechococcus, Nature 424 (2003) 1037–1042, https://doi.org/10.1038/nature01943.
[29] G. Rocap, D.L. Distel, J.B. Waterbury, S.W. Chisholm, Resolution of Prochlorococcus and Synechococcus ecotypes by using 16S-23S ribosomal DNA internal transcribed spacer sequences, Appl. Environ. Microbiol. 68 (2002) 1180–1191, https://doi.org/10.1128/AEM.68.3.1180-1191.2002.
[30] L.R. Moore, S.W. Chisholm, Photophysiology of the marine cyanobacterium Prochlorococcus: ecotypic differences among cultured isolates, Limnol. Oceanogr. 44 (1999) 628–638, https://doi.org/10.4319/lo.1999.44.3.0628.
[31] G. Rocap, F.W. Larimer, J. Lamerdin, S. Malfatti, P. Chain, N.A. Ahlgren, et al., Genome divergence in two Prochlorococcus ecotypes reflects oceanic niche differentiation, Nature 424 (2003) 1042–1047, https://doi.org/10.1038/nature01947.
[32] L. Moore, R. Goericke, S. Chisholm, Comparative physiology of Synechococcus and Prochlorococcus: influence of light and temperature on growth, pigments, fluorescence and absorptive properties, Mar. Ecol. Prog. Ser. 116 (1995) 259–275, https://doi.org/10.3354/meps116259.
[33] R.D. Smith, J.C. Walker, Plant protein phosphatases, Annu. Rev. Plant Physiol. Plant Mol. Biol. 47 (1996) 101–125, https://doi.org/10.1146/annurev.arplant.47.1.101.

[34] L.R. Moore, G. Rocap, S.W. Chisholm, Physiology and molecular phylogeny of coexisting Prochlorococcus ecotypes, Nature 393 (1998) 464–467, https://doi.org/10.1038/30965.
[35] L.R. Moore, A.F. Post, G. Rocap, S.W. Chisholm, Utilization of different nitrogen sources by the marine cyanobacteria *Prochlorococcus* and *Synechococcus*, Limnol. Oceanogr. 47 (2002) 989–996, https://doi.org/10.4319/lo.2002.47.4.0989.
[36] M.B. Sullivan, J.B. Waterbury, S.W. Chisholm, Cyanophages infecting the oceanic cyanobacterium Prochlorococcus, Nature 424 (2003) 1047–1051, https://doi.org/10.1038/nature01929.
[37] T. Mizuno, T. Kaneko, S. Tabata, Compilation of all genes encoding bacterial two-component signal transducers in the genome of the cyanobacterium, Synechocystis sp. strain PCC 6803, DNA Res. 3 (1996) 407–414. http://www.ncbi.nlm.nih.gov/pubmed/9097043. (Accessed 26 March 2019).
[38] M. Ohmori, M. Ikeuchi, N. Sato, P. Wolk, T. Kaneko, T. Ogawa, et al., Characterization of genes encoding multi-domain proteins in the genome of the filamentous nitrogen-fixing Cyanobacterium anabaena sp. strain PCC 7120, DNA Res. 8 (2001) 271–284. http://www.ncbi.nlm.nih.gov/pubmed/11858227. (Accessed 26 March 2019).
[39] J. Meeks, E. Campbell, M. Summers, F. Wong, Cellular differentiation in the cyanobacterium Nostoc punctiforme, Arch. Microbiol. 178 (2002) 395–403, https://doi.org/10.1007/s00203-002-0476-5.
[40] S.W. Chisholm, S.L. Frankel, R. Goericke, R.J. Olson, B. Palenik, J.B. Waterbury, et al., Prochlorococcus marinus nov. gen. nov. sp.: an oxyphototrophic marine prokaryote containing divinyl chlorophyll a and b, Arch. Microbiol. 157 (1992) 297–300, https://doi.org/10.1007/BF00245165.
[41] A. Dufresne, M. Salanoubat, F. Partensky, F. Artiguenave, I.M. Axmann, V. Barbe, et al., Genome sequence of the cyanobacterium Prochlorococcus marinus SS120, a nearly minimal oxyphototrophic genome, Proc. Natl. Acad. Sci. 100 (2003) 10020–10025, https://doi.org/10.1073/pnas.1733211100.
[42] G. Ralf, D.J. Repeta, The pigments of Prochlorococcus marinus: the presence of divinylchlorophyll a and b in a marine procaryote, Limnol. Oceanogr. 37 (1992) 425–433, https://doi.org/10.4319/lo.1992.37.2.0425.
[43] J. La Roche, G.W. van der Staay, F. Partensky, A. Ducret, R. Aebersold, R. Li, et al., Independent evolution of the prochlorophyte and green plant chlorophyll a/b light-harvesting proteins, Proc. Natl. Acad. Sci. U. S. A. 93 (1996) 15244–15248. http://www.ncbi.nlm.nih.gov/pubmed/8986795. (Accessed 26 March 2019).
[44] E.V. Armbrust, J.A. Berges, C. Bowler, B.R. Green, D. Martinez, N.H. Putnam, et al., The genome of the diatom Thalassiosira pseudonana: ecology, evolution, and metabolism, Science 306 (2004) 79–86, https://doi.org/10.1126/science.1101156.
[45] E. Derelle, C. Ferraz, S. Rombauts, P. Rouze, A.Z. Worden, S. Robbens, et al., Genome analysis of the smallest free-living eukaryote Ostreococcus tauri unveils many unique features, Proc. Natl. Acad. Sci. 103 (2006) 11647–11652, https://doi.org/10.1073/pnas.0604795103.
[46] C. Courties, A. Vaquer, M. Troussellier, J. Lautier, M.J. Chrétiennot-Dinet, J. Neveux, et al., Smallest eukaryotic organism, Nature 370 (1994) 255, https://doi.org/10.1038/370255a0.
[47] C. Courties, R. Perasso, M.-J. Chrétiennot-Dinet, M. Gouy, L. Guillou, M. Troussellier, Phylogenetic analysis and genome size of ostreococcus tauri (Chlorophyta, Prasinophyceae), J. Phycol. 34 (1998) 844–849, https://doi.org/10.1046/j.1529-8817.1998.340844.x.
[48] M.-J. Chrétiennot-Dinet, C. Courties, A. Vaquer, J. Neveux, H. Claustre, J. Lautier, et al., A new marine picoeucaryote: Ostreococcus tauri gen. et sp. nov. (Chlorophyta, Prasinophyceae), Phycologia 34 (1995) 285–292, https://doi.org/10.2216/i0031-8884-34-4-285.1.
[49] L. Guillou, W. Eikrem, M.-J. Chrétiennot-Dinet, F. Le Gall, R. Massana, K. Romari, et al., Diversity of picoplanktonic prasinophytes assessed by direct nuclear SSU rDNA sequencing of environmental samples and novel isolates retrieved from oceanic and coastal marine ecosystems, Protist 155 (2004) 193–214, https://doi.org/10.1078/143446104774199592.
[50] A.Z. Worden, J.K. Nolan, B. Palenik, Assessing the dynamics and ecology of marine picophytoplankton: the importance of the eukaryotic component, Limnol. Oceanogr. 49 (2004) 168–179, https://doi.org/10.4319/lo.2004.49.1.0168.
[51] P.D. Countway, D.A. Caron, Abundance and distribution of Ostreococcus sp. in the San Pedro Channel, California, as revealed by quantitative PCR, Appl. Environ. Microbiol. 72 (2006) 2496–2506, https://doi.org/10.1128/AEM.72.4.2496-2506.2006.

[52] A. Worden, Picoeukaryote diversity in coastal waters of the Pacific Ocean, Aquat. Microb. Ecol. 43 (2006) 165–175, https://doi.org/10.3354/ame043165.

[53] F. Rodríguez, E. Derelle, L. Guillou, F. Le Gall, D. Vaulot, H. Moreau, Ecotype diversity in the marine picoeukaryote Ostreococcus (Chlorophyta, Prasinophyceae), Environ. Microbiol. 7 (2005) 853–859, https://doi.org/10.1111/j.1462-2920.2005.00758.x.

[54] M. Matsuzaki, O. Misumi, T. Shin-i, S. Maruyama, M. Takahara, S. Miyagishima, et al., Genome sequence of the ultrasmall unicellular red alga Cyanidioschyzon merolae 10D, Nature 428 (2004) 653–657, https://doi.org/10.1038/nature02398.

[55] C. Six, A.Z. Worden, F. Rodríguez, H. Moreau, F. Partensky, New insights into the nature and phylogeny of prasinophyte antenna proteins: Ostreococcus tauri, a case study, Mol. Biol. Evol. 22 (2005) 2217–2230, https://doi.org/10.1093/molbev/msi220.

[56] M. Giordano, J. Beardall, J.A. Raven, CO_2 concentrating mechanisms in algae: mechanisms, environmental modulation, and evolution, Annu. Rev. Plant Biol. 56 (2005) 99–131, https://doi.org/10.1146/annurev.arplant.56.032604.144052.

[57] J. Raven, J.E. Kubler, New light on the scaling of metabolic rate with the size of algae, J. Phycol. 38 (2002) 11–16. https://onlinelibrary.wiley.com/doi/abs/10.1046/j.1529-8817.2002.01125.x; J. Reinfelder, A. Milligan, F. Morel, The role of the C4 pathway in carbon accumulation and fixation in a marine diatom, Plant Physiol. (2004) Available at: http://www.plantphysiol.org/content/135/4/2106.short.

[58] G. Bowes, S. Rao, G. Estavillo, J. Reiskind, C4 mechanisms in aquatic angiosperms: comparisons with terrestrial C4 systems, Funct. Plant Biol. 29 (2002) 379–392. CSIRO, http://www.publish.csiro.au/FP/PP01219.

[59] R.H. Wijffels, M.J. Barbosa, An outlook on microalgal biofuels, Science 329 (2010) 796–799, https://doi.org/10.1126/science.1189003.

[60] M.T. Guarnieri, A. Nag, S.L. Smolinski, A. Darzins, M. Seibert, P.T. Pienkos, Examination of triacylglycerol biosynthetic pathways via de novo transcriptomic and proteomic analyses in an unsequenced microalga, PLoS One 6 (2011), https://doi.org/10.1371/journal.pone.0025851.

[61] M.T. Guarnieri, L.M.L. Laurens, E.P. Knoshaug, Y.-C. Chou, B.S. Donohoe, P.T. Pienkos, Complex system engineering: a case study for an unsequenced microalga, in: Engineering Complex Phenotypes in Industrial Strains, John Wiley & Sons, Inc., Hoboken, NJ, USA, 2012, pp. 201–231, https://doi.org/10.1002/9781118433034.ch8

[62] M.J. Griffiths, R.P. van Hille, S.T.L. Harrison, The effect of nitrogen limitation on lipid productivity and cell composition in Chlorella vulgaris, Appl. Microbiol. Biotechnol. 98 (2014) 2345–2356, https://doi.org/10.1007/s00253-013-5442-4.

[63] H.G. Gerken, B. Donohoe, E.P. Knoshaug, Enzymatic cell wall degradation of Chlorella vulgaris and other microalgae for biofuels production, Planta 237 (2013) 239–253, https://doi.org/10.1007/s00425-012-1765-0.

[64] C. Zuñiga, C.-T. Li, T. Huelsman, J. Levering, D.C. Zielinski, B.O. McConnell, et al., Genome-scale metabolic model for the green alga *Chlorella vulgaris* UTEX 395 accurately predicts phenotypes under autotrophic, heterotrophic, and mixotrophic growth conditions, Plant Physiol. 172 (2016) 589–602, https://doi.org/10.1104/pp.16.00593.

[65] M.T. Guarnieri, J. Levering, C.A. Henard, J.L. Boore, M.J. Betenbaugh, K. Zengler, et al., Genome sequence of the oleaginous green alga, Chlorella vulgaris UTEX 395, Front. Bioeng. Biotechnol. 6 (2018) 37, https://doi.org/10.3389/FBIOE.2018.00037.

[66] R. Abdrabu, S.K. Sharma, B. Khraiwesh, K. Jijakli, D.R. Nelson, A. Alzahmi, R. Jagannathan, Single-cell characterization of microalgal lipid contents with confocal Raman microscopy, in: F.-G. Tseng, T.S. Santra (Eds.), *Essentials of Single-Cell Analysis*, Springer, 2016, pp. 363–382.

[67] W. Fu, A. Chaiboonchoe, B. Khraiwesh, D. Nelson, D. Al-Khairy, A. Mystikou, et al., Algal cell factories: approaches, applications, and potentials, Mar. Drugs 14 (2016) 225, https://doi.org/10.3390/md14120225.

[68] D.R. Nelson, B. Khraiwesh, W. Fu, S. Alseekh, A. Jaiswal, A. Chaiboonchoe, et al., The genome and phenome of the green alga Chloroidium sp. UTEX 3007 reveal adaptive traits for desert acclimatization, elife 6 (2017), https://doi.org/10.7554/eLife.25783.

[69] W. Gross, Ecophysiology of algae living in highly acidic environments, Hydrobiologia 433 (2000) 31–37, https://doi.org/10.1023/A:1004054317446.
[70] M.J. Ferris, K.B. Sheehan, M. Kühl, K. Cooksey, B. Wigglesworth-Cooksey, R. Harvey, et al., Algal species and light microenvironment in a low-pH, geothermal microbial mat community, Appl. Environ. Microbiol. 71 (2005) 7164–7171, https://doi.org/10.1128/AEM.71.11.7164-7171.2005.
[71] P.M. Novis, J.S. Harding, Extreme Acidophiles, (2007) pp. 443–463, https://doi.org/10.1007/978-1-4020-6112-7_24.
[72] H. Qiu, D.C. Price, A.P.M. Weber, V. Reeb, E. Chan Yang, J.M. Lee, et al., Adaptation through horizontal gene transfer in the cryptoendolithic red alga Galdieria phlegrea, Curr. Biol. 23 (2013) R865–R866, https://doi.org/10.1016/j.cub.2013.08.046.
[73] G. Schönknecht, W.-H. Chen, C.M. Ternes, G.G. Barbier, R.P. Shrestha, M. Stanke, et al., Gene transfer from bacteria and archaea facilitated evolution of an extremophilic eukaryote, Science 339 (2013) 1207–1210, https://doi.org/10.1126/science.1231707.
[74] S. Hirooka, Y. Hirose, Y. Kanesaki, S. Higuchi, T. Fujiwara, R. Onuma, et al., Acidophilic green algal genome provides insights into adaptation to an acidic environment, Proc. Natl. Acad. Sci. U. S. A. 114 (2017) E8304–E8313, https://doi.org/10.1073/pnas.1707072114.
[75] N.J. Wickett, S. Mirarab, N. Nguyen, T. Warnow, E. Carpenter, N. Matasci, et al., Phylotranscriptomic analysis of the origin and early diversification of land plants, Proc. Natl. Acad. Sci. 111 (2014) E4859–E4868, https://doi.org/10.1073/pnas.1323926111.
[76] Y. Xie, G. Wu, J. Tang, R. Luo, J. Patterson, S. Liu, et al., SOAPdenovo-Trans: de novo transcriptome assembly with short RNA-Seq reads, Bioinformatics 30 (2014) 1660–1666, https://doi.org/10.1093/bioinformatics/btu077.
[77] M.T.J. Johnson, E.J. Carpenter, Z. Tian, R. Bruskiewich, J.N. Burris, C.T. Carrigan, et al., Evaluating methods for isolating total RNA and predicting the success of sequencing phylogenetically diverse plant transcriptomes, PLoS One 7 (2012), https://doi.org/10.1371/journal.pone.0050226.
[78] S. Cheng, M. Melkonian, S.A. Smith, S. Brockington, J.M. Archibald, P.-M. Delaux, et al., 10KP: a phylodiverse genome sequencing plan, Gigascience 7 (2018), https://doi.org/10.1093/gigascience/giy013.
[79] D. Normile, Plant scientists plan massive effort to sequence 10,000 genomes, Science (2017), https://doi.org/10.1126/science.aan7165.
[80] H. Nozaki, et al., A 100%-complete sequence reveals unusually simple genomic features in the hot-spring red alga *Cyanidioschyzon merolae*, BMC Biol. 5 (2007) 28.
[81] I. Heilmann, C. Schnarrenberger, W. Gross, Mannose metabolizing enzymes from the red alga *Galdieria sulphuraria*, Phytochemistry 45 (1997) 903–906.
[82] B.L. de Groot, H. Grubmuller, Water permeation across biological membranes: mechanism and dynamics of aquaporin-1 and GlpF, Science 294 (2001) 2353–2357.
[83] D. Thomas, P. Bron, G. Ranchy, L. Duchesne, A. Cavalier, J.P. Rolland, et al., Aquaglyceroporins, one channel for two molecules, Biochim. Biophys. Acta 1555 (2002) 181–186.
[84] R.M. Stroud, L.J. Miercke, J. O'Connell, S. Khademi, J.K. Lee, J. Remis, et al., Glycerol facilitator GlpF and the associated aquaporin family of channels, Curr. Opin. Struct. Biol. 13 (2003) 424–431.
[85] D. Bhattacharya, D.C. Price, C.X. Chan, H. Qiu, N. Rose, S. Ball, et al., Genome of the red alga Porphyridium purpureum, Nat. Commun. 4 (2013) 1941.
[86] Y. Nakamura, N. Sasaki, M. Kobayashi, N. Ojima, M. Yasuike, et al., The first symbiont-free genome sequence of marine red alga, Susabi-nori (*Pyropia yezoensis*), PLoS One 8 (3) (2013).
[87] S.H. Brawley, A. Nicolas, N.A. Blouin, E. Ficko-Blean, G.L. Wheeler, M. Lohr, H.V. Goodson, et al., Insights into the red algae and eukaryotic evolution from the genome of *Porphyra umbilicalis* (Bangiophyceae, Rhodophyta), PNAS E6 (2017) 361–370.
[88] S. Bose, S.K. Herbert, D.C. Fork, Fluorescence characteristics of photoinhibition and recovery in a sun and a shade species of the red algal genus porphyra, Plant Physiol. 86 (1988) 946–950.
[89] J. Lee, E.C. Yang, L. Graf, J.H. Yang, H. Qiu, U. Zelzion, et al., Analysis of draft genome of the red seaweed Gracilariopsis chorda provides insights into genome size evolution in Rhodophyta, Mol. Biol. Evol. 35 (8) (2018) 1869–1886.

[90] J.S. Hawkins, H.R. Kim, J.D. Nason, R.A. Wing, J.F. Wendel, Differential lineage-specific amplification of transposable elements is responsible for genome size variation in *Gossypium*, Genome Res. 1610 (2006) 1252–1261.

[91] G. Barbier, C. Oesterhelt, M.D. Larson, R.G. Halgren, C. Wilkerson, R.M. Garavito, C. Benning, A.P. Weber, Comparative genomics of two closely related unicellular thermo-acidophilic red algae, *Galdieria sulphuraria* and *Cyanidioshyzon merolae*, reveals the molecular basis of the metabolic flexibility of *Galdieria sulpuraria* and significant differences in carbohydrate metabolism of both algae, Plant Physiol. 137 (2005) 460–474.

[92] B. Piegu, R. Guyot, N. Picault, A. Roulin, A. Sanyal, A. Saniyal, H. Kim, K. Collura, D.S. Brar, S. Jackson, et al., Doubling genome size without polyploidization: dynamics of retrotransposition-driven genomic expansions in *Oryza australiensis*, a wild relative of rice, Genome Res. 1610 (2006) 1262–1269.

[93] N. Phillips, R. Burrowes, F. Rousseau, B. de Reviers, G.W. Saunders, Resolving evolutionary relationships among the brown algae using chloroplast and nuclear genes, J. Phycol. 44 (2008) 394–405.

[94] J. Cock, L. Sterck, P. Rouze, D. Scornet, A. Allen, G. Amoutzias, V. Anthouard, F. Artiguenave, J. Aury, J. Badger, et al., The Ectocarpus genome and the independent evolution of multicellularity in brown algae, Nature 465 (2010) 617–621.

[95] N. Ye, X. Zhang, M. Miao, F. Xiao, Y. Zheng, D. Xu, et al., *Saccharina* genomes provide novel insight into kelp biology, *Nat. Commun.* 6 (2015) 6986.

[96] K. Nishitsuji, A. Arimoto, K. Iwai, Y. Sudo, K. Hisata, M. Fujie, N. Arakaki, T. Kushiro, T. Konishi, C. Shinzato, N. Satoh, E. Shoguchi, A draft genome of the brown alga, *Cladosiphon okamuranus,* S-strain: a platform for future studies of 'mozuku' biology, DNA Res. 23 (6) (2016) 561–570.

[97] H. Pavia, B.G. Torth, Inducible chemical resistance to herbivory in the brown seaweed Ascophyllum nodosum, Ecology 81 (2000) 3215–3225.

[98] H. Tettelin, V. Masignani, M.J. Cieslewicz, C. Donati, D. Medini, N.L. Ward, et al., Genome analysis of multiple pathogenic isolates of Streptococcus agalactiae: implications for the microbial "pan-genome", Proc. Natl. Acad. Sci. U. S. A. 102 (39) (2005) 13950–13955, https://doi.org/10.1073/pnas.0506758102.

Further reading

[99] E. Urbach, D.J. Scanlan, D.L. Distel, J.B. Waterbury, S.W. Chisholm, Rapid diversification of marine picophytoplankton with dissimilar light-harvesting structures inferred from sequences of Prochlorococcus and Synechococcus (Cyanobacteria), J. Mol. Evol. 46 (1998) 188–201. http://www.ncbi.nlm.nih.gov/pubmed/9452521. (Accessed 26 March 2019).

[100] A. Oren, Acidophiles, in: Encyclopedia of Life Sciences, 2010, https://scholar.google.com/scholar_lookup?author=A+Oren&title=Acidophiles&publication_year=2010&pages=192-206. (Accessed 29 March 2019).

CHAPTER 14

Pan-genomics of plants and its applications

Faiza Munir[a], Noor Ul Saba[a], Muneeba Arveen[a], Amnah Siddiqa[b], Jamil Ahmad[b], Rabia Amir[a]
[a]Department of Plant Biotechnology, Atta-ur-Rahman School of Applied Biosciences (ASAB), National University of Sciences and Technology (NUST), Islamabad, Pakistan
[b]Research Center for Modeling & Simulation (RCMS), National University of Sciences and Technology (NUST), Islamabad, Pakistan

1 Plant pan-genome concept

Most genomic investigations in plants emphasized on the single nucleotide polymorphisms (SNPs) due to their relative ease of identifications in different populations but with increasing knowledge of genome sequencing, the concept of single reference genome now seems insufficient. Most recent approach in this regard is pan-genome (originated from Greek term pan, which means whole) which came into existence with the development of an initial pan-genome, for a bacterial species *Streptococcus agalactiae* [1]. Total number of genes for a species are collectively classified as its pan-genome where its construction and annotation is a prerequisite for recognizing variations within different species [2]. Essentially, the concept of pan-genome is based on the fact that a single reference genome cannot represent all strains within the species due to various sequence variation causing mechanisms such as copy number variations (CNVs), presence/absence variations (PAVs), and SNPs. These variations are described comprehensively with special reference to plant genomes in the Section 2.

Increasing number of evidences recommend that a sole organism does not have capability to contain all the genetic information of a species due to different variations in the genomic sequences. Thus, the assembly of the pan-genome is necessary to determine the extent of genomic variation. A pan-genome signifies the entire count of genes for a specific species; it is made up of the core genes existing in every individual along with the variable genes and strain- (or isolate-) specific genes [3]. Apart from disparities at genetic level, many SNPs are also a significant type of genetic distinctions. There are evidences supporting the fact that CNVs and PAVs can be present even between individuals of the same species. Scientists have now agreed upon the fact that there is a share of genome which exists in every individual (core genome) of a species and other part which is from combined genomic content that occurs only in a subclass of the individuals (dispensable genome) and the addition of both of these constitutes the pan-genome of a species [4].

Pan-genomics: Applications, Challenges, and Future Prospects
https://doi.org/10.1016/B978-0-12-817076-2.00014-7
© 2020 Elsevier Inc.
All rights reserved.

The structure and dynamics of pan-genome is described in detail in the following sections.

A pan-genome is also classified as open or closed (limited). Open pan-genome approach refers to the condition when there is no fixed quantity of the genes in a species and whenever a new organism is included in the pan-genome analysis, there is a probability of identification of the novel genes. On the other hand, in closed pan-genome approach the gene pool is restricted and the pan-genome analyzed based on the definite number of individuals (or samples, strains) would not be extended in terms of gene numbers with the genome of additional individuals analyzed [5]. An example of open pan-genome is group B *Streptococcus* whose pan-genome is projected to rise on average of 33 novel genes whenever a novel strain is sequenced. An analogous analysis performed on five strains of *Streptococcus pyogenes* discovered a parallel genetic variety, demonstrating 27 definite genes for every novel genome included, indicating an 'open' pan-genome. An altered manner was perceived in examination of eight different *Bacillus anthracis* isolates. In this instance, the quantity of genes included in the pan-genome was found to promptly unite to zero when only fourth genome was included. Thus, *B. anthracis* species has 'closed' pan-genome and it can be entirely characterized by analyzing four genome sequences [6].

Thus, the number of genes increases with the number of additionally sequenced strains in the case of open pan-genomes. Usually, species like *Escherichia coli* which live in various environments of diverse microbial populations have numerous means of trading genetic material and hence constantly extent their total set of genes (open pan-genome). While in the case of closed pan-genome, analysis of further strains do not offer new genes to the species pan-genome after some initially sequenced strains. A closed or restricted pan-genome is distinctive for species (like *B. anthracis*) that are living in insulated places which has inadequate contact to the universal microbial gene pool. For all those species, a lesser number of sequenced strains already cover the entire pan-genome. The discovery of pan-genome in species *S. agalactiae* paved the path for same discoveries in other bacterial species as well [7, 8].

Additionally, it was revealed by the comparative sequencing of various grass genomes that the widespread discrepancies in both intergenic and the local genic content that is expressed in them is due to transposable elements between closely related species as well as among individuals within the same species. It is illustrated by these interpretations that a sole genome sequence may not signify the whole genomic complement in a species requiring the notion of the plant pan-genome which contains core genomic characteristics (mutual to all individuals) and the dispensable genome (comprised of moderately shared or no common DNA sequence elements). By understanding the nature of dispensable genome whether it is arrangement, origin or purpose, the procedures that create genetic variety and morphological diversity could be examined [9]. The measure of pan-genome with respect to the quantity of genes present in the classic strain is a symbol of the flexibility of the species and can indicate its prospective for settlement in a different environmental condition [10]. In the following section, a comprehensive discussion of the structure and dynamics of plant pan-genomes that have been previously reported in the literature is included to expand our comprehension.

2 Structure and dynamics of plant pan-genome

Last century witnessed a tremendous paradigm shift in plants genomics from comparisons of few genomes to meta-genomics-based pan-genome studies at various levels of resolution [2]. This is due to the continuous advancements in next-generation genomics technologies along with the reduced costs for resequencing of whole genomes allowing the large-scale genomes sequencing projects for characterizing not only model plants (*Arabidopsis thaliana*) but also many other crops (Sorghum, Tomato, Rice, Wheat, and Barley) [11, 12]. Such sequencing projects for delineating the structural variations in different plants genomes are at increase on a rate higher than ever before. Thus, the pan-genomic approaches are emerging as one of the promising tools to leverage and study the complex plants genomes from these sequencing projects in order to identify and characterize the genetic structures responsible for variations and different phenotypes among them. The concept is based on the realization of the increasing structural variation evidences from sequencing data making a single genome reference insufficient to depict the entire diversity of the genome of species. The genomes even among different individuals are variable from each other to different extents in species. Thus, the pan-genomes capture this diversity as a core genome (shared among all individuals in a species) and a variable genome (Fig. 1). The variable genome is further categorized as either dispensable genome (partially shared among individuals in a species) or species specific genome (not shared among different individuals).

The plant genomes are even more complex than many higher eukaryotic organisms mainly due to polyploidy and presence of repetitive genomic segments. Some of the plant genomes contain multiple copies of full chromosomes arising either from spontaneous duplication event (autopolyploidy) or through hybridization from other species (allopolyploidy). The repetitive genomic segments in plants genomes can vary from 10% (as present in Arabidopsis) to 80% (as present in wheat bread). Overall, the main sources of structural variations leading to diversity and complexity to plants genomes include SNPs, deletions, insertions, duplications, and genetic rearrangements [13]. These genomic structural variations with respect to studies of plants genomes can be broadly categorized into two types that is, CNVs and PAVs [13]. The CNVs correspond to the presence of multiple copies of a genetic segment/feature in an individual genome whereas PAVs correspond to the presence or absence of genomic segments or features in different individuals (Fig. 1B). Thus, the total genetic composition of plants genomes may highly vary among the individuals of same species which is dependent on the CNVs

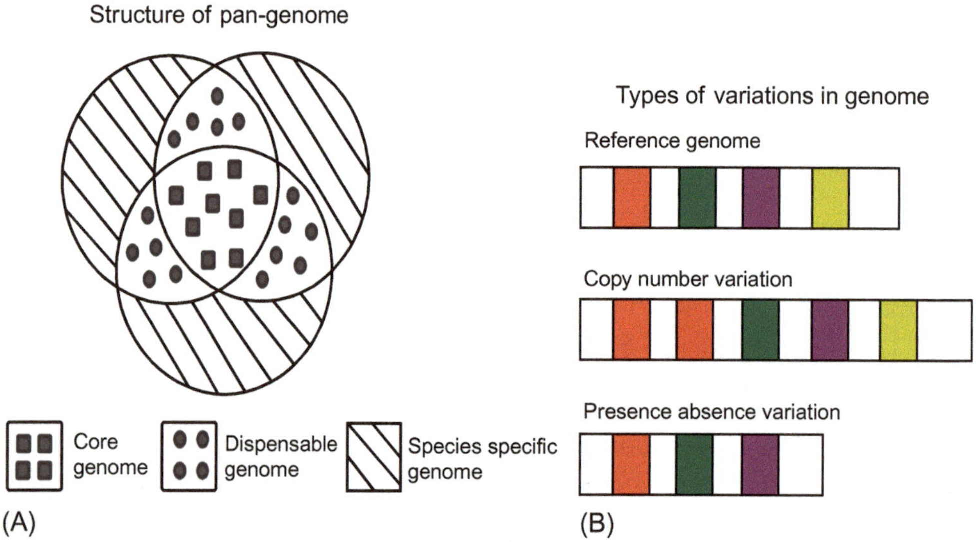

Fig. 1 (A) Structure of a pan-genome constitutes of core genome, dispensable genome, and species-specific genome. (B) Two different types of variations in the pan-genomes, that is, copy number variations (CPVs) and presence-absence variations (PAVs) are illustrated.

and PAVs. The CNVs particularly can affect the phenotypes of plants depending on their own role (being regulatory regions or gene themselves) and location of integration in the genome [14, 15]. On the basis of these changes, CNVs may contribute to cause changes in genes expression, structure, and identification of recessive alleles. These structural variations can be both beneficial and deleterious and their identification can increase our understanding of their roles in better adaptation to the environment and the diseases susceptibility of plants. Although plant pan-genome studies are still in infancy, many CNVs and PAVs have been identified in different plants species including arabidopsis [16, 17], maize [18], foxtail millet [19], opium [20], potato [21], tomato [14], rice [22], sorghum [23], soybean [24, 25], and wheat [26]. Some of these structural variations have been identified to be associated with different phenotypic traits such as temperature and biotic stress resistance in Arabidopsis [17], aluminum and boron toxicity tolerance in barley [27, 28], disease resistance and fruit shape in tomato [14, 29], and grain size in rice [22].

3 Plant pan-genome studies

3.1 Pan-genome of agronomically important crops

Pan-genome studies of agronomically important crops including Brassica, maize, wheat, rice, soya bean have been carried out in past. *Brassica rapa* is a plant comprising of numerous commonly grown subspecies like turnip and cabbage which are important vegetables. Various studies have reported the pan-genomes of *B. rapa*. In one of the studies, analyzing the pan-genome of Brassica plant 38,186 genes were reported as mutual because they existed in all of the three analyzed genomes (the pan-genome) while 1464 genes existed in only Chiffu, 1090 in turnip and 1118 genes were exclusive to rapid cycling. The study presented two new reference genomes and their observations signified different types of turnip and rapid cycling to the *B. rapa* population offering dependable models to study genetic discrepancy among these two types and the reference genome, Chiifu. Examination of the genetic information of the *B. rapa* pan-genome revealed that these three morphotypes diverged almost 5000–10,000 years ago [30].

Brassica is a perfect model to enhance knowledge of polyploid evolution. Pan-genome sequence is reported for *Brassica oleracea* where 20% of genes were reported to be influenced by PAVs. Also, there are two reference genomes available for *B. oleracea* [31] because a single reference was not able to present the whole gene count of the species due to structural variants like PAVs and CNVs. Its pan-genome includes 61,379 genes out of which 18.7% exhibit PAV as demonstrated in different studies. Numerous variable genes are explained to have functions that are linked with key agronomic attributes including resistance to different diseases, time of florescence, glucosinolate metabolism, and biosynthesis of various vitamins. This is indicative of the vital role of PAVs in breeding of upgraded Brassica crops [32]. The genome sequencing has additionally established a framework for the examination of the huge scope of morphological variety in *B. oleracea* and also supporting genome examination of the essential allotetraploid species.

In maize plants, an assessment through four arbitrarily selected genomic regions discovered a solid proof for PAV among the inbred lines B73 and Mo17. This comparison shows that only half number of the sequence are similar in both genotypes. Likewise, for the 8681 representative transcript assemblies (RTAs) which were recognized in this study, 83% were deficient in sequence support at the transcriptome level for presence in all 503 lines [4]. Another study found that only 70% of the entire pan-genome is represented by reference genome B73 [33]. Recently, an examination of pan-genome of maize was performed by resequencing six elite maize inbred lines which are essential for commercial hybrid production in China and various complete dispensable genes were recognized [34]. In two inbred lines, B73 and Mo17, the pan-genome is composed of the core genome signifying the half of the genome that is common in the two lines (having a size of 1.67 Gb) and dispensable genome having similar total size that is equally dispersed between these two lines [35]. The core genome contains single-copy sequences (containing most of the genes) along with transposable elements that are present in all organisms at a specific genomic locality. The nonessential (dispensable) genome is made up of mostly the transposable components of various sorts that, albeit present in different duplicates in every individual, can be found in a particular location just in some of them [9].

In another related study, a draft of wheat pan-genome was constructed and analyzed by a single reference and whole-genome sequencing statistics taken from 18 different cultivars. The pan-genome covers 128,656 expected genes of which 64.3% are acknowledged as core genes that exist in entire cultivars, whereas rest is variable and exhibit PAVs. Moreover, 12,150 genes were lacking in the Chinese Spring (CS) reference sequence but existed in all the other cultivars that were investigated. The pan-genome sequence is beneficial and significant source for researchers working in wheat genomics and breeding because knowledge of the variety of different genes is crucial for understanding their linkage with different agronomic characters. Spreading the genomic reference and SNP content to areas which are absent in CS offers a more complete source for genomics-based development of wheat crop [36]. The reference base which is used in wheat genetics and genomics is genotype CS. A large number of hexaploid wheat fragments are found to be possibly lost or considerably different in the CS genome. A finding after the evaluation of RNA sequences from 26 hexaploid wheat genotypes additionally strengthen the idea that dispensable genes are mutual in hexaploid wheat. Along with these widespread intra and interchromosomal readjustments within CS, presence of these dispensable genes is an additional feature emphasizing possible concerns that can occur by using one single reference genome for any species [37].

Soon after the publication of soybean genome, efforts were carried out to classify *Glycine soja* by resequencing of its genome [38]. In order to seizure most of the genetic variation between the species, newly created pan-genome is necessary that contains core genes existing in all organisms and dispensable genes which occurs only in some organisms. Furthermore, one particular genome is inadequate for signifying the genetic information of a mostly selfing (autogamous) species like *G. soja*, in which different variations like genetic exchange and recombination results in organisms that are different from each other [39]. When genomes of seven different varieties were examined and pan-genome was established, the resulting information revealed that most of the pan-genome (80%) existed in all organisms however the other 20% was distinct and exhibited advanced sequence variations in comparison to core genome [38]. In another study with 17 wild and 14 cultivated soybean genomes were examined, higher genetic deviations were revealed to exist in wild soybean accessions [40]. Pan-genome of soya bean displayed a range of new genes and alleles in wild lineages that can be introgressed in the crops that characteristically had less genetic variation as the consequence of domestication and breeding [41, 42].

Rice (*Oryza sativa* L.) is a principal diet crop and a perfect model for functional genomic investigations of monocots. The accessibility of the high-quality Nipponbare rice reference genome sequence has impressively enhanced gene cloning (international rice genome sequencing project); approximately 600 genes were cloned by the end of 2010 [43]. Nevertheless, a lot of genes governing vital qualities have been found to be lacking from the Nipponbare reference genome, such as GW5 [44], Sub1A [45], and Pikm-1 [46]. This specifies that one genome is not sufficient to seizure entire genetic variants. More genome sequences were required to achieve a more inclusive understanding of the pan-genome of rice. In order to explore the genetic diversity of rice several accessions have been restudied and sequenced again in recent times by the use of high-throughput sequencing technologies. Inspection of a 2.3 Mb homologous region of rice chromosome number four validated that two rice accessions *O. sativa* ssp. japonica and ssp. indica were different by 27 genes in this region, having greater gene density in japonica [47].

Numerous genome-specific loci lacking in the reference genome were acknowledged by newly created sequences of three different rice varieties and demonstrated the efficacy of these newly developed assemblies for examining natural innovation in rice. A pan-genome dataset was assembled for the *O. sativa-Oryza rufipogon* species complex, a source for comprehensive functional genomics studies and molecular breeding. By the pan-genome dataset, genome-wide assessments of the assemblies allowed the classification of several intricate variants that includes several large-effect coding variants along with various coding genes lacking in the reference genome sequence of rice that can be useful to identify the underlying dissimilarities in QTL cloning plus genome-wide association studies (GWAS) [48].

In addition to the abovementioned plant species, pan-genome discoveries have been made in some other plants including *Brachypodium distachyon*. This plant is commonly known as purple false brome or stiff brome. It is a model organism for functional genomic studies in grasses. A pan-genome of *B. distachyon* was constructed by analyzing 54 lines and it revealed that the size of its pan-genome sequence was almost twice the number of genes as compared to its individual genome. Genes that were common to all lines encode vital biological roles whereas genes that existed in some lines encoded occasional functions like defense and development. Phenotypic variations within species were due to differentially present genes that had a noteworthy impact on population genetics where transposable elements played a crucial role in pan-genome evolution [49].

Another plant whose pan-genome has been constructed is poplar (genus populous). It contains 35 species of trees and belongs to willow family (Salicaceae). A pan-genome of poplar is estimated to be of 497 Mb which is composed of core genome of size 401 Mb (80.7%) and dispensable genome of size 96 Mb (19.3%) [50].

4 Plant pan-genome analysis tools

Most of the software tools were developed initially for analysis of prokaryotic pan-genomes and have guided the workflows and pipelines being used in the current pan-genomics analysis tools [3]. However, all of the tools differ from each other in their functional performance including the methods used for orthologous genes identification, SNPs identification, phylogenetic analysis, and profiling of pan-genomes structures (core/dispensable/species specific genes identification) [3]. Summary of the available pan-genomics analysis tools is listed in Table 1. References of the tools are provided

Table 1 Pan-genome analysis Tools

Tool name	Weblink	Citations[a]	Reference
Harvest	https://github.com/marbl/harvest	295	[51]
Get_HOMOLOGUES	http://www.eead.csic.es/compbio/soft/gethoms.php	201	[52]
PGAP	http://pgap.sourceforge.net/	178	[53]
Panseq	https://lfz.corefacility.ca/panseq/	145	[54]
Spine and Agent	http://vfsmspineagent.fsm.northwestern.edu/index_age.html	61	[55]
PGAT	http://nwrce.org/pgat	48	[56]
PanGP	http://PanGP.big.ac.cn	47	[57]
ITEP	https://price.systemsbiology.net/itep	45	[58]
Micropan	https://cran.r-project.org/web/packages/micropan/index.html	24	[59]
PanCGHweb	http://bamics2.cmbi.ru.nl/websoftware/pancgh/pancgh_start.php	13	[60]
PANNOTATOR	http://bnet.egr.vcu.edu/pannotator/index.html	11	[61]
CAMBer	http://bioputer.mimuw.edu.pl/camber/index.html	9	[62]
PanCake	https://bitbucket.org/CorinnaErnst/pancake/wiki/Home	9	[63]

[a]As accessed on October 4, 2018.

in the Table 1 which will direct readers to gain deeper insights of the specific functionality of each tool.

4.1 Approaches to characterize plant pan-genomes

The field of pan-genomics is evolving at the moment because of continuous technological advancements in sequencing paradigms. Therefore, there is no particular protocol which can be applied to analyze the pan-genomes and different studies are utilizing different set of algorithms, tools, analysis pipelines for dealing with sequencing data, pan-genomes analysis, and its visualization. Initially all the pan-genomics analysis tools were developed for bacterial genomes which were much easier to handle than large eukaryotic genomes and also less complex in structural variations than most of the plant genomes. However, the general workflow of a pan genomic analysis follows the acquisition of the raw reads (sequencing) data for individual samples of same species. The quality of the obtained sequencing data is assessed using different established methods in order to remove technical sequencing biases [64].

The data could be further cleaned, trimmed, and normalized depending on the data quality. The next and most important step is to choose and apply the genome assembly approach. The genome assembly is the process of organizing the genomic reads or contigs data (coming from sequencing machines) back into the original order either in the presence or the absence of a reference genome [2]. The assembled data are again assessed for its quality of annotation. As a last step, pan-genome is calculated and analyzed to identify the structural variations followed by its visualization. Regardless of the approach used, pan-genome analysis allows us to estimate the size of the core genome, accessory genome, and the possibility of variation in both with the addition of new individuals.

To date, different methods have been used for the assembly of plant pan-genomes broadly categorized into *k*-mer-based approach, comparative de-novo assembly approach, and iterative assembly approach. Each of them is explained in detail below and also graphically illustrated in Fig. 2. However, for a deeper review of computational strengths and weaknesses of different tools based on these methods we refer our readers to Ref. [65].

4.1.1 k-mer-based approaches

The pan-genomes can be assembled using *k*-mer-based approaches in which the sequences are represented as a set of strings of length *k* (called *k*-mers). The sequences belonging to the whole genomes, contigs or sequencing reads are broken down into *k*-mers which can be then visualized using a special type of graph representation called as De Bruijn graphs. The nodes in this graph are the *k*-mers and the two *k*-mers are connected with each other based on the overlapping sequence relationship between them. *k*-mers are connected with an edge between them if they overlap by $k-1$ characters. Whole genomes can be assembled or reconstructed using this approach by taking into the account the information regarding nodes and their relationship to each other. Since, many reads/sequences can overlap with each other, the formed graph can be complex and many nodes can be connected to form loops like structures.

However, for a pan-genomics analyses in which we have to take into the account different genomes, the color code is used to identify the origin of *k*-mer. The nodes (*k*-mers) specific to one particular sample are assigned unique colors whereas the *k*-mer common to all of the samples are assigned single color. This information is then used to construct the pan-genome in the form of a colored *De Bruijn* graph. The edges between these colored nodes in a colored *De Bruijn* graph allows easier trace back of each genome along with the identification of common and unique sequences in the pan-genome [66, 67].

4.1.2 Comparative de novo assembly approach

A de novo assembly approach refers to the process of individual genome assembly without using a reference genome. The comparative de-novo gene assembly approach refers to the comparison of the de-novo assembled genomes of multiple samples/strains/individuals constructed from de-novo assembly approach using whole-genome

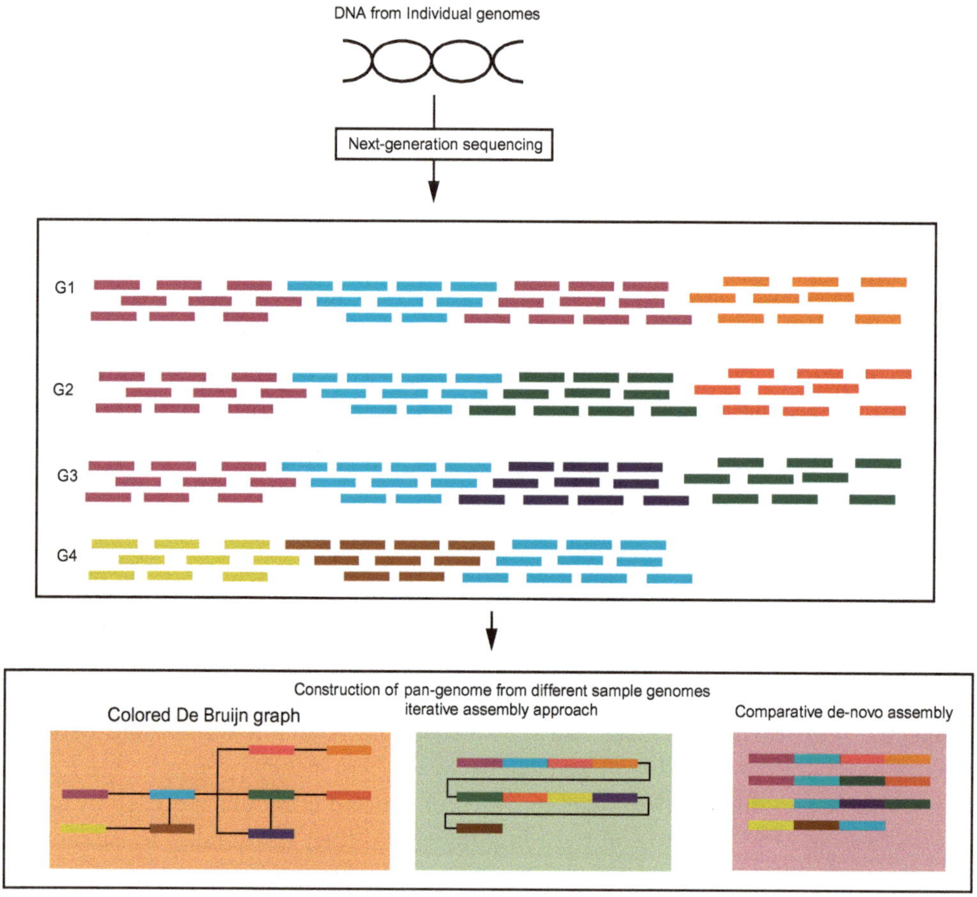

Fig. 2 Four different genomes (represented by G1, G2, G3, and G4) giving rise to a pan-genome by three different genome assembly approaches are shown. Each of these genomes is composed of different distinct DNA segments represented with different colors. Same color is used to represent the presence of same segment in all genomes. In the colored De Bruijn graph, any genome can be assembled back by tracing the edges of the graph. In the second approach, G1 is used as a reference genome and every other genome is compared with it and variable genome segments are then added to it sequentially. This forms a nonredundant pan-genome. In the *de novo* assembly approach, the genomes are assembled individually which are comparable to each other through whole-genome alignments.

alignments [38]. It allows for identification of variable and shared regions in the genomes along with orthologous gene clusters and core or variable gene families if followed by the annotations of genomes [38]. This approach is particularly useful for conducting a pan-genome analysis as it allows largest identification of the SVs and CVS when compared to other approaches. However, comparative *de novo* assembly approach is a computationally extensive approach and high sequencing coverage of individuals is a prerequisite for it in order to allow their individual assembly [2]. Velvet [68], SOAPdenovo [69], ALLPATHS [70], and MaSuRCA [71] are among some of the most cited tools for de novo genome assembly.

4.1.3 Iterative assembly approach

The iterative assembly approach is useful for low-coverage sequencing samples. A single whole-genome assembly is used as a reference genome and reads from all other genomes (samples) are mapped on it sequentially [5, 72]. One genome from each individual is mapped at a time on the reference genome afterwards. If the same genomic segment is present in the reference (starting genome) and the sample (being considered for mapping) genomes then the segment from reference genome will be kept in the pan-genome. If there are segments which are not present in the reference genome but only in sample genome then they are extracted and assembled. These newly assembled genomic sequences are then mapped back onto the reference genome. This whole process allows creation of one nonredundant pan-genome. This approach is computationally less intensive in terms of time and memory because whole-genome de-novo assemblies are not required for this process [5].

5 Applications of plant pan-genomics

5.1 Genetic mapping approaches and plant pan-genomics

Pan-genome analysis is the basis of assessing a variety of genome in large dataset and prediction of a number of additional whole-genome sequences that would be required to fully characterize the diversity of that genome sequence [73]. For pan-genome analysis and assembly there are several approaches [2]. The first, traditional approach that was applied in bacteria involved in assembly of whole-genome of all genotypes, individual annotation and comparison of the gene content is followed thereafter [1, 74].

Mendelian principles are used in genetic mapping that describe the relative closeness of DNA markers along the organismal chromosomes. Most botanical models and major crops now possess detailed reference genetic maps of specific populations as well as less detailed maps of many additional populations made for specific research group or crop improvement goals. Plants continuously evolve, diversify, and adapt so the variations in genes represent only one genome sequence which is insufficient to efficiently signify the presence of differences within a species [75]. Reference genomes are becoming extremely valuable as models for 'mapping' of data from further accessions that led to understanding of the history and structure of genetic variations within a crop plant and many other species [34]. The evolution of new genes inside a genus or species might not be apprehended in one genome sequence due to the effect of transposable elements.

Pan-genomes and pan transcriptomes have become efficient tools for capturing this added layer of deviation [75]. This approach involves sequencing of several genomes in a species like in maize [76] or soya bean [38] or even in whole genus (e.g., *Oryza*) [77] in a way that rearranged and diverged sequences can be mapped and are then analyzed. Pan-genome references of high quality include natural variants and unusual variations which are required for gene identification or regions that are associated with crop improvement and of adaptation to environmental conditions. Misassemble and split genes in assemblies are significant for downstream investigates such as pan-genomics and genome diversity examination. New approaches are provided by optical mapping and long read sequencing to increase contig length, reconstruction of repetitive regions and gaps in genome assemblies are filled [78].

5.2 Pan-genomics in crop diversity

A pan-genome study can offer the outline for accessing genomic diversity of data and provide background for understanding its observed attributes. Genomic diversity is studied by introducing the concept of pan-genome and the core-genome which is based on the genome analysis of maize or other organisms as maize is hypermutable, mainly due to retrotransposon activity, with one polymorphic insertion or deletion. In maize genome, range of diversity is incorporated by the pan-genome in a set of core, conserved components; this hyper variation is considered to be a major consequence of sequencing and the resequencing of cultivars [79].

As genome sequence information increases day by day, the genomic data of a single crop is not sufficient to represent the variety in the species. Population-level genotyping has given new insights to recognize the widespread genomic variant in species [80]. The examination of crop pan-genomes aims at effectively representing the diversity of genome in a species which has also provided a better understanding of intraspecific variation in crops [32]. By high-quality genome assemblies, precise description of genome diversity, and link of heritable agronomic traits with genotypes, stability of crop production, and environmental elasticity will be improved [80].

Natural and occasional variants are captured in high-quality pan-genome references which are central for the identification of genes or regions which are linked to adaptations toward environmental stresses and for crop development [75]. DivSeek and global crop diversity trust are the organizations working to coordinate the resequencing efforts of whole germplasm collections. The Chinese Academy of Sciences (CAS), Beijing Genomics Institute (BGI), and International Rice Research Institute (IRRI), also recently resequenced 3000 diverse rice accessions [81]. Notably, for Asian cultivated rice, only one high-quality reference genome is present, so a big amount of the resequencing data is unable to be mapped. High-quality pan-genome reference sets are needed for rice as well as for the majority of other crop plants [75].

5.3 Pan-genomics in adaptations to climate changes

Climate adapted crops are produced and breeding is accelerated through genome editing approaches. With growing volume of genomic data and progress in genome editing, genomics-assisted breeding will play a significant role in guaranteeing food security in the era of climate change [82]. Plant genomes commonly contain intraspecific copy number variants (CNVs) and presence-absence variants (PAVs) [74]. For this reason, a single crop reference genome offers a partial picture of crop's multiplicity or variations. In recent years, crop pan-genomes have been established for *B. rapa* [30] soybean [38], maize [4], and rice [74]. Presence-absence variation is central in understanding the diversity of genomic content in crops that have been shown to have an impact on climate-related agronomic characters like submergence tolerance and phosphorus uptake efficacy in rice [74] besides biotic stress reactions in various species such as muskmelon [83] and soybean [84]. In general pan-genomics helps in studying the multidimensional aspect of variation which will contribute in identification of genetic variation that includes several intricate agronomic qualities that can improve resistance to climate change.

Pan-genome investigations of different crops offer unique understanding of genetic range that exists in secondary crop gene pools. Latest studies on pan-genome sequencing in soya bean [38] and maize [85] define the valued involvement of pan-genomic variation to phenotypic differences for key adaptive behaviors. Identification and execution of such adaptive potential using high-resolution genotyping may be a key toward targeted renewal of exhausted phenotypic diversity in response to climate change. For instance, plant breeders increasingly use high-resolution genome information to illustrate germplasm, identifying genes that trigger vital agronomic qualities or they estimate the breeding values of individuals in breeding programs to speed up the choice of improved varieties [86]. Either with or without a completed reference sequence, the development of reduced-representation genotyping-by-sequencing (GBS) methods has paved the path for implementing NGS technologies for high-throughput genomic resequencing even for large plant populations, at a continuously reducing rate.

For instance, in maize [87] used the GBS method to produce almost 700,000 genome wide SNPs in a section of 2815 different inbred lines from universally dispersed breeding programs. Wide-ranging sequencing data sets of this type facilitate tremendously high-resolution assessment of genetic diversity and population structure, giving a complete understanding of the history of recombination and allelic diversity through various breeding pools [88, 89].

The wide effectivity of GBS procedures for genetic investigations has been effectively proved in several significant crops, for instance rice [90], soya bean [91], barley [92], wheat [93], and potato [94]. In segregating populations, along with quantitative phenotype investigations, NGS approaches also provide a potent basis for fast mapping and gene identification which underlie quantifiable characters [95].

5.4 Pan-genomics in plant breeding

A pan-genome approach provides several benefits over a single, linear reference genome sequence in various plant breeding applications. Pan-genome accessibility for a specified crop involves its remote lineages which offers a single synchronized system to anchor all known variation and phenotype information, and will permit for identification of new genes from already available germplasm which are absent in the reference genome(s). Sampling bias has been reduced as the trend toward crop pan-genomes in spite of single sample reference genomes as resource for molecular breeding has advanced because it allows a better illustration of genetic diversity [96].

The large quantity of SNPs, along with high-throughput discovery and detection approaches make them model aspirants for use in genetic investigations involving linkage mapping, quantitative trait loci (QTL) mapping, and map-based positional cloning. The identification of unusual variants that are linked with the QTLs for agronomic characters can aid in the development of cultivars through breeding in case of pan-genomes [5]. For example, in the pan-genome study of the *G. soja*, which is the wild relative of the soya bean *Glycine* max, most of the genes influenced by SNPs cause gain or loss in stop codon and frameshift are unusual events, which are normally present in only one of the seven *G. soja* accessions [38]. For instance, the gene *Gm*02g25230, which is one of the two homologs of Spiral2, a crucial microtubule gene for directional cell elongation that is involved in the right-handed helical growth in *Arabidopsis* [97] was found to harbor three indels in all *G. soja* accessions, but not in *G. max*. These indels were associated with the changes in amino acid in one of the Huntingtin, elongation factor 3 (EF3), and the yeast kinase TOR1 (HEAT)-repeat motifs, protein phosphatase 2A (PP2A) that cause the twisting growth habit in *G. soja* as compared to the upright growth found in *Glycine max*. This investigation also emphasized the prospect of using newly identified genetic variation confined within genomic regions that have been fixed in *G. max*. This data may be implemented to design crosses to define if the fixed regions are linked with phenotypes of agricultural importance, thus providing surplus candidate genes for the production of novel, better varieties.

5.5 Pan-genomics in production of desirable traits

Genomic hybridization of crops has been done to produce desirable traits in its offspring for over a century. Crop varieties are produced commercially by breeding genes that are found in wild varieties those with the improved properties of crops like, its appearance, resistance to certain pests or diseases, nutrient content of a particular crop, tolerance to various stresses such as drought, heat, salt, or wounding [96]. Large-scale genomics projects are already going on to completely describe the genetic diversity in plants, not only for the model plant *Arabidopsis thaliana* [11] but also for various other crop species. Resequencing of large number of varieties of rice [98], maize [99], sorghum [100], and tomato [101] has been done.

PAV is displayed in the poppy which contains a 10 gene cluster, existing in plants that produce noscapine, an antitumor alkaloid [20]. Studies on wheat genes revealed that photoperiod response (Ppd-B1 alleles) or vernalization necessity (Vrn-A1 alleles) are altered that regulate flowering because both of those genes display CNVs. An early flowering day neutral phenotype is conferred in alleles with an increased copy number of Ppd-B1, while for the increased requirement for vernalization plants have increased copy number of Vrn-A1, so that lengthier cold phases are essential to boost up flowering [102].

The Sub1A gene has role in submergence tolerance in rice, and is absent in nonsubmergence tolerant rice varieties [74]. Also in rice, a gene coding a protein kinase Pstol1 is involved in phosphorus acceptance efficiency phenotype, while in phosphorus starvation-intolerant varieties of rice that particular gene is absent [74].

Moreover, PAV is demonstrated in biotic stress response genes in a series of species [83, 103]. Numerous CNVs/PAVs affecting genes in plants were postulated which may not have great influence on the phenotype since the concerned genes that are influenced by CNVs/PAVs are considered the portions of large multigene families, where at least partial functional redundancy is expected between members [104]. CNVs/PAVs influence genes which belong to gene families, each gene contribute as a 'functional block' that provide partial to complete functionality for the family. Relatively minor effect on phenotype may be resulted due to the loss of a single member of a gene family as other members of the family produce compensatory effect in the function. However, the collective effect of lacking members in many gene families could cause reduced vigor [104]. Hybrids can alleviate the effect causing substantial hybrid vigor [104]. While in other case, it is also probable that exclusive genes exhibiting PAV are contributing to heterosis on a single gene rather than gene family level. In view of this, abundance of CNVs/PAVs in the whole population used for breeding may bound prospective for enhancement [104].

Upcoming projects are aiming to sequence many more assortments, for example, 100,000 varieties of rice. It is also a challenge because many of the plant genomes are huge, intricate, and are often polyploid [105]. A pan-genomic approach is required in excavating and leveraging the sequence data in such large-scale projects. In general pan-genomics provides the system for a more complex view of diversity which will play a significant role in recognizing genetic variation underlying many complex agronomic traits that can improve flexibility to environmental changes.

Genomic information acquired from various plant species feature CNV as an important source of genetic diversity. CNVs are linked with nucleotide-binding leucine-rich repeat (NB-LRR) genes and receptor-like kinase (RLK) genes, and have a role in plant defense-related approaches [13]. CNVs can be associated with variation in gene expression as well [106, 107]. Many disease resistance genes are found in CNV sections in rice [108]. These genes encode specific functions that are involved in cell death protein phosphorylation and defense mechanisms. Moreover, genetic approaches for CNV of resistance genes were inspected through phylogenetic analysis of resistance genes in the Cucurbitaceae family by Ref. [109].

CNV polymorphism was found to be linked with resistance to northern leaf blight on the basis of nested association mapping GWAS [110]. Gene ontology enrichment examination of the 672 genes located within CNV regions in soybean showed that genes associated with disease resistance response were considerably over-expressed [103]. It has also been stated that resistance gene function is modified to frequent rearrangements and CNVs [111]. CNVs linked to disease resistance have also been reported in numerous plant species, where disease resistance genes signify an important segment of genes in CNV regions and were greatly improved for resistance gene models [85, 112]. Bertioli et al. (2003) presented that in peanut and legumes *R*-genes have experienced extensive CNV. It is predictable that high copy number of resistance genes in plants is beneficial as it will provide better resistance against pathogens [109]. On the other hand, low copy number might be a result of less challenge from pathogens [53, 113]. This strengthens the hypothesis that CNV and the genes encoded within these regions have a role in disease resistance in plants through natural genome variation. CNV could permit gene diversification and development of novel resistance genes.

Pan-genome is constructed for several grass species as well. One of the most important species in this respect is *Setaria italica* (foxtail millet). Foxtail millet is narrowly associated with numerous biofuel grasses with complex genomes, for instance napier grass (*Pennisetum purpureum*), pearl millet (*Pennisetum glaucum*), switchgrass (*Panicum virgatum*), so pan-genome analysis of foxtail millet can be helpful in investigating its potential as biofuel in future [114]. Hence, the pan-genomic studies emphasize greatly on crop diversity and improvement. The above mentioned applications are summarized in Table 2.

6 Conclusions and future directions

The rapid improvements in sequencing methodologies have dramatically reduced the time and costs for the whole-genome analyses leading to greater comprehension of the mechanistic details in genome structures and the involved dynamics. Pan-genome analysis is one of the emerging theoretical paradigms based on the results from these large sequencing projects which contributed in our enhanced understanding regarding the insufficiency of a single reference genome for subsequent genomics analyses. This particularly allowed in comprehending the genomes structure and associated complexity of genotype to phenotype associations. The significance of pan-genome analyses has been specifically demonstrated in removing the reduced sample biases and ensuring the representation of genetic diversity. Most of the initial pan-genome studies have been restricted to bacterial and other smaller genomes. However, several plants species have been analyzed now with pan-genome methods and tools and have elucidated the role of CNVs and PAVs in different plant phenotypes such as flowering times, different stress resistance mechanisms, phenotypic traits association among strains, and the subsequent

Table 2 Applications of plant pan-genomics

Sr. No.	Applications	Attributes	References
1	Genetic mapping approaches and plant pan-genomics	• Assessment of the gene content. • Estimation of additional whole-genome sequences. • Optical mapping. • Long read sequencing. • Reestablishment of repetitive regions.	[73, 77, 78]
2	Pan-genomics in crop diversity	• Recognition of variants. • Basis for improvement of agronomic traits.	[79, 80]
3	Pan-genomics in adaptations to climate changes	• Production of climate adapted cultivars. • Food security. • Identification of stress resistant traits.	[82]
4	Pan-genomics in plant breeding	• Mapping of quantitative trait loci(QTLs). • Exploration of single nucleotide polymorphism (SNPs). • Investigation of candidate genes for releasing improved varieties.	[5]
5	Pan-genomics in production of desirable traits	• Production of hybrid cultivars. • Exploitation of PAVs/CNVs in crop breeding.	[20]

diversification. Such studies contributed in enhanced understanding to utilize and apply this genotypic information for increasing the crop production of better varieties in terms of sizes and flavors, increasing the abiotic stress and pathogens/disease resistances among many others reviewed above. However, there are some challenges regarding pan-genome studies including the availability of completed and well-annotated completed reference genomes. It is especially important for plant genomes because of the large number of repetitive sequences in them which could make it difficult to assemble the new plant genomes from shorter read fragments. In future, more sophisticated algorithms and methods for read alignment and assembly are required to allow the full exploitation of pan-genome analyses for practical and industrial applications regarding plant genomes.

References

[1] H. Tettelin, et al., Genome analysis of multiple pathogenic isolates of *Streptococcus agalactiae*: implications for the microbial "pan-genome" Proc. Natl. Acad. Sci. 102 (39) (2005) 13950–13955.

[2] A.A. Golicz, J. Batley, D. Edwards, Towards plant pangenomics, Plant Biotechnol. J. 14 (4) (2016) 1099–1105.

[3] J. Xiao, et al., A brief review of software tools for pangenomics, Genomics Proteomics Bioinformatics 13 (1) (2015) 73–76.
[4] C.N. Hirsch, et al., Insights into the maize pan-genome and pan-transcriptome, Plant Cell 26 (2014) 121–135.
[5] B. Hurgobin, D.J.B. Edwards, SNP discovery using a pangenome: has the single reference approach become obsolete? Biology (Basel) 6 (1) (2017) E21.
[6] D. Medini, et al., The microbial pan-genome, Curr. Opin. Genet. Dev. 15 (6) (2005) 589–594.
[7] C. Donati, et al., Structure and dynamics of the pan-genome of *Streptococcus pneumoniae* and closely related species, Genome Biol. 11 (10) (2010) R107.
[8] R. Baddam, et al., Genome dynamics and evolution of Salmonella Typhi strains from the typhoid-endemic zones, Sci. Rep. 4 (2014) 7457.
[9] M. Morgante, E. De Paoli, S. Radovic, Transposable elements and the plant pan-genomes, Curr. Opin. Plant Biol. 10 (2) (2007) 149–155.
[10] L. Snipen, T. Almøy, D.W. Ussery, Microbial comparative pan-genomics using binomial mixture models, BMC Genomics 10 (1) (2009) 385.
[11] D. Weigel, R. Mott, The 1001 genomes project for *Arabidopsis thaliana*, Genome Biol. 10 (5) (2009) 107.
[12] D. Weigel, R. Mott, The 1001 genomes project for *Arabidopsis thaliana*, Genome Biol. 10 (5) (2009) 107.
[13] R.K. Saxena, D. Edwards, R.K. Varshney, Structural variations in plant genomes, Brief. Funct. Genomics 13 (4) (2014) 296–307.
[14] B. Cong, L.S. Barrero, S.D. Tanksley, Regulatory change in YABBY-like transcription factor led to evolution of extreme fruit size during tomato domestication, Nat. Genet. 40 (6) (2008) 800–804.
[15] D.M. Bickhart, et al., Copy number variation of individual cattle genomes using next-generation sequencing, Genome Res. 22 (4) (2012) 778–790.
[16] X. Gan, et al., Multiple reference genomes and transcriptomes for *Arabidopsis thaliana*, Nature 477 (7365) (2011) 419–423.
[17] S. DeBolt, Copy number variation shapes genome diversity in *Arabidopsis* over immediate family generational scales, Genome Biol. Evol. 2 (2010) 441–453.
[18] N.M. Springer, et al., Maize inbreds exhibit high levels of copy number variation (CNV) and presence/absence variation (PAV) in genome content, PLoS Genet. 5 (11) (2009) e1000734.
[19] H. Bai, et al., Identifying the genome-wide sequence variations and developing new molecular markers for genetics research by re-sequencing a landrace cultivar of foxtail millet, PLoS One 8 (9) (2013).
[20] T. Winzer, et al., A *Papaver somniferum* 10-gene cluster for synthesis of the anticancer alkaloid noscapine, Science 336 (6089) (2012) 1704–1708.
[21] M. Iovene, et al., Copy number variation in potato—an asexually propagated autotetraploid species, Plant J. 75 (1) (2013) 80–89.
[22] Y. Wang, et al., Copy number variation at the GL7 locus contributes to grain size diversity in rice, Nature 47 (8) (2015) 944.
[23] L.-Y. Zheng, et al., Genome-wide patterns of genetic variation in sweet and grain sorghum (*Sorghum bicolor*), Genome Biol. 12 (11) (2011) R114.
[24] W.J. Haun, et al., The composition and origins of genomic variation among individuals of the soybean reference cultivar Williams 82, Plant Physiol. 155 (2) (2010) 645–655.
[25] H.-M. Lam, et al., Resequencing of 31 wild and cultivated soybean genomes identifies patterns of genetic diversity and selection, Nat. Genet. 42 (12) (2010) 1053.
[26] A. Díaz, et al., Copy number variation affecting the Photoperiod-B1 and Vernalization-A1 genes is associated with altered flowering time in wheat (*Triticum aestivum*), Plos One 7 (3) (2012).
[27] T. Sutton, et al., Boron-toxicity tolerance in barley arising from efflux transporter amplification, Science 318 (5855) (2007) 1446–1449.
[28] M. Fujii, et al., Acquisition of aluminium tolerance by modification of a single gene in barley, Nat. Commun. 3 (2012) 713.
[29] H. Xiao, et al., A retrotransposon-mediated gene duplication underlies morphological variation of tomato fruit, Science 319 (5869) (2008) 1527–1530.

[30] K. Lin, et al., Beyond genomic variation-comparison and functional annotation of three *Brassica rapa* genomes: a turnip, a rapid cycling and a Chinese cabbage, BMC Genomics 15 (1) (2014) 250.
[31] S. Liu, et al., The *Brassica oleracea* genome reveals the asymmetrical evolution of polyploid genomes, Nat. Commun. 5 (2014) 3930.
[32] A.A. Golicz, et al., The pangenome of an agronomically important crop plant *Brassica oleracea*, Nat. Commun. 7 (2016) 13390.
[33] M.A. Gore, et al., A first-generation haplotype map of maize, Science 326 (5956) (2009) 1115–1117.
[34] J. Lai, et al., Genome-wide patterns of genetic variation among elite maize inbred lines, Nat. Genet. 42 (11) (2010) 1027.
[35] M. Lee, et al., Expanding the genetic map of maize with the intermated B73 × Mo17 (IBM) population, Plant Mol. Biol. 48 (5-6) (2002) 453–461.
[36] J.D. Montenegro, et al., The pangenome of hexaploid bread wheat, Plant J. 90 (5) (2017) 1007–1013.
[37] M. Liu, et al., Chromosome-specific sequencing reveals an extensive dispensable genome component in wheat, Sci. Rep. 6 (2016) 36398.
[38] Y.-h. Li, et al., ' assembly of soybean wild relatives for pan-genome analysis of diversity and agronomic traits, Nat. Biotechnol. 32 (10) (2014) 1045.
[39] R. Li, et al., De novo assembly of human genomes with massively parallel short read sequencing, Genome Res. 20 (2) (2010) 265–272.
[40] H.-M. Lam, et al., Resequencing of 31 wild and cultivated soybean genomes identifies patterns of genetic diversity and selection, Nat. Genet. 42 (12) (2010) 1053.
[41] D.L. Hyten, et al., Impacts of genetic bottlenecks on soybean genome diversity, Proc. Natl. Acad. Sci. 103 (45) (2006) 16666–16671.
[42] J.F. Doebley, B.S. Gaut, B.D. Smith, The molecular genetics of crop domestication, Cell 127 (7) (2006) 1309–1321.
[43] Y. Jiang, et al., Rice functional genomics research: progress and implications for crop genetic improvement, Biotechnol. Adv. 30 (5) (2012) 1059–1070.
[44] A. Shomura, et al., Deletion in a gene associated with grain size increased yields during rice domestication, Nat. Genet. 40 (8) (2008) 1023.
[45] K. Xu, et al., Sub1A is an ethylene-response-factor-like gene that confers submergence tolerance to rice, Nature 442 (7103) (2006) 705.
[46] I. Ashikawa, et al., Two adjacent NBS-LRR class genes are required to confer Pikm-specific rice blast resistance, Genetics 180 (4) (2008) 2267–2276.
[47] B. Han, Y. Xue, Genome-wide intraspecific DNA-sequence variations in rice, Curr. Opin. Plant Biol. 6 (2) (2003) 134–138.
[48] Q. Zhao, et al., Pan-genome analysis highlights the extent of genomic variation in cultivated and wild rice, Nat. Genet. 50 (2) (2018) 278.
[49] S.P. Gordon, et al., Extensive gene content variation in the *Brachypodium distachyon* pan-genome correlates with population structure, Nat. Commun. 8 (1) (2017) 2184.
[50] S. Pinosio, et al., Characterization of the poplar pan-genome by genome-wide identification of structural variation, Mol. Biol. Evol. 33 (10) (2016) 2706–2719.
[51] T.J. Treangen, et al., The harvest suite for rapid core-genome alignment and visualization of thousands of intraspecific microbial genomes, Genome Biol. 15 (11) (2014) 524.
[52] B. Contreras-Moreira, P. Vinuesa, GET_HOMOLOGUES, a versatile software package for scalable and robust microbial pan-genome analysis, Appl. Environ. Microbiol. 79 (24) (2013) 7696–7701.
[53] Y. Zhao, et al., PGAP: pan-genomes analysis pipeline, Bioinformatics 28 (3) (2011) 416–418.
[54] C. Laing, et al., Pan-genome sequence analysis using Panseq: an online tool for the rapid analysis of core and accessory genomic regions, BMC Bioinformatics 11 (1) (2010) 461.
[55] E.A. Ozer, J.P. Allen, A.R.J.B.g. Hauser, Characterization of the core and accessory genomes of *Pseudomonas aeruginosa* using bioinformatic tools Spine and AGEnt, BMC Genomics 15 (1) (2014) 737.
[56] M.J. Brittnacher, et al., PGAT: a multistrain analysis resource for microbial genomes, Bioinformatics 27 (17) (2011) 2429–2430.
[57] Y. Zhao, et al., PanGP: a tool for quickly analyzing bacterial pan-genome profile, Bioinformatics 30 (9) (2014) 1297–1299.

[58] M.N. Benedict, et al., ITEP: an integrated toolkit for exploration of microbial pan-genomes, BMC Genomics 15 (1) (2014) 8.
[59] L.-G. Snipen, H.L. Kristian, micropan: an R-package for microbial pan-genomics, BMC Bioinformatics 16 (1) (2015).
[60] J.R. Bayjanov, R.J. Siezen, S.A.J.B. van Hijum, PanCGHweb: a web tool for genotype calling in pangenome CGH data, Bioinformatics 26 (9) (2010) 1256–1257.
[61] A. Santos, et al., PANNOTATOR: an automated tool for annotation of pan-genomes, Genet. Mol. Res. 12 (3) (2013) 2982–2989.
[62] M. Wozniak, L. Wong, J. Tiuryn, CAMBer: an approach to support comparative analysis of multiple bacterial strains, BMC Genomics 12 (2) (2011) 121–126.
[63] Ernst, C. and S. Rahmann. PanCake: a data structure for pangenomes. in Open Access Series in Informatics. 2013. Schloss Dagstuhl-Leibniz-Zentrum fuer Informatik.
[64] Computational Pan-Genomics Consortium, Computational pan-genomics: status, promises and challenges, Brief. Bioinform. 19 (1) (2016) 118–135.
[65] B. Wajid, E. Serpedin, Do it yourself guide to genome assembly, Brief. Funct. Genomics 15 (1) (2014) 1–9.
[66] Z. Iqbal, et al., De novo assembly and genotyping of variants using colored de Bruijn graphs, Nat. Genet. 44 (2) (2012) 226.
[67] S. Marcus, H. Lee, M.C. Schatz, SplitMEM: a graphical algorithm for pan-genome analysis with suffix skips, Bioinformatics 30 (24) (2014) 3476–3483.
[68] D. Zerbino, E. Birney, Velvet: algorithms for de novo short read assembly using de Bruijn graphs, Genome Res. 18 (5) (2008) 821–829.
[69] R. Li, et al., De novo assembly of human genomes with massively parallel short read sequencing, Genome Res. 20 (2) (2010) 265–272.
[70] J. Butler, et al., ALLPATHS: de novo assembly of whole-genome shotgun microreads, Genome Res. 18 (5) (2008) 810–820.
[71] A.V. Zimin, et al., The MaSuRCA genome assembler, Bioinformatics 29 (21) (2013) 2669–2677.
[72] A.A. Golicz, J. Batley, D. Edwards, Towards plant pangenomics, Plant Biotechnol. J. 14 (4) (2016) 1099–1105.
[73] G. Vernikos, et al., Ten years of pan-genome analyses, Curr. Opin. Microbiol. 23 (2015) 148–154.
[74] M.C. Schatz, et al., Whole genome de novo assemblies of three divergent strains of rice, *Oryza sativa*, document novel gene space of aus and indica, Genome Biol. 15 (11) (2014) 506.
[75] J.F. Wendel, et al., Evolution of plant genome architecture, Genome Biol. 17 (1) (2016) 37.
[76] C.N. Hansey, et al., Maize (*Zea mays* L.) genome diversity as revealed by RNA-sequencing, PLoS One 7 (3) (2012).
[77] J. Jacquemin, et al., The International Oryza Map Alignment Project: development of a genus-wide comparative genomics platform to help solve the 9 billion-people question, Curr. Opin. Plant Biol. 16 (2) (2013) 147–156.
[78] S. Koren, A.M. Phillippy, One chromosome, one contig: complete microbial genomes from long-read sequencing and assembly, Curr. Opin. Microbiol. 23 (2015) 110–120.
[79] P. SanMiguel, et al., The paleontology of intergene retrotransposons of maize, Nat. Genet. 20 (1) (1998) 43.
[80] M. Abberton, et al., Global agricultural intensification during climate change: a role for genomics, Plant Biotechnol. J. 14 (4) (2016) 1095–1098.
[81] J.-Y. Li, J. Wang, R.S. Zeigler, The 3,000 rice genomes project: new opportunities and challenges for future rice research, GigaScience 3 (1) (2014) 8.
[82] A. Scheben, Y. Yuan, D. Edwards, Advances in genomics for adapting crops to climate change, Curr. Plant Biol. 6 (2016) 2–10.
[83] V.M. González, et al., High presence/absence gene variability in defense-related gene clusters of *Cucumis melo*, BMC Genomics 14 (1) (2013) 782.
[84] D.E. Cook, et al., Copy number variation of multiple genes at Rhg1 mediates nematode resistance in soybean, Science 338 (6111) (2012) 1206–1209.

[85] F. Lu, et al., High-resolution genetic mapping of maize pan-genome sequence anchors, Nat. Commun. 6 (2015) 6914.
[86] R.K. Varshney, R. Terauchi, S.R. McCouch, Harvesting the promising fruits of genomics: applying genome sequencing technologies to crop breeding, PLoS Biol. 12 (6) (2014).
[87] M.C. Romay, et al., Comprehensive genotyping of the USA national maize inbred seed bank, Genome Biol. 14 (6) (2013) R55.
[88] L. Qian, W. Qian, R.J. Snowdon, Sub-genomic selection patterns as a signature of breeding in the allopolyploid *Brassica napus* genome, BMC Genomics 15 (1) (2014) 1170.
[89] K. Voss-Fels, et al., Subgenomic diversity patterns caused by directional selection in bread wheat gene pools, Plant Genome 8 (2) (2015).
[90] J. Spindel, et al., Bridging the genotyping gap: using genotyping by sequencing (GBS) to add high-density SNP markers and new value to traditional bi-parental mapping and breeding populations, Theor. Appl. Genet. 126 (11) (2013) 2699–2716.
[91] D. Jarquín, et al., Genotyping by sequencing for genomic prediction in a soybean breeding population, BMC Genomics 15 (1) (2014) 740.
[92] R.J. Elshire, et al., A robust, simple genotyping-by-sequencing (GBS) approach for high diversity species, PLoS One 6 (5) (2011).
[93] J. Poland, et al., Genomic selection in wheat breeding using genotyping-by-sequencing, Plant Genome 5 (3) (2012) 103–113.
[94] J.G. Uitdewilligen, et al., A next-generation sequencing method for genotyping-by-sequencing of highly heterozygous autotetraploid potato, PLoS One 8 (5) (2013).
[95] K. Schneeberger, et al., SHOREmap: simultaneous mapping and mutation identification by deep sequencing, Nat. Methods 6 (8) (2009) 550.
[96] Computational Pan-Genomics Consortium, Computational pan-genomics: status, promises and challenges, Brief. Bioinform. 19 (1) (2016) 118–135.
[97] T. Shoji, et al., Plant-specific microtubule-associated protein SPIRAL2 is required for anisotropic growth in Arabidopsis, Plant Physiol. 136 (4) (2004) 3933–3944.
[98] X. Huang, et al., A map of rice genome variation reveals the origin of cultivated rice, Nature 490 (7421) (2012) 497.
[99] Y. Jiao, et al., Genome-wide genetic changes during modern breeding of maize, Nat. Genet. 44 (7) (2012) 812.
[100] E.S. Mace, et al., Whole-genome sequencing reveals untapped genetic potential in Africa's indigenous cereal crop sorghum, Nat. Commun. 4 (2013) 2320.
[101] S. Aflitos, et al., Exploring genetic variation in the tomato (Solanum section Lycopersicon) clade by whole-genome sequencing, Plant J. 80 (1) (2014) 136–148.
[102] A. Díaz, et al., Copy number variation affecting the Photoperiod-B1 and Vernalization-A1 genes is associated with altered flowering time in wheat (*Triticum aestivum*), PLoS One 7 (3) (2012).
[103] L.K. McHale, et al., Structural variants in the soybean genome localize to clusters of biotic stress response genes, Plant Physiol. 159 (4) (2012) 1295–1308.
[104] R.A. Swanson-Wagner, et al., Pervasive gene content variation and copy number variation in maize and its undomesticated progenitor, Genome Res. 20 (12) (2010) 1689–1699.
[105] M.G. Claros, et al., Why assembling plant genome sequences is so challenging, Biology 1 (2) (2012) 439–459.
[106] L.D. Orozco, et al., Copy number variation influences gene expression and metabolic traits in mice, Hum. Mol. Genet. 18 (21) (2009) 4118–4129.
[107] M. Ortiz-Estevez, et al., Segmentation of genomic and transcriptomic microarrays data reveals major correlation between DNA copy number aberrations and gene–loci expression, Genomics 97 (2) (2011) 86–93.
[108] P. Yu, et al., Genome-wide copy number variations in *Oryza sativa* L, BMC Genomics 14 (1) (2013) 649.
[109] X. Lin, et al., Frequent loss of lineages and deficient duplications accounted for low copy number of disease resistance genes in *Cucurbitaceae*, BMC Genomics 14 (1) (2013) 335.

[110] T.M. Jamann, et al., Unraveling genomic complexity at a quantitative disease resistance locus in maize, Genetics 198 (1) (2014) 333–344.
[111] D. Leister, et al., Rapid reorganization of resistance gene homologues in cereal genomes, Proc. Natl. Acad. Sci. 95 (1) (1998) 370–375.
[112] X. Xu, et al., Resequencing 50 accessions of cultivated and wild rice yields markers for identifying agronomically important genes, Nat. Biotechnol. 30 (1) (2012) 105.
[113] D. Bertioli, et al., A large scale analysis of resistance gene homologues in *Arachis*, Mol. Genet. Genomics 270 (1) (2003) 34–45.
[114] G. Zhang, et al., Genome sequence of foxtail millet (*Setaria italica*) provides insights into grass evolution and biofuel potential, Nat. Biotechnol. 30 (6) (2012) 549.

CHAPTER 15

Pan-cancer analysis and applications

Dipali Dhawan
Baylor Genetics, Houston, TX, United States

1 Introduction

Since early 1970s it has been known that cancers carry genetic abnormalities which can be specific as well as recurrent [1–3]. Thereafter, researchers have been elucidating the list of genes and mutations that give rise to cancer and cancer progression, for four decades. Technological advancements have fueled this research by enabling in genomic analysis, with chromosome banding techniques for analysis of chromosome structure [2] as well as other cytogenetic techniques, positional cloning of cancer genes [4], capillary sequencing [5] which is the most popular technique, comparative genomic hybridization (CGH) [6], and the latest being massively parallel whole genome sequencing [7–11].

With the advancing technology, the global cancer genomics community established The International Cancer Genome Consortium (ICGC) in 2008 for systematically analyzing and documenting the somatic mutations observed in 25,000 samples covering the most common types of tumors [12]. Researchers from The Cancer Genome Atlas (TCGA) in the United States as well as those from 17 countries in Asia, Europe, and America form the ICGC. The main reasons for the formation of this global consortium include there is a huge scope of cancer analysis, duplication of effort if there are independent cancer genome studies, different technologies used worldwide could pose a challenge in comparing datasets across studies, different types of cancers at various frequencies around the world, and availability of data to the scientific fraternity. The samples analyzed are carefully prescreened by ICGC histopathologists and clinicians to ensure the quality of the sample and thereby ensure the accuracy of the diagnosis. In order to meet the minimum coverage and quality requirements, sequencing of both tumor and matched constitutional DNA are necessary. The data generated from ICGC are released rapidly to the scientific fraternity with utmost care to protect the ethical and regulatory aspects [13]. Fig. 1 illustrates the facts and figures of the TCGA project.

The extensive amount of data generated from the ICGC projects have enabled in discovering new cancer genes and their pathways [14–18], in elucidating mutational processes operative in human cancers [19–23], in delineation of the patterns of tumor heterogeneity (which is majorly observed in most cancers) and evolution of clonality [24–27], for cancer prevention using genomics [28] and clinical management of cancer

Pan-genomics: Applications, Challenges, and Future Prospects
https://doi.org/10.1016/B978-0-12-817076-2.00015-9
© 2020 Elsevier Inc.
All rights reserved.

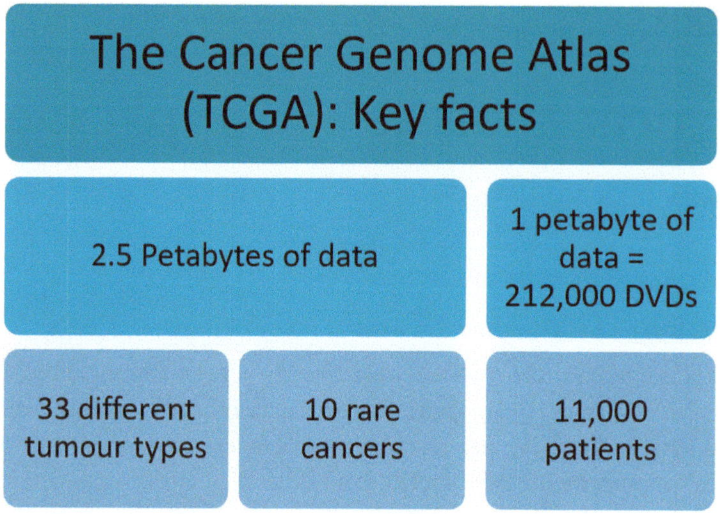

Fig. 1 Facts and figures of the TCGA project. *(Adapted from NIH: The Cancer Genome Atlas.)*

[29–31]. The enormous amount of data was analyzed using novel computational and statistical algorithms specifically designed to accurately identify genomic alterations and enable in elucidating new insights in cancer.

The gene expression level data generated by TCGA has been made available for tumors as well as normal tissues of 10 types of cancer which includes breast cancer, kidney renal papillary cell carcinoma, lung adenocarcinoma, colon adenocarcinoma, low-grade glioma, glioblastoma, ovarian carcinoma, lung squamous cell carcinoma, uterine corpus endometrioid carcinoma, and rectum adenocarcinoma [32]. This dataset is of utmost importance to understand the biological mechanisms underlying cancer and in the identification of targets for new therapy.

2 Methods in pan-cancer analysis

The numerous projects involved in pan-cancer analysis generated huge volumes of data using various technologies including high-end molecular genetics and cytogenetics techniques. Various web tools have been developed and used to interpret the large amount of data generated by the pan-cancer projects (summarized in Table 1). The various tools complement each other and help in the analysis of various datasets and enable in generating the pan-cancer atlas.

3 Pan-cancer analysis findings and applications

ICGC and TCGA studies have identified commonalities as well as differences in somatic genomic makeup across different tumor types. It was observed that while some cancer

Table 1 Web tools for pan-cancer studies

Name of tool and purpose	Weblink	Reference
IntOGen-mutations: Identification of cancer drivers across various cancer types	http://www.intogen.org/mutations	Gonzalez-Perez et al. [78]
CancerMiner: Identification of recurring miRNA-mRNA associations across various cancer types	http://cancerminer.org	Jacobsen et al. [79]
Synapse: In collaboration with TCGA pan-cancer group, sharing and updating data and results	https://www.synapse.org/	Omberg et al. [80]
TCGA: Enables researchers to search, download and analyze data generated by TCGA	http://cancergenome.nih.gov/	Weinstein et al. [81]
TCPA: Provides access to cancer proteomics datasets	http://bioinformatics.mdanderson.org/main/TCPA:Overview	Li et al. [82]
UCSC Cancer GenomicsBrowser: Provides interactive exploration of genomic and clinical data	https://genome-cancer.ucsc.edu	Cline et al. [83]

Adapted from Z. Liu, S. Zhang., Toward a systematic understanding of cancers: a survey of the pan-cancer study, Front. Genet. 5 (2014) 194.

genes are mutated in many different types of tumors, others might be more specific to a certain subtype of cancer [33, 34]. There is a coexistence of certain combinations of driver mutations observed in individual patients whereas common tumor types have few frequently mutated genes [35–37]. There is a huge diversity observed in terms of evolution of mutations, like some evolve by large-scale restructuring of chromosomes [38], others by mutations in tumor suppressor genes predominantly [39] and some others with an increased frequency of driver mutations activating the oncogenes [40].

Phosphorylation which is involved in important processes like proliferation and oncogenic kinase signaling has been suspected to play a key role in cancer. Spatially specific mutations at protein sites have been observed to be included in cellular phosphorylation signaling in cancer [41]. The same group of researchers studied 3185 tumor genomes in 12 different tumor types and identified 54 cancer-specific drivers and 82 genes only observed in pan-cancer analysis [42]. It has been speculated that the temporal relationship of somatic genetic alterations may provide newer insights into identification of driver oncogenes. Further, the timing of the most important mutation might be related to occurrence of metastasis. Early occurring genetic alterations are most important as therapeutic targets for early and most promising intervention [43].

The extensive amount of data of copy number profiles for various different cancer types will help in distinguishing which chromosomal alterations are clinically relevant and provide a better understanding of cancer progression. One of the studies by Zack and colleagues analyzed the high-resolution copy number profiles generated by TCGA and have identified common patterns of somatic copy number alterations (SCNA) in various different tumor types [44]. Many other researchers are trying to elucidate the functionality of these SCNA, however, due to the intrinsic complexity of cancer genomics, it is crucial to have more powerful algorithmic approaches for evaluating large-scale copy number alterations.

MicroRNAs (miRNA) have been observed to play important roles in gene regulation and are being studied for their roles in tumor occurrence and progression. Hamilton and colleagues studied the miRNA regulatory aspects and identified pan-cancer miRNA drivers of tumor by integrating TCGA Pan-cancer miRNA, mRNA, exome sequencing data, and copy number variation (CNV) across 12 different cancer types with a miRNA target atlas [45]. Other pan-cancer studies have used gene expression signatures to identify stromal and immune cell fractions in tumor samples [46] and also generate a virus-tumor map by interpreting the transcriptome data [47].

It has also been observed that DNA methylation, which has an important regulatory role in chromatin complexes, is dysregulated in a lot of cancers. One of the large-scale studies on DNA methylation analyzed 82 human cell lines and tissues to enable in generating an atlas of DNA methylation and its functional role in regulation of genes and disease progression [48]. Further studies are needed to understand the association of DNA methylation and gene expression levels with other phenotypic and molecular observations across various cancers.

It has also been possible to understand the order in which alterations occur in the different pathways during cancer progression and is understood to be nonrandom because of gene–gene interactions [49]. The genomic 'scars' [50] that result due to accumulation of endogenous and environmental exposures, mutations involved in cellular processes, DNA-repair defects in the pathways over many years enable in the identification of disease cause in particular individual. The analysis of data has helped in identifying previously unknown mutational processes which are suspected to be key players in cancer development and progression of certain cancers [19]. For example, germline variants have enabled in identifying genes that predispose to familial malignancies like *PALB2* and pancreatic cancer [51, 52].

Tissue-specific regulatory regions have been analyzed by various studies including ENCODE [53], Blueprint [54], and Epigenome Roadmap [55]. The interplay between transcription factors and enhancers, silencers, and other elements is responsible for cell-specific regulatory responses and its role in cancer [14, 56–60]. Epigenetic marks are also associated with cancer in terms of mutation densities [61, 62]. Fig. 2 summarizes the major findings of the pan-cancer project.

The TCGA projects also used multiomics technologies to understand the effect of genetic variants (including somatic variants) on transcription, mainly in cases of breast

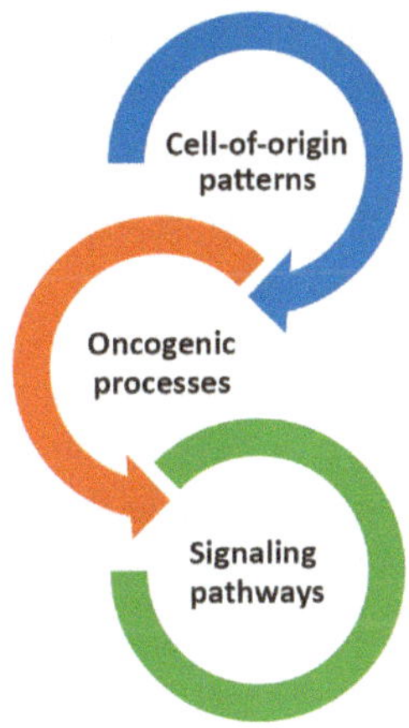

Fig. 2 Major findings of the pan-cancer project.

and liver cancers [63, 64]. It was observed that cancer cells were transcriptionally more active, and the genomic variants of cancers showed an increased expression which triggered alterations in transcription including over- or underexpression of genes [65, 66]. With progress in omics technologies it has become possible to do a comprehensive analysis of multiple cancer types for understanding cancer [67–69]. One of the studies identified 11 subtypes based on expression profiles by integrating different omics datasets including mRNA-Seq, miRNA-Seq, reverse-phase protein arrays, structural copy number alterations, DNA methylation, and somatic mutations from 12 different cancer types [68]. Studies like these have used various different omics technologies to complement the data generated by each technique and identify key players of different pathways. Pan-cancer transcriptome analysis has revealed differentially expressed genes (DEGs) between tumor and normal expression [70, 71]. The identification of therapeutic targets has been possible by using high-throughput screens of RNA interference [72].

4 Limitations of analysis

There have been several limitations faced by the researchers during analysis of different tumor types. One of the major challenges included the integration of data from different platforms as versions may get updated for different or same platforms regularly. During pan-cancer studies a lot of changeover has been observed to higher density DNA methylation arrays, different technologies for exome capture, RNA sequencing for characterizing microarray-based RNA, and lastly an increase in the number of antibodies and their quality for reverse-phase proteomic arrays (RPPAs) [73]. Further work is needed in order to improve the analysis while minimizing batch effects and maintaining biological data.

The differences in the quality and nature of clinical data generated affects comparison of demographic information, histopathological characterization, and clinical outcomes for various types of cancer. Gene expression profiles and patterns of coaberrations might show different consequences though similar expected results due to tumor lineage. However, in spite of the challenges faced during analysis in these projects the pan-cancer data represents a landmark in elucidating the common and contrasting cancer biologies.

5 Future prospects

Through the many studies conducted as part of the pan-cancer project, a lot of efforts have been put in by researchers in understanding the molecular landscape of cancer with the large number of samples of various tumor types being investigated. Further studies with more number of samples included in each tumor type will enable in identifying the rare driver mutations in the tumor samples, well known to be heterogeneous. The utilization of technologies like laser capture microdissection as well as cell sorting will help in distinguishing if the signals are received from the malignant or stromal cells, thus increasing accuracy of the data generated. Future studies may incorporate analyzing primary tumors paired with metastasized tumors which will enable in identifying characteristics of primary versus metastasized tumors which might be different as speculated [73].

Another challenge that needs to be overcome is the development of clinical trial strategies to connect tumor subsets from diverse tissue types (diverse molecular signatures). Some studies conducted on pharmacological profiling experiments across cancer cell lines have elucidated the fact that some common genetic variations may predict response to a particular therapy across various cell lineages [74–77]. Clinical trials designed on the basis of biomarkers can increase statistical significance, and decrease the expense, size, and duration of the trials conducted. With the lowering cost of genomic sequencing it is anticipated that in the near future cancer genomes might be routinely sequenced as part of the clinical management of cancer patients.

6 Summary

Pan-cancer analysis has enabled in identifying molecular aspects underlying cancer thereby benefiting diagnosis, prevention, and therapy for patients. One of the major applications of the pan-cancer data is for drug development by ranking drug targets that can be further exploited to develop targeted therapies for cancer. Further analysis of the data is needed for understanding gene–gene interactions and roles of genetic variants affecting pathways. An alternative route for drug development could be by pharmacological modulation of an important cancer pathway. In the coming years when there is more insight available from the data analyzed we might have information on additional tumor classifications [12], other categories of mutations like those found in the noncoding regions as well as functionally annotated genome (The [53]).

References

[1] E.P. Reddy, R.K. Reynolds, E. Santos, M. Barbacid, A pointmutation is responsible for the acquisition of transforming propertiesby the T24 human bladder carcinoma oncogene, Nature 300 (1982) 149–152.

[2] J.D. Rowley, A new consistent chromosomal abnormality in chronicmyelogenous leukaemia identified by quinacrine fluorescence and Giemsa staining, Nature 243 (1973) 290–293.

[3] C.J. Tabin, et al., Mechanism of activation of a human oncogene, Nature 300 (1982) 143–149.

[4] S.H. Friend, et al., A human DNA segment with properties of the gene that predisposes to retinoblastoma and osteosarcoma, Nature 323 (1986) 643–646.

[5] T. Sjoblom, et al., The consensus coding sequences of human breast and colorectal cancers, Science 314 (2006) 268–274.

[6] R. Beroukhim, et al., The landscape of somatic copy-number alteration across human cancers, Nature 463 (2010) 899–905.

[7] P.J. Campbell, et al., Identification of somatically acquired rearrangements in cancer using genome-wide massively parallel paired-end sequencing, Nat. Genet. 40 (2008) 722–729.

[8] J.O. Korbel, et al., Paired-end mapping reveals extensive structural variation in the human genome, Science 318 (2007) 420–426.

[9] T.J. Ley, et al., DNA sequencing of a cytogenetically normal acute myeloid leukaemia genome, Nature 456 (2008) 66–72.

[10] E.D. Pleasance, et al., A comprehensive catalogue of somatic mutations from a human cancer genome, Nature 463 (2010) 191–196.

[11] E.D. Pleasance, et al., A small-cell lung cancer genome with complex signatures of tobacco exposure, Nature 463 (2010) 184–190.

[12] The International Cancer Genome Consortium, International network of cancer genome projects, Nature 464 (2010) 993–998.

[13] Y. Joly, E.S. Dove, B.M. Knoppers, M. Bobrow, D. Chalmers, Data sharing in the post-genomic world: the experience of the International Cancer Genome Consortium (ICGC) Data Access Compliance Office (DACO), PLoS Comput. Biol. 8 (2012) e1002549.

[14] P.A. Northcott, C. Lee, T. Zichner, A.M. Stütz, S. Erkek, D. Kawauchi, D.J. Shih, V. Hovestadt, M. Zapatka, D. Sturm, D.T. Jones, M. Kool, M. Remke, F.M. Cavalli, S. Zuyderduyn, G.D. Bader, et al., Enhancer hijacking activates GFI1 family oncogenes in medulloblastoma, Nature 511 (2014) 428–434.

[15] E. Papaemmanuil, et al., Somatic SF3B1 mutation in myelodysplasia with ring sideroblasts, N. Engl. J. Med. 365 (2011) 1384–1395.

[16] X.S. Puente, et al., Whole-genome sequencing identifies recurrent mutations in chronic lymphocytic leukaemia, Nature 475 (2011) 101–105.

[17] J. Richter, et al., Recurrent mutation of the ID3 gene in Burkittlymphoma identified by integrated genome, exome and transcriptome sequencing, Nat. Genet. 44 (2012) 1316–1320.

[18] The Cancer Genome Atlas Research Network, Comprehensive molecular portraits of human breast tumours, Nature 490 (2012) 61–70.

[19] L.B. Alexandrov, et al., Signatures of mutational processes in human cancer, Nature 500 (2013) 415–421.

[20] N.J. Haradhvala, et al., Mutational strand asymmetries in cancer genomes reveal mechanisms of DNA damage and repair, Cell 164 (2016) 538–549.

[21] T. Rausch, et al., Genome sequencing of pediatric medulloblastoma links catastrophic DNA rearrangements with TP53 mutations, Cell 148 (2012) 59–71.

[22] The Cancer Genome Atlas Research Network, Comprehensive molecular characterization of clear cell renal cell carcinoma, Nature 499 (2013) 43–49.

[23] Y. Totoki, et al., Trans-ancestry mutational landscape of hepatocellular carcinoma genomes, Nat. Genet. 46 (2014) 1267–1273.

[24] C.S. Cooper, et al., Analysis of the genetic phylogeny of multifocal prostate cancer identifies multiple independent clonal expansions in neoplastic and morphologically normal prostate tissue, Nat. Genet. 47 (2015) 367–372.

[25] S. Nik-Zainal, et al., The life history of 21 breast cancers, Cell 149 (2012) 994–1007.
[26] A.M. Patch, et al., Whole–genome characterization of chemoresistant ovarian cancer, Nature 521 (2015) 489–494.
[27] C.S. Ross-Innes, et al., Whole-genome sequencing provides newinsights into the clonal architecture of Barrett's esophagus and esophageal adenocarcinoma, Nat. Genet. 47 (2015) 1038–1046.
[28] G. Scelo, et al., Variation in genomic landscape of clear cell renal cell carcinoma across Europe, Nat. Commun. 5 (2014) 5135.
[29] H. Davies, et al., HR Detect is a predictor of BRCA1 and BRCA2deficiency based on mutational signatures, Nat. Med. 23 (2017) 517–525.
[30] N. Waddell, et al., Whole genomes redefine the mutational landscape of pancreatic cancer, Nature 518 (2015) 495–501.
[31] Y. Yuan, et al., Assessing the clinical utility of cancer genomic and proteomic data across tumour types, Nat. Biotechnol. 32 (2014) 644–652.
[32] R. Neapolitan, C.M. Horvath, X. Jiang, Pan-cancer analysis of TCGA data reveals notable signaling pathways, BMC Cancer 15 (2015) 516.
[33] C. Kandoth, et al., Mutational landscape and significance across 12major cancer types, Nature 502 (2013) 333–339.
[34] M.S. Lawrence, et al., Discovery and saturation analysis of cancer genes across 21 tumour types, Nature 505 (2014) 495–501.
[35] M.D.M. Leiserson, et al., Pan-cancer network analysis identifies combinations of rare somatic mutations across pathways and protein complexes, Nat. Genet. 47 (2014) 106–114.
[36] The Cancer Genome Atlas Research Network, Comprehensive genomic characterization of squamous cell lung cancers, Nature 489 (2012) 519–525.
[37] The Cancer Genome Atlas Research Network, Comprehensive molecular characterization of human colon and rectal cancer, Nature 487 (2012) 330–337.
[38] G. Ciriello, et al., Emerging landscape of oncogenic signatures across human cancers, Nat. Genet. 45 (2013) 1127–1133.
[39] The Cancer Genome Atlas Research Network, Integrated genomic characterization of endometrial carcinoma, Nature 497 (2013) 67–73.
[40] The Cancer Genome Atlas Research Network, Comprehensive molecular profiling of lung adenocarcinoma, Nature 511 (2014) 543–550.
[41] J. Reimand, G.D. Bader, Systematic analysis of somatic mutations in phosphorylation signalling predictsnovelcancerdrivers, Mol. Syst. Biol. 9 (2013) 637.
[42] J. Reimand, O. Wagih, G.D. Bader, Themutational landscape of phosphorylation signalling in cancer, Sci. Rep. 3 (2013) 2651.
[43] B. Vogelstein, N. Papadopoulos, V.E. Velculescu, S. Zhou, L.A. Diaz, K.W. Kinzler, Cancergenome landscapes, Science 339 (2013) 1546–1558.
[44] T.I. Zack, S.E. Schumacher, S.L. Carter, A.D. Cherniack, G. Saksena, B. Tabak, et al., Pan-cancer patterns of somatic copy number alteration, Nat. Genet. 45 (2013) 1134–1140.
[45] M.P. Hamilton, K. Rajapakshe, S.M. Hartig, B. Reva, M.L. MD, C. Kandoth, et al., Identification of apan-canceroncogenic micro RNA super-family anchored by acentralcoreseed motif, Nat. Commun. 4 (2013) 2730.
[46] K. Yoshihara, M. Shahmoradgoli, E. Martínez, R. Vegesna, H. Kim, W. Torres-Garcia, et al., Inferring tumor purity and stromal and immune cell admixture from expression data, Nat. Commun. 4 (2013) 2612.
[47] K.W. Tang, B. Alaei-Mahabadi, T. Samuelsson, M. Lindh, E. Larsson, The landscape of viral expression and host gene fusion and adaptation in human cancer, Nat. Commun. 4 (2013) 2513.
[48] K.E. Varley, J. Gertz, K.M. Bowling, S.L. Parker, T.E. Reddy, F. Pauli-Behn, et al., Dynamic DNA methylation across diverse human cell lines and tissues, Genome Res. 23 (2013) 555–567.
[49] A. Ashworth, C.J. Lord, J.S. Reis-Filho, Genetic interactions in cancer progression and treatment, Cell 145 (2011) 30–38.
[50] C.J. Lord, A. Ashworth, The DNA damage response and cancer therapy, Nature 481 (2012) 287–294.

[51] S. Jones, et al., Core signaling pathways in human pancreatic cancers revealed byglobal genomic analyses, Science 321 (2008) 1801–1806.

[52] S. Jones, et al., Exomic sequencing identifies PALB2 as a pancreatic cancer susceptibility gene, Science 324 (2009) 217.

[53] ENCODE Project Consortium, An integrated encyclopedia of DNA elements in the human genome, Nature 489 (2012) 57–74.

[54] H.G. Stunnenberg, International Human Epigenome Consortium, M. Hirst, The International Human Epigenome Consortium: a blueprint for scientific collaboration and discovery, Cell 167 (2016) 1145–1149.

[55] Roadmap Epigenomics Consortium, A. Kundaje, W. Meuleman, J. Ernst, M. Bilenky, A. Yen, A. Heravi-Moussavi, P. Kheradpour, Z. Zhang, J. Wang, et al., Integrative analysis of 111 reference human epigenomes, Nature 518 (2015) 317–330.

[56] D. Hnisz, A.S. Weintraub, D.S. Day, A.L. Valton, R.O. Bak, C.H. Li, J. Goldmann, B.R. Lajoie, Z.P. Fan, A.A. Sigova, J. Reddy, D. Borges-Rivera, T.I. Lee, R. Jaenisch, M.H. Porteus, J. Dekker, R.A. Young, Activation of protooncogenes by disruption of chromosome neighborhoods, Science 351 (2016) 1454–1458.

[57] S. Horn, A. Figl, P.S. Rachakonda, C. Fischer, A. Sucker, A. Gast, S. Kadel, I. Moll, E. Nagore, K. Hemminki, D. Schadendorf, R. Kumar, TERT promoter mutations in familial and sporadic melanoma, Science 339 (2013) 959–961.

[58] F.W. Huang, E. Hodis, M.J. Xu, G.V. Kryukov, L. Chin, L.A. Garraway, Highly recurrent TERT promoter mutations in human melanoma, Science 339 (2013) 957–959.

[59] E. Rheinbay, P. Parasuraman, J. Grimsby, G. Tiao, J.M. Engreitz, J. Kim, M.S. Lawrence, A. Taylor-Weiner, S. Rodriguez-Cuevas, M. Rosenberg, J. Hess, C. Stewart, Y.E. Maruvka, P. Stojanov, M. L. Cortes, S. Seepo, C. Cibulskis, A. Tracy, T.J. Pugh, J. Lee, Z. Zheng, L.W. Ellisen, A.J. Iafrate, J.S. Boehm, S.B. Gabriel, M. Meyerson, T.R. Golub, J. Baselga, A. Hidalgo-Miranda, T. Shioda, A. Bernards, E.S. Lander, G. Getz, Recurrent and functional regulatory mutations in breast cancer, Nature 547 (7661) (2017) 55–60, https://doi.org/10.1038/nature22992.

[60] J. Weischenfeldt, T. Dubash, A.P. Drainas, B.R. Mardin, Y. Chen, A.M. Stütz, S.M. Waszak, G. Bosco, A.R. Halvorsen, B. Raeder, T. Efthymiopoulos, S. Erkek, C. Siegl, H. Brenner, O.T. Brustugun, S.M. Dieter, P.A. Northcott, I. Petersen, S.M. Pfister, M. Schneider, S. K. Solberg, E. Thunissen, W. Weichert, T. Zichner, R. Thomas, M. Peifer, A. Helland, C.R. Ball, M. Jechlinger, R. Sotillo, H. Glimm, J.O. Korbel, Pan-cancer analysis ofsomatic copy-number alterations implicates IRS4 and IGF2 inenhancer hijacking, Nat. Genet. 49 (2017) 65–74.

[61] P. Polak, R. Karlić, A. Koren, R. Thurman, R. Sandstrom, M.S. Lawrence, A. Reynolds, E. Rynes, K. Vlahoviček, J.A. Stamatoyannopoulos, S.R. Sunyaev, Cell-of-origin chromatin organization shapes the mutationallandscape of cancer, Nature 518 (2015) 360–364.

[62] B. Schuster-Böckler, B. Lehner, Chromatin organization is a major influence on regional mutation rates in human cancer cells, Nature 488 (2012) 504–507.

[63] Y. Shiraishi, et al., Integrated analysis of whole genome and transcriptome sequencing reveals diverse transcriptomic aberrations driven by somatic genomic changes in liver cancers, PLoS One 9 (2014) e114263.

[64] A. Shlien, et al., Direct transcriptional consequences of somatic mutation in breast cancer, Cell Rep. 16 (2016) 2032–2046.

[65] C. Suo, et al., Integration of somatic mutation, expression and functional data reveals potential driver genes predictive of breast cancer survival, Bioinformatics 31 (2015) 2607–2613.

[66] J. Zhang, Z. Abrams, J.D. Parvin, K. Huang, Integrative analysis of somatic mutations and transcriptomic data to functionally stratify breast cancer patients, BMC Genomics 17 (2016) 513.

[67] R. Akbani, et al., A pan-cancer proteomic perspective on The Cancer Genome Atlas, Nat. Commun. 5 (2014) 3887.

[68] K.A. Hoadley, et al., Multiplatform analysis of 12 cancer types reveals molecular classification within and across tissues of origin, Cell 158 (2014) 929–944.

[69] Z. Liu, S. Zhang, Tumor characterization and stratification by integrated molecular profilesreveals essential pan-cancer features, BMC Genomics 16 (2015) 503.

[70] C.R. Cabanski, et al., Pan-cancer transcriptome analysis reveals long noncoding RNAs with conserved function, RNA Biol. 12 (2015) 628–642.
[71] Z. Cao, S. Zhang, An integrative and comparative study of pan-cancer transcriptomes reveals distinct cancer common and specific signatures, Sci. Rep. 6 (2016).
[72] L. Chin, J.W. Gray, Translating insights from the cancer genome into clinical practice, Nature 452 (2008) 553–563.
[73] The Cancer Genome Atlas Research Network, J.N. Weinstein, E.A. Collisson, G.B. Mills, et al., The Cancer Genome Atlas Pan-Cancer analysis project, Nat. Genet. 45 (2013) 1113–1120.
[74] J. Barretina, G. Caponigro, N. Stransky, K. Venkatesan, A.A. Margolin, et al., The Cancer Cell Line Encyclopedia enables predictive modelling of anticancer drug sensitivity, Nature 483 (2012) 603–607.
[75] M.J. Garnett, E.J. Edelman, S.J. Heidorn, C.D. Greenman, A. Dastur, et al., Systematic identification of genomic markers of drug sensitivity in cancer cells, Nature 483 (2012) 570–575.
[76] L.M. Heiser, A. Sadanandam, W.L. Kuo, S.C. Benz, T.C. Goldstein, et al., Subtype and pathway specific responses to anticancer compounds in breast cancer, Proc. Natl. Acad. Sci. U. S. A. 109 (2012) 2724–2729.
[77] J.N. Weinstein, Drug discovery: cell lines battle cancer, Nature 483 (2012) 544–545.
[78] A. Gonzalez-Perez, C. Perez-Llamas, J. Deu-Pons, D. Tamborero, M.P. Schroeder, A. Jene-Sanz, et al., Int OGen-mutations identifies cancer drivers across tumor types, Nat. Methods 10 (2013) 1081–1082.
[79] A. Jacobsen, J. Silber, G. Harinath, J.T. Huse, N. Schultz, C. Sander, Analysis of micro RNA-target interactions across diverse cancer types, Nat. Struct. Mol. Biol. 20 (2013) 1325–1332.
[80] L. Omberg, K. Ellrott, Y. Yuan, C. Kandoth, C. Wong, M.R. Kellen, et al., Enabling transparent and collaborative computational analysis of 12tumor types within the cancer genomeatlas, Nat. Genet. 45 (2013) 1121–1126.
[81] J.N. Weinstein, E.A. Collisson, G.B. Mills, S. KRM, B.A. Ozenberger, K. Ellrott, et al., The cancer genomeatlaspan-canceranalysis project, Nat. Genet. 45 (2013) 1113–1120.
[82] J. Li, A.R. LuY, R.P.L. JuZ, W. Liu, et al., TCPA: a resource for cancer functional proteomics data, Nat. Methods 10 (2013) 1046–1047.
[83] M.S. Cline, B. Craft, T. Swatloski, M. Goldman, S. Ma, D. Haussler, et al., Exploring TCGA pan-cancer data at the UCSC cancergenomics browser, Sci. Rep. 3 (2013) 2652.

Further reading

[84] J.S. Boehm, S.B. Gabriel, M. Meyerson, T.R. Golub, J. Baselga, A. Hidalgo-Miranda, T. Shioda, A. Bernards, E.S. Lander, G. Getz, Recurrent and functional regulatory mutations in breast cancer, Nature 547 (2017) 55–60.
[85] Z. Liu, S. Zhang, Toward a systematic understanding of cancers: a survey of the pan-cancer study, Front. Genet. 5 (2014) 194.
[86] NIH (n.d.): The Cancer Genome Atlas (https://cancergenome.nih.gov/abouttcga).

CHAPTER 16

Reverse vaccinology and drug target identification through pan-genomics

Anam Naz, Ayesha Obaid, Fatima Shahid, Hamza Arshad Dar, Kanwal Naz, Nimat Ullah, Amjad Ali
Department of Plant Biotechnology, Atta-ur-Rahman School of Applied Biosciences (ASAB), National University of Sciences and Technology (NUST), Islamabad, Pakistan

1 Introduction and goals of pan-genomics and reverse vaccinology

Publication of first complete genome of the bacterium *Haemophilus influenzae* in 1995 [1] opened new ventures and challenges for biologists. It leads to the series of genome sequencing which till now has turned into a huge mass of genomic data consisting of thousands of "reference genomes." These reference genomes, in the beginning, were quite helpful to discover and analyze genome data sets. But rapid next-generation sequencing, resequencing, and genome analysis workflows have gifted us with a great wealth of evolutionary genomics, genome alterations, functional annotations, genomic variants, homologs, and many more relations within genomes. This advent of rapid genome data generation led scientists to rethink the genome analysis platform and change the concept of reference genomes by introducing pan-genomes, which entails a paradigm shift from a single genome representation to the full complement of genes in a clade. Tettelin et al. presented the first microbial pan-genome as the collection of "core" and "dispensable" genomes [2]. Core genome represents the genes common to all species, thus conserved by the nature for the species, whereas the dispensable genome also known as flexible or accessory genome consists of genome-specific or partially shared genes.

Insights into genomic evolution, diversity, and pathogenic potential revealed by pan-genome can then be utilized to access the core gene repertoire in order to design a broad spectrum and effective drug/vaccine candidates. Reverse vaccinology (RV), a competent genome-based strategy has allowed researchers to comprehensively analyze the antigenic repository of a particular organism using its genome sequence. This term was first introduced by Rappuoli in 2000 [3]. RV employs cloning and expression of computationally identified exoproteome and secretome of the pathogen followed by screening and selection based on the antibody response in suitable animal models [4]. This revolutionary strategy was first successfully employed to identify potential antigens against serogroup B meningococcus and effective vaccines were designed [5]. RV mainly targets outer membrane proteins (OMPs) for vaccine and drug candidate identification and has reportedly been able to discover various novel antigens as potential vaccines [6–9]. It is an

Pan-genomics: Applications, Challenges, and Future Prospects
https://doi.org/10.1016/B978-0-12-817076-2.00016-0
© 2020 Elsevier Inc.
All rights reserved.

integrated computational approach which employs various bioinformatics tools to target antigenic proteins. More explanation of the methodology and tools used for RV and pan-genome analysis has been given in the upcoming sections of this chapter.

1.1 RV vs conventional vaccinology

The term "vaccine" was first introduced in 1796 by Edward Jenner when he successfully developed the smallpox vaccine by isolating the vaccinating molecules from a cow. That was the beginning of a new era in the field of medicine and the successful use of vaccines against certain diseases widespread the idea to combat pathogens by live, attenuated, dead pathogens, or pathogenic materials. Rules of vaccinology were then set by Louis Pasteur and followed by various other scientists to design effective vaccines against polio, measles, mumps, rubella, diphtheria, tetanus, etc. [10]. Thus, it remained the most powerful technique since the first revolution of vaccinology which was based on genetic engineering to design vaccines. This revolution did not bring much-appreciated success till the second revolution which was made in the 20th century with the advent of computational genomics which leads to the modern vaccinology or immunoinformatic approaches [11].

Classical vaccinology almost takes 10–15 years to develop an effective vaccine against any microbe or a disease, but advances in molecular biology and genomic techniques have introduced attractive approaches to design efficient and novel vaccines rapidly [12]. Identification of antigenic epitopes from potentially virulent proteins is an emerging technique nowadays, which focuses on structural, functional, and immunogenic properties of protein epitopes [12]. The rapid and cheap whole-genome sequencing has made it more convenient to determine and explore the antigens against a pathogen and thus can be easily expressed. For this purpose, all genes/proteins of the pathogen are screened to determine potential antigenic and virulent proteins/adjuvants that can successfully elicit the immune response within the host and can be determined as active compounds for a vaccine. All this can be done within hours now with the computational power and then the prioritized candidates can be easily checked for their activity against the pathogen experimentally. This whole process employing computational or RV usually takes 3–5 years which is significantly less time taken by the conventional vaccinology to design effective vaccines. A brief comparison of conventional and RV has been shown in Table 1.

2 Outcomes of pan-genomics and RV

The pan-genomic analysis of a pathogenic species delivers substantial data for putative vaccine candidate identification. Theoretically, the core genome is comprised of the most desired set of target genes for the purpose of broad-spectrum vaccines construction. However, these may be less immunogenic in nature. Conversely, the accessory genome containing jumping genes, plasmids, and acquired genes that may vary from pathogen to

Table 1 Comparison of properties, targets, and outcomes of conventional and reverse vaccinology

Properties	Conventional vaccinology	Reverse vaccinology
Targeting every single antigen	X	✓
Targeting nonculturable pathogens	X	✓
Screening of nonabundant antigens	X	✓
Targeting antigens immunogenic during infection only	✓	X
Identifies proteins only	X	✓
Only structural proteins are considered	✓	X
Time consuming	✓	X

pathogen, majorly manifests pathogenesis and might be able to elicit a strong immunogenic response. A combinational strategy incorporating both essential, core and accessory genome may be employed for better results. This strategy was employed in the case of group B streptococcus (GBS) vaccine development [4].

A substantial rise in the antibiotic resistance exhibited by bacterial species has lately imperiled health-care facilities. The group of "ESKAPE" pathogens that contains antibiotic-resistant strains poses a huge risk to the modern world. Resistant determinants encoded by accessory genes form certain genomic islands that confer discernable resistance, while core-resistant genes confer substantial intrinsic resistance. Intrinsic resistance is responsible for modulating the permeability of the outer membrane, mediating broad-spectrum drug efflux pumps and enhancing stress enduring ability [13,14].

Employing pan-genomics for interspecies analysis of pathogens that can cause similar infections can refine the process of drug target prediction, because species with similar characteristics and survival requirements colonize in the same niche and possess virulence factors. Virulence factors can be a potent target for new drug design against a group of bacteria [15]. Thus, predicting some potent virulent factors from the pan-genome of a species can encounter some specific broad-spectrum targets and can help to design a universal vaccine which can potentially be comprised of biologically cross-protective antigens against various strains, serovars, or pathovars of a single pathogen.

In contrast to classical vaccinology, the RV is the most practiced methodology since the discovery of the menB vaccine [5,16]. It involves the use of several computational analyses to predict potential immunogenic antigens from the genomes or proteomes of pathogens and provides a comprehensive view of the pathogen overall genome/pan-genome, essential and pathogen-specific pathways, virulence-determining factors, and protein-protein interactions between pathogenic and host proteins. Thus, having the influencing advantages of speed and preferable cost, RV holds a very important role in modern vaccinology to predict potential vaccine candidates. This eventually answers many questions which usually remain unaddressed experimentally. Thus, the contribution of pan-genome analysis and RV has led to the identification of many novel antigenic and potential drug/vaccine candidates and reduced the burden of classical vaccinology and drug resistance patterns.

3 Core genome as the basis of the novel and broad-spectrum drugs and vaccine candidates

A set of genes that remains significantly unchanged throughout a species or genus is regarded as the core genome. Naturally, the core genome testifies the concept that genomes belonging to closely related bacteria must have some common features [17] and they constitute the basis of all the essential elements required for the genome. Thus, the core genome prediction of pathogenic bacteria allows innovative advancements in epidemiology, diagnostics, antigen identification, and vaccine and drug design. Bacterial genomes are diverse in nature hence targeting accessory genome is not prolific to obtain broad-spectrum therapeutics. When pan-genomics is employed in combination with RV, it reveals a set of highly conserved surface exposed and secretory proteins (encoded by core genes) that provide antigens to facilitate broad coverage vaccine design. Once these antigens pass the screening steps, they have the potential to show immunogenic responses in animal models and fight infections [18].

The recent spread of antibiotic resistance needs to be addressed with all the knowledge of transcriptomics, proteomics, and genomics. Pan-genome profiling is instrumental as it distinguishes core and accessory genes. It elucidates broad-spectrum core targets that are intrinsically encoded and have eventually been evolved to confer selective resistance as well as potent resistance in different strains of the same species [19]. Hence, it is necessary to explore and bioinformatically catalog significantly conserved core genomes that can assist in the taxonomic marking of microbial clades, to find similarities among several phylogenies produced from diverse core genes, and to interpret the biological functions augmented within core genome [20]. This information can then be used to combat unprecedentedly increasing bacterial infections. The detailed overview of the methodology and tools employing pan-genomics and RV has been discussed in the following sections.

4 Pan-genomics

Pan-genome is the global gene repertoire of a given set of genomes, at the species level or genus level and consists of the core genome, dispensable genome, and species- or strain-specific genes [13,21]. The core genome represents those genes that are present in all genomes under study and can be determined by comparing the different genomes. Normally, genes in the core genome are associated with the maintenance of the elementary aspects of the organism and mostly focused on basic processes such as translation, replication, and cellular homeostasis [2,22]. Due to the essential nature of functions associated with the core genome, significant selective pressure is applied to these genes, which minimize the chances of drastic events in core genome composition [23]. The number of genes in the core genome indicates the genetic diversity among the genomes [24]. So, there is a tendency for phylogenetically related genomes to share more genes and present a larger core genome.

The genes present in some organisms, but not all of them are considered to be a part of the dispensable genome [23]. These genes impart specific functions to the organisms that are important for survival in different environments. Usually, these functions are linked with virulence or resistance to antibiotics [25,26]. The dispensable genome arises from variations in gene sequences that can lead to new functions from the genes [23]. It is believed that the dispensable genome is formed by horizontal gene transfer and paraphyletic evolution. Strain divergence may occur due to these genetic changes [27,28].

Species-specific genes are exclusively present in a single species at the interspecies level, while strain-specific genes are unique to a single strain at the intraspecies level [29]. These genes are typically acquired by horizontal gene transfer among species and may confer an adaptive advantage over strains that lack these genes. Moreover, these genes are associated with virulence or pathogenicity in pathogenic organisms [30,31]. Whereas, in nonpathogenic organisms, these genes maybe important for metabolism perspective and could be metabolic islands that are acquired by horizontal gene transfer [32].

4.1 Drug targets identification employing pan-genomics

The pan-genome analysis is usually followed by subtractive proteomics approach to determine putative drug targets. It is the stepwise filtration of a bacterial genome for target prioritization scheme [33]. For this purpose, core proteome is first computed and then subjected to various sequential filtration steps to retain only the attributes of drug targets identified through a literature study. A detailed scheme is provided below.

4.1.1 Determination of nonhomologous protein sequences to the human proteome

Core proteome is subjected to BLASTp against human proteome using NCBI at a suitable threshold value, that is, *E*-value 10^{-3}. Consequently, the bacterial proteins found to be homologous to any human protein are discarded from further analysis. This step is carried out to avoid targeting human proteins through drug action.

4.1.2 Identification of virulence factors and essential proteins

The nonhuman homologous proteins are checked for similarity to proteins from database of essential genes (DEG) using BLASTp search with *E*-value 10^{-5} [34]. The sequences showing significant similarity to any protein in the DEG database are considered essential for the bacteria. Virulence factors involved in pathogenesis and disease development are determined from selected proteins using BLASTp against microbial virulence database

(MvirDB) and virulence factors database (VFDB) [35,36]. Targeting these crucial proteins involved in disease progression through drug therapy will help in the control of the bacterial infection.

4.1.3 Metabolic pathway analysis

Comparative pathway analysis is conducted on nonhuman homolog virulent and essential bacterial proteins from the previous step using KEGG Automated Annotation Server (KAAS) [37]. This is a crucial step to identify proteins associated with cellular pathways unique to bacteria and absent in humans so that pathogen-specific pathways can be specifically targeted with minimum disruption to human pathways [38].

4.1.4 Prediction of subcellular localization

The unique bacterial proteins are later checked for their subcellular localization using different methods. For this, PSORTb is commonly used [39]. PSORTb provides a prediction for different subcellular localization, namely extracellular, outer membrane, cytoplasm, cytoplasmic membrane, cell wall, and unknown. The option of cytoplasmic proteins is selected as cytoplasmic proteins are considered good drug targets [40]. CELLO, another localization prediction program, can also be used to confirm the location of proteins provided by PSORTb [41].

4.1.5 Assessment of protein involvement in conferring antibiotic resistance

Pathogenic bacteria typically contain numerous antibiotic resistance genes in their genome [42]. The proteins are checked for their possible involvement in conferring antimicrobial resistance. BLASTp is performed against ARDB (antibiotic resistance genes database) and CARD (comprehensive antibiotic resistance database) and proteins with at least 30% sequence identity are retained for further scrutinization [43,44].

4.1.6 Druggability potential of shortlisted sequences

The druggability of shortlisted protein targets is evaluated through BLASTp search against DrugBank database which contains a collection of FDA-approved drugs as well as their drug targets [45]. Druggability is the ability of drug targets to bind drug or drug-like molecule [46]. A default E-value of 10^{-3} can be used for this purpose.

4.2 Advantages and success stories

The first successful experiment of cumulative pan-genomic and RV approach was the GBS vaccine development against variable genomes. This lead to the vital discovery of pilus-like structures in group A streptococcus as well as *S. pneumoniae* [47,48]. Later, it was experimentally validated that immunization with GAS pilus vaccine conferred significant protection against the highly virulent GAS pathogen. This could be the first step toward a vaccine against GAS [47]. Below are mentioned some efforts employing either the combinations of pan-genomics, subtractive proteomics, and RV.

In 2010, Deng et al. studied *Listeria monocytogenes*, a renowned food born pathogen by focusing on its core proteome to unveil its resistance mechanisms as well as its adaptation strategies to evade the digestive system of the host. This provided all-encompassing coverage of the entire species with precision [49]. Similarly, Hassan et al. followed the same lead in 2016, by applying pan-genomics and immunoproteomics approaches simultaneously. This resulted in the proposition of 13 highly antigenic core proteins in *Acinetobacter baumannii*. These proteins (P pilus assembly protein, AdeK, PonA, OmpA, peptidoglycan-associated lipoprotein, and peptidyl-prolyl *cis-trans* isomerase GspD, FhuE receptor, type VI secretion system OmpA/MotB, TonB-dependent siderophore receptor, GspD, OMP, and pili assembly chaperone) had vital roles in the survival and pathogenesis of the bacterium. Protein scrutinization was followed by identification of promiscuous, surface exposed, 9mer T-cell epitopes that could be potential targets to eradicate the infections caused by this nosocomial bacterium [7].

Uddin et al. wisely merged pan-genomics and subtractive proteomics to identify novel therapeutic targets against the ominous human pathogen *Salmonella enterica*. The pan-genome analysis was carried out on a total of 42 strains. Scrutinized core proteins were further checked for their involvement in pathogen specific pathways, essentiality, and the absence of human host homology. This study predicted 49 novel putative drug targets against *S. enterica* [50]. Yang et al. also studied 42 complete *Brucella* genomes in 2016 and identified 1710 core genes among them. Only 1210 out of these mediated some essential functions of the pathogen. Additional analysis was performed by comparing the core essential genes of Brucella to *Mycobacterium tuberculosis*, *F. novicida*, and *Burkholderia* spp. Results revealed that these 4 species possess 340 common essential genes that can be targeted for therapeutic purposes [51].

In 2017, Ibrahim et al. carried out a detailed pan-genome analysis of eight diverse *Prevotella* species involved in oral infections like periodontitis. These include *P. details*, *P. denticola*, P. sp. oral taxon 299, *P. intermedia* 17, *P. intermedia* 17-2, *P. enoeca*, *P. melaninogenica*, and *P. fusca*. The study revealed a highly conserved set of core genes among these all. Further annotation deciphered that these species had a potential to secrete cysteine proteases as virulence factors that damage nutrient acquisition proteins of the host and sequester the pathogen from hosts' immune response. Additionally, the core genome contained oxidative stress regulatory proteins to counter the stress during host inflammation. These were comprised of certain glutathione peroxidases, some of the genes from SUF pathway as well as aerotolerance regulators. Certain proteins shared by these eight species and some other oral pathogens were checked for human host homology and broad-spectrum drug targets were predicted against oral infections [52].

Most recently, Uddin et al. implemented a similar strategy as an effort to tackle pan-drug resistance exhibited by *Pseudomonas aeruginosa*. A total of 68 strains of *P. aeruginosa* were subjected to pan-genome analysis and the core genes were extracted and further analyzed via subtractive genomic approach. Once the bacterial essential nonhuman homologous proteins were shortlisted on the basis of their presence in suitable subcellular vicinity, their druggability scores were measured and functional annotation was performed by String analysis. Finally, eight proteins were proposed as potential drug targets [53].

4.3 Bioinformatics tools to determine pan-genome

Various tools have been developed for pan-genome analysis. EDGAR, a web-tool software can be used that performs homology analyses based on a defined threshold value that is automatically adjusted to the query data [54]. The orthology analysis to calculate the core genome, pan-genome, and singletons are performed using BLAST score ratio values. PGAT another web-based tool that has been developed compares multiple strains of the given input species to predict genetic differences [55]. Its analyses include pan-genome, synteny, the presence or the absence of genes in a dataset, comparison of SNPs in orthologous genes, metabolic pathway analysis, and the development of functional annotation. PGAP is a stand-alone tool that conducts analysis related to pan-genome, genetic variation, evolution, and functional aspects of gene clusters [56]. The software uses two methods to calculate all of the analyses: (i) the GF method to detect homologous genes and (ii) the MP method to detect orthologous genes.

Core genome estimation is usually done on the basis of BLAST, which determines similarities between the genomes under study as per the 50/50 rule [57,58]. In accordance with this rule, BLAST results are analyzed and if in any two genes, minimum 50% identity over minimum 50% of the longest gene is observed, the two genes are considered conserved and assumed to belong to the same gene family. Adopting this strategy, similar genes can be grouped into gene families. Genes that do not fit into any gene family are allocated into their unique gene family. Gene families containing at least one common gene are gathered into the core genome, while the remaining genes are considered part of the species or genus pan-genome.

Another strategy can be adopted for pan-genome analysis using Roary [59]. Roary requires assembled annotated genomes in GFF3 format, which in turn can be obtained by Prokka [60]. All the coding regions are taken from the submitted input, converted to protein sequences and screened to retain complete sequences. Next, preclusters are formed using CD-HIT [61]. The filtration steps thus lead to a reduced set of protein sequences on which further analysis is performed. A comprehensive comparison involving all-against-all method is carried out with BLASTp on the selected sequences with a user-specified percentage sequence identity threshold (default threshold is 95%). Clustering is done on the sequences with MCL and the resulting information is integrated with the preclustering results of CD-HIT [62]. Homologous paralog-containing groups are divided into groups of true orthologs. As per the sequence of input sequences, a graph is built describing the relationships of the clusters and, therefore, enabling the ordering of clusters and providing gene contexts. Considering the presence of genes in the accessory genome, clustering of isolates is carried out, with cluster size depicting the contribution of individual isolates to the graph. Command line-based tools are then used to assess the dataset obtained to provide information about the union, complement, and intersection.

5 Reverse vaccinology

5.1 RV methodology to predict potential vaccine candidates

RV is an approach in which in silico sequential filtering (Fig. 1) of the entire proteome of the pathogenic bacteria is conducted to prioritize a select few antigenic vaccine candidates which could be tested in wet laboratory conditions for validation [9].

Some steps are common to both vaccine and drug target prioritization which will be indicated where applicable; steps are detailed below.

5.1.1 Selection of host nonhomologous, essential, and virulent proteins

Nonhuman homologs from the core proteome are selected to minimize chances of autoimmunity. This step is performed through BLASTp against human proteome as

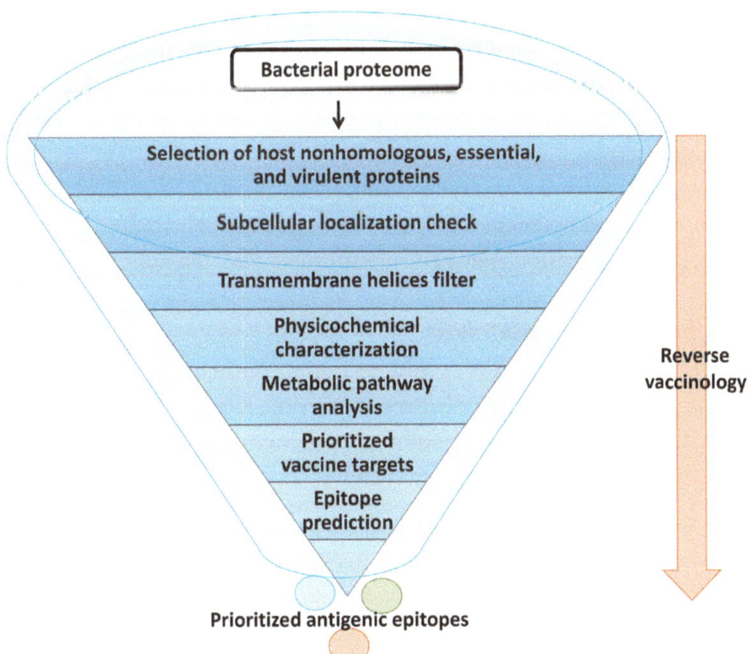

Fig. 1 Flow diagram representing filtration steps of reverse vaccinology.

explained earlier. Similarly, proteins are subjected to BLASTp against DEG for assessing essentiality [34]. Further, virulence factors are searched using MvirDB and VFDB [35,36]. Essentiality check is crucial for the identification of vaccine candidates as essential genes control major cellular functions of microbes, while virulence is also important to study pathogenesis, hence both parameters are applied for target prioritization [6]. These steps are common in both vaccine and drug target prioritization.

5.1.2 Subcellular localization check

The selected proteins from previous steps are subjected to subcellular localization similarly as discussed in pan-genomics section. This is important as vaccine candidates are likely to be exoproteome or secretome in nature [6].

5.1.3 Transmembrane helices filter

Proteins possessing more than one transmembrane helices are difficult to express and purify for vaccine development studies [63]. Hence, proteins are checked for the number of transmembrane helices using different online tools, that is, TMHMM version 2.0 and HMMTOP version 2.0 [64,65]. Consequently, the proteins containing at most one transmembrane helix are retained for further scrutinization.

5.1.4 Physicochemical characterization

Physicochemical features such as pI (isoelectric point), molecular weight, hydropathicity, and aliphatic index of selected proteins can be computed through ProtParam tool [66]. Proteins with a low molecular weight (<110 kDa) are considered good vaccine candidates as they can be easily purified for vaccine development [6]. The higher aliphatic index indicates the stability of proteins at several temperatures, while negative GRAVY values show that the protein is hydrophilic in nature and thus displays significant interactions with water molecules rendering it favorable for vaccine [67]. This step is checked for both vaccine and drug target prioritization.

5.1.5 Metabolic pathway analysis

Comparative pathway analysis is conducted on nonhuman homolog virulent and essential bacterial proteins from the previous step using KAAS [37]. This is a crucial step to identify proteins associated with cellular pathways unique to bacteria and absent in humans so that pathogen-specific pathways can be specifically targeted with minimum disruption to human pathways [38].

5.1.6 Epitope prediction

It is one of the main steps of the RV techniques, where prioritized targets are screened for antigenic determinants present on the surface, which can be easily detected by the antibody. Majorly, the immune system of the host is controlled by B- and T-cell lymphocytes by recognizing the epitopes and regulating cell immunity. Epitope prediction can be based on the sequence, structure of the antigen-antibody complex, or mimotope analysis where a macromolecule mimics the structure of the epitope. Various tools by today are now available to predict B- and T-cell epitopes among which most comprehensive epitope mapping algorithm is EpiMatrix. It has now also been aided with various computational tools to aid the conformational epitope prediction [68]. Other B- and T-cell prediction tools include ABCPred [69] and to check their affinity with MHC I and II alleles ProPred1 [70] and ProPred [71] are commonly used.

5.2 Tools available for RV

Many bioinformatics tools have been developed that adopt RV approach for the identification of potential vaccine candidates. These tools are available online as well as stand-alone and are used by researchers working on vaccine development. Some of them are discussed:

NERVE (New Enhanced Reverse Vaccinology Environment), a Perl-implemented pipeline for *the identification* of putative vaccine candidates, is a modular-based stand-alone software that generates results through text interface configuration [72]. But its main limitation is that it only focuses on adhesion proteins, whereas several nonadhesion proteins can also participate in host-pathogen interactions (such as flagellin, porins, invasin, etc.), and most of them are pathogenic as well as antigenic [73,74]. Vacceed, a Linux-based platform designed for in silico identification of vaccine candidates in eukaryotes [75]. Vacceed is able to reduce false vaccine candidates that are selected for laboratory validation, but this program provides inadequate information about functional annotation of prioritized targets and pathogenicity. Vaxign, an online RV tool, provides precomputed results of 400 genomes, however, only of 30 organisms (mostly bacterial species) [76]. The dynamic mode can be utilized to query proteins from other organisms; however, only up to 500 protein sequences are permitted which is not suitable to search vaccine candidates in bacteria as bacteria typically encode thousands of proteins. JennerPredict, a webserver performs vaccine candidate identification using bacterial proteomes [77]. This is by taking into account known functional domains from protein classes belonging to adhesins as well as nonadhesins. The server generated better results than NERVE, Vaxign, etc.

Web-based pipelines are easy to use; however, their drawbacks include time delays and limited input file size. A recently designed pipeline named VacSol addresses the issue and is a dynamic, flexible stand-alone pipeline fully functional on Ubuntu Linux environment and windows operating system [78]. It is a highly scalable and flexible software designed to predict potential vaccine candidates quickly and efficiently. VacSol performs the target identification as well as conducts epitope mapping in identified vaccine candidates. This tool is a valuable addition to the RV tools developed earlier. Majority of the attributes of potential vaccine candidates, that is, nonhomology filter, essentiality and virulence check, <2 transmembrane helices, exoproteome, and secretome localization, functional annotation along with epitope mapping can be performed easily through Vac-Sol [78]. Although the pipeline has been designed for vaccine candidate identification, with some modifications it can be used for drug target prioritization too, especially in the initial stages.

5.3 Currently available effective RV vaccines

The RV strategy exclusively filters out pathogenic determinants from the whole bacterial genome that can behave as putative vaccine candidates. Generally, surface-exposed proteins are chosen and later evaluated for their immunogenic potential. A significantly fruitful story of the development of subunit MenB vaccine resulted in the identification of a majority of vaccine candidates in merely 18 months and the number of identified candidates exceeds the total number of vaccine candidates identified via conventional vaccinology in the past 40 years combined together. The genome of MenB strain MC58 was screened by implementing RV and out of 2158 ORFs, 570 were computationally analyzed to be associated with exoproteome or secretome and showed a tendency to elicit an immune response. Further filtration was based on the capability of the antigen to be actively expressed in *E. coli* and later flow cytometry assay to confirm the status of the potential antigen is on the cell surface of the bacterial cell and bactericidal assay or passive immunization to ensure the cell-killing potential. These steps gradually narrowed down 570 candidates to 350 and then to 91 and eventually, 28 shortlisted candidates were checked for their conservation among all similar meningococcal strains. As a result of this process, the famous MenB vaccine has formed that move into phase III clinical trials in 2008. This vaccine is now commercially available under the name BEXSERO (Novartis Vaccines) [79]. Several other vaccines predicted by RV have been under consideration for animal testing and clinical trials against zika virus [80,81], *Helicobacter pylori* [6,82,83], *S. aureus* [84], Tuberculosis [85], and many more.

6 Limitations of RV and pan-genomics

Apart from the robustness of this technology, it is also associated with various limitations. First of all, the RV-based predicted vaccines consist of only proteins which are encoded by the genome. However, other antigenic factors such as carbohydrates, polysaccharide, lipid antigens, and other factors cannot be identified through this technique.

Moreover, the vaccine candidates predicted on the basis of pan-genome analysis and RV are mere predictions, which need to be validated in vivo. Also, when subjected to clinical analysis, these predictions might fail to evoke a proper immunogenic response. The reason being that as these vaccine candidates are predicted solely based on algorithms and tools designed to extract the best antigenic signatures based on the prior knowledge of vaccinology. Once these candidates are inside the living system, the holistic dynamics of the living cells are beyond human control. There are numerous factors at work at a time which might interfere with the immunogenic response of that vaccine.

The generation of a large amount of data through the latest sequencing and other robust technologies has been highly beneficial for vaccine prediction. However, this data are mostly heterogeneous and are needed to be efficiently integrated and harmonized together. For that, a number of analysis tools are currently available. Still, there is a limitation in terms of huge data analysis simultaneously which also poses a big challenge both in terms of algorithms and software development. This type of analysis requires distributed and parallel computing and will require rigorous efforts from the scientific community to develop such algorithms which are able to handle larger genomes and big data sets.

7 Future prospects

Development of new strategies to tackle various obstacles and challenges is at its peak in today's world. Apart from improving the available technology for the development of vaccines, there are some future aspects of vaccinology which should be pondered upon. The commercially available vaccines mostly depend on stimulating B cell responses and strong antibody repertoire production. There is a need to analyze these signatures of antibodies against a particular antigen to propose proper antigens. This will aid in the structural design and selection of new, novel, and effective vaccines in the future. There is a need to investigate effective alternate delivery systems for new vaccines. Most of the predicted vaccine antigens are mere epitopes, which require proper adjuvants and a delivery system. New approaches are being designed which will be able to increase the half-life, stability, and effectiveness of such vaccines. Apart from infectious diseases, it is the need of the hour that noninfectious chronic diseases are also subjected to new vaccinology approaches such as cancer. Another future prospect of the new approach is the "personalized vaccinology." As every individual is different and the response to a vaccine is also different because of the difference in immune signatures. The future holds a promise in which "systems vaccinology" will garner much attention in which knowledge from all the "-omics" technologies along with "vaccinomics" will be applied to design personalized vaccines. In which various formulations of the vaccine can be used depending on the individual variations.

References

[1] R.D. Fleischmann, M.D. Adams, O. White, et al., Whole-genome random sequencing and assembly of Haemophilus influenzae Rd, Science 269 (5223) (1995) 496–512.

[2] H. Tettelin, V. Masignani, M.J. Cieslewicz, et al., Genome analysis of multiple pathogenic isolates of Streptococcus agalactiae: implications for the microbial "pan-genome", Proc. Natl. Acad. Sci. U. S. A. 102 (39) (2005) 13950–13955.

[3] R. Rappuoli, Reverse vaccinology, Curr. Opin. Microbiol. 3 (5) (2000) 445–450.
[4] C.D. Rinaudo, J.L. Telford, R. Rappuoli, et al., Vaccinology in the genome era, J. Clin. Invest. 119 (9) (2009) 2515–2525.
[5] M. Pizza, V. Scarlato, V. Masignani, et al., Identification of vaccine candidates against serogroup B meningococcus by whole-genome sequencing, Science 287 (5459) (2000) 1816–1820.
[6] A. Naz, F.M. Awan, A. Obaid, et al., Identification of putative vaccine candidates against Helicobacter pylori exploiting exoproteome and secretome: a reverse vaccinology based approach, Infect. Genet. Evol. 32 (2015) 280–291.
[7] A. Hassan, A. Naz, A. Obaid, et al., Pangenome and immuno-proteomics analysis of Acinetobacter baumannii strains revealed the core peptide vaccine targets, BMC Genomics 17 (1) (2016) 732.
[8] M.-H. Chiang, W.-C. Sung, S.-P. Lien, et al., Identification of novel vaccine candidates against Acinetobacter baumannii using reverse vaccinology, Hum. Vaccin. Immunother. 11 (4) (2015) 1065–1073.
[9] A. Sette, R. Rappuoli, Reverse vaccinology: developing vaccines in the era of genomics, Immunity 33 (4) (2010) 530–541.
[10] A.M. Kanampalliwar, R. Soni, A. Girdhar, et al., Reverse vaccinology: basics and applications, J. Vaccines Vaccin. 4 (6) (2013) 194–198.
[11] D.R. Flower, et al., Computational vaccine design, in: D.R. Flower (Ed.), Drug Design: Cutting Edge Approaches, The Royal Society of Chemistry, 2002, , pp. 136–180.
[12] S. Bambini, R. Rappuoli, The use of genomics in microbial vaccine development, Drug Discov. Today 14 (5) (2009) 252–260.
[13] L. Rouli, V. Merhej, P.-E. Fournier, et al., The bacterial pangenome as a new tool for analysing pathogenic bacteria, New Microbes New Infect. 7 (2015) 72–85.
[14] L.A. Gallagher, S.A. Lee, C. Manoil, Importance of core genome functions for an extreme antibiotic resistance trait, MBio 8 (6) (2017).
[15] D.T. Hung, E.A. Shakhnovich, E. Pierson, et al., Small-molecule inhibitor of Vibrio cholerae virulence and intestinal colonization, Science 310 (5748) (2005) 670–674.
[16] M.M. Giuliani, J. Adu-Bobie, M. Comanducci, et al., A universal vaccine for serogroup B meningococcus, Proc. Natl. Acad. Sci. U. S. A. 103 (29) (2006) 10834–10839.
[17] A.O. Kislyuk, B. Haegeman, N.H. Bergman, et al., Genomic fluidity: an integrative view of gene diversity within microbial populations, BMC Genomics 12 (1) (2011) 32.
[18] D.G. Moriel, L. Tan, K.G. Goh, et al., A novel protective vaccine antigen from the core *Escherichia coli* genome, mSphere 1 (6) (2016).
[19] F.R. Fields, S.W. Lee, M.J. McConnell, Using bacterial genomes and essential genes for the development of new antibiotics, Biochem. Pharmacol. 134 (2017) 74–86.
[20] N. Segata, C. Huttenhower, Toward an efficient method of identifying core genes for evolutionary and functional microbial phylogenies, PLoS One 6 (9) (2011).
[21] L. Carlos Guimaraes, L. Benevides de Jesus, M. Vinicius Canario Viana, et al., Inside the pan-genome-methods and software overview, Curr. Genomics 16 (4) (2015) 245–252.
[22] D. Medini, C. Donati, H. Tettelin, et al., The microbial pan-genome, Curr. Opin. Genet. Dev. 15 (6) (2005) 589–594.
[23] P. Lapierre, J.P. Gogarten, Estimating the size of the bacterial pan-genome, Trends Genet. 25 (3) (2009) 107–110.
[24] J.G. Lawrence, H. Hendrickson, Genome evolution in bacteria: order beneath chaos, Curr. Opin. Microbiol. 8 (5) (2005) 572–578.
[25] A. Mira, A.B. Martín-Cuadrado, G. D'Auria, et al., The bacterial pan-genome: a new paradigm in microbiology, Int. Microbiol. 13 (2) (2010) 45–57.
[26] T.D. Read, D.W. Ussery, Opening the pan-genomics box, Curr. Opin. Microbiol. 9 (5) (2006) 496–498.
[27] D. Croll, B.A. McDonald, The accessory genome as a cradle for adaptive evolution in pathogens, PLoS Pathog. 8 (4) (2012).
[28] C.J. Grim, M.L. Kotewicz, K.A. Power, et al., Pan-genome analysis of the emerging foodborne pathogen Cronobacter spp. suggests a species-level bidirectional divergence driven by niche adaptation, BMC Genomics 14 (1) (2013) 366.

[29] A. Muzzi, V. Masignani, R. Rappuoli, The pan-genome: towards a knowledge-based discovery of novel targets for vaccines and antibacterials, Drug Discov. Today 12 (11–12) (2007) 429–439.
[30] I.K. Jordan, K.S. Makarova, J.L. Spouge, et al., Lineage-specific gene expansions in bacterial and archaeal genomes, Genome Res. 11 (4) (2001) 555–565.
[31] S.C. Soares, V.A. Abreu, R.T. Ramos, et al., PIPS: pathogenicity island prediction software, PLoS One 7 (2) (2012) e30848.
[32] K. Penn, C. Jenkins, M. Nett, et al., Genomic islands link secondary metabolism to functional adaptation in marine Actinobacteria, ISME J. 3 (10) (2009) 1193.
[33] Z. Khalid, S. Ahmad, S. Raza, et al., Subtractive proteomics revealed plausible drug candidates in the proteome of multi-drug resistant Corynebacterium diphtheriae, Meta Gene 17 (2018) 34–42.
[34] H. Luo, Y. Lin, F. Gao, et al., DEG 10, an update of the database of essential genes that includes both protein-coding genes and noncoding genomic elements, Nucleic Acids Res. 42 (D1) (2013) D574–D580.
[35] L. Chen, J. Yang, J. Yu, et al., VFDB: a reference database for bacterial virulence factors, Nucleic Acids Res. 33 (Suppl. 1) (2005) D325–D328.
[36] C. Zhou, J. Smith, M. Lam, et al., MvirDB—a microbial database of protein toxins, virulence factors and antibiotic resistance genes for bio-defence applications, Nucleic Acids Res. 35 (Suppl. 1) (2006) D391–D394.
[37] Y. Moriya, M. Itoh, S. Okuda, et al., KAAS: an automatic genome annotation and pathway reconstruction server, Nucleic Acids Res. 35 (Suppl. 2) (2007) W182–W185.
[38] V. Solanki, V. Tiwari, Subtractive proteomics to identify novel drug targets and reverse vaccinology for the development of chimeric vaccine against Acinetobacter baumannii, Sci. Rep. 8 (1) (2018) 9044.
[39] N.Y. Yu, J.R. Wagner, M.R. Laird, et al., PSORTb 3.0: improved protein subcellular localization prediction with refined localization subcategories and predictive capabilities for all prokaryotes, Bioinformatics 26 (13) (2010) 1608–1615.
[40] T.M. Bakheet, A.J. Doig, Properties and identification of antibiotic drug targets, BMC Bioinformatics 11 (1) (2010) 195.
[41] C.S. Yu, Y.C. Chen, C.H. Lu, et al., Prediction of protein subcellular localization, Proteins 64 (3) (2006) 643–651.
[42] S.J. Patel, L. Saiman, Antibiotic resistance in neonatal intensive care unit pathogens: mechanisms, clinical impact, and prevention including antibiotic stewardship, Clin. Perinatol. 37 (3) (2010) 547–563.
[43] B. Jia, A.R. Raphenya, B. Alcock, et al., CARD 2017: expansion and model-centric curation of the comprehensive antibiotic resistance database, Nucleic Acids Res. 45 (2016) D566–D573.
[44] B. Liu, M. Pop, ARDB—antibiotic resistance genes database, Nucleic Acids Res. 37 (Suppl. 1) (2008) D443–D447.
[45] D.S. Wishart, Y.D. Feunang, A.C. Guo, et al., DrugBank 5.0: a major update to the DrugBank database for 2018, Nucleic Acids Res. 46 (D1) (2017) D1074–D1082.
[46] P. Schmidtke, X.J.J.o.m.c. Barril, Understanding and predicting druggability. A high-throughput method for detection of drug binding sites, J. Med. Chem. 53 (15) (2010) 5858–5867.
[47] M. Mora, G. Bensi, S. Capo, et al., Group A Streptococcus produce pilus-like structures containing protective antigens and Lancefield T antigens, Proc. Natl. Acad. Sci. U. S. A. 102 (43) (2005) 15641–15646.
[48] M. Barocchi, J. Ries, X. Zogaj, et al., A pneumococcal pilus influences virulence and host inflammatory responses, Proc. Natl. Acad. Sci. U. S. A. 103 (8) (2006) 2857–2862.
[49] X. Deng, A.M. Phillippy, Z. Li, et al., Probing the pan-genome of Listeria monocytogenes: new insights into intraspecific niche expansion and genomic diversification, BMC Genomics 11 (1) (2010) 500.
[50] R. Uddin, M. Sufian, Core proteomic analysis of unique metabolic pathways of Salmonella enterica for the identification of potential drug targets, PLoS One 11 (1) (2016).
[51] X. Yang, Y. Li, J. Zang, et al., Analysis of pan-genome to identify the core genes and essential genes of Brucella spp, Mol. Gen. Genomics 291 (2) (2016) 905–912.
[52] M. Ibrahim, A. Subramanian, S. Anishetty, Comparative pan genome analysis of oral Prevotella species implicated in periodontitis, Funct. Integr. Genomics 17 (5) (2017) 513–536.

[53] R. Uddin, F. Jamil, Prioritization of potential drug targets against P. aeruginosa by core proteomic analysis using computational subtractive genomics and protein-protein interaction network, Comput. Biol. Chem. 74 (2018) 115–122.
[54] J. Blom, S.P. Albaum, D. Doppmeier, et al., EDGAR: a software framework for the comparative analysis of prokaryotic genomes, BMC Bioinformatics 10 (1) (2009) 154.
[55] M.J. Brittnacher, C. Fong, H. Hayden, et al., PGAT: a multistrain analysis resource for microbial genomes, Bioinformatics 27 (17) (2011) 2429–2430.
[56] Y. Zhao, J. Wu, J. Yang, et al., PGAP: pan-genomes analysis pipeline, Bioinformatics 28 (3) (2011) 416–418.
[57] P. Leekitcharoenphon, O. Lukjancenko, C. Friis, et al., Genomic variation in *Salmonella enterica* core genes for epidemiological typing, BMC Bioinformatics 13 (1) (2012) 88.
[58] O. Lukjancenko, D.W. Ussery, T.M.J.M.e. Wassenaar, Comparative genomics of Bifidobacterium, Lactobacillus and related probiotic genera, Microb. Ecol. 63 (3) (2012) 651–673.
[59] A.J. Page, C.A. Cummins, M. Hunt, et al., Roary: rapid large-scale prokaryote pan genome analysis, Bioinformatics 31 (22) (2015) 3691–3693.
[60] T.J.B. Seemann, Prokka: rapid prokaryotic genome annotation, Bioinformatics 30 (14) (2014) 2068–2069.
[61] Y. Huang, B. Niu, Y. Gao, et al., CD-HIT Suite: a web server for clustering and comparing biological sequences, Bioinformatics 26 (5) (2010) 680–682.
[62] A.J. Enright, S. Van Dongen, C.A. Ouzounis, An efficient algorithm for large-scale detection of protein families, Nucleic Acids Res. 30 (7) (2002) 1575–1584.
[63] Z. Ni, Y. Chen, E. Ong, et al., Antibiotic resistance determinant-focused Acinetobacter baumannii vaccine designed using reverse vaccinology, Int. J. Mol. Sci. 18 (2) (2017) 458.
[64] A. Krogh, B. Larsson, G. Von Heijne, et al., Predicting transmembrane protein topology with a hidden Markov model: application to complete genomes, J. Mol. Biol. 305 (3) (2001) 567–580.
[65] G.E. Tusnady, I.J.B. Simon, The HMMTOP transmembrane topology prediction server, Bioinformatics 17 (9) (2001) 849–850.
[66] E. Gasteiger, C. Hoogland, A. Gattiker, et al., Protein identification and analysis tools on the ExPASy server, in: The Proteomics Protocols Handbook, Springer, 2005, , pp. 571–607.
[67] D. Droppa-Almeida, E. Franceschi, F.F.J.B. Padilha, et al., Immune-informatic analysis and design of peptide vaccine from multi-epitopes against Corynebacterium pseudotuberculosis, Bioinform. Biol. Insights 12 (2018).
[68] D. Sirskyj, F. Diaz-Mitoma, A. Golshani, et al., Innovative bioinformatic approaches for developing peptide-based vaccines against hypervariable viruses, Immunol. Cell Biol. 89 (1) (2011) 81–89.
[69] S. Saha, G. Raghava, Prediction of continuous B-cell epitopes in an antigen using recurrent neural network, Proteins 65 (1) (2006) 40–48.
[70] H. Singh, G. Raghava, ProPred1: prediction of promiscuous MHC class-I binding sites, Bioinformatics 19 (8) (2003) 1009–1014.
[71] H. Singh, G. Raghava, ProPred: prediction of HLA-DR binding sites, Bioinformatics 17 (12) (2001) 1236–1237.
[72] S. Vivona, F. Bernante, F. Filippini, NERVE: new enhanced reverse vaccinology environment, BMC Biotechnol. 6 (1) (2006) 35.
[73] F. Dorner, J.L. McDonel, Bacterial toxin vaccines, Vaccine 3 (2) (1985) 94–102.
[74] T.M. Wizemann, J.E. Adamou, S. Langermann, Adhesins as targets for vaccine development, Emerg. Infect. Dis. 5 (3) (1999) 395.
[75] S.J. Goodswen, P.J. Kennedy, J.T.J.B. Ellis, Vacceed: a high-throughput in silico vaccine candidate discovery pipeline for eukaryotic pathogens based on reverse vaccinology, Bioinformatics 30 (16) (2014) 2381–2383.
[76] Y. He, Z. Xiang, H.L. Mobley, Vaxign: the first web-based vaccine design program for reverse vaccinology and applications for vaccine development, J. Biomed. Biotechnol. 2010 (2010).
[77] V. Jaiswal, S.K. Chanumolu, A. Gupta, et al., Jenner-predict server: prediction of protein vaccine candidates (PVCs) in bacteria based on host-pathogen interactions, BMC Bioinformatics 14 (1) (2013) 211.
[78] M. Rizwan, A. Naz, J. Ahmad, et al., VacSol: a high throughput in silico pipeline to predict potential therapeutic targets in prokaryotic pathogens using subtractive reverse vaccinology, BMC Bioinformatics 18 (1) (2017) 106.

[79] D. Serruto, R. Rappuoli, Post-genomic vaccine development, FEBS Lett. 580 (12) (2006) 2985–2992.
[80] A. Alam, S. Ali, S. Ahamad, et al., From ZikV genome to vaccine: in silico approach for the epitope-based peptide vaccine against Zika virus envelope glycoprotein, Immunology 149 (4) (2016) 386–399.
[81] H. Dar, T. Zaheer, M.T. Rehman, et al., Prediction of promiscuous T-cell epitopes in the Zika virus polyprotein: an in silico approach, Asian Pac. J. Trop. Med. 9 (9) (2016) 844–850.
[82] W.-Y. Zhou, Y. Shi, C. Wu, et al., Therapeutic efficacy of a multi-epitope vaccine against Helicobacter pylori infection in BALB/c mice model, Vaccine 27 (36) (2009) 5013–5019.
[83] L. Guo, R. Yin, K. Liu, et al., Immunological features and efficacy of a multi-epitope vaccine CTB-UE against H. pylori in BALB/c mice model, Appl. Microbiol. Biotechnol. 98 (8) (2014) 3495–3507.
[84] N. Hajighahramani, N. Nezafat, M. Eslami, et al., Immunoinformatics analysis and in silico designing of a novel multi-epitope peptide vaccine against Staphylococcus aureus, Infect. Genet. Evol. 48 (2017) 83–94.
[85] N. Chatterjee, R. Ojha, N. Khatoon, et al., Scrutinizing Mycobacterium tuberculosis membrane and secretory proteins to formulate multiepitope subunit vaccine against pulmonary tuberculosis by utilizing immunoinformatic approaches, Int. J. Biol. Macromol. 118 (2018) 180–188.

Further reading

[86] F. Sigaux, Cancer genome or the development of molecular portraits of tumors, Bull. Acad. Natl. Med. 184 (7) (2000) 1441–1447 (discussion 1448–1449).

CHAPTER 17

Pan-metagenomics: An overview of the human microbiome

Madangchanok Imchen, Ravali Krishna Vennapu, Ranjith Kumavath
Department of Genomic Science, School of Biological Sciences, Central University of Kerala, Kasaragod, India

17.1 Introduction

Metagenomics is the study of the microbial community without culturing the microbes residing in the given sample. Initially, metagenomic sequence were cloned and the inserts were sequenced with Sanger sequencing technology [1]. However, with the advent of next-generation sequencing (NGS) and the drop in cost per gigabyte (Gb) data has made its application very common. Hence, metagenomic studies have generated a huge amount of data, which have found applications in many fields such as in industry, pharmaceuticals, ecology, agriculture, etc. Pan-metagenome is the collective study of all or several metagenomes from all possible units belonging to a particular type of ecosystem or host. The pan-microbiome composition can be classified into core, accessory, and unique microbiome (Fig. 17.1.). The core microbiome would include only those microbiotas that are common to all sets of samples, hence, presumably essential to maintain the community structure irrespective of geographical locations or other biogeochemical parameters. On the other hand, the accessory microbiome would shine light on the microbiomes that are essential in two or more sets of sample. This could also provide information regarding the similarity of biogeochemical properties between the samples. Finally, the unique microbiome would be those that are present only in a particular sample. This subdivision of pan-microbiome would allow us to narrow down the huge data that are specific or common between samples. Similar to pan-genome studies, a pan-metagenome could be classified into either closed or open type [2, 3]. A closed-type pan-metagenome would have a limited amount of associated microbiota. On the other hand, an open-type pan-metagenome would have an unlimited (infinite) amount of associated microbiota. Therefore, one of the factors of difficulty in obtaining a complete set of pan-microbiota would be dependent on the closed or open type of pan-metagenome. Pan-microbiome studies have huge potential in health care, industry, ecology, and animal husbandry. For instance, according to a large-scale study by Xue et al. [4], the lactation performance of cows is dictated to some extent by the rumen pan- and core-microbiome. The authors further detail that lactate-producing bacterial

Pan-genomics: Applications, Challenges, and Future Prospects
https://doi.org/10.1016/B978-0-12-817076-2.00017-2
© 2020 Elsevier Inc.
All rights reserved.

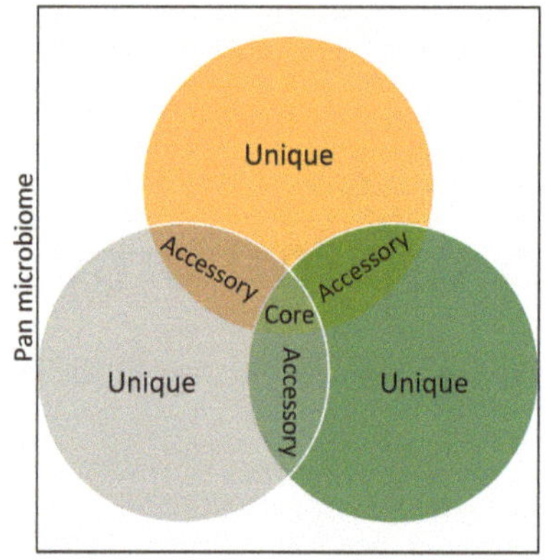

Fig. 17.1 Components of the pan-metagenome visualized through a Venn diagram would provide the core, accessory, and unique microbiome among the datasets.

genera such as *Sharpea*, *Bifidobacterium*, and *Lactobacillus* could alter the rumen pH and have significant role in metabolic dysfunction [5].

17.2 Gut pan-metagenome

The human gut microbiome holds the highest amount of microbial biomass accounting for about 10^{11} microbial cells per ml [6]. Gut microbiome studies are mainly focused on the large intestine, which is accessed through the stool as a proxy [6, 7]. It represents a wide composition of microbes that varies from one to another individual. This large variation is a major hurdle in cataloguing the core gut microbiome. Gut microbiome from industrialized and nonindustrialized samples have been seen to vary distinctly with the former having about 30% fewer species than the later [8–10], leading to the "disappearing human microbiota" hypothesis [6, 11]. Core microbiome is the set of OTUs, which are present in all individual. Hence, an understanding of the pan-microbiome would be a prior requirement before coming up with core gut microbiome. A large-scale study of 2243 metagenomic 16S rRNA human stool samples by de Cárcer [12] revealed that the core groups were dominated by members of phyla Firmicutes and Bacteroidetes. The study also found that the human gut pan-microbiome is constituted with discrete units present in all samples. However, the study noted that the composition of core units varied according to the prevalence threshold. The human gut pan-microbiome could also be influenced by the individual's social behavior. A study on the pan-microbiome on chimpanzee by Moeller et al. [13] shows that the gut microbiome could exhibit a vertical inheritance from mother to offspring. However, the authors also noted that after several generations the social behavior of the chimpanzees under the study had a higher significant effect on horizontal transfer of microbiota rather than the vertical inheritance. For instance, the variability between the individuals was negatively correlated to the individual's social life while species richness was positively correlated. This suggested that the pan microbiota from social interactions enriches the microbiota of an individual through dispersal of symbionts and provides a room for selection of species richness. Similarly, pan-microbiomes of ponies were also found to be linked to social interactions [14]. Mammals have been shown to harbor distinct gut microbiome based on the host species [15]; yet, it can be anticipated that the social behavior could play a significant part in altering human gut microbiota.

17.3 Pan-microbiome and its built environment

The pan-microbiome of build environments provides a strong foundation to the health of the occupants. Hanski and his research group [16] established that atopic patients had lower diversity of microbiome surrounding their home (build environment) compared to the healthy counterpart. The authors also found that the build environment pan-microbiome of healthy subjects have higher abundance of Gram-negative Gammaproteobacteria genus *Acinetobacter* on the skin. This genus is linked to the expression of anti-inflammatory cytokine IL-10 in peripheral blood mononuclear cells (PBMCs). Overall, the atopic patients had a lower association with the microbial community. Such effects of environmental pan-microbiome has clinical significance as build environments pan-microbiome studies at a global scale could link us with the clues to variations and correlations on the distribution and prevalence of autoimmune diseases [17, 18]. The built environment pan-microbiome is affected by several factors such as the outdoor environment (landscape, climate, nearby build environments, etc.), occupants and households [19]. Hence, the study of the surrounding environment would be vital to build up a strong pan-microbiome of any environment. A well-studied example of environmental influence on the BinhPhuoc Anopheles microbiota was recently reported by Ngo et al. [20]; several bacterial genera were found in the *Anopheles*, which were not yet described in *Anopheles* pan-microbiota suggesting that the microbiota would be dependent on the sampling locations rather than host. The northeast Indian female *Aedes aegypti* and *Aedes albopictus* showed the presence of similar microbiota between both species [21] using a traditional cultivable technique. Similarly, several mosquito species from the United States were dominated by a few selected bacterial groups [22]. The presence of such common microbiota regionally in the pan-microbiota of interspecies indicates the influence of environmental factors [23]. However, the interindividual variations were hypothesized to have risen from difference in food [24], and infection status [25]. In addition, the core sets of common microbiota were found in different species regardless of development stages further leading to the environmental acquisition hypothesis [23]. Pan-microbiotas of mosquito have a huge diversity but low richness per specimen (*Anopheles* mosquito), which they concluded that the microbiotas of *Anopheles* were linked to the sampling location rather than the host [20, 26]. In another study, the human skin microbiome samples from the same individual have a higher similarity compared to samples from other individuals in spite of cohabiting in the same build environment [27]. However, the authors also noted that the individuals from sample households had higher similar microbiome compared to individuals from other households. Such studies indicate the importance of understanding our surrounding pan-microbiome.

17.4 Pan-microbiome in pharmacokinetic studies

Pan-genomes have been fairly progressive in several microbial studies. However, the pan-metagenome is relatively less explored both in human and environmental aspects [28]. Core and pan metagenome has a promising potential in the development of herd as well as personalized medicine [28, 29]. An interesting example of the role of the human microbiome in pharmacokinetics was validated by Haiser et al. [30] revealing that the human cardiac drug digoxin is metabolized by the gut bacterium *Eggerthella lenta* belonging to the Actinobacterium phylum. They found that arginine downregulates the expression of the *cgr* operon, which is essential for drug inactivation. However, when the *Eggerthella lenta* was cocultured with the fecal microbiome, abundance of the bacterium as well as the drug inactivation rate was increased significantly. They hypothesized that the mixed community could enhance *Eggerthella lenta* growth through the release of several growth-promoting factors and also by competing for arginine. Thus, the role of the microbial community and the diet plays a significant part in pharmacokinetics. Another example of microbiome-linked complications is *Clostridium difficile* infections (CDI), which are most commonly acquired from healthcare units. They are mostly treated by vancomycin as first-line treatment with high efficacy; however, they are highly recurring (up to 65%) once the antibiotic treatment is stopped leading to chronic complication and frequent hospitalization, even to death [31]. An antibiotic treatment for CDI affects the whole microbial community. However, the spore-forming nature of *Clostridium difficile* renders it unaffected, which leads to the high recurring rates. Fecal microbiota transplantation (FMT) has been very successful in the treatment of CDI with a single treatment by FMT [32]. This works by correcting the imbalance in gut microbiota by reintroducing the healthy donor fecal samples to restore the microbiome community. The interplay between the pharmacokinetics of a drug and the microbiome of the host is essential to understand in order to provide the highest efficacy of the drug. However, a variation in the microbiota of the host is a major hurdle. For instance, palms microbiota tend to differ based on location and racial background [27]. The microbiome of an individual has also been linked to the genetic makeup of the host [33]. The uniqueness of microbiome at the individual level indicates the significance of cataloguing a pan-microbiome map, which could have subsequent applications in pharmaceuticals (Fig. 17.2) [34].

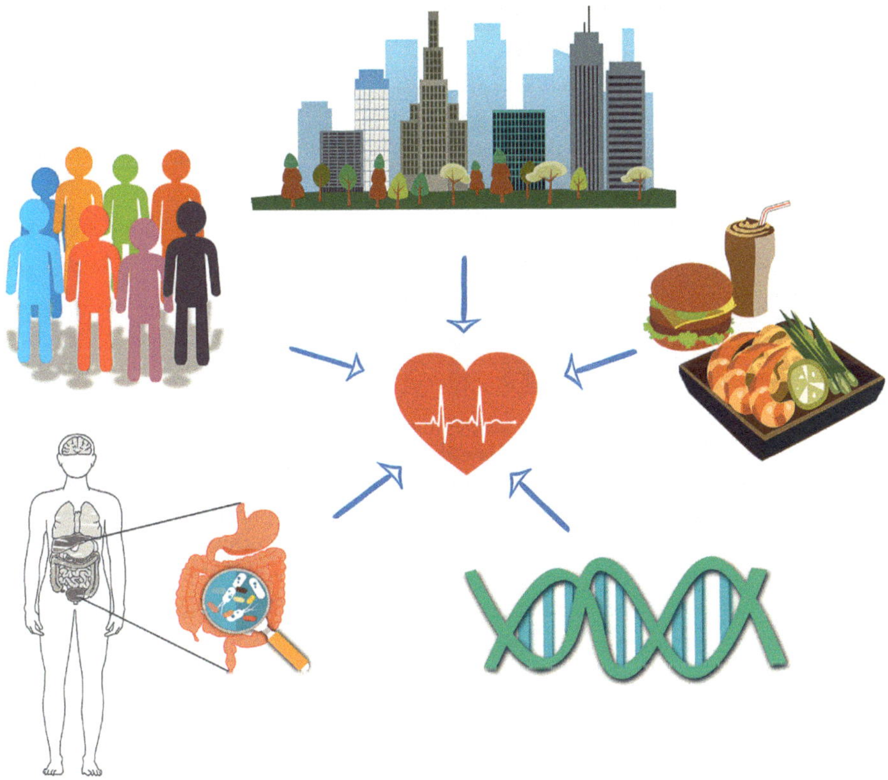

Fig. 17.2 Factors that shapes microbiome: several factors are involved in shaping the human pan-microbiome such as the host genetic makeups, food, build environments, social life, etc. Understanding the pan-microbiome is essential to develop diagnosis measure and efficient drugs. *The images used are under Creative Commons CC0 license from https://pixabay.com*

17.5 The large-scale microbiome projects

The NIH Human Microbiome Project (HMP, https://hmpdacc.org; [35]) established in 2008 and ended in 2013 have focused on several body parts such as the nasal passages, oral cavity, skin, gastrointestinal tract, and urogenital tract from 300 healthy individuals generating over 14.23 terabytes of data. The main aim of the HMP is to understand how the microbiome impacts our health and disease. Similarity, the expanded Human Oral Microbiome Database (eHOMD, http://www.ehomd.org; [36]) provides curated information on the microbial community present in the human aerodigestive tract, upper digestive tract, lower respiratory tract oral cavity, pharynx, nasal passages, sinuses, and esophagus. The project also houses genomes of 475 taxas. Recently, the American Gut Project (AGP; http://www.americangut.org) was initiated as a citizen scientist-driven project where the samples are deposited by the crowd. It is also one of the largest crowd-sourced project in the United States. According to its website, it has more samples

Fig. 17.3 Large-scale microbiome projects: Several microbiome projects have been initiated targeting human microbiome as well as the built environment microbiota. Projects such as the Earth Microbiome project comprises hundreds of collaborators while the home microbiome project and the American Gut Project (AGP) are crowd sourced. Such helping hands strengthen the sampling diversity and amount, which ultimately leads to a wider and deeper view of the microbiome.

than other microbiome projects such as the HMP (Fig. 17.3). The project is not only aimed at the gut microbiome but also skin and oral sites of the human body. The project has led to interesting findings such as the diversity of the gut microbiome is proportional to varieties of plants consumed and age of the host but indirectly proportional to antibiotic usage.

In addition to human microbiome projects, there are natural and build environment microbiome projects such as the Home Microbiome Project (http://homemicrobiome.com), which aim to uncover how microbiome shapes in a new home. The project aims to collect samples when a resident shifts to a new home and monitor for 6 weeks. This would provide the answer to whether the human microbiome shapes the build environment or vice versa. Similarly, the build environment Subways and Urban Biomes are relatively less explored. The MetaSUB (http://metasub.org) project aims to map the microbiome of cities around the globe specially targeting the public transport system, which is prone to maximum interaction by millions of different individuals on a daily basis. There are also projects dedicated to diverse environments such as The Earth Microbiome Project ([37]; EMP, http://earthmicrobiome.org), which was founded in 2010. The project has aimed to achieve 200,000 samples using various omics technologies with the aim to come up with a global "Gene Atlas" (Fig. 17.3).

17.6 Conclusion

Besides the human gut microbiome, little attention has been focused on other species and the environment [6]. A pan-metagenome study at a global scale is of immediate need to build a strong platform on the abundance and distribution of both harmful and beneficial microbes. In addition, studies on the build environment and its surrounding habitat including food and water would help us to elucidate a strong hint on the evolutionary history between microbiota and the host through the pan-metagenome [38]. On the other hand, pan-metagenome could reveal common adaptations to the environmental factors as well as co-evolution of interactions and genetic variations.

References

[1] J. Handelsman, Metagenomics: application of genomics to uncultured microorganisms, Microbiol. Mol. Biol. Rev. 68 (4) (2004) 669–685.

[2] T. Lefebure, P.D. Pavinski Bitar, H. Suzuki, M.J. Stanhope, Evolutionary dynamics of complete campylobacter pan-genomes and the bacterial species concept, Genome Biol. Evol. 2 (2010) 646–655.

[3] L. Rouli, V. Merhej, P.E. Fournier, D. Raoult, The bacterial pangenome as a new tool for analysing pathogenic bacteria, New Microbes New Infections 7 (2015) 72–85.

[4] M. Xue, H. Sun, X. Wu, J. Liu, Assessment of rumen microbiota from a large dairy cattle cohort reveals the pan and core bacteriomes contributing to varied phenotypes, Appl. Environ. Microbiol. 84 (19) (2018) e00970-18.

[5] T.G. Nagaraja, C.J. Newbold, C.J. Van Nevel, D.I. Demeyer, Manipulation of ruminal fermentation, in: The Rumen Microbial Ecosystem, Springer, Dordrecht, 1997, pp. 523–632.

[6] E.R. Davenport, J.G. Sanders, S.J. Song, K.R. Amato, A.G. Clark, R. Knight, The human microbiome in evolution, BMC Biol. 15 (1) (2017) 127.

[7] J. Walter, R. Ley, The human gut microbiome: ecology and recent evolutionary changes, Annu. Rev. Microbiol. 65 (2011) 411–429.

[8] J.C. Clemente, E.C. Pehrsson, M.J. Blaser, K. Sandhu, Z. Gao, B. Wang, O. Lander, The microbiome of uncontacted Amerindians, Sci. Adv. 1 (2015).

[9] I. Martínez, J.C. Stegen, M.X. Maldonado-Gómez, A.M. Eren, P.M. Siba, A.R. Greenhill, J. Walter, The gut microbiota of rural Papua new guineans: composition, diversity patterns, and ecological processes, Cell Rep. 11 (4) (2015) 527–538.

[10] T. Yatsunenko, F.E. Rey, M.J. Manary, I. Trehan, M.G. Dominguez-Bello, M. Contreras, A.C. Heath, Human gut microbiome viewed across age and geography, Nature 486 (7402) (2012) 222.

[11] M.J. Blaser, S. Falkow, What are the consequences of the disappearing human microbiota? Nat. Rev. Microbiol. 7 (12) (2009) 887.

[12] D.A. de Cárcer, The human gut pan-microbiome presents a compositional core formed by discrete phylogenetic units, Sci. Rep. 8 (1) (2018).

[13] A.H. Moeller, S. Foerster, M.L. Wilson, A.E. Pusey, B.H. Hahn, H. Ochman, Social behavior shapes the chimpanzee pan-microbiome, Sci. Adv. 2 (1) (2016).

[14] R.E. Antwis, J.M. Lea, B. Unwin, S. Shultz, Gut microbiome composition is associated with spatial structuring and social interactions in semi-feral Welsh Mountain ponies, Microbiome 6 (1) (2018) 207.

[15] R.E. Ley, M. Hamady, C. Lozupone, P.J. Turnbaugh, R.R. Ramey, J.S. Bircher, J.I. Gordon, Evolution of mammals and their gut microbes, Science 320 (5883) (2008) 1647–1651.

[16] I. Hanski, L. von Hertzen, N. Fyhrquist, K. Koskinen, K. Torppa, T. Laatikainen, E. Vartiainen, Environmental biodiversity, human microbiota, and allergy are interrelated, Proc. Natl. Acad. Sci. 109 (21) (2012) 8334–8339.

[17] M.H. Leung, P.K. Lee, The roles of the outdoors and occupants in contributing to a potential pan-microbiome of the built environment: a review, Microbiome 4 (1) (2016) 21.

[18] P.M. Salo, S.J. Arbes Jr., M. Sever, R. Jaramillo, R.D. Cohn, S.J. London, D.C. Zeldin, Exposure to Alternariaalternata in US homes is associated with asthma symptoms, J. Allergy Clin. Immunol. 118 (4) (2006) 892–898.

[19] M. Täubel, H. Rintala, M. Pitkäranta, L. Paulin, S. Laitinen, J. Pekkanen, A. Nevalainen, The occupant as a source of house dust bacteria, J. Allergy Clin. Immunol. 124 (4) (2009) 834–840.

[20] C.T. Ngo, S. Romano-Bertrand, S. Manguin, E. Jumas-Bilak, Diversity of the bacterial microbiota of Anopheles mosquitoes from BinhPhuoc Province, Vietnam, Front. Microbiol. 7 (2016) 2095.
[21] K.K. Yadav, A. Bora, S. Datta, K. Chandel, H.K. Gogoi, G.B.K.S. Prasad, V. Veer, Molecular characterization of midgut microbiota of *Aedesalbopictus* and *Aedesaegypti* from Arunachal Pradesh, India, Parasite Vectors 8 (1) (2015) 641.
[22] E.J. Muturi, J.L. Ramirez, A.P. Rooney, C.H. Kim, Comparative analysis of gut microbiota of mosquito communities in Central Illinois, PLoS Negl. Trop. Dis. 11 (2) (2017).
[23] M. Guégan, K. Zouache, C. Démichel, G. Minard, P. Potier, P. Mavingui, C.V. Moro, The mosquito holobiont: fresh insight into mosquito-microbiota interactions, Microbiome 6 (1) (2018) 49.
[24] G. Gimonneau, M.T. Tchioffo, L. Abate, A. Boissière, P.H. Awono-Ambéné, S.E. Nsango, I. Morlais, Composition *of Anopheles coluzzii* and *Anopheles gambiae* microbiota from larval to adult stages, Infect. Genet. Evol. 28 (2014) 715–724.
[25] K. Zouache, R.J. Michelland, A.B. Failloux, G.L. Grundmann, P. Mavingui, Chikungunya virus impacts the diversity of symbiotic bacteria in mosquito vector, Mol. Ecol. 21 (9) (2012) 2297–2309.
[26] J. Osei-Poku, C.M. Mbogo, W.J. Palmer, F.M. Jiggins, Deep sequencing reveals extensive variation in the gut microbiota of wild mosquitoes from Kenya. Mol. Ecol. 21 (2012) 5138–5150, https://doi.org/10.1111/j.1365-294X.2012.05759.x.
[27] M.H. Leung, D. Wilkins, P.K. Lee, Insights into the pan-microbiome: skin microbial communities of Chinese individuals differ from other racial groups, Sci. Rep. 5 (2015).
[28] A. Shafquat, R. Joice, S.L. Simmons, C. Huttenhower, Functional and phylogenetic assembly of microbial communities in the human microbiome, Trends Microbiol. 22 (5) (2014) 261–266.
[29] M. Imchen, R. Kumavath, Metagenomics of antimicrobial resistance in gut microbiome. in: Metagenomics for Gut Microbes, 2018 https://doi.org/10.5772/intechopen.76214.
[30] H.J. Haiser, D.B. Gootenberg, K. Chatman, G. Sirasani, E.P. Balskus, P.J. Turnbaugh, Predicting and manipulating cardiac drug inactivation by the human gut bacterium *Eggerthellalenta*, Science 341 (6143) (2013) 295–298.
[31] J.S. Bakken, T. Borody, L.J. Brandt, J.V. Brill, D.C. Demarco, M.A. Franzos, T.A. Moore, Treating Clostridium difficile infection with fecal microbiota transplantation, Clin. Gastroenterol. Hepatol. 9 (12) (2011) 1044–1049.
[32] K. Rao, V.B. Young, Fecal microbiota transplantation for the management of Clostridium difficile infection, Infect. Dis. Clin. 29 (1) (2015) 109–122.
[33] J.K. Goodrich, J.L. Waters, A.C. Poole, J.L. Sutter, O. Koren, R. Blekhman, T.D. Spector, Human genetics shape the gut microbiome, Cell 159 (4) (2014) 789–799.
[34] T. Johnson, B. Gómez, M. McIntyre, M. Dubick, R. Christy, S. Nicholson, D. Burmeister, The cutaneous microbiome and wounds: new molecular targets to promote wound healing, Int. J. Mol. Sci. 19 (9) (2018) 2699.
[35] J. Peterson, S. Garges, M. Giovanni, P. McInnes, L. Wang, J.A. Schloss, V. Bonazzi, et al., The NIH human microbiome project, Genome Res. 19 (12) (2009) 2317–2323.
[36] I.F. Escapa, T. Chen, Y. Huang, P. Gajare, F.E. Dewhirst, K.P. Lemon, New insights into human nostril microbiome from the expanded Human Oral Microbiome Database (eHOMD): a resource for species-level identification of microbiome data from the aerodigestive tract, bioRxiv (2018).
[37] L.R. Thompson, J.G. Sanders, D. McDonald, A. Amir, J. Ladau, K.J. Locey, R.J. Prill, et al., A communal catalogue reveals Earth's multiscale microbial diversity, Nature 551 (7681) (2017).
[38] A.H. Moeller, Y. Li, E.M. Ngole, S. Ahuka-Mundeke, E.V. Lonsdorf, A.E. Pusey, H. Ochman, Rapid changes in the gut microbiome during human evolution, Proc. Natl. Acad. Sci. USA 111 (46) (2014) 16431–16435.

CHAPTER 18

Pan-transcriptomics and its applications

Attya Bhatti, Faisal Sheraz Shah, Jahanzaib Azhar, Shahbaz Ahmad, Peter John
Atta-ur-Rahman School of Applied Biosciences, National University of Sciences and Technology, Islamabad, Pakistan

1 Introduction

The year 1995 marked the initiation of the genomic era with the sequencing of the first ever genome of a microbe followed by genome sequencing of many other organisms and ultimately to the sequencing of human genome by 2001. During the most recent years, the substantial increase in genome sequencing has changed our thoughts regarding microorganism diversity. From a single gene to entire community DNA studies, we have discovered that nature's largest gene repository resides in bacteria. Omics has become a routine word in molecular biology research. Genomics has provided us the prodigious wealth of knowledge about genes and their functional product. Likewise, investigations of single-nucleotide polymorphisms (SNPs) have revealed the genetic diversity among the species. Moreover, genome-wide association studies (GWAS) provides important information regarding genotype-phenotype correlation. However, our understanding about genome functionality is still limited. The functionality of the genome is not only dependent on protein-coding genes but also on the noncoding RNAs which are involved in the regulation of gene expression. From this point of investigation about genome functionality, the science of transcriptomics emerged. Therefore, transcriptomics can be described as the molecular biology of gene expression on an extensive scale or it can also be described as the study of all the RNA transcripts available in a cell or tissue at a given time.

Everything relatable to RNAs comes under the domain of transcriptomics including the transcription, expression, functions, location, developmental stage, and degradation of RNAs. Other aspects that are explored in transcriptomics include the structures of the transcripts along with their splicing patterns and other post-transcriptional modifications [1]. In transcriptomics, all the different types of RNAs are studied such as messenger RNAs, microRNAs, and other noncoding RNAs [2]. Useful information regarding gene regulation is obtained when the expression profiling of an organism's genome is carried out at its different developmental stages and in the case of eukaryotes in its many different tissues at varying conditions. An organism's biology can be understood in detail through transcriptomics as it can help in identifying the functions of identified but un-annotated genes. This has also enabled us to understand the changing patterns of gene expression

Pan-genomics: Applications, Challenges, and Future Prospects
https://doi.org/10.1016/B978-0-12-817076-2.00018-4
© 2020 Elsevier Inc.
All rights reserved.

between the different organisms whether prokaryotes or eukaryotes along with an understanding of human diseases. The most important aspect that transcriptomics cover is the identification of broad coordinated trends through analyzing varying gene expression, which is not possible otherwise [3].

Every cell has a fate, a developmental plan and in case of disease it has a proper pathway for its progression. In order to understand the underlying pathways for these, transcriptomic studies provide a base to form associations between the genotypes and their phenotypic response [4]. In disease diagnosis and profiling of eukaryotes, RNA-Seq has helped in the identification of various important sites that play their role in disease progression such as transcription start sites, splicing sites, and alternative promoter sites among others. Apart from these other factors that contribute in disease onset and progression such as SNPs, fused genes, and allele specific expressions have also been categorized to get a better understanding of variations and their role in disease. However, prokaryote gene expression studies, which are also a part of transcriptome analysis, have helped greatly in studying the pathogens and their expression profiling has helped to understand the role of specific genes in pathogen-mediated disease. This also helps in determining the drug targets to develop optimized infection control strategies and targeted treatments. Apart from all that, whenever a new sequence is discovered its function is also determined by transcriptomic studies; so, functional annotation is another aspect covered by transcriptomics [3].

The past few years have brought some technological advances such as high-throughput genomic studies across various species and that enabled researchers to use the information obtained for healthcare and agricultural benefits [5]. These newly developed technologies have provided an enormous amount of data in the study areas of genomics and transcriptomics of a variety of organisms, which was previously limited to only some of them [6]. As a result, the transcriptomic studies have scaled up to the pan-transcriptomic level which allowed researchers to do comparative analyses of genomes among the species. One of the approaches toward the identification of novel transcripts in comparison to reference genomes and building a pan-transcriptome, RNA sequencing has become quite successful so far because of its efficiency to remove redundant sequences, robustness and cost-effectiveness. Although, whole-genome sequencing was the suitable option to study microbial genomics, but with complex genomes employing a pan-transcriptomic approach will enable us to understand the role of differential expression of genes within the species. The data obtained from this approach can be integrated with other datasets for a better understanding of complex organisms. Diversity between the species and in a single specie due to structural variation lead to a conclusion that current reference transcriptome is incomplete to represent the total gene repertoire of the species and virulence factor of the microorganisms [7]. Due to large variation in gene expression, there is much phenotypic diversity and as day-by-day sequencing is becoming easier we are surprised with sequence variation of complex genomes with reference genomes and transcriptomes. Therefore, building a pan-transcriptome is necessary for studying and analyzing complex genomes and variation between species.

The pan-transcriptome can be defined by recalling the concept of the pan-genome. It reflects the set of all the RNA molecules present in a specie or in an organism containing core, accessory, and unique transcriptomes (specific to one specie or organism). Building a pan-transcriptome will lead us to study genome organization and genetic association of different species. Systematically, the functional consecution and regulating mechanism of pan-transcriptomics is undetermined yet. Phenotypic evolution is highly influenced by differences in transcript abundance, which leads to allelic variation [8,9] in transcriptomes, which regulates both genetic and epigenetic factors at the genome-wide level [10].

There is a new concept of pan-regulon emerging, which aims at identifying the role of regulons in the genome expression by identifying all the putative targets of a given transcription factor in all the available genomes that are under study. This approach helps in understanding the similarities and differences between genomes and to see whether the observed expression of genes is because of the presence or absence of a gene or because of the polymorphism in regulons [11]. Therefore, the term pan-regulon has been defined as the "sum of all putative targets for a given transcription factor in all the genomes considered" [12]. Like the pan-genome and pan-transcriptome, the pan-regulon of specie can also be divided into three parts that is, the core pan-regulon which includes the shared target genes in all the genomes under study, accessory pan-regulon which includes target genes in some of the genomes under study and unique pan-regulon which is specific to only one genome under study, for a given transcription factor [13]. Building the pan-transcriptome of species will be useful in determining functionally dispensable and unique cloud genes. While identifying the pan-regulon at the same time will enable researchers to make sure that it's actually the gene that is missing and not the regulon due to which variability in pan-transcriptome occurs. So far, only a few pan-transcriptomic studies have been carried out and many more studies are required to understand about accessory and unique gene at the functional genomics level.

2 Methodologies in transcriptomics

Data gathering is an important first step in transcriptomic and pan-transcriptomic studies and for this purpose two basic principle methods are used, that is, the first approach is to sequence the transcripts which are to be studied by using high-throughput sequencing such as NGS, while the other approach is the microarray which uses hybridization techniques in which an ordered array of nucleotide probes is used to hybridize the transcripts with it.

2.1 Microarray

The microarray is one of the most widely used techniques in transcriptomics for analyzing the expression of many genes simultaneously in a single reaction. This chip technology was introduced in early 1990s where thousands of nucleotide probes are immobilized on a small surface area [14]. While developing this technology the most important goal was to achieve maximum sensitivity even from a limited amount of sample. In this regard, some parameters were set that included the amount of DNA to be fixed on the array, concentration of fluorescently labeled RNA, activity of labeling, and duration of hybridization and finally the duration of exposure of array to the phosphor imager shield. The technology thus developed opened new vistas of research due to their cost effectiveness and efficiency; this allowed researchers, especially the academic ones, to perform numerous expression profiling studies even when there is limited sample [15]. For the analysis, the sample mRNA is reverse transcribed to synthesize cDNA, which is also fluorescently labeled. This product is then hybridized and the fluorescence intensity is measured as it is proportional to the amount of mRNA present [16].

The intensity of fluorescence at a specific probe location indicates the expression level of that particular transcript [17]. Gene function is a key factor that bio-scientists want to examine or understand about and microarray did brought that revolution in the research. The reliability of the microarray is a hot topic for discussion since the research started in the field of transcriptomics and through several studies it became clear that many factors are involved in determining that starting from the fabrication of chips to the experimental conditions that were provided with several other parameters as well. Overall, the microarray is considered effective in studying and understanding gene functionality at the genomic level, which further helps in differing molecular mechanisms in normal and dysfunctional biological processes [18,19]. Further, the use of this technology in pan-transcriptomic studies can help in measuring the expression levels of transcripts in between all the genomes under study, which will help in establishing pan-transcriptomic maps for the genomes under study along with the characterization of studies transcripts into core, accessory, or unique pan-transcriptomes. This will especially be helpful in identifying targets in case of a disease outbreak or a newly detected mutation.

2.2 Next-generation sequencing

For quite some time, the gold standard for identifying differential gene expression was the microarray technique which allowed understanding and assessing the complex biological processes in a tissue or cell based upon its transcriptome. However, in the recent years DNA sequencing technologies have been developed and these advanced technologies have been used to completely profile a cell's transcriptome. The next-generation sequencing technologies have helped a great deal in advancing our knowledge of molecular biology. These sequencing technologies also enabled us to completely characterize the complete transcriptome at various genomic and exomic levels [20,21] along with an ability to identify new genes, gene fusions, and rare transcripts [1,22].

RNA-seq works by utilizing high-throughput sequencing and combining it with computational methods to get information about transcripts in the RNA extract [23]. This works by converting the extracted RNA into cDNA and the nucleotide sequences generated are usually 100 bp but can be extended up to 10,000 bp with the lower limit being as low as 30 bp [3]. This sequencing technology can identify the activity of genes at some particular point in time and an expression profile over the different stages or points of time can be obtained which can be used to understand a biological process or a disease progression and much more. It can also be helpful in identifying new genes or mutations that might cause the overexpression of a gene readout involved in some physiological function or malfunction through generating datasets for many organisms of the same species thus building a pan-transcriptomic map. The technology can also be used to confirm the presence or absence of a gene in the genome and to find whether it's the gene responsible or a regulon which will make it a lot easier to categorize between accessory and unique pan-transcriptome from the core pan-transcriptome. One of the most important advantages of high-throughput sequencing technology over microarray is that it has the ability to identify new and rare transcripts unlike microarrays which depend only on an existing genomic sequence. Another important advantage is that this method has no chance of cross-hybridization unlike the microarray and this helps in reducing the background levels [1]. The huge amount of data generated through these high-throughput sequencing techniques requires the development of robust bioinformatics tools that can handle such massive amounts of data and at the same time be efficient enough to search for the useful biological information [2].

3 Computational framework of pan-transcriptomics

Interpretation of quickly expanding number of genomes has become a challenge for numerous disciplines ranging from human genetics and cancer studies to plant rearing, virology, and microbiology. In case of humans, the quantity of sequenced genomes will approach several thousands in the following couple of years. Just scaling up set-up bioinformatics pipelines will not be adequate for leveraging the full potential of such rich genomic datasets. Rather, novel, quantitatively unique computational techniques and ideal models are required. We will observe the quick expansion of computational pan-genomes, another subregion of research in computational science.

However, the field is in its early developmental process yet and few tools are available that are currently being used to analyze and assemble pan-transcriptomic data. Some of which have been discussed briefly below.

4 Prokaryotic data analysis software

4.1 WoPPER

To determine differentially expressed chromosomal regions in bacteria, a web tool integrating gene expression and genomic annotations WoPPER is used. RNA-sequencing or microarray-based gene expression data are provided as input, along with gene annotations [24].

4.2 JCoast

JCoast consolidates all capacities required for the mining, annotation, and interpretation of (meta) genomic information. The lightweight programming software enables the client to effectively exploit progressed back-end database structures by giving a programming and graphical user interface (UI) to respond to biological inquiries [25].

4.3 EDGAR

EDGAR is intended to automatically perform genome correlations in a high-throughput approach. The product bolsters a speedy review of evolutionary connections and disentangles the way toward getting new natural bits of knowledge into the differential gene substance of related genomes [26].

4.4 Trimmomatic

Trimmomatic utilizes a pipeline-based design, permitting individual "steps" (adapter expulsion, quality sifting, and so forth) to be connected to each peruse/read pair, in the request indicated by the client. Each progression can chip away at the peruses in seclusion, or work on the joined pair, as fitting. The apparatus tracks read matching and stores "combined" and "single" peruses independently [27].

4.5 Roary

Roary is software that quickly manufactures huge-scale pan-genomes, recognizing pan- and accessory genomes. Roary makes the development of the pan-genome of thousands of prokaryote tests conceivable on a standard desktop without settling on the precision of results. [28] (Table 18.1).

5 Eukaryotic data analysis software

5.1 AGAPE

AGAPE comprises of three principle parts: assembly, annotation, and variety calls. Given the raw sequences peruses of a given genome, a reference genome succession, and reference genome annotations, the pipeline creates all over again get-together platforms and contigs, ORF comments including non-reference ORFs, and arrangement

Table 18.1 Prokaryotic data analysis softwares

Software name	Prokaryotes/ eukaryotes	Data analysis type	Limitations	References
WoPPER	Prokaryotes	Microarray	Yet, this kind of examination is prescribed just for genomes having in excess of 2000 genes, as on littler genomes the smoothing and stage ventures of WoPPER would be found on couple of information focuses, along these lines yielding noisy measurements.	[24]
JCoast	Prokaryotes	RNA sequencing	The GenDB framework underpins the investigation of eukaryotic information too, yet taking care of the extra data for such undertakings is not executed in JCoast right now.	[25]
EDGAR	Prokaryotes	RNA sequencing	Sometimes, score ratio values (SRV) dispersion does not demonstrate the normal bimodal shape. This is for the most part the case if there is a high variety inside the genomes of a class.	[26]
Trimmomatic	Prokaryotes	RNA sequencing	It is not easy to run lots of commands in a row or to run the same command again.	[27]
Roary	Prokaryotes	RNA sequencing	Laborious to use.	[28]

variety calls, for example, extra recently embedded sequences in the genome (not present in the reference genome) just as SNPs in respect to the reference [29].

5.2 SHOE

SHOE empowers clients to investigate potential interactions between translation factors and target genes by means of various information sees, find interpretation factor restricting themes over gene co-articulation, and picture genes as a system of gene and translation factors on its local device GeneViz, the CellDesigner pathway analyzer, and the Reactome database to look through the pathways included [30].

5.3 TaxMapper

TaxMapper is intended to look sequences peruses against remotely comparable hits in compiled database and to sift through hits of flawed sureness. It comprises of five modules (seek, map, filter, count, and plot) that can be run independently with client-characterized parameters or as a solitary advance with default settings [31] (Table 18.2).

6 Applications

Studying pan-transcriptomes and pan-regulons is the gateway toward many different applications. Various disorders that occurred due to problem in the regulation of transcription and transcription factors can be characterized or profiled by pan-transcriptome and pan-regulon analysis, which will lead to predict novel drug targets to alleviate

Table 18.2 Eukaryotic data analysis softwares

Software name	Prokaryotes/ eukaryotes	Data analysis	Limitations	References
AGAPE	Eukaryotes	RNA sequencing	AGAPE should be adjusted to consider complex gene models and increasingly advanced assembly strategies to explore genomes wealthy in redundant sequences.	[29]
SHOE	Eukaryotes	Microarray	The current investigative requests require in excess of a basic yield of putative transcription factor binding sites.	[30]
TaxMapper	Eukaryotes	RNA sequencing	Tool was not planned for metagenome investigations. It is relied upon to execute well on protein-coding locales, yet because of the protein reference sequences, the ordered and functional task will come up short for noncoding and intronic regions.	[31]

different disorders. Antimicrobial-resistant (AMR) pathogens greatly undermine people's ability to control pathogens and cure diseases. The ultrafast mutation rates of these microbes render our existing drugs useless against superbugs, and existing classes of antibiotics are probably the best we will ever have. Pan-transcriptome analysis of antibiotic-resistant strains will lead us to evaluate those core and dispensable genes that confer the ability to resist antibiotics. Likewise, we will be able to differentiate between the features of different bacterial strains, especially the virulent factor of microbes and their ability to cope with biotic and abiotic stress. The accessory pan-transcriptome and pan-regulon is greatly associated with microbial evolution with different traits. A few studies conducted on pan-transcriptome and pan-regulon in both prokaryotes and eukaryotes are discussed below.

7 Prokaryotic examples

7.1 View of the pan-genome and pan-regulon of *Dickeya solani*

Dickeya solani is emerging plant pathogen that is very harmful to crops and ornamentals [32]. Recently, a study has been conducted on this pathogen to identify diverse virulent factors through concepts of comparative genomics [12]. Ten available genomes and four de novo sequenced genomes were used to build the structure of the *Dickeya solani* pan-genome. In the pan-genome, 74.8% genes were found in the core genome while 25.2% genes were comprised of the accessory genome. Further, in the pan-regulon analysis of *Dickeya solani* binding sites for four transcription factors including Kdgr, Fur, CRP, and Pecs were predicted to identify the regulons of these mentioned virulence regulators. It was found that the potential virulent factors were predicted to belong to the accessory regulon of Kgdr, Pecs, and CRP and it concludes that there is variation in gene expression between the strains of *Dickeya solani*. Moreover, the comparison between high virulent and less virulent strain of *Dickeya solani* disclose significant differences in the production of virulence factors and these strains have variation in size and number of prophages contained in their genome. The mobility factor was considered to be responsible for more virulence or aggressiveness.

7.2 Pan-regulon study of *Listeria monocytogenes* σB

Listeria monocytogenes is divided into different phylogenetic lineages and has the ability to transmit in humans [33]. The alternative sigma factor σB is demonstrated to be involved in stress response to the environment and virulence in different strains of *Listeria monocytogenes* [34]. It was hypothesized that σB provides the differential abilities to different strains of *Listeria monocytogenes* for survival and adaptation in multiple niches. Oliver et al. [35] conducted a study to evaluate the contribution of σB to stress response in different strains of *Listeria monocytogenes* belonging to different lineages [35]. Δ*sigB* mutations were introduced in strains belong to I, II, IIIB, and IIIA lineages. These strains were demonstrated for their survival under oxidative stress conditions. The Δ*sigB* mutant and parent strain were analyzed and compared through whole-genome expression microarrays. Results had shown that the alternative sigma factor σB has a small core regulon and is responsible for virulence in all strains of different lineages but its contribution toward stress tolerance are different in different strains belonging to different lineages.

8 Eukaryotic examples

8.1 Insight into maize pan-transcriptome

Maize possesses a high level of variation in phenotype and to artificial and natural selection due to its outcrossing nature during the increasing population expansion globally [36]. Jin et al. [7] constructed a maize pan-transcriptome and measured quantitative trait variation, which gives insight into the genome complexity of maize.

They selected 368 maize diverse inbred lines and analyzed them by RNA-seq data. Results show that showed that more than 1/3 of genes had variation in expression. These genes were in fact an essential component of regulation in the cell and were managed by expression of quantitative trait loci (eQTLs). Expression presence and absence variation (ePAV) was utilized in a straightforward manner to be "genotype" to do genome-wide association studies for about 15 phenotypic appearances in agronomics as well as for about 526 qualities related to metabolism for effectively investigating the relationship among variations in transcriptomics and phenomics. With the help of an efficient and updated assembly technique, more than 2000 high-certainty new sequences were discovered missing inside the reference genome. The total length of the novel sequence was 1.9 Mb. A simulation model shows that pan-transcriptome analysis of whole kernel maize is moving toward a greatest estimation of about 63,000 genes and by establishing two test crosses for the populations and looking over a few most vital yield characteristics, the accessory genes appeared to add to heterosis [7]. Establishment of the pan-transcriptomics for the maize plant is better and more efficient than the pan-genomic results of the same plant in light of fact that the greater extent of the presence of such repetitive elements in the genome complicates the total assembly.

8.2 Pan-transcriptome analysis of barley

Barley has been used as food for humans and animals and is an important crop containing lot of diversity. Barley is a type of inbreeding species, which is diploid in hereditary and contains a massive genetic complement of about 5.1 Gb. About 80% genome of barley contains repetitive sequences and huge pericentromeric regions that lack the ability of meiotic recombination [37,38]. Even with the advent of next-generation sequencing, it is still difficult to construct the complete genome assembly of barley due to its diversity. After over a time of devoted endeavors, a worldwide group comprising of more than 70 scientists have succeeded in sequencing the whole hereditary material of barley using the cultivar Morex [38].

It appears to be far-fetched that high-quality genome assemblies for barley genotypes can be gotten soon. In this way, a pan-genome of barley may need to pause. However, catching most of the expressed genes in a barley genotype is now moderately simple and reasonable. The pan-transcriptome analysis of barley has been done by de novo assembling 288 sets of RNA-seq data from 63 genotypes. The pan-transcriptome of barley comprises of 756,632 transcripts with a normal N50 length of 1240 bp [39]. Around 38% of transcripts related to the recently organized pan-transcriptome were not located in the reference genome (Morex) (Fig. 18.1). New transcripts contain genomic sequences related to reaction to abiotic and biotic burdens.

Comparing the genetic complement of the wild and cultivated barley at the pan-transcriptomic level revealed that genes which are involved in creating resistance have increased greater in the wild species than cultivated ones and also underwent greater selective pressures during barley domestication [39].

8.3 Pan-transcriptome reconstruction in *Phaeoacremonium minimum*

Phaeoacremonium minimum is an ascomycete fungus, which is the primary causative agent of Esca. Esca is a boundless and harming grapevine trunk infection. Diversity among virulent *Phaeoacremonium minimum* isolates was known, but still, the basic genomic premise about the phenotypic diversity stays obscure. There is so much genetic diversity within the strains of *Phaeoacremonium minimum*; so, reference transcriptome is not sufficient to report the virulence factor of the entire collection of isolates. Therefore, a study was conducted by Massonnet et al. [40] to assemble the reference pan-transcriptome of

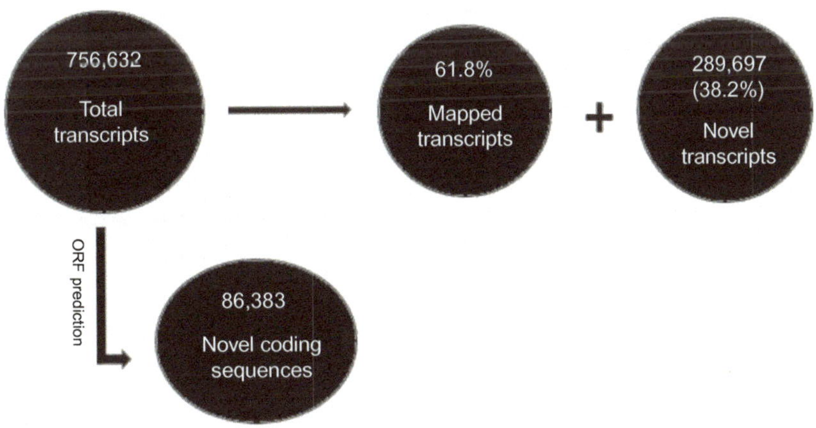

Fig. 18.1 Novel CDS and transcripts not reported in the reference transcriptome. Around 289,697 of RNA sequences from recently managed pan-transcriptome was not identified for reference genetics (Morex). Approximately, 86,383 new coding sequences were assembled. *Image adapted from Ma, Y., Liu, M., Stiller, J., & Liu, C. (2019). A pan-transcriptome analysis shows that disease resistance genes have undergone more selection pressure during barley domestication. BMC Genomics, 20(1), 12.*

Phaeoacremonium minimum [40]. The pan-transcriptome reference contained about 15,000 nonredundant coding sequences.

Utilizing naturally infected field samples expressing Esca symptoms, mapping of meta-transcriptomics data on a multispecies reference that incorporated the *Phaeoacremonium minimum* pan-transcriptome permits the profiling of virulent factors, incorporating variable genes related to the cell transport as well as metabolism.

9 Conclusion and future prospects

The genomic era brought great advancements in understanding the molecular biology of organisms, both prokaryotes and eukaryotes, in terms of their evolution with respect to pathogenesis, survival, proliferation, and adaptation. A decade after the beginning of the genomic era, the question of how genomics can describe variation between species has not been fully addressed. Experimental data have shown that in some species new genes are discovered after sequencing the genomes of several strains. Computational modeling predicts that new genes will be discovered even after sequencing hundreds of genomes per species. Therefore, the pan-genome era started, which revealed the concept of core, accessory, and unique genome of an organism. This new concept aimed at assembling the whole repertoire of genes present in specie. After the concept of the pan-genome, scientists had seen that there is lot of variation in gene expression among the species. So, they started to focus on the gene readouts and their regulatory factors that are involved in species evolution and their survival under environmental stress. From here, the concept of the pan-transcriptome and pan-regulon arises. In this chapter, we have discussed the transcriptomic technologies such as RNA-seq and microarray and their utilization in evaluating pan-transcriptome and pan-regulon. The data that is being generated through these technologies is enormous and needs to be analyzed using computational techniques which can extract useful information. Currently, various software tools for prokaryotic and eukaryotic data analysis are being used to generate maps of pan-transcriptomes and pan-regulons but these software have certain limitations that need to be addressed. So far, only few pan-transcriptomic studies have been carried out and more research studies should be conducted so that the behavior of organisms at the molecular level could be assessed.

References

[1] Z. Wang, M. Gerstein, M. Snyder, RNA-Seq: a revolutionary tool for transcriptomics, Nat. Rev. Genet. 10 (1) (2009) 57.

[2] T. Raiol, D.P. Agustinho, K.C.R. Simi, C.M. de Souza Silva, M.E. Walter, I. Silva-Pereira, M. Brígido, Transcriptome analysis throughout RNA-seq, in: G. A. Passos (Ed.), Transcriptomics in Health and Disease, Springer, 2014, pp. 49–68.

[3] R. Lowe, N. Shirley, M. Bleackley, S. Dolan, T. Shafee, Transcriptomics technologies, PLoS Comput. Biol. 13 (5) (2017).

[4] Y. Ruan, P. Le Ber, H.H. Ng, E.T. Liu, Interrogating the transcriptome, Trends Biotechnol. 22 (1) (2004) 23–30.
[5] Y. Han, S. Gao, K. Muegge, W. Zhang, B. Zhou, Advanced applications of RNA sequencing and challenges, Bioinform. Biol. Insights *9* (2015) 29–46.
[6] A.A. Pandit, R.A. Shah, A.M. Husaini, Transcriptomics: a time-efficient tool with wide applications in crop and animal biotechnology, J. Pharmacogn. Phytochem. 7 (2) (2018) 1701–1704.
[7] M. Jin, H. Liu, C. He, J. Fu, Y. Xiao, Y. Wang, …J. Yan, Maize pan-transcriptome provides novel insights into genome complexity and quantitative trait variation, Sci. Rep. 6 (2016) 18936.
[8] F.W. Albert, L. Kruglyak, The role of regulatory variation in complex traits and disease, Nat. Rev. Genet. 16 (4) (2015) 197.
[9] H. Liu, X. Wang, M.L. Warburton, W. Wen, M. Jin, M. Deng, et al., Genomic, transcriptomic, and phenomic variation reveals the complex adaptation of modern maize breeding, Mol. Plant 8 (6) (2015) 871–884.
[10] J. Fu, Y. Cheng, J. Linghu, X. Yang, L. Kang, Z. Zhang, et al., RNA sequencing reveals the complex regulatory network in the maize kernel, Nat. Commun. 4 (2013) 2832.
[11] M. Galardini, A. Mengoni, M. Brilli, F. Pini, A. Fioravanti, S. Lucas, et al., Exploring the symbiotic pangenome of the nitrogen-fixing bacterium *Sinorhizobium meliloti*, BMC Genomics 12 (1) (2011) 235.
[12] M. Golanowska, M. Potrykus, A. Motyka-Pomagruk, M. Kabza, G. Bacci, M. Galardini, et al., Comparison of highly and weakly virulent *Dickeya solani* strains, with a view on the pangenome and panregulon of this species, Front. Microbiol. 9 (2018) 1940.
[13] M. Galardini, M. Brilli, G. Spini, M. Rossi, B. Roncaglia, A. Bani, et al., Evolution of intra-specific regulatory networks in a multipartite bacterial genome, PLoS Comput. Biol. 11 (9) (2015).
[14] R.C.D. Duran, S. Menon, J. Wu, The analyses of global gene expression and transcription factor regulation, in: J. Wu (Ed.), Transcriptomics and Gene Regulation, Springer, 2016, pp. 1–35.
[15] F. Bertucci, K. Bernard, B. Loriod, Y.-C. Chang, S. Granjeaud, D. Birnbaum, et al., Sensitivity issues in DNA array-based expression measurements and performance of nylon microarrays for small samples, Hum. Mol. Genet. 8 (9) (1999) 1715–1722.
[16] R.E. Petty, R.M. Laxer, C.B. Lindsley, L. Wedderburn, Textbook of Pediatric Rheumatology, Elsevier Health Sciences, 2015.
[17] I. Barbulovic-Nad, M. Lucente, Y. Sun, M. Zhang, A.R. Wheeler, M. Bussmann, Bio-microarray fabrication techniques—a review, Crit. Rev. Biotechnol. 26 (4) (2006) 237–259.
[18] F.M. Bareyre, M.E. Schwab, Inflammation, degeneration and regeneration in the injured spinal cord: insights from DNA microarrays, Trends Neurosci. 26 (10) (2003) 555–563.
[19] A. Jaerve, F. Kruse, K. Malik, H.-P. Hartung, H.W. Müller, Age-dependent modulation of cortical transcriptomes in spinal cord injury and repair, PLoS One 7 (12) (2012).
[20] S. Anders, A. Reyes, W. Huber, Detecting differential usage of exons from RNA-seq data, Genome Res. 22 (10) (2012) 2008–2017.
[21] V.M. Kvam, P. Liu, Y. Si, A comparison of statistical methods for detecting differentially expressed genes from RNA-seq data, Am. J. Bot. 99 (2) (2012) 248–256.
[22] Y. Katz, E.T. Wang, E.M. Airoldi, C.B. Burge, Analysis and design of RNA sequencing experiments for identifying isoform regulation, Nat. Methods 7 (12) (2010) 1009.
[23] F. Ozsolak, P.M. Milos, RNA sequencing: advances, challenges and opportunities, Nat. Rev. Genet. 12 (2) (2011) 87.
[24] S. Puccio, G. Grillo, F. Licciulli, M. Severgnini, S. Liuni, S. Bicciato, et al., WoPPER: web server for position related data analysis of gene expression in prokaryotes, Nucleic Acids Res. 45 (W1) (2017) W109–W115.
[25] M. Richter, T. Lombardot, I. Kostadinov, R. Kottmann, M.B. Duhaime, J. Peplies, F.O. Glöckner, JCoast–a biologist-centric software tool for data mining and comparison of prokaryotic (meta) genomes, BMC Bioinformatics 9 (1) (2008) 177.
[26] J. Blom, S.P. Albaum, D. Doppmeier, A. Pühler, F.-J. Vorhölter, M. Zakrzewski, A. Goesmann, EDGAR: a software framework for the comparative analysis of prokaryotic genomes, BMC Bioinformatics 10 (1) (2009) 154.
[27] A.M. Bolger, M. Lohse, B. Usadel, Trimmomatic: a flexible trimmer for Illumina sequence data, Bioinformatics 30 (15) (2014) 2114–2120.

[28] A.J. Page, C.A. Cummins, M. Hunt, V.K. Wong, S. Reuter, M.T. Holden, et al., Roary: rapid large-scale prokaryote pan genome analysis, Bioinformatics 31 (22) (2015) 3691–3693.
[29] G. Song, B.J. Dickins, J. Demeter, S. Engel, B. Dunn, J.M. Cherry, AGAPE (Automated Genome Analysis PipelinE) for pan-genome analysis of *Saccharomyces cerevisiae*, PLoS One 10 (3) (2015).
[30] N. Polouliakh, P. Horton, K. Shibanai, K. Takata, V. Ludwig, S. Ghosh, H. Kitano, Sequence homology in eukaryotes (SHOE): interactive visual tool for promoter analysis, BMC Genomics 19 (1) (2018) 715.
[31] D. Beisser, N. Graupner, L. Grossmann, H. Timm, J. Boenigk, S. Rahmann, TaxMapper: an analysis tool, reference database and workflow for metatranscriptome analysis of eukaryotic microorganisms, BMC Genomics 18 (1) (2017) 787.
[32] J. Mansfield, S. Genin, S. Magori, V. Citovsky, M. Sriariyanum, P. Ronald, et al., Top 10 plant pathogenic bacteria in molecular plant pathology, Mol. Plant Pathol. 13 (6) (2012) 614–629.
[33] M.J. Gray, R.N. Zadoks, E.D. Fortes, B. Dogan, S. Cai, Y. Chen, et al., *Listeria monocytogenes* isolates from foods and humans form distinct but overlapping populations, Appl. Environ. Microbiol. 70 (10) (2004) 5833–5841.
[34] S. Raengpradub, M. Wiedmann, K.J. Boor, Comparative analysis of the σB-dependent stress responses in *Listeria monocytogenes* and *Listeria innocua* strains exposed to selected stress conditions, Appl. Environ. Microbiol. 74 (1) (2008) 158–171.
[35] H. Oliver, R. Orsi, M. Wiedmann, K. Boor, Listeria monocytogenes σB has a small core regulon and a conserved role in virulence but makes differential contributions to stress tolerance across a diverse collection of strains, Appl. Environ. Microbiol. 76 (13) (2010) 4216–4232.
[36] J. Yan, M. Warburton, J. Crouch, Association mapping for enhancing maize (*Zea mays* L.) genetic improvement, Crop Sci. 51 (2) (2011) 433–449.
[37] P. Langridge, B. Shi, G. Fincher, A physical, genetic and functional sequence assembly of the barley genome, Nature 491 (7426) (2012) 711–716.
[38] M. Mascher, H. Gundlach, A. Himmelbach, S. Beier, S.O. Twardziok, T. Wicker, et al., A chromosome conformation capture ordered sequence of the barley genome, Nature 544 (7651) (2017) 427.
[39] Y. Ma, M. Liu, J. Stiller, C. Liu, A pan-transcriptome analysis shows that disease resistance genes have undergone more selection pressure during barley domestication, BMC Genomics 20 (1) (2019) 12.
[40] M. Massonnet, A. Morales-Cruz, A. Minio, R. Figueroa-Balderas, D.P. Lawrence, R. Travadon, et al., Whole-genome resequencing and pan-transcriptome reconstruction highlight the impact of genomic structural variation on secondary metabolite gene clusters in the grapevine esca pathogen *Phaeoacremonium minimum*, Front. Microbiol. 9 (2018).

CHAPTER 19

Pan-proteomics: Technologies, applications, and challenges

Wanderson Marques da Silva[a], Nubia Seyffert[b]
[a]Institute of Agrobiotechnology and Molecular Biology, INTA-CONICET, Buenos Aires, Argentina
[b]Biology Institute, Federal University of Bahia, Salvador, Brazil

1 Introduction

In the past years, with molecular biology advances various information about the molecular basis of biological systems was uncovered, this is due to a variety of genomic projects that were performed. However, these results reveal little information about how proteins of a given organism might operate individually or together to perform their functions. In this context, the large-scale study of proteins through proteomic studies becomes important to elucidate this information. Unlike conventional biochemical studies that focus only on one protein or simple sets of macromolecules, the proteomic has an approach more comprehensive and systematic to the investigation of the biological systems [1]. Proteomic analysis is a discipline that has been utilized to understate the global protein dynamic in one cell, where the proteins are identified by mass spectrometry (MS). The proteomic study is often used to evaluate the posttranslational modification (PTM) protein, abundance or relative expression in different conditions or during different steps of the cell growth, thus providing important information about the activities of protein components or protein networks and routes. In addition, these generated datasets contribute to understanding biological function and the properties of the cell as well as how cells respond to environmental changes or metabolic stresses [2].

Proteomics techniques like the classical two-dimensional gel electrophoresis (2-DE) and liquid-chromatography coupled to mass spectrometry (LC-MS) have been contributing to proteomic success. Currently, with introduction of new technologies like multidimensional protein identification technology (MudPIT) that combines several two-dimensional chromatographic methodologies and MS, researchers are able perform studies with highly complex protein samples [3]. To follow these technological advances new software and algorithm has been developed what contributed to increasing proteomic power. Nowadays, with advances in proteomic techniques, it has been possible to establish a bridge between our understanding of the genome sequence and cellular behavior which encompasses the functional genomics; through the result obtained by the proteomics, it is possible to validate genes, and establish the correlation between the

Pan-genomics: Applications, Challenges, and Future Prospects
https://doi.org/10.1016/B978-0-12-817076-2.00019-6
© 2020 Elsevier Inc.
All rights reserved.

genotype and phenotype. Hence, proteomics has become a powerful approach to study the functional genome of a given organism at the protein level in different areas, including human medicine, veterinary, and plant and animal science.

Quantitative proteomics studies have been promoted several insights into proteomic, this type of study can be classified in relative or absolute quantification in global proteome analysis [2]. The relative quantification aim detected differentially expressed proteins between distinct sample groups, for example, tumor vs healthy, microorganisms under stress condition vs laboratory condition. In this study, the difference observed between abundance ratio or fold change between different samples is used to quantify the relative expression [4]. On the other hand, absolute quantification is often utilized to determine the absolute concentration of one or more protein in a given sample. To obtain a more accurate result and avoid a possible variability inside this complex analysis, an external standard with known concentration is spiked into the sample [4].

Due to all these technological advances, the proteomic study has contributed to broaden our knowledge about drug discovery, and microbial meta-proteomics in distinct environmental, plant proteomic, host-pathogen interaction, human or animal health. More recently, the proteomic studies have been utilized in pan-proteomic studies. This strategy aim compares all proteins of several organisms with genetic variation, enabling through qualitative or/and quantitative strategies identify the entire proteome and promote information attributions about the biology of organisms within a given specie [5]. Here in this chapter, we provide information and discussion about the contribution of the pan-proteomic study into different areas of knowledge such as microbiology, botany and zoology to elucidate important factors related to basic life processes. In addition, we also discuss about proteomics technologies and bioinformatics strategies developed to increase the proteomic power.

2 Pan-proteomics concept and proteomics technologies used in pan-proteomics

Pan-proteomics is an approach that aims to characterize and compare the proteome across the intraspecific diversity of organisms inside a species [5]. Like the pan-genomics, the pan-proteome can be divided into: (i) the core proteome, which is composed of proteins present in all individuals and (ii) the dispensable proteome or accessory proteome, which comprises proteins absent in some individuals. The pan-proteomics complements both pan-genomics and pan-transcriptomics, which makes it possible to identify not only genetic variants, but also the presence of homologous sequence of an organism at the protein level [5].

As in any proteomic study, in pan-proteomics some steps should be also considered, such as: (i) sample preparation, what type of protein fraction will be evaluated; since each one present a physical-chemical property, the choice of the extraction buffer or precipitation method must be optimized to obtain the maximum proteome information; (ii) identification method, the technique utilized must have a high performance to obtain the maximum of qualitative as well as quantitative information. Thus, the utilization of high-throughput proteomics techniques like MudPIT is very suitable to pan-proteomics studies due to enhanced sensitivity and dynamic range and proteome coverage [5, 6]; (iii) protein sequence database, a pan-proteomic database must be constructed with curated protein sequences and must contain the proteome of all species or strains utilized in this study. Thus, this database will be composed of sequences of a reference proteome and unique sequences that are present in other species or strains, but absent in the reference proteome [7].

Unlike gel-based techniques like the classical 2-DE, gel-free proteomics based in LC-MS allow the continuous separation of thousands of proteins promoting a high throughput in pan-proteomic studies. On the other hand, with the introduction of quantitative studies several insights have been promoted in the proteomic study [2]. In gel-free proteomics, the quantitative analysis can be subdivided in two approaches: (i) label-free, where the quantification can be obtained through spectral counting (relative quantitation of the total number of MS/MS spectra for a protein across multiple LC-MS runs) or by precursor signal intensity from extracted ion chromatogram (determination of the area under the curve of the chromatographic peak on the MS1 level); (ii) stable isotope label-based, this approach can be divided into two strategies that distinguish themselves in the way of incorporation of the stable isotope-enriched labels. In the postharvest tagging strategies, peptides are chemically labeled after the cells have been lysed, and some techniques were developed for chemical labeling, such as: isobaric tags for relative and absolute quantification (iTRAQ), isotope-coded affinity tag (ICAT), and tandem mass tag (TMT). In turn, in metabolic labeling strategy the incorporation of stable isotopes is made endogenously in metabolically active cells; this strategy is based on techniques such as uniform ^{15}N labeling, amino acid-coded mass tagging/stable isotope labeling with amino acids in cell culture (AACT/SILAC), neutron encoding (NeuCode), and cell-selective labeling with amino acid precursors (CTAP) [4].

3 Bioinformatics strategies/tools used in proteomics

Over the years, several methodologies have been developed to increase the analytical power of proteomic studies, which consequently generates a large amount of MS data. Thus, in order to keep up with this technological advance, it is necessary to develop robust software that is capable of processing and transforming these raw data from MS into quantitative and qualitative data useful for the knowledge of biological systems. Some softwares and bioinformatics tools that are utilized in proteomics studies are visualized in Table 1. Some characteristics should be considered in the development of this software such as the ability to process a high number of peptides, evaluate the ions

Table 1 Softwares and bioinformatics tools utilized in proteomics studies

Softwares/ database	Description	Link
MaxQuant	Platform for quantitative (stable isotope label and label-free quantification) and qualitative proteomics	http://www.maxquant.org
Progenesis	Absolute quantification, label-free quantification, ion-mobility separations, analysis of Waters MSE, and HDMSE data-independent (DIA) data	http://www.waters.com
ProteinLynx Global SERVER (PLGS)	Platform for quantitative and qualitative proteomics research, label-free quantification	http://www.waters.com
MASCOT	Peptide mass fingerprint and MS/MS database searches, quantitative and qualitative proteomics research	http://www.matrixscience.com/
UniProt	Protein sequence databases	http://www.uniprot.org/
PDB	Protein Data Bank	https://www.rcsb.org/
Proteome Discoverer	Quantitation methods including label-free quantitation, TMT, and SILAC	http://www.thermofisher.com
Scaffold	Visualize and validate complex MS/MS proteomics experiments, quantitative studies	http://www.proteomesoftware.com/
PRIDE	Proteomics data repositories	http://www.proteomexchange.org/
GO	Gene Ontology Consortium	http://geneontology.org/
Cytoscape	Protein-protein interactions network databases	https://cytoscape.org/
STRING	Protein-protein interactions network databases	https://string-db.org/
KEGG	Kyoto Encyclopedia of Genes and Genomes	https://www.genome.jp/kegg/
IPA	Ingenuity pathways analysis	http://www.qiagen.com/
ExPASy	EXpert Protein Analysis System	https://www.expasy.org/
PFAM	Protein Families Database	http://pfam.xfam.org/
DrugBank	Bioinformatics and cheminformatics resource that combines detailed drug data with comprehensive drug target information	https://www.drugbank.ca/
Spectrum Mill	Platform for quantitative and qualitative proteomics research	https://www.agilent.com
ProteinScape	Platform for quantitative (stable isotope label and label-free quantification) and qualitative proteomics research	https://www.bruker.com/
SCIEX Software	Software for high-throughput proteomics	https://sciex.com

intensity, and determine the peptide and protein ratio. Due to mass spectrometer diversity from different companies, the software must be able to process different types of data. MaxQuant software allows an efficient analysis of raw proteomic data with high peptide identification rates, individualized range mass accuracies, and protein quantification [8]. Progenesis QI software performs quantitative and qualitative analysis in complex samples using the advantages of label-free analysis. This software supports LC-MS data from Waters, Thermo, AB Sciex, Agilent, and Bruker [9].

In the case of quantitative studies, the software should be versatile in order to be configured to analyze the different labeling methods to evaluate isotopes and charge states, and calculation of statistical significance [10]. Both label-based and label-free quantification of PTM can result in a more difficult analysis due to their short biological lifetime and/or labile nature. In turn, when evaluating data from label-free quantification, noise reduction, peak picking, and retention time alignment are challenges that need to be overcome [11]. However, several computational tools (i) commercial: Mascot Distiller, Proteome Discoverer, SIEVE, PEAKSQ, QuanLynx, ProteinPilot, ProteinScape, and Spectrum Mill and (ii) free: APEX, ASAP Ratio, MaxQuant, MSQuant, OpenMS, ProteoSuite, and XPRESS are utilized to overcome these difficulties and improve the peptide and protein quantification [11].

However, the proteomic analysis does not end only with the identification and quantification of the proteins; additional analysis is necessary to correlate the identified proteins with biological pathways and cellular function. Thus, proteins whether grouped or not, in hierarchical clustering are subject to functional analysis by bioinformatics tools to identify the gene ontology (GO), and biological pathway and network analysis. The identification of enriched functional categories in a high-throughput analysis increases the likelihood to identify biological processes most relevant [12]. The utilization of database platform also contributes to enhance the comprehensiveness and performance of the functional analysis. For example, the Kyoto Encyclopedia of Genes and Genomes (KEGG) is an integrated database collection developed for the biological interpretation of high throughput from genomics, transcriptomics, and proteomics. In the KEGG Genes database are deposited more than 4000 complete genomes annotated with KEGG Orthology (KO); these datasets can be utilized for the reconstruction of KEGG pathways and molecular networks [13]. The Database for Annotation, Visualization, and Integrated Discovery (DAVID) provides a comprehensive set of functional annotation and enrichment tools for high-throughput data set analysis [12, 14]. Already, to identify the relationships between these proteins as colocalization patterns and protein-protein interaction some platforms are available, such as Cytoscape [15] and STRING [16]. These databases allow data integration, visualization, and protein-protein interaction network. Currently, several softwares, bioinformatics tools, databases, and computational algorithms are available, which are capable of storing, disseminating, and analyzing these high-throughput proteomic data are available for proteomic study [17].

4 Pan-proteomics applications and outcomes in microbes

In the microbiological field, proteomic study has contributed to broadening our knowledge about differences related to environmental stress, host-pathogen interaction, microbial meta-proteomics, and more recently microbial pan-proteomics. In microbiology, the pan-proteomics approach makes it possible to compare an unlimited number of bacterial strains and all this information provides insights about the biological pathways that are extremely important to understand the biology of a given microorganism.

Once the elucidation of molecules and mechanisms underlying bacterial pathogenesis is one of the main focuses of microbiological research, pan-proteomic studies have been applied to characterize the proteomes of different bacterial pathogens (Table 2). The proteomic analysis of four epidemics *Salmonella paratyphi A* strains reveals a core proteome highly conserved among the strains, composed mainly of proteins involved in energy metabolism [18]. On the other hand, the quantitative proteomic analysis of seven enterotoxigenic *Escherichia coli* (ETEC) shows in all strains the induction of proteins related to iron acquisition, maltose metabolism, and acid resistance [19]. A comparative proteomic study among avirulent, virulent, and clinical strains of *Mycobacterium tuberculosis* revealed differences in the expression of proteins involved in virulence and responsible for drug resistance and lipid metabolism. These observed variations might contribute to the specific biology of each pathogen [20]. *Streptococcus agalactiae* (GBS) is a major pathogen of Nile tilapia; infection by this pathogen causes outbreaks of septicemia and meningoencephalitis causing significant economic losses in the aquaculture sector. Comparative analysis among seven *S. agalactiae* strains revealed a pan-proteome composed of 1065 proteins. Functional analysis identified proteins related to stress response and in the regulation of gene expression, metabolism, and virulence, suggesting that these proteins might contribute to both adaptive processes and pathogenesis [21].

Due to the important role of the secreted and surface proteins in the bacterial pathogenesis, a proteomic study was conducted to characterize the secretome of the Big-six group (O26, O45, O103, O111, O121, and O145) of enterohemorrhagic *Escherichia coli* (EHEC) non-O157 strains, which are reported in food outbreaks. Quantitative proteomic analysis showed relative difference among proteins related to EHEC virulence. In addition, strain-specific proteins were also identified; these proteins could represent potential markers for identification of these strains. These variations observed among the proteome of these strains could be related to the degree of pathogenicity of each strain [22]. The pan-surface proteome of 54 uropathogenic *Escherichia coli* (UPEC) isolates showed the induction of adhesins and iron-acquisition proteins in the majority of the strains, which are proteins involved in the UPEC pathogenesis [23].

Pan-proteomics studies conducted in nonpathogenic bacteria have contributed in the identification of factors that contribute to their adaptive and physiological processes. The characterization of the core proteome of three mollicute species was composed mainly of

Table 2 Pan-proteomics studies in prokaryotic

Organism	Number of organism	Cellular fraction	Approach	Technique	Type of study	Basis of quantifications	Reference
Salmonella Paratyphi A	4	Whole bacterial lysates	Gel-based	2-DE/MALDI	Quantitative/qualitative	Spot abundance	[18]
ETEC	7	Whole bacterial lysates	Gel-free	LC/MS	Quantitative/qualitative	Label-free	[19]
M. tuberculosis	4	Whole bacterial lysates	Gel-free	LC/MS	Quantitative/qualitative	Label-free	[20]
S. agalactiae	7	Whole bacterial lysates	Gel-free	LC/MS	Quantitative/qualitative	Label-free quantification	[21]
EHEC non-O157:H7	6	Secreted proteins	Gel-free	LC/MS	Quantitative/qualitative	TMT labeling	[22]
UPEC	44	Surface proteins	Gel-free	LC/MS	Qualitative	–	[23]
Mollicutes (mycoplasmas)	3	Whole bacterial lysates	Gel-based	2-DE/MALDI 1D-PAGE/LC-MS	Qualitative	–	[24]
L. lactis	4	Whole bacterial lysates	Gel-free	LC/MS	Quantitative/qualitative	Label-free quantification	[25]
Roseobacters	11	Secreted proteins	Gel-based/Gel-free	1D-PAGE/LC-MS	Quantitative/qualitative	Label-free quantification	[26]

proteins related to replication, transcription, translation, and minimal metabolism, which are extremely necessary to biology [24]. The proteomic analysis of four biotechnological strains of *Lactococcus lactis* showed a core proteome with proteins involved in the resistance to different stress conditions as well as related to the probiotic characteristic of *L. lactis*. In turn, in the accessory proteome were identified proteins related to the metabolic pathway exclusive for the different strains [25]. *Roseobacters* are a group of generalist bacteria found in the oceans. To promote insights in the factors that might contribute in their interaction with the extracellular milieu, a study pan-exoproteome of 11 bacteria of the Roseobacter clade was performed. This comparative proteomic analysis reveals that the most abundant proteins are related to transporter, adhesion, motility, and toxin-like proteins, suggesting that might be the main categories involved in the adaptive process of Roseobacter bacteria [26].

5 Pan-proteomics applications and outcomes in plants

Unlike prokaryotic organisms, plants are more complex organisms because they have different types of cells, tissues, and organs that interact with each other. The leaves and seeds are widely used to obtain protein samples, and consequently, for proteomic analyses, due to photosynthetic activity, active growth and ease of obtainment. However, the stem, roots, and inflorescences and other specialized structures of plants can also be studied at the proteomic level [27–29].

The proteomic methodologies have been an important part of research on plant breeding, identification of food allergens, identification of nutritional value, traceability of agronomic characteristics, among others [30–32]. The field of research is broad and has been explored since the development of the first techniques used in protein analysis.

For many years, conventional 2-DE has been employed to separate plant proteins followed by matrix-assisted laser desorption ionization-time of flight-mass spectrometry (MALDI-TOF-MS). Similar to the proteomics experiments performed on microorganisms and animals, these methodologies have limitations related to the amount of sample, sensitivity, and specificity of the analyses [33].

2-DE and MALDI-TOF-MS were used in a comparative study of the mitochondria of *Triticum aestivum* with the purpose of verifying dysfunctions related to the sterility of wheat. In all, 71 proteins were differentially expressed in the samples evaluated, which enabled a complete view of the pathways involved in abortion and anther defects [34].

2D-DIGE and MALDI-TOF-MS were used to analyze the leaf sheaths of two rice lines, one resistant and another susceptible to the parasite *Laodelphax striatellus*. Two lines showed 138 differentially expressed proteins after determined periods of *L. striatellus* infestation. These proteins are involved in response to stresses, photosynthetic processes, cellular metabolism, cell wall proteins, and transcriptional regulators [35].

With the development of free-gel methodologies, a new era of plant proteomic research has resurfaced with high-precision technologies such as LC-MS [36]. For example, in a study with *Arabidopsis thaliana* in response to the hormone strigolactone that stimulates seed germination and development of this plant, 2095 proteins were identified through LC-MS. Significant expression differences were also observed in 37 proteins, which will certainly contribute in studies that involve the molecular mechanisms of action of this hormone in plants [37].

Researchers have extensively explored proteomics tools to study plant responses to different environmental stresses with traditional and advanced omics. These studies are essential because usually the plants are susceptible to different environmental conditions and need to adapt to the soil, variations in humidity, and adverse temperatures [38]. Through these studies it has been possible to find unique proteins, expression differences between lines, signal transduction pathways, PTMs, and interaction networks [39].

Currently, there is a necessity for the integration of the omics data performed on plant species and their varieties to support research in the area of botany. Plantomics will be important not only for basic studies of pan-genomics and systems biology, but also to increase the biological significance of studies already conducted in plant biotechnology.

6 Pan-proteomics applications and outcomes in animals

Protein characterization of the animal organs and secretions are important ways to study biological systems and modifications that may occur during life. However, there are difficulties in correlating proteomics experiments from samples of epithelial, bone, nervous, connective, and blood animal tissues. Thousands of proteins may be identified in these tissues through LC-MS, necessary being the elaboration of software to integrate these data. An extensive study in silico of a healthy human skin proteome has been performed and a total of approximately 3000 proteins were correlated through an automated literature review [40].

Currently, in the world there are several research groups that aim to study specific human organs. The EyeOme Project uses methodologies to identify, quantify, and describe modifications and interactions of proteins from various components of the human eye. More than 16,000 proteins have been identified in the ciliary body, retina, iris, retrobulbar optic nerve, and sclera [41].

LC-MS-based clinical proteomics platforms allow studying the dynamics of protein concentrations in bodily fluids and quantification of candidate biomarker proteins. 357 proteins differentially expressed were identified in the semen of men with type 2 diabetes mellitus and reduced fertility comparing to control [42].

A human proteome project has been developed to understand the expression of the 20,000 protein-coding genes to understand human biology in health and disease.

The idea of the research group is to create several proteomic atlas regarding cells, tissues, and cytology [43].

Plasma proteins are biological indicators or biomarkers important for diagnosing various diseases, including cancer. The complexity of biological fluid proteomes and extensive heterogeneity among diseases are obstacles to successful research. Serum samples from 70 patients with gastric cancer were analyzed and only the biomarkers SERPINA1 and ENOSF1 were identified through MS and related to gastric cancer [44]. Zhong et al. [45] isolated 31 membrane proteins in pancreatic cancer cells and observed that the expression level of Prohibitin-1 is correlated with pancreatic carcinoma differentiation. Proteomic analysis of lung cancer cells revealed higher expression of Stanniocalcin-2 (STC2) compared to normal cells; other cell biology experiments have confirmed STC2 as a good biomarker for cancer [46]. In the case of breast cancer, Mucin-1 was identified as a biomarker protein used to monitor the progression of the patients with metastasis [47]. These are just a few examples, as there are many tumor protein markers already described in the literature for different types of cancers.

A pan-cancer proteome has been developed for correlating data analysis and emphasize the discovery of pathways, determining proteins for the prognostic and therapeutic of the functional proteome. For this purpose, the Clinical Proteomic Tumor Analysis Consortium has used MS to analyze the human proteome and determinate tumors [48].

Similarly, proteomics analyses have been applied in animal production to identify proteins of the fluid, cell, tissue, organs, and pathogens. However, these analyses have not been very efficient for the quantitative analysis of less abundant proteins in complex biological samples, such as bovine milk [49]. Different strategies have been used to characterize disease status, determine the origin, reproduction, and general characteristics of products of animal origin. A consortium dedicated to the applications of proteomics to animal production and health was created to facilitate the contact between researchers in the world, involving 31 countries [50]. The emergence of pan-proteomics will be important in the development of engineered constructs and offer insight into disease treatments human and animal.

7 Conclusions and future prospective

Over the years, proteomic study has been applied in different areas of knowledge to uncover the functional bases of the genome of a given organism. The introduction of new technologies has substantially contributed to enhance the proteomic study. Thus, the integration of pan-proteomics studies with new technologies and bioinformatics tools is a strategy to understating about factors related to genetic variation into given specie. The emergence of pan-proteomics will be important in the development of engineered constructs and offer insight into human and animal disease treatments. In addition, this strategy opens perspectives new to the systematic study of physiology and pathogenesis of several pathogens at the protein level. Although, pan-proteomics can promote insights in the biology of several organisms; its association with genomics, transcriptomics and metabolomics are powerful tools for obtaining information on gene function and regulatory networks inside highly related organisms, besides of promote impact across health, industry, and the environment.

References

[1] M.J. Han, S.Y. Lee, S.T. Koh, S.G. Noh, W.H. Han, Biotechnological applications of microbial proteomes, J. Biotechnol. 15 (2010) 341–349.
[2] T.C. Chao, N. Hansmeier, The current state of microbial proteomics: where we are and where we want to go, Proteomics 12 (2012) 638–650.
[3] J.R. Yates, C.I. Ruse, A. Nakorchevsky, Proteomics by mass spectrometry: approaches, advances, and applications, Annu. Rev. Biomed. Eng. 11 (2009) 49–79.
[4] J.A. Ankney, A. Muneer, X. Chen, Relative and absolute quantitation in mass spectrometry-based proteomics, Annu. Rev. Anal. Chem. (Palo Alto, Calif.) 12 (2018) 49–77.
[5] J.A. Broadbent, D.A. Broszczak, I.U. Tennakoon, F. Huygens, Pan-proteomics, a concept for unifying quantitative proteome measurements when comparing closely-related bacterial strains, Expert Rev. Proteomics 13 (2016) 355–365.
[6] Z. Zhang, S. Wu, D.L. Stenoien, L. Pasa-Tolic, High-throughput proteomics, Annu. Rev. Anal. Chem. (Palo Alto, Calif.) 7 (2014) 427–454.
[7] The UniProt Consortium, UniProt: the universal protein knowledgebase, Nucleic Acids Res. 46 (2018) 2699.
[8] J. Cox, M. Mann, MaxQuant enables high peptide identification rates, individualized p.p.b.-range mass accuracies and proteome-wide protein quantification, Nat. Biotechnol. 26 (2008) 1367–1372.
[9] http://www.waters.com.
[10] S. Lemeer, H. Hahne, F. Pachl, B. Kuster, Software tools for MS-based quantitative proteomics: a brief overview, Methods Mol. Biol. 893 (2012) 489–499.
[11] I.M. Lazar, Bioinformatics resources for interpreting proteomics mass spectrometry data, Methods Mol. Biol. 1647 (2017) 267–295.
[12] D.W. Huang, B.T. Sherman, R.A. Lempicki, Bioinformatics enrichment tools: paths toward the comprehensive functional analysis of large gene list, Nucleic Acids Res. 37 (2009) 1–13.
[13] M. Kanehisa, Y. Sato, M. Kawashima, M. Furumichi, M. Tanabe, KEGG as a reference resource for gene and protein annotation, Nucleic Acids Res. 4 (2016) D457–D462.
[14] D.W. Huang, B.T. Sherman, R.A. Lempicki, Systematic and integrative analysis of large gene lists using DAVID bioinformatics resources, Nat. Protoc. 4 (2009) 44–57.
[15] P. Shannon, A. Markiel, O. Ozier, N.S. Baliga, J.T. Wang, D. Ramage, N. Amin, B. Schwikowski, T. Ideker, Cytoscape: a software environment for integrated models of biomolecular interaction networks, Genome Res. 13 (2003) 2498–2504.
[16] C. von Mering, M. Huynen, D. Jaeggi, S. Schmidt, P. Bork, B. Snel, STRING: a database of predicted functional associations between proteins, Nucleic Acids Res. 1 (2003) 258–261.
[17] S. Keerthikumar, An introduction to proteome bioinformatics, Methods Mol. Biol. 1549 (2017) 1–3.
[18] L. Zhang, D. Xiao, B. Pang, Q. Zhang, H. Zhou, L. Zhang, J. Zhang, B. Kan, The core proteome and pan proteome of *Salmonella* Paratyphi A epidemic strains, PLoS One 24 (2014).
[19] V.K. Pettersen, H. Steinsland, H.G. Wiker, Comparative proteomics of enterotoxigenic *Escherichia coli* reveals differences in surface protein production and similarities in metabolism, J. Proteome Res. 5 (2018) 325–336.
[20] G.D. Jhingan, S. Kumari, S.V. Jamwal, H. Kalam, D. Arora, N. Jain, L.K. Kumaar, A. Samal, K.V. Rao, D. Kumar, V.K. Nandicoori, Comparative proteomic analyses of avirulent, virulent, and clinical strains of *Mycobacterium tuberculosis* identify strain-specific patterns, J. Biol. Chem. 1 (2016) 14257–14273.

[21] G.C. Tavares, F.L. Pereira, G.M. Barony, C.P. Rezende, W.M. da Silva, G.H.M.F. de Souza, T. Verano-Braga, V.A. de Carvalho, C.A.G. Leal, H.C.P. Figueiredo, Delineation of the pan-proteome of fish-pathogenic *Streptococcus agalactiae* strains using a label-free shotgun approach, BMC Genomics 7 (2019) 11.
[22] R.S. Nirujogi, B. Muthusamy, M.S. Kim, G.J. Sathe, P.T. Lakshmi, O.N. Kovbasnjuk, T.S. Prasad, M. Wade, R.E. Jabbour, Secretome analysis of diarrhea-inducing strains of *Escherichia coli*, Proteomics 17 (2017) 6.
[23] D.J. Wurpel, D.G. Moriel, M. Totsika, D.M. Easton, M.A. Schembri, Comparative analysis of the uropathogenic *Escherichia coli* surface proteome by tandem mass-spectrometry of artificially induced outer membrane vesicles, J. Proteome 6 (2015) 93–106.
[24] G.Y. Fisunov, D.G. Alexeev, N.A. Bazaleev, V.G. Ladygina, M.A. Galyamina, I.G. Kondratov, N.A. Zhukova, M.V. Serebryakova, I.A. Demina, V.M. Govorun, Core proteome of the minimal cell: comparative proteomics of three mollicute species, PLoS One 6 (2011) e21964.
[25] W.M. Silva, C.S. Sousa, L.C. Oliveira, S.C. Soares, G.F.M.H. Souza, G.C. Tavares, C.P. Resende, E.L. Folador, F.L. Pereira, H. Figueiredo, V. Azevedo, Comparative proteomic analysis of four biotechnological strains *Lactococcus lactis* through label-free quantitative proteomics, Microb. Biotechnol. 12 (2019) 265–274.
[26] J.A. Christie-Oleza, J.M. Piña-Villalonga, R. Bosch, B. Nogales, J. Armengaud, Comparative proteogenomics of twelve *Roseobacter* exoproteomes reveals different adaptive strategies among these marine bacteria, Mol. Cell. Proteomics 11 (2012).
[27] W. Albertin, O. Langella, J. Joets, L. Négroni, M. Zivy, C. Damerval, H. Thiellement, Comparative proteomics of leaf, stem, and root tissues of synthetic *Brassica napus*, Proteomics 9 (2009) 793–799.
[28] U. Mathesius, M.A. Djordjevic, M. Oakes, N. Goffard, F. Haerizadeh, G.F. Weiller, M.B. Singh, P.L. Bhalla, Comparative proteomic profiles of the soybean (*Glycine max*) root apex and differentiated root zone, Proteomics 11 (2011) 1707–1717.
[29] N. Ahsan, S.E. Stevenson, Proteomic mapping for legume nodule organogenesis, Proteomics 14 (2014) 153–154.
[30] A. Molassiotis, G. Tanou, P. Filippou, V. Fotopoulos, Proteomics in the fruit tree science arena: new insights into fruit defense, development, and ripening, Proteomics 13 (2013) 1871–1884.
[31] R. Pedreschi, S. Lurie, M. Hertog, B. Nicolaï, J. Mes, E. Woltering, Post-harvest proteomics and food security, Proteomics 13 (2013) 1772–1783.
[32] P. Šotkovský, M. Hubálek, L. Hernychová, P. Novák, M. Havranová, I. Šetinová, A. Kitanovičová, M. Fuchs, J. Stulík, L. Tučková, Proteomic analysis of wheat proteins recognized by IgE antibodies of allergic patients, Proteomics 8 (2008) 1677–1691.
[33] M.R. Roe, T.J. Griffin, Gel-free mass spectrometry based high throughput proteomics: tools for studying biological response of proteins and proteomes, Proteomics 6 (2006) 4678–4687.
[34] S. Wang, G. Zhang, Y. Zhang, Q. Song, Z. Chen, J. Wang, J. Guo, N. Niu, J. Wang, S. Ma, Comparative studies of mitochondrial proteomics reveal an intimate protein network of male sterility in wheat (*Triticum aestivum* L.), J. Exp. Bot. 66 (2015) 6191–6203.
[35] Y. Dong, X. Fang, Y. Yang, G.P. Xue, X. Chen, W. Zhang, X. Wang, C. Yu, J. Zhou, Q. Mei, W. Fang, C. Yan, J. Chen, Comparative proteomic analysis of susceptible and resistant rice plants during early infestation by small brown planthopper, Front. Plant Sci. 8 (2017) 1744.
[36] J. Grossmann, B. Fischer, K. Baerenfaller, J. Owiti, J.M. Buhmann, W. Gruissem, S. Baginsky, A workflow to increase the detection rate of proteins from unsequenced organisms in high-throughput proteomics experiments, Proteomics 7 (2007) 4245–4254.
[37] Z. Li, O. Czarnecki, K. Chourey, J. Yang, G.A. Tuskan, G.B. Hurst, C. Pan, J.G. Chen, Strigolactone-regulated proteins revealed by iTRAQ-based quantitative proteomics in *Arabidopsis*, J. Proteome Res. 13 (2014) 1359–1372.
[38] E. Gemperline, C. Keller, L. Li, Mass spectrometry in plant-omics, Anal. Chem. 88 (2016) 3422–3434.
[39] J.V. Jorrín, D. Rubiales, E. Dumas-Gaudot, G. Recorbet, A. Maldonado, M.A. Castillejo, M. Curto, Proteomics: a promising approach to study biotic interaction in legumes. A review, Euphytica 147 (2006) 37–47.

[40] S.A. Hibbert, M. Ozols, C.E. Griffiths, R.E. Watson, M. Bell, M.J. Sherratt, Defining tissue proteomes by systematic literature review, Sci. Rep. 8 (2018) 546.
[41] R.D. Semba, J.J. Enghild, V. Venkatraman, T.F. Dyrlund, J.E. Van Eyk, The Human Eye Proteome Project: perspectives on an emerging proteome, Proteomics 13 (2013) 2500–2511.
[42] T. An, Y.F. Wang, J.X. Liu, Y.Y. Pan, Y.F. Liu, Z.C. He, B.H. Lv, S. Gao, G. Jiang, Comparative analysis of proteomes between diabetic and normal human sperm: insights into the effects of diabetes on male reproduction based on the regulation of mitochondria-related proteins, Mol. Reprod. Dev. 85 (2018) 7–16.
[43] P.J. Thul, C. Lindskog, The human protein atlas: a spatial map of the human proteome, Protein Sci. 27 (2018) 233–244.
[44] J. YangXiong, X. Wang, B. Guo, K. He, C. Huang, Identification of peptide regions of SERPINA1 and ENOSF1 and their protein expression as potential serum biomarkers for gastric cancer, Tumor Biol. 36 (2015) 5109–5118.
[45] N. Zhong, Y. Cui, X. Zhou, T. Li, J. Han, Identification of prohibitin 1 as a potential prognostic biomarker in human pancreatic carcinoma using modified aqueous two-phase partition system combined with 2D-MALDI-TOF-TOF-MS/MS, Tumor Biol. 36 (2015) 1221–1231.
[46] S.S. Na, M.B. Aldonza, H.J. Sung, Y.I. Kim, Y.S. Son, S. Cho, J.Y. Cho, Stanniocalcin-2 (STC2): a potential lung cancer biomarker promotes lung cancer metastasis and progression, Biochim. Biophys. Acta Proteins Proteom. 1854 (2015) 668–676.
[47] Y. Liu, Y. Liao, L. Xiang, K. Jiang, S. Li, M. Huangfu, S. Sun, A panel of autoantibodies as potential early diagnostic serum biomarkers in patients with breast cancer, Int. J. Clin. Oncol. 22 (2017) 291–296.
[48] R. Akbani, P.K. Shing Ng, H.M.J. Werner, M. Shahmoradgoli, F. Zhang, Z. Ju, W. Liu, J. Yang, K. Yoshihara, J. Li, S. Ling, E.G. Seviour, P.T. Ram, J.D. Minna, L. Diao, P. Tong, J.V. Heymach, S.M. Hill, F. Dondelinger, N. Städler, L.A. Byers, F. Meric-Bernstam, J.N. Weinstein, B.M. Broom, R.G.W. Verhaak, H. Liang, S. Mukherjee, Y. Lu, G.B. Mills, A pan-cancer proteomic perspective on The Cancer Genome Atlas, Nat. Commun. 5 (2014) 3887.
[49] J. Boehmer, J. Ward, R. Peters, K. Shefcheck, M. McFarland, D. Bannerma, Proteomic analysis of the temporal expression of bovine milk proteins during coliform mastitis and label-free relative quantification, J. Dairy Sci. 93 (2010) 593–603.
[50] A.M.D. Almeida, A. Bassols, E. Bendixen, M. Bhide, F. Ceciliani, S. Cristobal, M. McLaughlin, Animal board invited review: advances in proteomics for animal and food sciences, Animal 9 (2015) 1–17.

CHAPTER 20

Pan-metabolomics and its applications

Li Bao[a,b], **Xiaofeng Liu**[a,b]
[a]National Clinical Research center for Cancer, Tianjin Medical University Cancer Institute and Hospital, Tianjin, China
[b]Key Laboratory of Cancer Prevention and Therapy, Tianjin, China

1 Introduction

Metabolomics is one of many omics, which has developed with the development of life sciences. Unlike other omics, metabolomics is a discipline that studies biological systems by investigating the changes of metabolites or their changes overtime after stimulation or perturbation of biological systems (cells, tissues, or organisms), such as mutation of a particular gene or environmental change [1]. Metabolome is the downstream and final product of genome [2]. It is a collection of small molecular compounds involved in metabolism, maintenance of normal function, and growth of organisms, mainly small endogenous molecules with relative molecular weight less than 1000 [3]. The number of metabolites in the metabolome varies greatly with biological species. It is estimated that the number of metabolites in the plant kingdom is more than 200,000 [4]. The number of metabolites in a single plant is between 5000 and 25,000. Even *Arabidopsis thaliana* produces about 5000 metabolites, far more than the metabolites in microorganisms (about 1500 types) and animals (about 2500 types) [5]. In fact, in humans and animals, due to the coexistence of microbial metabolism and the re-degradation of food and its metabolites themselves, it is not possible to estimate how many metabolites there are, and the concentration distribution ranges from 7 to 9 orders of magnitude. Therefore, the study of metabolomics is facing many challenges in terms of analysis platform, data processing, and biological interpretation [2].

Life sciences study the nature, characteristics, occurrence, and development of life phenomena and activities, as well as the relationship between various organisms and between organisms and environment. Since Watson and Crick established the DNA double helix structure model in 1953, the life sciences research has taken a new look [6]. The development of molecular biology on this basis enables the basic problems of life, such as genetics, development, disease, and evolution, to be interpreted from the molecular mechanism. Biological research has entered the stage of quantitative description of life phenomena. The rapid development of molecular biology has greatly promoted people's understanding of biological systems at the level of molecular composition. The genome project shows the composition of model organisms including *Escherichia coli*, yeast,

Pan-genomics: Applications, Challenges, and Future Prospects
https://doi.org/10.1016/B978-0-12-817076-2.00020-2
© 2020 Elsevier Inc.
All rights reserved.

nematodes, fruit flies, mice, and all the genetic information of human beings [7]. The mystery of life lies in these sequences.

Owing to technological breakthroughs the acquisition of genomic data is no longer a difficult task in life sciences. The basic completion of the Human Genome Project marks the arrival of the post-genome era [7, 8]. In this period, genome functional analysis has become the main task of life sciences. The core idea is to look at the material groups in organisms from a connected point of view, to study how genetic information is transferred from gene transcription to functional protein, how gene functions are expressed by product proteins, and so on [9]. Metabolomics appeared following the emergence of genome, transcriptome, and proteome one after another, and accordingly formed "omics" theory, such as transcriptomics, proteomics, and so on. However, the relationship between genes and functions is very complex, and transcriptome and proteome cannot be used to express all the functions of organisms. There are very complete and elaborate regulatory systems and complex metabolic networks in organisms, which are responsible for the generation and regulation of substances and energy required for life activities [10]. In this complex system, not only carbohydrates, fats, and their intermediate metabolites are directly involved in material and energy metabolism, but also substances that play an important role in regulating metabolism [11]. These substances form an interrelated metabolic network in the body. Gene mutation, diet, environmental factors, and so on can cause changes in some metabolic pathways in this network. The changes in these substances can reflect the state of the body [11].

Metabolites that play a regulatory role include neurotransmitters, hormones and cell signal transduction molecules in physiological function, and polypeptides, amino acids and their derivatives, amines, lipids, and metal ions in chemical composition. Most of these regulators are small molecular substances, and there are a large number of secondary metabolites in plants and microorganisms. These molecules are widely distributed in the body and have a wide range of regulatory effects on a variety of physiological activities. Only a small number of them can have a very strong biological effect. Molecules with different activities interact with each other by synergy, antagonism, or modification, forming complex networks in biological effects, signal transduction, and gene expression regulation, undertaking the important mission of maintaining the homeostasis of the organism, and being the material basis of neuroendocrine and immune network regulation and the most important component of homeostasis regulation [12, 13]. It is difficult to cover these very active and very important bioactive substances by the study of transcriptome and proteome. However, if we do not fully understand the physiological and pathophysiological significance of these substances, it is impossible to really clarify the essence of life function activities.

Traditional research methods are based on physiological and pharmacological experimental methods, and in the lack of high-throughput research technology, it is difficult to establish a research model for complex systems of small biological molecules.

In this case, metabolome and metabolomics emerged as the times require, and became an important breakthrough in systems biology [14]. Metabolism occurs at the end of the regulation of life activities, so metabolomics is closer to phenotype than genomics and proteomics [15].

In the broad sense of metabolomics, the history of metabolomics is quite long. Some target compounds in biological samples have been analyzed long ago to understand the state of living organisms. Currently, some technical platforms use in metabolomics, such as nuclear magnetic resonance (NMR), chromatography, and mass spectrometry, have a long application history [16]. Metabolomics in the strict sense (qualitative and quantitative analysis of all metabolic components in a specific biological sample under limited conditions) has been proposed for only a few years ago. It is generally believed that metabolomics originates from metabolic profiling analysis, which embodies the germination of the concept of "analyzing metabolites in biological samples as much as possible" [16].

In the early 1970s, Baylor Medical College published papers on metabolic profiling analysis. In their work, gas chromatography-mass spectrometry (GC-MS) was used to analyze the metabolites of various steroids, organic acids, and medicines for urinary problems [17]. This multicomponent analysis method was called metabolic profiling analysis, which pioneered the metabolic profiling analysis of complex samples. Metabolic profiling analysis has been widely used in qualitative and quantitative analysis of metabolites in blood and urine samples for screening and diagnosis of diseases. GC-MS is still used to diagnose diseases in clinics [17]. Subsequently, the emphasis was placed on the automation of analysis and the GC method was applied in the analysis of other types of compounds. In the 1980s, people began to use high-performance liquid chromatography (HPLC) and NMR to analyze metabolic profiles [18]. For example, in 1982, van der Greef from the Netherlands Institute of Applied Sciences (TNO) first used mass spectrometry to study metabolic fingerprints in urine [19]. In 1983, Sadler, Buckingham, and Nicholson published the first ^{1}H-NMR spectra of whole blood and plasma [20]. In 1986, chromatography magazine Journal of Chromatography published an album on metabolic profiling analysis. In the 1990s, metabolic profiling analysis technology has been developing steadily, with 10–15 papers published every year. However, during this period, people focused more on specific target compounds [21]. In the early 1990s, Sauter et al. studied the effects of different herbicides on barley by using GC-MS metabolic profiling analysis. The idea of using metabolic profiling analysis to study the effects of various factors on biological functions was then recognized [21]. In 1997, Steven Oliver's team proposed to assess the genetic function and redundancy of yeast genes by quantifying and characterizing metabolites, and pioneered the concept of metabolome [22]. In 1999, Nicholson and others put forward the concept of metabolomics [23], and have done a lot of fruitful work in disease diagnosis and drug screening [24–26]. Then, the scientists of Max-Planck-Institut in Germany began the study of plant metabolomics [27], which greatly enriched it.

Metabolomics is characterized by:

(1) A focus on endogenous compounds.

(2) Qualitative and quantitative studies of small molecular compounds in biological systems.

(3) Upregulation and downregulation of these compounds which indicate the effects of disease, toxicity, gene modification, or environmental factors.

(4) Knowledge of these endogenous compounds which can be used for disease diagnosis and drug screening.

Compared with transcriptomics and proteomics, metabolomics has the following advantages:

(1) Small changes in gene and protein expression can be amplified on metabolites, making detection easier.

(2) Metabolomics does not require genome-wide sequencing and the number of expressed sequence tags (ESTs) database.

(3) The types of metabolites are much smaller than the number of genes and proteins (about 10^3 orders of magnitude per tissue, even in the smallest bacterial genome there are thousands of genes).

(4) The technology used in the study is more general because the given metabolites are the same in every tissue.

In several common omics studies, genomics mainly studies the genetic structure of biological systems, that is, the sequence and expression of DNA. Proteomics studies proteins expressed by biological systems and differences caused by external stimuli. Metabolomics is an extension of genomics and proteomics, which studies the changes in all metabolites produced by external stimuli in biological systems (cells, tissues, or organisms). With the deepening of these studies, scientists have come to realize that changes in the genome are not necessarily expressed, and thus do not have a substantial impact on the system. Concentrations of some proteins may increase due to changes in external conditions, but this protein may not be active and thus has no impact on the system [28]. At the same time, due to the functional compensation of genes or proteins, the loss of one gene or protein will be compensated by the presence of other genes or proteins, and the net result of the final reaction is zero. The production and metabolism of small molecules is the final result of these events, which can more accurately reflect the state of biological systems [29]. Therefore, the study of systems biology should cover genomics, transcriptomics, proteomics, and metabolomics. Any single genomics study is incomplete for understanding biological problems.

Systematic biology is a science that studies various molecules with different structures and functions and their interactions at the level of cells, tissues, organs, and organisms as a whole, and quantitatively describe and predict biological functions, phenotypes, and behaviors through computational biology [30]. Systems biology starts with genome sequence and completes the research from the life code to life process [31]. If life is regarded as a metabolic network composed of innumerable interrelated biochemical reactions regulated by genes, then systems biology will identify the various molecules and their interactions at each reaction node, from local to global, and finally complete the road map of the whole life activity [31, 32]. The main technical platforms of systems biology are genomics, transcriptomics, proteomics, metabolomics, interatomics, and phenomics. These "omics" detects and identifies various molecules at the level of DNA, RNA, protein, and metabolites, and studies their functions and interrelationships among various molecules [33]. Furthermore, the pathways and networks of biochemical reactions are found, the biological modules are constructed, and the interaction maps of organisms are drawn based on the interaction of modules. The combination of metabolomics and other omics is of great significance in elucidating the mysteries of life [34, 35].

2 Methodologies of pan-metabolomics

Metabolomics research generally includes sample collection and preparation, metabolomic data collection, data preprocessing, multivariate data analysis, marker identification, and pathway analysis [36]. Biological samples can be urine, blood, tissues, cells, and cultures. After collection, biological reactions are inactivated and pretreated. Then the types, contents, states and changes of metabolites are detected by NMR, MS, or GC, and metabolic profiles or metabolic fingerprints are obtained. Then, multivariate data analysis method is used to determine the dimensionality reduction and information on the obtained multidimensional complex data, identify the metabolic markers with significant changes, and study the metabolic pathways and changes involved in order to elaborate the response mechanism of organisms to the corresponding stimuli, so as to achieve the purpose of classification and discovery of biomarkers [37].

Oliver Fiehn has divided the analysis of metabolites in biological systems into four levels according to different objects and purposes of the study [27].

(1) Metabolite target analysis: Analysis of a specific component or components. In this level, it is necessary to adopt certain pretreatment technology to remove interferences in order to improve the sensitivity of detection.

(2) Metabolic profiling analysis: Quantitative analysis of a few presupposed metabolites, for example, a class of structure-related compounds (such as amino acids, cis-diols), and all intermediates of a metabolic pathway or marker components of multiple metabolic pathways. In metabolic profiling analysis, we can make full use of the unique physical and chemical properties of these compounds, and use specific techniques to complete the pretreatment and detection of samples.

(3) Metabolomics: Qualitative and quantitative analysis of all endogenous metabolic components in specific biological samples under limited conditions. In metabolomics research, sample pretreatment and detection techniques must meet the requirements of high sensitivity, high selectivity, and high throughput for all metabolic components, and the matrix interference should be small. Metabolomics involves a large amount of data, so it needs chemometric techniques that can analyze its data.

(4) Metabolic fingerprint analysis: Instead of identifying a single component, samples are quickly classified by comparing the differences in the fingerprints of metabolites (e.g., phenotypic identification).

Strictly speaking, only the third level is the true meaning of metabolomics research. Currently, the ultimate goal of metabolomics is still an impossible task, because no real metabolomics technology has been developed to cover all metabolites regardless of the size and nature of the molecules. However, it differs significantly from metabolic profile (spectrum) analysis. In specific experiments, metabolomics try to analyze all the visible peaks, so the characteristics of metabolomics can also be expressed as it will try to analyze as many metabolic components as possible.

2.1 Sample collection and preparation

Sample collection and preparation is one of the first and most important steps in metabolomics, which requires strict experimental design. First, we need to collect a sufficient number of representative samples to reduce the impact of individual differences in biological samples on the analysis results. In the design of the experiment, the time, location, type, and group of samples should be fully considered. In the study of human samples, many factors such as diet, sex, age, day, and night, and geography should also be taken into account. In addition, strict quality control is needed in the analysis process, such as sample repeatability, analysis accuracy, blank, and so on [38].

Sample extraction and pretreatment methods are different according to different research objects, purposes, and analytical techniques. If the NMR technology platform is used, the sample can be analyzed with less pretreatment. For the analysis of body fluids, in most cases, only buffer or water can be used to control pH and reduce viscosity. When using MS for "full" component analysis, the sample processing method is relatively simple, but there is no universal standardization method. The principle of "similarity and cleanliness" is still the basis. After deproteinization, metabolites are usually extracted with water or organic solvents (such as methanol, hexane, etc.) to obtain water extracts and organic solvent extracts, thus separating the nonpolar phase from the polar phase [39]. For metabolic profiling or target analysis, more complex pretreatments are needed, such as solid-phase microextraction, solid-phase extraction (SPE), affinity chromatography, and so on [39]. When using gas chromatography or GC-MS, derivatization is often needed to increase the volatility of samples [40]. Because the specific extraction conditions are usually only suitable for some compounds, there is no suitable extraction method for all metabolites at present. Different extraction methods should be selected depending on different compounds, and the extraction conditions should be optimized [41].

Because metabolomics analyzes many samples at a time, it is impossible to collect samples in 1 day. Therefore, the problem of sample preservation should also be noted, preferably at −80°C. The COMET project showed that when urine samples were kept in a refrigerator at −40°C, no change was found for at least 9 months. But after 18 months, slight changes were observed in the intermediate products of tricarboxylic acid (TCA) cycle. The plasma was stored at −80°C for 6 months, and no significant changes were found in NMR spectra.

2.2 Data collection

After sample collection and pretreatment, the metabolites in samples need to be determined by appropriate methods. Metabolomics analysis methods require high sensitivity, high throughput, and unbiased characteristics. Unlike the existing methods, which only analyze specific types of compounds, the size, number, functional groups, volatility, chargeability, electromobility, polarity, and other physical and chemical parameters of the objects analyzed by metabolomics vary greatly. Due to the complexity of metabolites and biological systems, up to now, there exists no metabolomics analysis technology that can meet all the above requirements. The existing analysis techniques have their own advantages and application scope [42–45]. It is preferable to use a combination of techniques and multiple methods for comprehensive analysis. Separation and analysis methods such as chromatography, mass spectrometry, NMRM, capillary electrophoresis, infrared spectroscopy, electrochemical detection, and their combinations all appear in metabolomics research [46–50]. Among them, GC-MS has high separation, high throughput and universality, and high sensitivity and specificity of mass spectrometry. NMR, especially ^{1}H-NMR, has become the most important analytical tool because of its universality for hydrogen-containing metabolites.

NMR: NMR is the main technology in metabolomics research. The advantage of NMR is that it can detect samples noninvasively and unbiased. It has good objectivity and reproducibility. Samples do not need tedious processing. It has high throughput and low detection cost per sample. In addition, ^{1}H-NMR is responsive to hydrogen-containing compounds. It can detect most of the compounds in the samples and meet the target of detecting as many compounds as possible in metabolomics. Although NMR can be used for nondestructive analysis of complex samples such as urine and blood, compared with mass spectrometry, its disadvantages are as follow: its detection sensitivity is relatively low (using the mature ultralow temperature probe technology, its detection sensitivity is at the level of nanogram), its dynamic range is limited, and it is difficult to simultaneously determine metabolites coexisting in biological systems with large concentration differences. However, the instrument is purchased, hence the investment needed is also large.

In order to improve the sensitivity of NMR technology, researchers have adopted the methods for increasing field strength, using cryogenic probes and microprobes. To solve the problem of resolution, multidimensional NMR technology and liquid chromatography-nuclear magnetic resonance (LC-NMR) were used. Daykin et al. used chromatographic techniques to detect the metabolites of lipoproteins in patients with cardiovascular diseases by LC-NMR [51]. Nicholson's team used the magic angle spinning (MAS) technology, which has been developed in recent years, to rotate the sample to 54.17 degrees in the direction of magnetic field, thus overcoming the anisotropy of line broadening and chemical displacement caused by dipolar coupling [52–54]. Using MAS technology, researchers can obtain high-quality NMR spectra. The sample only needs a small amount of D2O without pretreatment. The sample volume is only about 10 mg. Metabolomics based on NMR technology has been widely used in the clinical diagnosis of drug toxicity, gene function, and disease [54–59].

Mass spectrometry: Compared with the weaknesses of low sensitivity and narrow detection dynamic range of NMR, MS has high sensitivity and specificity, and can achieve simultaneous rapid analysis and identification of multiple compounds. With the development of mass spectrometry and its combination technology, more and more researchers have applied GC-MS to metabolomics research [60, 61]. The main advantages of GC-MS method include high resolution and detection sensitivity, and standard spectrogram library for reference and comparison, which can be used for qualitative analysis of metabolites. However, GC cannot directly obtain the information on most metabolic components which are difficult to volatilize in the system. For metabolites with lower volatility, derivatization is needed, and the pretreatment process is cumbersome. GC-MS is often used to analyze the metabolic fingerprints of plants and microorganisms [62, 63]. For example, Fiehn et al. used GC-MS to study the genotype and phenotype relationship of Arabidopsis [27], and Styczynski et al. analyzed the metabolic products of *E. coli* in detail [62]. LC-MS avoids the complicated sample pretreatment in GC-MS. Because of its high sensitivity and wide dynamic range, LC-MS has been increasingly used in metabolomics research [64, 65]. It is highly suitable for the detection of complex metabolites in biological samples and the identification of potential markers. Metabolomics of LC-MS usually uses reverse packing and gradient elution procedure. However, in body fluid samples, especially urine samples, there are a large number of hydrophilic metabolites, which are not retained or very weak in reversed-phase chromatography. Recently, researchers have used hydrophilic interaction chromatography (HILIC) to solve the problem of weak retention of hydrophilic substances [66]. New analytical techniques, such as ultrahigh-performance liquid chromatography/high-resolution time-of-flight mass spectrometry [67], capillary liquid chromatography-mass spectrometry (LC-MS) [64], and Fourier transform ion cyclotron resonance [68], have also been used in metabolomics to improve the sensitivity and flux of metabolites.

2.3 Data analysis platform

Metabolomics obtains a great deal of multidimensional information. In order to fully tap the potential information in the obtained data, a series of chemometrics methods are needed. In metabolomics, most of them are based on the information about the detected metabolites, including the response before and after gene mutation, the discrimination and classification of two or more types (such as metabolites between different phenotypes) [69], and the discovery of biomarkers [70, 71]. The main methods used in data analysis are pattern recognition technology, including unsupervised learning method and supervised learning (SL) method.

The unsupervised learning method is used to classify the samples from the original spectral information or preprocessed information, and the corresponding visualization technology is used to express the samples intuitively without any background information about the sample classification. This method compares the classified information with the original information on these samples (such as drug action sites or disease types), establishes the relationship between metabolites and the original information, screens the markers related to the original information, and then examines the metabolic pathways. The method used for this purpose has no training samples available for learning, so it is called unsupervised learning method, such as principal component analysis (PCA) [72], non-linear mapping [73], cluster analysis [74], and so on. The SL is used to establish mathematical models between classes to maximize the separation between samples, and to predict unknown samples using the established multiparameter model. Such methods are called SL because there are training samples available for learning when building models. This method often requires the establishment of validation sets for the validation of sample classification (to prevent over-fitting) and test sets for testing model performance. The main methods used in this field are the improved methods based on PCA, partial least squares (PLS), and neural network. The commonly used methods are analog soft independent modeling and PLS-discriminant analysis (PLS-DA) [75], orthogonal (O)-PLS [76]. As a nonlinear pattern recognition method, artificial neural network (ANN) technology has also been widely used. PCA and PLS-DA are the most commonly used pattern recognition methods in metabolomics. These two methods usually use a score plot to obtain information about sample classification and a load plot to obtain variables contributing to classification and their contribution size, so as to find variables that can be used as biomarkers. In addition, in the various stages of data processing and analysis, the quality control of data and validation of model also need to be paid enough attention [77, 78].

It should be emphasized that the metadata derived from the above-mentioned analytical instruments cannot be directly used for pattern recognition analysis, but need to be preprocessed to transform the metadata into a data form suitable for multivariate analysis (mainly pattern recognition), so that the same metabolites are represented by the same variable in the generated data matrix, and all samples have the same number variables. Finally, the data used for pattern recognition is in the form of a two-dimensional (2D) matrix data. The rows represent the number of samples or experiments and list the corresponding individual measurement indicators (usually the signal intensity of metabolites, etc.). Microwave fluctuation of instrument and changes in sample pH and matrix will cause changes in chemical displacement in NMR [79]. The difference in retention time is often caused by the composition of mobile phase, the slight change in column temperature, the reproducibility of gradient, and the change in the state of column surface in LC-MS. Before pattern recognition, peak matching (or peak alignment) is needed to compare the data of each sample correctly. The main data preprocessing includes noise filtering, deconvolution, peak alignment, peak matching, standardization, and normalization. In practice, not all these steps need to be carried out, but according to the actual situation, only a few of them are pretreated. In contrast, the retention time repeatability of HPLC is worse than that of GC, and peak matching is relatively difficult. The peak alignment algorithm developed by us can be used not only for HPLC, but also for peak matching in GC metabolomics [80].

2.4 Database of metabolomics

Metabolomics analysis is inseparable from various metabolic pathways and biochemical databases. Compared with genomics and proteomics, there is no complete database with similar functions in metabolomics. Some biochemical databases can be used for structural identification of unknown metabolites or for biological function interpretation of known metabolites, such as Connections Map DB, Kyoto Encyclopedia of Genes and Genomes (KEGG), METLIN, HumanCyc, EcoCycmetacyc, BRENDA, MetaCyc, UMBBD, EMP, IRIS, AraCyc, Ex-PASy, main metabolic pathways (MMP) on the Internet, Dr. Duke's database of Phytochemistry and Ethnobotany, and The University of Arizona's database of natural products.

Ideal metabolomics databases should include metabolomic information on various organisms and quantitative data of metabolites, as in the Human Metabolomics Database (http://www.hmdb.ca). But, in fact, this information is very scarce. Some public databases are also very useful for structural identification of metabolites in various biological samples, such as PubMed Compound Library and ChemSpider Database, which contain structural information about 16.5 million compounds and can be retrieved online (Table 1).

3 Application of pan-metabolomics

3.1 Drug discovery

In the field of drug research and development, especially in the western countries, targeted R&D strategy is mainly used to "make 90% of drugs effective only for 30%–50% of

Table 1 Database for searching metabolomes

Name of database	Address
KEGG	http://www.genome.jp/kegg/ligand.html
HumanCyc	https://biocyc.org/
METLIN database	https://metlin.scripps.edu/
Tumor metabolome database	https://www.metabolic-database.com/
LIPID MAPS	http://www.lipidmaps.org/data/index.html
SphinGOMAP	http://sphingomap.org/
LipidBank	http://lipidbank.jp/
The Human Metabolome Database	http://www.hmdb.ca/
PubChem Compound	https://www.ncbi.nlm.nih.gov/pubmed/
NIST Mass Spectrometry Data Center	https://chemdata.nist.gov/
ChemSpider	http://www.chemspider.com/

patients." That is to say, 50%–70% of patients do not benefit from the received drug treatment and bear its side effects as well. A consensus has been reached to identify effective physiological and clinical markers as a cheap and fast method for screening effective or toxic drugs for specific populations [81]. In fact, as the cost of drug development increases, the consumption of new drug discovery-development links has become one of the great challenges facing the pharmaceutical industry. Any tool that can quickly, economically, and effectively predict the potential toxicity of drugs to a particular population will undoubtedly be the focus of attention.

Metabolomics has been widely used in the confirmation of diseases (including genetically modified animals), drug screening, evaluation of efficacy and toxicity, mechanism of action, and clinical evaluation in animal models [82–85]. Nicholson's team has done profound work on toxicity assessment of drugs using NMR based metabolomics techniques. Their research shows that metabolomics can be used to judge the organs and tissues affected by toxicity, to speculate the mechanism of drug-related action, and to identify potential biomarkers related to toxicity. Based on this, an expert system for toxicity prediction and the time-varying trajectory of endogenous metabolites affected by toxicants in animals can be established. In the COMET research project, hepatorenal toxicity of 147 typical drugs was studied. After analyzing the NMR spectra of metabolites in body fluid and tissues of normal and poisoned rats and mice, and combining with the pathological effects of known toxic substances, the first expert system of rat liver and kidney toxicity was established. The expert system can be divided into three independent levels to realize normal/abnormal discrimination, toxicity or disease identification of unknown specimens, and pathological biomarkers identification. Currently, the goal of COMET project is to study the molecular mechanism of standard toxic drugs and establish predictable structure-activity relationships [86]. Mally et al. EM have studied the feasibility of 1H-NMR metabolomics for markers of renal function damage. They studied the feasibility of 4-hydroxy-2(E)-nonylaldehyde-mercaptouric acid as a marker of renal function damage in mice models induced by FeNTA or potassium bromide. The results showed that ^{1}H-NMR metabolomics could be used to indicate renal function damage, but had no specificity for oxidative stress. HNE-MA and other markers of phospholipid peroxidation were well correlated. The types of markers were related to pathological conditions, and no universal markers of oxidative stress were found.

3.2 Disease research

Due to the pathological changes in the body, the metabolites also undergo some corresponding changes. The analysis of metabolites caused by diseases, that is, metabolomics analysis, can help people better understand the process of disease and metabolic pathways of substances in the body, as well as the discovery of biomarkers of diseases and assist in clinical diagnosis. For example, Brin-die et al. used ^{1}H-NMR technology to analyze the metabolomics of 36 patients with severe cardiovascular disease and 30 patients with cardiovascular atherosclerosis. Combining PCA, SIMCA (soft independent modeling of class analogy), PL DA, OSC-PLS (orthogonal signal correction-partial least squares) and other pattern recognition techniques, we realized the identification of cardiovascular disease and its severity, and obtained more than 90% sensitivity and specificity.

The application of metabolomics in disease research mainly includes the discovery of pathological markers, diagnosis, treatment, and prognosis judgment [87, 88]. The most extensive application is to find metabolic markers (groups) related to disease diagnosis and treatment. The related markers obtained by metabolite spectrum analysis are the basis of disease classification, diagnosis, and treatment. Currently, many literatures have reported the application of metabolomics in disease research, such as neonatal metabolic disorders [89], coronary heart disease [90], cystitis [91], hypertension [92], and mental system diseases [93].

The established metabolomics has been applied to study the pathological markers in major diseases (such as cancer, type 2 diabetes, and severe hepatitis). A normal-phase liquid chromatography/electrospray ionization linear ion-trap mass spectrometry method has been developed for the analysis of phospholipid metabolic profiles in body fluids. The method was applied to classify type 2 diabetes mellitus and healthy people. Four possible molecular biomarkers of phospholipids were identified to study the effects of n-3 polyolefin fatty acid eicosapentaenoic acid (EPA) and docosahexaenoic acid (DHA) on several major phospholipid compositions in membrane lipid raft and soluble membrane region of human JurkatE-6-lT cells. The results showed that EPA or DHA could significantly increase the phospholipid content of n-3 polyunsaturated fatty acid (PUFA), which revealed the molecular mechanism of PUFA immunosuppression [94, 95].

In addition, the method of nucleoside metabolic profiling in body fluid based on SPE-HPLC has been established and used in cancer research. The level and pattern of nucleoside emission in urine of normal people and cancer patients have been established. The difference in emission levels between different cancers has been compared, and the sensitivity of detecting cancer is significantly higher than that of tumor markers currently used in clinical practice. However, it has good value in differentiating benign and malignant tumors, monitoring the effect of surgery and chemotherapy, and predicting the recurrence of tumors. At the same time, the metabolomics platform based on LC-MS method was applied to study liver diseases, which realized the effective distinction between patients with different liver diseases and normal people. The false positive rate of hepatitis and cirrhosis patients was only 7.4% in the diagnosis of liver cancer; when it was applied to study the acute episode of chronic hepatitis B, the diagnostic accuracy was 100%. A traditional marker and four new markers were identified [96].

3.3 Plant metabolomics

Many studies of plant metabolomics have focused on the relatively independent branch of cell metabolomics. In order to further understand the metabolic pathway of plants, we mainly study the relationship between genotype and phenotype and reveal the function of silent genes by studying the corresponding changes in metabolome in plant cells after gene variation or changes in environmental factors. Metabolomics studies in plants mostly focus on metabolite fingerprinting or metabolite profiling. According to different objects, the research of plant metabolomics mainly includes: (1) metabolomics of some specific species. This kind of research usually takes a plant as an object, chooses an organ or tissue, and carries out the qualitative and the quantitative analysis of the metabolite in it. (2) Metabolomic phenotypes of different genotypes. Generally, two or more plants of the same species (including normal control and genetically modified plants) are needed, and then metabolomics is used to compare and identify the plants of different genotypes studied [97]. (3) Metabolomics of some ecotypes. This kind of research usually chooses the same plant in different ecological environments to study the effects of growth environment on plant metabolites. (4) Plant autoimmune response to external stimulation.

With the rapid development of plant cell metabolomics, people have begun to make use of this technology. The establishment of metanomics is a typical example. Their goal is to find key genes in plant metabolism, such as those that make plants resistant to cold. The idea is to follow the method of metabolomics, after changing plant genes, carry out plant metabolic analysis or record metabolites, to grasp the information about plant metabolic pathways more quickly.

About 80% of the secondary metabolites found so far come from plants. Plant secondary metabolites contain many functional components, which can be used as drugs (such as artemisinin, taxol, three terpenoidsaponins, etc.), insecticides, dyes, flavors, and fragrances [98]. Although plants can synthesize hundreds of thousands of low-molecular-weight organic compounds (secondary metabolites), many of which are valuable for use, the enormous synthetic ability of plant cells has not been well utilized. More importantly, the content of important secondary metabolites is very low, such as artemisinin, which was first isolated from *Artemisia annua* and has outstanding therapeutic effects on falciparum malaria, such as cerebral malaria. The content of *A. annua* is less than 1%, which is far from people's expectation [98]. Up to now, the secondary metabolic network of plants has not been well characterized, and the functional genome maps related to biosynthesis are far from complete. These are very important to break the bottleneck of low productivity in plant or plant cell culture. A method for the separation and analysis of volatile oil from *Artemisia annua L.* based on full 2D gas chromatography-time-of-flight mass spectrometry (GC × GC-TOFMS) was established [99]. The constituents of volatile oil from *Artemisia annua L.* were analyzed. The results showed that the volatile oil was mainly composed of alkanes, monoterpenes, monoterpene oxygen derivatives, sesquiterpenes, and sesquiterpene oxygen derivatives. More than 300 compounds can be identified from the volatile oil of *A. annua* by 2D GC × GC-TOFMS, and important intermediates in the metabolic pathway of artemisinin have been identified. The samples of transgenic *A. annua* were analyzed and qualitatively analyzed by GC × GC-TOFMS. Nearly 100 terpenoids were identified and the differences in metabolites between common and transgenic plants were found. The metabolic fingerprints of *A. annua* at different growth stages were studied by gas chromatography-flame ionization detector (GC-FID) and GC-MS [99]. The five growth stages (seedling stage, germination stage, pre-budding stage, budding stage, and blooming stage) of *A. annua* could be well distinguished, and the bottleneck in the production pathway of artemisinin was confirmed.

3.4 Microbial research

The first microbial metabolomics literature reported microbial contamination during the fermentation of glucan by *Leuconostoc mesenterica* through the analysis of fatty acids, amino acids, and sugars by GC-MS [100]. Currently, metabolomics technology has been applied to microbial phenotypic classification [101], mutant screening [102], metabolic pathway and microbial metabolic engineering [103], monitoring and optimization of fermentation process [104], and microbial degradation of environmental pollutants [105].

Buchholz et al. have combined rapid sampling technology with other analytical techniques to achieve rapid and high-frequency quantification of intracellular metabolites, which can be used for dynamic detection of fermentation process. This technology helps to study the effects of various factors on fermentation, thus increasing the yield of bioengineering. Dalliige et al. used liquid chromatography tandem mass spectrometry to monitor the amino acids in the fermentation process. Through analysis, it was confirmed that one of their subsets reflect the fermentation state. Grivet et al. made a detailed review of the application of NMR in microbial metabolomics research [106]; Ishii et al. made a detailed review of computer simulation in microbial cells [107]. Based on the changes and characteristics of intracellular and extracellular metabolites under different gene modification, and comparing the differences in fingerprints of intracellular metabolites between different substrates and strains, we have investigated the effects of environment on microbial metabolism, and obtained the relationship between intracellular amino acid changes of alkalophilic lactic acid bacteria and lactic acid production in fermentation broth under different conditions. The changes in the metabolic pools of main organic acid metabolites in TCA cycle and glycolysis pathway of *Pseudomonas aeruginosa* and *E. coli* under the action of different antibiotics were studied. It was found that the effect of quinolones on TCA metabolites in *P. aeruginosa* was negatively correlated with that of beta-lactams. The correlation was positive in antimicrobial activity. The succinic acid production of wild strains, ACK, and SDH gene-modified strains of *E. coli* was analyzed. It was found that when glucose and fructose were used as carbon sources, the biomass of ACK strains was significantly lower than that of the other two strains. Under the same conditions, when fructose was the only carbon source, the biomass of ACK strain was higher than that of glucose. Through the analysis of intracellular metabolic flux, it was found that the metabolic flux of ACK strain changed significantly.

3.5 Nutritional research

Nutrition is a science that studies the rules of human nutrition and measures to improve it. Nutrition has always played an important role in the medical system as a discipline to study the relationship between the food and the health. Reasonable dietary allocation is not only beneficial to health, but also can prevent the occurrence of some diseases in the early stage, so as to improve people's quality of life. With the continuous development of modern analytical techniques, metabolomics not only provides an important research platform for understanding the metabolic pathway and its regulation mechanism in body fluid or tissue, but also provides a new research method for nutrition, that is, to study the intake of various nutrients and bioactive substances in food by metabolomics [108]. The influence and intervention of metabolic pathways in the body can also be assessed by metabolomics methods, including the role of intestinal flora and the health effects of environmental and behavioral factors.

The content of traditional nutritional research includes the nutritional requirements of the body, the biological function of dietary nutrients, and the relationship between nutrition and health. Among them, the function of nutrients and their effects on health are mostly studied by in vitro experiments with pure nutrients. However, our daily intake of food nutrients are not only diverse, but also complex; these substances meet the different needs of the body: energy intake, the acquisition of various trace elements, the absorption of various substances that cannot be synthesized in the body, and are closely related to the occurrence and development of diseases [109], for example, the effects on cell differentiation, apoptosis and cell cycle, DNA repair, hormone secretion, carcinogen metabolism, and inflammation response. However, due to the complexity of the body's internal environment, we have not yet formed a complete understanding of the function of nutrients intake, the role of nutrients in the body and the interaction between them [109]. With the development of systemic biomedicine, modern nutrition pays more attention on how to balance the internal environment of the body through dietary regulation in order to prevent the occurrence of diseases. Therefore, while studying the functions of some precise nutrients in food independently, we also need a research method that can understand the interaction between food and organism from a holistic and systematic perspective. In the post-genomics era, the research platforms of transcriptomics, proteomics, and metabolomics all provide new technological platforms for nutrition research at different levels to explore the effects of bioactive substances in food on gene phenotypes [110].

Genomic studies have shown that the diversity of genes among individuals is the decisive factor leading to different nutritional needs of different individuals. Genes determine the potential ability of different individuals to absorb, transform, and metabolize various nutrients. Genomic differences are usually expressed in single nucleotide polymorphisms, that is, changes in a single base in a DNA sequence. Genetic polymorphism is the root cause of the differences among various life forms. The expression of gene polymorphism in nutrition-related diseases has been reported. Because the regulation of gene on life process is ultimately achieved by the protein it expresses, proteomics can explore the interaction between nutrients and organism and its regulation mechanism through the study of protein expression status, which is an important platform for nutritional research. Compared with genomics and proteomics, metabolomics research clearly reflects the changes in metabolic pathways in the body by characterizing the changes in metabolites concentration in body fluids and tissues. It has more advantages in reflecting the development of diseases, especially in studying the metabolism of nutrients. Many multi-pathogenic diseases (such as diabetes mellitus, cancer, etc.) occur and develop over a long period of time [111]. The changes in genes and gene expression products and their interaction with the environment lead to the occurrence and development of diseases. The above-mentioned omics can monitor the process of disease occurrence at different levels, so as to achieve the goals of disease prevention, early detection, and intervention. For nutritional research, metabolic processes such as absorption, transformation, and degradation of bioactive substances in food are the main research objectives, especially some small molecular nutrients, such as amino acids, lipids, vitamins, and so on. These are consistent with the contents of metabolomics. It can be said that metabolomics provides a systematic biological perspective and a new research platform for nutrition research [112]. Metabolomics research methods have been widely used in the research and practice of nutrition to evaluate the relationship between individual dietary habits, nutritional status, and different food components and the occurrence of chronic diseases from a holistic point of view, and to explore the changes of metabolic pathways in vivo [113].

Among the four different levels of metabolomics, "whole component analysis" of metabolomics is the most widely used method in nutritional research, and there are many related reports on the analysis of specific target compounds and metabolic profiling analysis of a certain class of substances. In nutritional research, systematic study of small molecules in metabolic pathways has always been the main research direction. Earlier, due to technical constraints, nutritionists could only analyze the active ingredients of a small number of nutrients to determine their effects on cell function and the possible relationship between disease occurrences, and so on. With the development of modern analytical methods, it is possible to analyze a large number of small molecular compounds in body fluids or tissues at the same time. In the past 10 years, through continuous measurements of complex metabolite concentrations in body fluids, metabolomics has achieved and well completed the assessment of the role of food and nutrition in complex internal environment. Many scientists have explored the relationship between metabolism and food in human or animal bodies. In this way, scientists can explain the relationship between food nutrition and human health from the perspective of metabolism [114].

3.6 Environmental sciences

With the deterioration of environmental pollution, many chemicals have reached the alert level that threatens the existence various organisms. These chemicals are the main causes of biochemical, genetic, structural, or physiological damage in the process of life. It is necessary to elucidate the nature and mechanism of the physiological and toxicological interaction of these substances in order to elucidate the pathogenic mechanism of various environmental diseases and put forward effective treatment schemes [115]. As a new technology, metabolomics takes the final metabolites of organisms as the research object, quantitatively analyzes the changes in endogenous metabolites in biological system, and studies the regulation and response mechanism of organisms to external stimuli from the perspective of systematic biochemical spectrum [116]. In the past several years, this emerging discipline has developed rapidly, and has been widely applied to molecular pathology, toxicology, functional genomics, clinical medicine, and environmental science [116].

Since 1999, metabolomics has developed rapidly as a new technology of omics, and the papers on metabolomics have increased exponentially. Its application has affected many aspects, such as basic life science, drug research and development, disease physiology, nutrition and plant pharmacy, environmental science, and so on [117]. As one of the important research methods of environmental safety, metabolomics provides a powerful weapon for early detection of pathophysiological changes caused by environmental toxic substances. Compared with other omics (such as transcriptomics and proteomics), the advantage of metabolomics is that all exogenous stimuli (such as drugs, food, environment, etc.) can promote the regulation of biological system and lead to changes in the metabolome, which is the ultimate result of biological processes [118]. Many changes in organisms that cannot be reflected from the transcriptome and proteome can be reflected through the metabolome. The metabolic pathways of different organisms are relatively stable even though there are great differences in metabolic rates. Therefore, biomarkers found in experimental studies are often interoperable among different species. Moreover, the techniques used in metabolic studies of any species can be used in metabolic studies of other species at the same time. In addition, metabolomics has the advantages of being noninvasive, quantifiable, high throughput, and low cost [116, 119]. These characteristics lay a solid foundation for the application of metabolomics in environmental safety assessment.

In the field of environmental science, metabolomics is mainly used to study and elucidate the physiological and biochemical reactions of organisms after exposure to toxic chemicals, and to study the long-term mechanism of action of environmental chemicals. Its indicators can be used as a reasonable standard for predicting the safety of existing global chemical mixtures. Metabolism and physiological and biochemical reactions of organisms stimulated by external environment (such as cold, heat, hunger, etc.) can also be described by metabolomics. In addition, metabolomics has great potential in the evaluation and prediction of biological health. The National Institute of Environmental Health Sciences (NIEHS) has carried out metabolomics research on the interaction between potential environmental inputs and diseases, and attached great importance to it [120].

4 Development of pan-metabolomics in the future

In general, metabolomics is still in the developing stage. It faces great challenges in both methods and applications. It needs the cooperation and intersection of other disciplines.

In terms of technology platform and methodology, due to the complexity of biological samples, metabolomics research requires higher sensitivity, resolution, dynamic range, and flux of analytical techniques. The development of metabolomics research benefits from the development of analytical techniques, such as high-resolution mass spectrometry, ultrahigh performance LC-MS, capillary LC-MS, multidimensional LC-MS, and multidimensional NMR technology. Structural identification of biomarkers is also one of the key and difficult problems in metabolomics research. The application of LC-MS technology in metabolomics research has been restricted to a certain extent due to the lack of a standard and universal mass spectrometry database [121]. In theory, LC-MS$\sim$NMR can provide better information about component structure, but the instrument is complex, complicate to operate, where sensitivity and flux need to be improved urgently. More and more attention has been paid to the construction of a well-functioning metabolite database and the standardization of metabolomics research (http://msi-workgroups.sourceforge.net/) [122]. Like in other omics, important factors such as how to overcome the bottleneck and find specific biomarkers (especially low abundance biomarkers) from a large number of metabolites determine whether this technology can be widely used in the field of medicine and clinical. The advent of capillary electrophoresis technology has promoted the development of genomics. A 2D gel electrophoresis and 2D liquid chromatography mass spectrometry technology have promoted the development of proteomics. Currently, there is no similar and versatile new technology in metabolomics. Nowadays, the integration of various analytical techniques is the main technology platform. For example, in the study of "insulin resistance population" markers, a set of new methods for the detection and identification of biomarkers were proposed, including LC-MS fingerprint analysis, multivariate data analysis, detection of possible biomarkers, FT-MS (Fourier transform-mass spectrometry) determination of accurate quality, micro-preparation, MS/MS (tandem mass spectrometry) fragment information, gas chromatography retention index, literature retrieval and synthesis of isotope compounds, identification of metabolic markers, etc. Based on this, not only possible biomarkers can be found, but also their structures can be determined and isomers can be distinguished. This method has a strong role in promoting the development of metabolomics based on LC-MS.

Since the information of all metabolites cannot be obtained by one analysis, biological problems need to be understood from many aspects. Therefore, the integration of different fields should be emphasized, including the integration of the data and knowledge of cell biology and animal models, the integration of different metabolomic methods [such as ultrahigh-performance liquid chromatography-time-of-flight mass spectrometry (UPLC-TOF-MS) and NMR] [123], the integration of metabolomic data of different samples (urine, blood, tissue, etc.) [124], the integration of transcriptome, genome, pro teome, and several omics data [125–128], the integration of metabolomics and computational biology [107], the construction of metabolic networks and mathematical models of metabolic flow dynamics [129–132]. It has broad prospects in the field of metabolomics, and is also the focus of future research.

In terms of application, if metabolomics is to survive and develop, it must have its own characteristics. It is necessary to answer the biological questions that other omics cannot answer from the phenotype. Metabolites represent the total analysis of immediate in vitro and in vivo stimuli. The phenotypes of drugs in patients (drug distribution, efficacy, treatment failure, and toxicity) can be monitored by metabolic profiles of urine, drug metabolites, or thousands of endogenous small molecule compounds. Specific patterns can show individual susceptibility to drug toxicity before clinical effects occur. In this way, metabolomics may help doctors achieve personalized treatment of patients, avoid poisoning, and reduce adverse drug reactions. Doctors can also analyze the course of a patient's illness and formulate treatment plans according to the patient's phenotype. Therefore, there is great room for the development of drug metabolomics in personalized drug therapy and other medical fields [133–135].

References

[1] J.K. Nicholson, J. Connelly, J.C. Lindon, E. Holmes, Metabonomics: a platform for studying drug toxicity and gene function, Nat. Rev. Drug Discov. 1 (2002) 153–161.
[2] C.H. Johnson, F.J. Gonzalez, Challenges and opportunities of metabolomics, J. Cell. Physiol. 227 (2012) 2975–2981.
[3] B. Misra, Individualized metabolomics: opportunities and challenges, Clin. Chem. Lab. Med. (2019).
[4] L. Casadei, M. Valerio, C. Manetti, Metabolomics: challenges and opportunities in systems biology studies, Methods Mol. Biol. 1702 (2018) 327–336.
[5] K.M. Oksman-Caldentey, K. Saito, Integrating genomics and metabolomics for engineering plant metabolic pathways, Curr. Opin. Biotechnol. 16 (2005) 174–179.
[6] S. Kundu, S.N. Karmakar, Localization phenomena in a DNA double-helix structure: a twisted ladder model, Phys. Rev. E Stat. Nonlinear Soft Matter Phys. 89 (2014).
[7] W.F. Dietrich, The origin and implications of the Human Genome Project: scientific overview, Natl. Cathol. Bioeth. Q. 1 (2001) 489–495.
[8] L. Uzych, The Human Genome project: an overview of ethical issues and public policy concerns, Nurs. Outlook 44 (1996) 150–151.
[9] D.R. Bentley, The Human Genome Project—an overview, Med. Res. Rev. 20 (2000) 189–196.
[10] F. He, E. Murabito, H.V. Westerhoff, Synthetic biology and regulatory networks: where metabolic systems biology meets control engineering, J. R. Soc. Interface 13 (2016).
[11] I. Sanchez-Osorio, F. Ramos, P. Mayorga, E. Dantan, Foundations for modeling the dynamics of gene regulatory networks: a multilevel-perspective review, J. Bioinforma. Comput. Biol. 12 (2014) 1330003.
[12] D. Chen, X. Liu, Y.P. Yang, H.J. Yang, P. Lu, Systematic synergy modeling: understanding drug synergy from a systems biology perspective, BMC Syst. Biol. 9 (2015).
[13] H. Modell, W. Cliff, J. Michael, J. McFarland, M.P. Wenderoth, A. Wright, A physiologist's view of homeostasis, Adv. Physiol. Educ. 39 (2015) 259–266.
[14] W. Weckwerth, Metabolomics in systems biology, Annu. Rev. Plant Biol. 54 (2003) 669–689.
[15] C.M. Metallo, M.G. Vander Heiden, Understanding metabolic regulation and its influence on cell physiology, Mol. Cell 49 (2013) 388–398.
[16] S. Rochfort, Metabolomics reviewed: a new "Omics" platform technology for systems biology and implications for natural products research, J. Nat. Prod. 68 (2005) 1813–1820.
[17] D.J. Beale, et al., Review of recent developments in GC-MS approaches to metabolomics-based research, Metabolomics 14 (2018).
[18] I.N. Acworth, The application of HPLC with coulometric electrochemical array detection to the study of natural products and botanicals: from targeted analyses to metabolomics, Planta Med. 77 (2011) 1248.
[19] X. Zhao, et al., Metabonomic fingerprints of fasting plasma and spot urine reveal human pre-diabetic metabolic traits, Metabolomics 6 (2010) 362–374.
[20] S.A. Sansone, et al., The metabolomics standards initiative, Nat. Biotechnol. 25 (2007) 844–848.
[21] A.R. Fernie, R.N. Trethewey, A.J. Krotzky, L. Willmitzer, Innovation—metabolite profiling: from diagnostics to systems biology, Nat. Rev. Mol. Cell Biol. 5 (2004) 763–769.
[22] S.G. Oliver, From gene to screen with yeast, Curr. Opin. Genet. Dev. 7 (1997) 405–409.
[23] J.K. Nicholson, J.C. Lindon, E. Holmes, 'Metabonomics': understanding the metabolic responses of living systems to pathophysiological stimuli via multivariate statistical analysis of biological NMR spectroscopic data, Xenobiotica 29 (1999) 1181–1189.
[24] J.T. Brindle, et al., Rapid and noninvasive diagnosis of the presence and severity of coronary heart disease using 1H-NMR-based metabonomics, Nat. Med. 8 (2002) 1439–1444.

[25] E. Holmes, et al., Chemometric models for toxicity classification based on NMR spectra of biofluids, Chem. Res. Toxicol. 13 (2000) 471–478.
[26] E. Holmes, J.K. Nicholson, G. Tranter, Metabonomic characterization of genetic variations in toxicological and metabolic responses using probabilistic neural networks, Chem. Res. Toxicol. 14 (2001) 182–191.
[27] O. Fiehn, J. Kopka, P. Dormann, T. Altmann, R.N. Trethewey, L. Willmitzer, Metabolite profiling for plant functional genomics, Nat. Biotechnol. 18 (2000) 1157–1161.
[28] B.H. Toyama, M.W. Hetzer, OPINION Protein homeostasis: live long, won't prosper, Nat. Rev. Mol. Cell Biol. 14 (2013) 55–61.
[29] X.J. Liu, J.W. Locasale, Metabolomics: a primer, Trends Biochem. Sci. 42 (2017) 274–284.
[30] A. Trewavas, A brief history of systems biology. "Every object that biology studies is a system of systems." Francois Jacob (1974), Plant Cell 18 (2006) 2420–2430.
[31] A. Spivey, Systems biology—the big picture, Environ. Health Perspect. 112 (2004) A938–A943.
[32] L. Hood, L. Rowen, The human genome project: big science transforms biology and medicine, Genome Med. 5 (2013).
[33] K.J. Karczewski, M.P. Snyder, Integrative omics for health and disease, Nat. Rev. Genet. 19 (2018) 299–310.
[34] E. Fukusaki, Application of metabolomics for high resolution phenotype analysis, Mass Spectrom. (Tokyo) 3 (2014) S0045.
[35] R. Chen, M. Snyder, Promise of personalized omics to precision medicine, Wiley Interdiscip. Rev. Syst. Biol. Med. 5 (2013) 73–82.
[36] M. Yan, G. Xu, Current and future perspectives of functional metabolomics in disease studies—a review, Anal. Chim. Acta 1037 (2018) 41–54.
[37] K. MacDonald, A. Krishnan, E. Cervenka, G. Hu, E. Guadagno, Y. Trakadis, Biomarkers for major depressive and bipolar disorders using metabolomics: a systematic review, Am. J. Med. Genet. B Neuropsychiatr. Genet. 180 (2019) 122–137.
[38] M.K. Townsend, Y. Bao, E.M. Poole, K.A. Bertrand, P. Kraft, B.M. Wolpin, C.B. Clish, S.S. Tworoger, Impact of pre-analytic blood sample collection factors on metabolomics, Cancer Epidemiol. Biomark. Prev. 25 (2016) 823–829.
[39] Y.H. Yang, C. Cruickshank, M. Armstrong, S. Mahaffey, R. Reisdorph, N. Reisdorph, New sample preparation approach for mass spectrometry-based profiling of plasma results in improved coverage of metabolome, J. Chromatogr. A 1300 (2013) 217–226.
[40] G. O'Maille, et al., Metabolomics relative quantitation with mass spectrometry using chemical derivatization and isotope labeling, Spectroscopy 22 (2008) 327–343.
[41] H. Miyagawa, T. Bamba, Comparison of sequential derivatization with concurrent methods for GC/MS-based metabolomics, J. Biosci. Bioeng. 127 (2019) 160–168.
[42] K. Dettmer, P.A. Aronov, B.D. Hammock, Mass spectrometry-based metabolomics, Mass Spectrom. Rev. 26 (2007) 51–78.
[43] B.L. Ackermann, J.E. Hale, K.L. Duffin, The role of mass spectrometry in biomarker discovery and measurement, Curr. Drug Metab. 7 (2006) 525–539.
[44] W.B. Dunn, N.J. Bailey, H.E. Johnson, Measuring the metabolome: current analytical technologies, Analyst 130 (2005) 606–625.
[45] K. Hollywood, D.R. Brison, R. Goodacre, Metabolomics: current technologies and future trends, Proteomics 6 (2006) 4716–4723.
[46] R. Ramautar, G.W. Somsen, G.J. de Jong, Direct sample injection for capillary electrophoretic determination of organic acids in cerebrospinal fluid, Anal. Bioanal. Chem. 387 (2007) 293–301.
[47] P. Britz-McKibbin, S. Terabe, On-line preconcentration strategies for trace analysis of metabolites by capillary electrophoresis, J. Chromatogr. A 1000 (2003) 917–934.
[48] P. Lasch, L. Chiriboga, H. Yee, M. Diem, Infrared spectroscopy of human cells and tissue: detection of disease, Technol. Cancer Res. Treat. 1 (2002) 1–7.
[49] J. Schmitt, M. Beekes, A. Brauer, T. Udelhoven, P. Lasch, D. Naumann, Identification of scrapie infection from blood serum by Fourier transform infrared spectroscopy, Anal. Chem. 74 (2002) 3865–3868.

[50] P.H. Gamache, D.F. Meyer, M.C. Granger, I.N. Acworth, Metabolomic applications of electrochemistry/mass spectrometry, J. Am. Soc. Mass Spectrom. 15 (2004) 1717–1726.
[51] C.A. Daykin, O. Corcoran, S.H. Hansen, I. Bjornsdottir, C. Cornett, S.C. Connor, J.C. Lindon, J.K. Nicholson, Application of directly coupled HPLC NMR to separation and characterization of lipoproteins from human serum, Anal. Chem. 73 (2001) 1084–1090.
[52] P. Krishnan, N.J. Kruger, R.G. Ratcliffe, Metabolite fingerprinting and profiling in plants using NMR, J. Exp. Bot. 56 (2005) 255–265.
[53] J.L. Griffin, L.A. Walker, S. Garrod, E. Holmes, R.F. Shore, J.K. Nicholson, NMR spectroscopy based metabonomic studies on the comparative biochemistry of the kidney and urine of the bank vole (Clethrionomys glareolus), wood mouse (Apodemus sylvaticus), white toothed shrew (Crocidura suaveolens) and the laboratory rat, Comp. Biochem. Physiol. B Biochem. Mol. Biol. 127 (2000) 357–367.
[54] J.L. Griffin, M.E. Bollard, Metabonomics: its potential as a tool in toxicology for safety assessment and data integration, Curr. Drug Metab. 5 (2004) 389–398.
[55] J.L. Griffin, Metabonomics: NMR spectroscopy and pattern recognition analysis of body fluids and tissues for characterisation of xenobiotic toxicity and disease diagnosis, Curr. Opin. Chem. Biol. 7 (2003) 648–654.
[56] I. Pelczer, High-resolution NMR for metabomics, Curr. Opin. Drug Discov. Dev. 8 (2005) 127–133.
[57] C.L. Gavaghan, E. Holmes, E. Lenz, I.D. Wilson, J.K. Nicholson, An NMR-based metabonomic approach to investigate the biochemical consequences of genetic strain differences: application to the C57BL10J and Alpk:ApfCD mouse, FEBS Lett. 484 (2000) 169–174.
[58] G.L. Jones, et al., A functional analysis of mouse models of cardiac disease through metabolic profiling, J. Biol. Chem. 280 (2005) 7530–7539.
[59] J.R. Marchesi, E. Holmes, F. Khan, S. Kochhar, P. Scanlan, F. Shanahan, I.D. Wilson, Y. Wang, Rapid and noninvasive metabonomic characterization of inflammatory bowel disease, J. Proteome Res. 6 (2007) 546–551.
[60] M. Glinski, W. Weckwerth, The role of mass spectrometry in plant systems biology, Mass Spectrom. Rev. 25 (2006) 173–214.
[61] C. Denkert, et al., Mass spectrometry-based metabolic profiling reveals different metabolite patterns in invasive ovarian carcinomas and ovarian borderline tumors, Cancer Res. 66 (2006) 10795–10804.
[62] M.P. Styczynski, J.F. Moxley, L.V. Tong, J.L. Walther, K.L. Jensen, G.N. Stephanopoulos, Systematic identification of conserved metabolites in GC/MS data for metabolomics and biomarker discovery, Anal. Chem. 79 (2007) 966–973.
[63] E.M. Lenz, I.D. Wilson, Analytical strategies in metabonomics, J. Proteome Res. 6 (2007) 443–458.
[64] I.D. Wilson, R. Plumb, J. Granger, H. Major, R. Williams, E.M. Lenz, HPLC-MS-based methods for the study of metabonomics, J. Chromatogr. B Anal. Technol. Biomed. Life Sci. 817 (2005) 67–76.
[65] S. Wagner, K. Scholz, M. Sieber, M. Kellert, W. Voelkel, Tools in metabonomics: an integrated validation approach for LC-MS metabolic profiling of mercapturic acids in human urine, Anal. Chem. 79 (2007) 2918–2926.
[66] S.U. Bajad, W. Lu, E.H. Kimball, J. Yuan, C. Peterson, J.D. Rabinowitz, Separation and quantitation of water soluble cellular metabolites by hydrophilic interaction chromatography-tandem mass spectrometry, J. Chromatogr. A 1125 (2006) 76–88.
[67] P. Yin, X. Zhao, Q. Li, J. Wang, J. Li, G. Xu, Metabonomics study of intestinal fistulas based on ultraperformance liquid chromatography coupled with Q-TOF mass spectrometry (UPLC/Q-TOF MS), J. Proteome Res. 5 (2006) 2135–2143.
[68] A. Oikawa, et al., Clarification of pathway-specific inhibition by Fourier transform ion cyclotron resonance/mass spectrometry-based metabolic phenotyping studies, Plant Physiol. 142 (2006) 398–413.
[69] M. Katajamaa, M. Oresic, Data processing for mass spectrometry-based metabolomics, J. Chromatogr. A 1158 (2007) 318–328.
[70] M. van Doorn, J. Vogels, A. Tas, E.J. van Hoogdalem, J. Burggraaf, A. Cohen, J. van der Greef, Evaluation of metabolite profiles as biomarkers for the pharmacological effects of thiazolidinediones in Type 2 diabetes mellitus patients and healthy volunteers, Br. J. Clin. Pharmacol. 63 (2007) 562–574.

[71] D. Morvan, A. Demidem, Metabolomics by proton nuclear magnetic resonance spectroscopy of the response to chloroethylnitrosourea reveals drug efficacy and tumor adaptive metabolic pathways, Cancer Res. 67 (2007) 2150–2159.

[72] A.M. Weljie, J. Newton, P. Mercier, E. Carlson, C.M. Slupsky, Targeted profiling: quantitative analysis of 1H NMR metabolomics data, Anal. Chem. 78 (2006) 4430–4442.

[73] E. Holmes, H. Antti, Chemometric contributions to the evolution of metabonomics: mathematical solutions to characterising and interpreting complex biological NMR spectra, Analyst 127 (2002) 1549–1557.

[74] H.C. Keun, T.M. Ebbels, M.E. Bollard, O. Beckonert, H. Antti, E. Holmes, J.C. Lindon, J.K. Nicholson, Geometric trajectory analysis of metabolic responses to toxicity can define treatment specific profiles, Chem. Res. Toxicol. 17 (2004) 579–587.

[75] U. Lutz, R.W. Lutz, W.K. Lutz, Metabolic profiling of glucuronides in human urine by LC-MS/MS and partial least-squares discriminant analysis for classification and prediction of gender, Anal. Chem. 78 (2006) 4564–4571.

[76] S. Wiklund, E. Johansson, L. Sjostrom, E.J. Mellerowicz, U. Edlund, J.P. Shockcor, J. Gottfries, T. Moritz, J. Trygg, Visualization of GC/TOF-MS-based metabolomics data for identification of biochemically interesting compounds using OPLS class models, Anal. Chem. 80 (2008) 115–122.

[77] O. Teahan, S. Gamble, E. Holmes, J. Waxman, J.K. Nicholson, C. Bevan, H.C. Keun, Impact of analytical bias in metabonomic studies of human blood serum and plasma, Anal. Chem. 78 (2006) 4307–4318.

[78] A. Craig, O. Cloarec, E. Holmes, J.K. Nicholson, J.C. Lindon, Scaling and normalization effects in NMR spectroscopic metabonomic data sets, Anal. Chem. 78 (2006) 2262–2267.

[79] S. Kuehn, S.A. Hickman, J.A. Marohn, Advances in mechanical detection of magnetic resonance, J. Chem. Phys. 128 (2008).

[80] P.G. Boswell, D. Abate-Pella, J.T. Hewitt, Calculation of retention time tolerance windows with absolute confidence from shared liquid chromatographic retention data, J. Chromatogr. A 1412 (2015) 52–58.

[81] J. Jin, G.E. Sklar, V. Min Sen Oh, S. Chuen Li, Factors affecting therapeutic compliance: a review from the patient's perspective, Ther. Clin. Risk Manag. 4 (2008) 269–286.

[82] D.G. Robertson, K. Datta, D. Wells, L. Egnash, L. Robosky, M. Manning, C. Rohde, M.D. Reily, Metabonomic evaluation of metabolic dysregulation in rats induced by PF 376304, a novel inhibitor of phosphoinositide 3-kinase, Chem. Res. Toxicol. 20 (2007) 1871–1877.

[83] H.C. Keun, Metabonomic modeling of drug toxicity, Pharmacol. Ther. 109 (2006) 92–106.

[84] D.B. Kell, Systems biology, metabolic modelling and metabolomics in drug discovery and development, Drug Discov. Today 11 (2006) 1085–1092.

[85] J.C. Lindon, E. Holmes, J.K. Nicholson, Metabonomics and its role in drug development and disease diagnosis, Expert. Rev. Mol. Diagn. 4 (2004) 189–199.

[86] J.C. Lindon, et al., Contemporary issues in toxicology the role of metabonomics in toxicology and its evaluation by the COMET project, Toxicol. Appl. Pharmacol. 187 (2003) 137–146.

[87] J.L. Griffin, Understanding mouse models of disease through metabolomics, Curr. Opin. Chem. Biol. 10 (2006) 309–315.

[88] L.K. Schnackenberg, R.D. Beger, Monitoring the health to disease continuum with global metabolic profiling and systems biology, Pharmacogenomics 7 (2006) 1077–1086.

[89] Z. Pan, H. Gu, N. Talaty, H. Chen, N. Shanaiah, B.E. Hainline, R.G. Cooks, D. Raftery, Principal component analysis of urine metabolites detected by NMR and DESI-MS in patients with inborn errors of metabolism, Anal. Bioanal. Chem. 387 (2007) 539–549.

[90] F. Fava, J.A. Lovegrove, R. Gitau, K.G. Jackson, K.M. Tuohy, The gut microbiota and lipid metabolism: implications for human health and coronary heart disease, Curr. Med. Chem. 13 (2006) 3005–3021.

[91] Q.N. Van, et al., The use of urine proteomic and metabonomic patterns for the diagnosis of interstitial cystitis and bacterial cystitis, Dis. Markers 19 (2003) 169–183.

[92] J.T. Brindle, J.K. Nicholson, P.M. Schofield, D.J. Grainger, E. Holmes, Application of chemometrics to 1H NMR spectroscopic data to investigate a relationship between human serum metabolic profiles and hypertension, Analyst 128 (2003) 32–36.
[93] J.K. Yao, R.D. Reddy, Metabolic investigation in psychiatric disorders, Mol. Neurobiol. 31 (2005) 193–203.
[94] Q. Li, L. Tan, C. Wang, N. Li, Y. Li, G. Xu, J. Li, Polyunsaturated eicosapentaenoic acid changes lipid composition in lipid rafts, Eur. J. Nutr. 45 (2006) 144–151.
[95] Q. Li, M. Wang, L. Tan, C. Wang, J. Ma, N. Li, Y. Li, G. Xu, J. Li, Docosahexaenoic acid changes lipid composition and interleukin-2 receptor signaling in membrane rafts, J. Lipid Res. 46 (2005) 1904–1913.
[96] Z.Z. Fang, F.J. Gonzalez, LC-MS-based metabolomics: an update, Arch. Toxicol. 88 (2014) 1491–1502.
[97] H. Rischer, K.M. Oksman-Caldentey, Unintended effects in genetically modified crops: revealed by metabolomics? Trends Biotechnol. 24 (2006) 102–104.
[98] M. Wink, Modes of action of herbal medicines and plant secondary metabolites, Medicines (Basel) 2 (2015) 251–286.
[99] C.F. Ma, H.H. Wang, X. Lu, H.F. Li, B.Y. Liu, G.W. Xu, Analysis of Artemisia annua L. volatile oil by comprehensive two-dimensional gas chromatography time-of-flight mass spectrometry, J. Chromatogr. A 1150 (2007) 50–53.
[100] I. Elmroth, A. Fox, O. Holst, L. Larsson, Detection of bacterial contamination in cultures of eucaryotic cells by gas chromatography-mass spectrometry, Biotechnol. Bioeng. 42 (1993) 421–429.
[101] J.G. Bundy, T.L. Willey, R.S. Castell, D.J. Ellar, K.M. Brindle, Discrimination of pathogenic clinical isolates and laboratory strains of Bacillus cereus by NMR-based metabolomic profiling, FEMS Microbiol. Lett. 242 (2005) 127–136.
[102] L.M. Raamsdonk, et al., A functional genomics strategy that uses metabolome data to reveal the phenotype of silent mutations, Nat. Biotechnol. 19 (2001) 45–50.
[103] A. Buchholz, J. Hurlebaus, C. Wandrey, R. Takors, Metabolomics: quantification of intracellular metabolite dynamics, Biomol. Eng. 19 (2002) 5–15.
[104] M. Dauner, J.E. Bailey, U. Sauer, Metabolic flux analysis with a comprehensive isotopomer model in Bacillus subtilis, Biotechnol. Bioeng. 76 (2001) 144–156.
[105] M.G. Boersma, I. Solyanikova, W.J. Van Berkel, J. Vervoort, L. Golovleva, I.M. Rietjens, 19F NMR metabolomics for the elucidation of microbial degradation pathways of fluorophenols, J. Ind. Microbiol. Biotechnol. 26 (2001) 22–34.
[106] J.P. Grivet, A.M. Delort, J.C. Portais, NMR and microbiology: from physiology to metabolomics, Biochimie 85 (2003) 823–840.
[107] N. Ishii, M. Robert, Y. Nakayama, A. Kanai, M. Tomita, Toward large-scale modeling of the microbial cell for computer simulation, J. Biotechnol. 113 (2004) 281–294.
[108] F. Savorani, M.A. Rasmussen, M.S. Mikkelsen, S.B. Engelsen, A primer to nutritional metabolomics by NMR spectroscopy and chemometrics, Food Res. Int. 54 (2013) 1131–1145.
[109] J.B. German, M.A. Roberts, S.M. Watkins, Personal metabolomics as a next generation nutritional assessment, J. Nutr. 133 (2003) 4260–4266.
[110] S.H. Zeisel, et al., The nutritional phenotype in the age of metabolomics, J. Nutr. 135 (2005) 1613–1616.
[111] E.M.S. McNiven, J.B. German, C.M. Slupsky, Analytical metabolomics: nutritional opportunities for personalized health, J. Nutr. Biochem. 22 (2011) 995–1002.
[112] L. Brennan, Metabolomics and nutritional applications, Ann. Nutr. Metab. 63 (2013) 11–12.
[113] H. Gibbons, A. O'Gorman, L. Brennan, Metabolomics as a tool in nutritional research, Curr. Opin. Lipidol. 26 (2015) 30–34.
[114] M. Suarez, A. Caimari, J.M. del Bas, L. Arola, Metabolomics: an emerging tool to evaluate the impact of nutritional and physiological challenges, Trac Trends Anal. Chem. 96 (2017) 79–88.
[115] T.J. Phelps, A.V. Palumbo, A.S. Beliaev, Metabolomics and microarrays for improved understanding of phenotypic characteristics controlled by both genomics and environmental constraints, Curr. Opin. Biotechnol. 13 (2002) 20–24.

[116] M.R. Viant, Recent developments in environmental metabolomics, Mol. BioSyst. 4 (2008) 980–986.
[117] M.J. Simpson, J.R. McKelvie, Environmental metabolomics: new insights into earthworm ecotoxicity and contaminant bioavailability in soil, Anal. Bioanal. Chem. 394 (2009) 137–149.
[118] J. Kikuchi, K. Ito, Y. Date, Environmental metabolomics with data science for investigating ecosystem homeostasis, Prog. Nucl. Magn. Reson. Spectrosc. 104 (2018) 56–88.
[119] M.A. Garcia-Sevillano, T. Garcia-Barrera, J.L. Gomez-Ariza, Environmental metabolomics: biological markers for metal toxicity, Electrophoresis 36 (2015) 2348–2365.
[120] M.R. Viant, U. Sommer, Mass spectrometry based environmental metabolomics: a primer and review, Metabolomics 9 (2013) S144–S158.
[121] M. Berg, M. Vanaerschot, A. Jankevics, B. Cuypers, R. Breitling, J.C. Dujardin, LC-MS metabolomics from study design to data-analysis—using a versatile pathogen as a test case, Comput. Struct. Biotechnol. J. 4 (2013).
[122] A.L. Castle, O. Fiehn, R. Kaddurah-Daouk, J.C. Lindon, Metabolomics Standards Workshop and the development of international standards for reporting metabolomics experimental results, Brief. Bioinform. 7 (2006) 159–165.
[123] D.J. Crockford, E. Holmes, J.C. Lindon, R.S. Plumb, S. Zirah, S.J. Bruce, P. Rainville, C.L. Stumpf, J.K. Nicholson, Statistical heterospectroscopy, an approach to the integrated analysis of NMR and UPLC-MS data sets: application in metabonomic toxicology studies, Anal. Chem. 78 (2006) 363–371.
[124] N.J. Waters, E. Holmes, A. Williams, C.J. Waterfield, R.D. Farrant, J.K. Nicholson, NMR and pattern recognition studies on the time-related metabolic effects of alpha-naphthylisothiocyanate on liver, urine, and plasma in the rat: an integrative metabonomic approach, Chem. Res. Toxicol. 14 (2001) 1401–1412.
[125] E. Davidov, et al., Methods for the differential integrative omic analysis of plasma from a transgenic disease animal model, OMICS 8 (2004) 267–288.
[126] C.D. Davis, J. Milner, Frontiers in nutrigenomics, proteomics, metabolomics and cancer prevention, Mutat. Res. 551 (2004) 51–64.
[127] L. Eriksson, et al., Using chemometrics for navigating in the large data sets of genomics, proteomics, and metabonomics (gpm), Anal. Bioanal. Chem. 380 (2004) 419–429.
[128] K.C. Verhoeckx, S. Bijlsma, S. Jespersen, R. Ramaker, E.R. Verheij, R.F. Witkamp, J. van der Greef, R.J. Rodenburg, Characterization of anti-inflammatory compounds using transcriptomics, proteomics, and metabolomics in combination with multivariate data analysis, Int. Immunopharmacol. 4 (2004) 1499–1514.
[129] J. van der Greef, S. Martin, P. Juhasz, A. Adourian, T. Plasterer, E.R. Verheij, R.N. McBurney, The art and practice of systems biology in medicine: mapping patterns of relationships, J. Proteome Res. 6 (2007) 1540–1559.
[130] H.W. Ma, A.P. Zeng, The connectivity structure, giant strong component and centrality of metabolic networks, Bioinformatics 19 (2003) 1423–1430.
[131] J. Ricard, Reduction, integration and emergence in biochemical networks, Biol. Cell. 96 (2004) 719–725.
[132] B.M. Lange, M. Ghassemian, Comprehensive post-genomic data analysis approaches integrating biochemical pathway maps, Phytochemistry 66 (2005) 413–451.
[133] T.A. Clayton, et al., Pharmaco-metabonomic phenotyping and personalized drug treatment, Nature 440 (2006) 1073–1077.
[134] P.K. Yeung, Metabolomics and biomarkers for drug discovery, Metabolites 8 (2018).
[135] D.C. Douillet, et al., Metabolomics and proteomics identify the toxic form and the associated cellular binding targets of the anti-proliferative drug AICAR, J. Biol. Chem. 294 (2019) 805–815.

CHAPTER 21

Pan-interactomics and its applications

Gyan P. Srivastava[a], Neelam Yadav[a], Bhupendra N.S. Yadav, Rajiv K. Yadav, Dinesh K. Yadav
Molecular Biology and Genetic Engineering Laboratory, Department of Botany, University of Allahabad, Allahabad, India

1 Introduction

Cells are intricate chemical factories which sink and swim with fine orchestration of numerous biomolecules to manifest life. Every single cellular event requires a channelized regulation of biomolecules that form well-organized and fine-tuned ultra-network in cell. Cellular integrity is regulated by a series of various biochemical reactions and the entire cellular events can be understood at various levels viz genomics (study of complete genetic makeup), transcriptomics (study of total RNA), proteomics (study of whole proteins), metabolomics (study of all metabolites of a cell), and many more aspects of knowledge about an organism. The literal meaning of life appears in the cell through signal generation, perception, and communication between all aforesaid aspects through all possible intermolecular interactions for the execution of signal and response. In molecular biology, the term 'interactome' refers to complete set of molecular interactions in a cell. It specifically attributes to all possible physical interactions between biomolecules and consequences of these interactions. Interactomics deals with the study of the intersection of bioinformatics and biology of complex interaction web formed after physical and functional interactions between all forms of biomolecules (DNA, RNA, proteins, lipids, ions, metabolites, etc.) leading to various physiological events in a cell. Thus, different cellular activities occurring under different physiological conditions are outcome of set(s) of interactions that regulate such events. Pan-interactomics information includes DNA-protein, protein-RNA, protein-protein interactions (PPIs), protein-ions, and many more possible interactions between biomolecules. As proteins are workhorses of the cell, the aforementioned PPIs serve as an epicenter in interactomics. PPIs are studied using in vitro, in vivo, and in silico approaches (Fig. 1). Interactome mapping of cellular events can be accomplished by large-scale Y_2H or other label-free and/or label-based techniques followed by sequencing and computational analysis. Pan-interactomics has ingrained its application in predicting pathogenicity, establishing evolutionary relationship between two or more species, unveiling metabolic pathways, predicting function of

[a] Authors have contributed equally.

Pan-genomics: Applications, Challenges, and Future Prospects
https://doi.org/10.1016/B978-0-12-817076-2.00021-4
© 2020 Elsevier Inc.
All rights reserved.

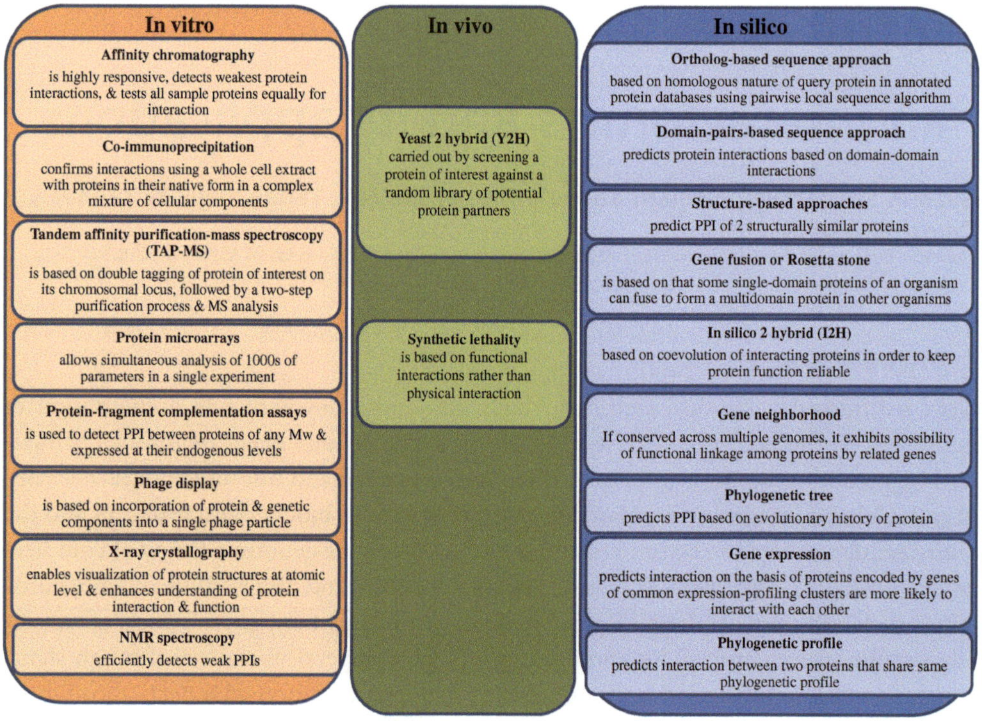

Fig. 1 Methods for the detection of PPIs.

single or multiple genes, and mutational studies. Thus, interactomics is of utmost importance for precise understanding of cellular events.

Apoplastic proteome has suggested biological functions in cell-to-cell interaction, extra/intracellular signal relays, appropriate cellular response to environmental stimuli, and regulation of host defense system against invading pathogen [1]. Apoplastic leaderless secretome proteins have been reported to play a significant role in various biotic and abiotic stresses. The discrimination of true secretome and secretome under various stress conditions imposes greater challenge and necessitates its comprehensive study. Moreover, much data is not available on extracellular PPI [2]. Majority of eukaryotic and one-third of biopharmaceutical proteins are glycoproteins and comprehensives understanding of role of glycans in biological processes including their interaction with carbohydrates, proteins, and nucleic acid is equally significant [3]. Protein-carbohydrate interactions between pathogen and host cell are significant in establishing attachment and invading host cell [4]. Hence comprehensive knowledge of host, vector, and pathogen glycoproteins and their interactome may provide better insights into pathogenesis [5], host-pathogen or vector-pathogen interaction, and disease management. Interactomics, a growing area of study, provides a compendious understanding of whole molecular interactions occurring across the cell. In biochemical and biomedical research it may accelerate insights into cellular regulatory mechanism, disintegrating mechanisms, biomolecule markers, drug designing, drug delivery mechanisms, and many more.

Comprehensive insight of molecular interactions within and outside of a cell was called interactomics, during the development of FlyNets database to decode complete molecular interactions in *Drosophila melanogaster* [6]. Ever since then interactomics has gained limelight in biological sciences, well depicted by continuous enrichment of publications in this area (Fig. 2).

Due to the involvement of diverse nature of biochemical interactions, that is, temporary or permanent, positive or negative, and obligate or non-obligate, interactomics has become highly dynamic. Interactomics has passed its adolescence and is expanding additional approach-based molecular biology involving experimental and/or computational methods. The information gained from these experiments complement each other. During computational studies, information obtained from in vitro experiments are analyzed, formatted, and stored in a retrievable format for users [7]. Fig. 3 depicts relationship between computational and experimental data development, and various processes involved in interactomics.

Recent researchers have developed many interactomics databases for prokaryotes (*Escherichia coli*, *Mesorhizobium loti*, and *Mycobacterium tuberculosis*), viruses [*human varicella-zoster virus* (HVZV), *Epstein-Barr virus* (EBV), *hepatitis E virus* (HEV)], and eukaryotic interactomes (*Schizosaccharomyces pombe*, *Caenorhabditis elegans*, and *Homo sapiens*). Proteins play a central role in the regulation of multitude of cellular pathways in coordination with other biomolecules (DNA, RNA). Different organisms synthesize similar proteins during same type of cellular events that encompass almost all signaling

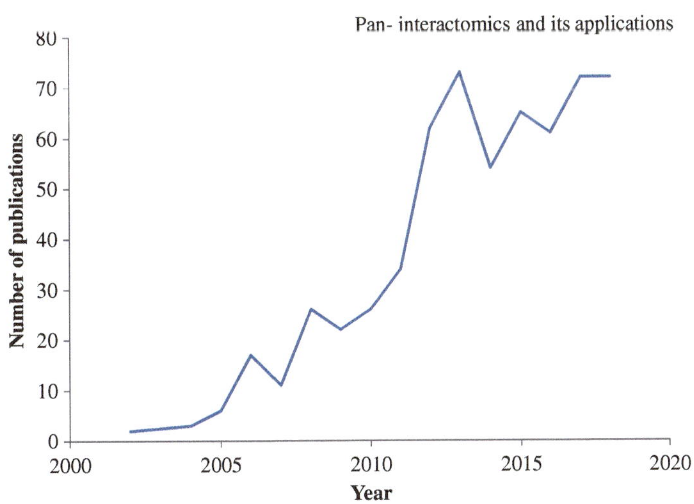

Fig. 2 Graphical representation of number of interactomics researches published in the 21st century (https://www.ncbi.nlm.nih.gov/pubmed/).

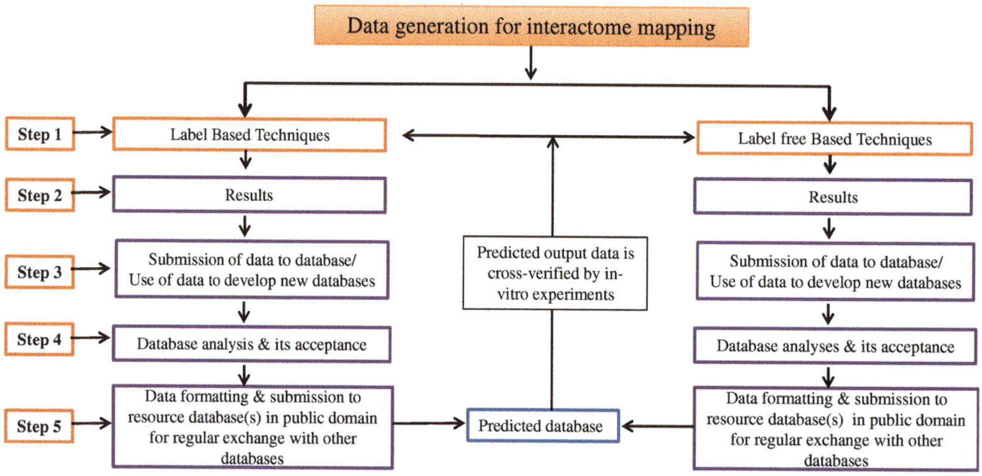

Fig. 3 Workflow representing interrelationship of computational and experimental approaches: Steps 1 and 2: data generation through experiments. Step 3: use of experimental data to create new databases or update existing databases. Step 4: data analyses and acceptance by databases after rigorous evaluation. Step 5: after evaluation data submitted to databases are placed in the public domain.

events; similarly through various channelized pathways numerous proteins are co-synthesized [8].

Mapping of co-expression data is one of the most effective ways to identify interactomes of any unknown species [9]. Few attempts were made to improve and optimize the co-expression-based mapping network [10, 11]. Likewise, it is possible to predict the expression pattern of genes related to a pathway in real time when a single regulatory gene of that pathway is knocked out [12]. It enlightens the significance of knockout genes as well as expresses its co-expression impact on other genes related to the respective pathway.

The combination of various instrumentations provides a better opportunity to elucidate molecular interactions under different physiological states of the cell. Recently evolved comparative and predicted interactomics are based on the combination of different interactomics databases and various criteria of their mining, like gene ontology, co-expression, and occurrence. The comparative interactome refers to the mapping of interacting networks between species, commonly host-pathogen interactions, for understanding the pathogenicity at molecular level. The species whose interactomes have not been unveiled so far can be mapped for their orthologous genes within a pre-mapped interactome database(s). Such mapping is referred as predicted interactomics.

The prediction of molecular networks in life sciences is gradually growing through decoding the evolutionary sketch as well as functional mapping of the organisms. Predicted interactomes have been reported for many plants viz *Oryza sativa*, *Xanthomonas oryzae, Arabidopsis thaliana, Brassica rapa, Zea mays, Populus trichocarpa*, etc. The hypotheses that are generated via predicted databases require experimental verification.

In this chapter we have discussed the present status of the interactomics, the techniques involved in the creation of the databases, and their applications in day to day life.

2 Computational analysis of interactome

PPIs are key to molecular functioning of cells and information pertaining to such interactions can be utilized to assign function to functionally orphan proteins. Common approaches used in the study and identification of proteins participating in complex biological functions include direct protein interaction analyses by binary systems viz split ubiquitin, yeast-two-hybrid (Y2H), or purification of protein complexes by tandem affinity purification systems [13] followed by MS for the identification of the protein.

Although various high-throughput techniques are available to unveil PPIs yet these data are quite small. Validation of these interactions is large, exhaustive, and time consuming even if done using high-throughput methods.

Moreover, bioinformatics tools/software can be used to identify putative protein partners of complex biological process or mechanisms which otherwise could manage to escape critical detection procedures. Such in silico interactome analyses have speeded the identification of suitable targets for drug design and discovery and crop improvement which could be used effectively after proper validation via wet lab experiments. Some commonly used computational platforms for interactome analyses are as follows:

Significance Analysis of INTeractome (SAINT; saint-apms.sourceforge.net/) uses a series of software tools to assign confidence scores to bait-prey interactions based on quantitative proteomics data in affinity purification-mass spectrometry experiments. SAINT is available in three versions viz *SAINT v2* relies on time-consuming sampling-based interface [14, 15], *SAINTexpress* was introduced to overcome limitation of v2, it only allows analysis of data sets with control purifications [16], *SAINTq* allows users to directly use the reproducibility information in transitions/peptides to score each protein as an interaction partner. It can also be used to score protein- or peptide-level intensity data [17].

Integrated Interactome System (IIS; http://www.lge.ibi.unicamp.br/lnbio/IIS/) is a free online integrative platform with a web-based interface for annotation, analysis, and visualization of interactions of target proteins/genes, metabolites, and drugs [18]. It works on four modules: *submission module*, which receives raw data derived from Sanger sequencing (two-hybrid system); *search module*, which enables user to search for the processed reads to be assembled into contigs/singlets, or for lists of desired proteins/genes, metabolites and drugs, and add them to the project; *annotation module*, which assigns annotations from several databases for contigs/singlets or lists of proteins/genes, generating tables with automatic annotation for manual curation; and *interactome module*, which maps contigs/singlets or uploaded lists to entries in the integrated database, building networks that gather novel identified interactions, protein and metabolite expression/concentration levels, subcellular localization and computed topological metrics, GO biological processes, and enrich KEGG pathways. The XGMML file generated in this module could be imported into Cytoscape or be visualized directly on the web.

Thus, IIS being an integration of diverse databases allows systematic analysis of physical, genetic, and chemical-genetic interactions. It was validated with Y2H, proteomics, and metabolomics datasets and is extendable to other datasets.

POINeT is an integrated online (http://poinet.bioinformatics.tw/) available software for PPI search, analysis, and visualization. It presents combinatorial analyses of PPI and tissue-specific expression data from multiple resources including published literatures. So, it allows identification and ranking of potential novel genes involved in a subnetwork [19].

GENE INFINITY uses informational resources and is a collection of tools and calculators to facilitate analysis of biological data, including online resources and software to analyze *interactions* between proteins, small molecules, and nucleic acids (http://www.geneinfinity.org/).

Struct2Net is a web server that works on structure-based approach ahead of homology modeling to predict protein interactions. Struct2Net only requires the sequence information on the proteins being queried. This freely available (http://struct2net.csail.mit.edu) web service allows multiple querying options, aimed at maximizing flexibility. Protein interactions of organisms such as fly, human and yeast have been precomputed for instantaneous retrieval. For proteins from other species, users have an option of having a quick-but-approximate result or perform a full-blown computation.

Cytoscape is an open source bioinformatics platform for integrating biomolecular interaction networks with high-throughput expression data and other molecular states into a unified conceptual framework [20]. It is applicable to varied range of biological molecular components and interactions. It becomes very powerful tool when used in conjunction with databases of protein-protein, protein-DNA, and genetic interactions, available for humans and model organisms. *Cytoscape Core* provides basic functionality to layout and queries the network; to visually integrate the network with expression profiles, phenotypes, and other molecular states; and to link the network to databases of functional annotations. This is a collection of protocols, tutorials, and workflows for Cytoscape and Cytoscape Apps for basic and advanced analysis. Most of its Apps are freely available on the Cytoscape App Store.

Protein Interaction Network Analysis (PINA) is a web-based platform which integrates PPI data from six manually curated public databases and provides a set of built-in tools for network construction, filtering, analysis, and visualization. PINA v2.0 has enhanced utility in PPI network analyses as it includes multiple collections of interaction modules identified by different interactome for six model organisms [21]. All the identified modules are fully annotated by enriched GO terms, KEGG pathways, Pfam domains, and chemical and genetic perturbations collection from MSigDB. It is freely accessible at http://cbg.garvan.unsw.edu.au/pina/.

Interactome INSIDER is a tool to link genomic variant information with structural protein-protein interactomes. It involves application of machine learning algorithm to predict protein interaction interfaces for 185,957 protein interactions with previously unresolved interfaces in human and seven model organisms, including entire experimentally determined human binary interactome. Predicted interfaces exhibit functional properties similar to those of known interfaces, including enrichment for disease mutations and recurrent cancer mutations. Interactome INSIDER (http://interactomeinsider.yulab.org) enables users to identify whether variants or disease mutations are enriched in known and predicted interaction interfaces at various resolutions. Users may explore known population variants, disease mutations, and somatic cancer mutations, or they may upload their own set of mutations [22].

Inter-Tools is a useful tool for interactome analysis to discover disease pathways. It allows users to search databases in the PSI-MITAB standard format for interactions on specified criteria and combines them into a single, species-specific interactome dataset. It produces graph files of interactome networks and performs basic analysis on user-input gene sets. This toolkit reduces the barriers to preliminary interactome investigation and hypothesis generation [23].

3 Databases and its types

First completely sequenced genome of *Haemophilus influenzae* gave new aspiration to molecular and computational biologists to predict the molecular mechanism of infection in the host cell. Genome sequencing data are stored in newly developed specialized databases which hold information in prescribed format and allow swapping of information with other available databases. These specialized databases are placed in the public domain.

Bioinformatics serves as an effective tool to utilize cumulated information from databases for interactome mapping. Certain bioinformatics tools used in interactome networking are listed in Table 1.

Interactome databases store information on interacting networks in all possible dimensions occurring in a cell. Some interactome databases are detailed below.

3.1 Viral interactomes

The viral interactomes studies were pioneered due to their small genome size facilitating analysis of interactions of viral proteome using available confined resources. Y2H serves as a promising tool for studying PPIs allowing prediction of significant interacting partners of viral components. Interactome of *bacteriophage lambda*, a dsDNA phage was mapped using Y2H which screened a sum of 97 interactions. A total of 18 new interactions of functionally related proteins were identified from

Table 1 List of bioinformatics tools used in interactomics

Databases	Description	Web address
3D-footprint	Collection of PWM extracted from structure	http://floresta.eead.csic.es/3dfootprint/
3DNA	Tool for extracting DNA structure parameters	http://x3dna.org/
AANT	Collection of amino acid-nucleotide interactions derived from experimentally determined protein nucleic acid structures	http://aant.icmb.utexas.edu/
BindN	SVM-based tool for prediction of DNA and RNA-binding residues in protein	http://bioinformatics.ksu.edu/bindn/
BIPA	Database for protein and nucleic acid interactions in 3D structures. It provides various features of protein-nucleic acid interfaces	http://www-cryst.bioc.cam.ac.uk/bipa
Curves+	Tool for extracting DNA structure parameters	https://bisi.ibcp.fr/tools/curves_plus/
iDBPs	Web server for identification of DNA-binding residues in protein sequences	http://idbps.tau.ac.il/
DP-Bind	Web server that takes a sequence of a DNA-binding protein and predicts residue positions involved in interactions with DNA	http://lcg.rit.albany.edu/dp-bind/
hPDI	Database that holds experimental protein and DNA interaction data for humans identified by protein microarray assays	http://bioinfo.wilmer.jhu.edu/PDI/
IAlign	Software to align protein-DNA interfaces based on a matrix score	http://cssb.biology.gatech.edu/iAlign.
JASPAR	Open-access database for eukaryotic transcription factor-binding profiles	http://jaspar.genereg.net/
NDB	The nucleic acid database	http://ndbserver.rutgers.edu
NPIDB	Database containing information derived from structures of DNA-protein and RNA-protein complexes extracted from PDB	http://monkey.belozersky.msu.ru/NPIDB/
NUCPLOT	Programme to generate schematic diagrams of protein-nucleic acid interactions	https://www.ebi.ac.uk/thornton-srv/software/NUCPLOT/
PDA	Automatic programme for the analysis of protein-DNA complex structures	http://bioinfozen.uncc.edu/webpda
PDBSum	Web-based database of summaries and analyses of all PDB structures	http://www.biochem.ucl.ac.uk/bsm/pdbsum
ProNIT	Database that collects experimentally observed binding data from the literature. It contains several important thermodynamic data for protein-nucleic acid binding	http://www.rtc.riken.go.jp/jouhou/pronit/pronit.html

Table 1 List of bioinformatics tools used in interactomics—cont'd

Databases	Description	Web address
ProNuC complex.	Database containing structural data of protein-nucleic acid complex. The data are classified according to the recognition motif of proteins and DNA forms involved in the protein-nucleic acid complex	http://npidb.belozersky.msu.ru/
ProtNA-ASA	Database that combines the data on conformational parameters of nucleic acids and accessible surface area of nucleic acid atoms in protein-DNA/RNA complexes	http://www.protna.bio-page.org
Tfmodeller	Programme for comparative modeling of protein-DNA complexes	http://www.ccg.unam.mx/tfmodeller
TRANSFAC	Private database for eukaryotic transcription factor-binding profiles	http://generegulation.com/pub/databases.html
ZifBase	Collection of various natural and engineered zinc finger proteins, containing sequence features and linked to their structures	http://web.iitd.ac.in/~sundar/zifbase

these results [24]. A study conducted on three members of *Herpes viruses* family viz *herpes simplex virus*-1 (HSV-1), *Murine cytomegalovirus*, and *EBV* involved comparison of interactomes using Y2H, identified ~735 common interactions suggesting that interaction between HSV-1 UL33 with nuclear egress proteins UL31/UL34 has less conserved sequences similarity [25]. It also reported that some orthologous proteins show least sequence similarities due to rapid evolution while exhibiting functional similarities [25].

Y2H followed by GST pull down and ELISA enabled investigation of interactions in *Chandipura virus* and identified four viral proteins showing eight interactions out of which five were newly cited in *Chandipura virus* [26]. The identification of cellular interactome of *Ebola virus* nucleoprotein (NP) revealed cellular chaperones, including HSP70, as its interaction partners which provide stability to NP. Also it was reported that disruption of stability of NP adversely affects viral RNA synthesis [27]. The interactome of *Hepatitis C virus* (HCV) that causes chronic liver diseases was mapped using a flow-cytometry-coupled FRET (fluorescence resonance energy transfer) assay in living cells. It showed ~20 interactions, out of which 7 were novel interactions specifying the role of p7 in HCV assembly [28]. The combination of immunoprecipitation assay with liquid chromatography tandem mass spectrometry (LC-MS-MS) has proven a new powerful tool for mapping host-parasite interactions. This study showed the significance of PA protein in replication and pathogenicity of *Influenza* A *virus* (IAV). It established 78 human cellular proteins as interaction partner of PA of IAV strain H5N1. Ontology of these host genes suggested their involvement in viral translation, replication and organismal injury, abnormalities, cell death, and survival. Co-immunoprecipitation and co-localization assays identified two proteins as nucleolin and eukaryotic translation elongation factor 1-alpha, out of 78 proteins. It established the role of PA in regulating viral life cycle in human cells [29].

Interactome mapping studies on RNA viruses have been under taken since past few decades. An investigation involving molecular interactions of five RNA viruses identified interaction of 44 autophagy-related proteins with 83 viral proteins. Among the identified host proteins, the expression of immunity-associated GTPase family M was shown to prevent virus-induced autophagy by measles virus, HCV, and HIV-1 by preventing viral replication [30]. Interactome mapping of *Zika virus* (ZIKV) proteins for topology network identified 3033 interactions, among which 1224 were unique human polypeptides. The identified interaction partners are involved in quality control, vesicle trafficking, RNA processing, and lipid metabolism. However, more than 60% of network components have been previously reported in other viral infections. This study also unveiled the role of peroxisomes in ZIKV infection [31]. *Human-Nipah virus* PPIs mapping identified 101 interactions including 88 novel interactions, which suggest PRP19 complex and miRNA to be main host target [32]. Like PPIs, RNA-protein interaction also plays a significant role in virus infection and replication. Comparative RNA-protein interactome mapping of eukaryotic PCBP2 protein was separately performed in human Huh-7 cell and human Huh-7 cells infected with HCV stain JFH-1. Interaction map in Huh-7 cells identified its interaction with mRNA and noncoding RNA suggesting its role in various biological functions. PCBP2 interaction with 5′-UTR and 3′-UTR regions of viral RNA suggests its role in viral replication via viral genome circularization. For accurate findings fully automated and standardized integrated cross-linking immunoprecipitation (FAST-iCLIP) with improved CLIP biochemistry was used to provide comprehensive analysis across coding, noncoding, repetitive, retroviral, and nonhuman transcriptomes [33].

3.2 Bacterial interactomes

The identification of molecular interactions is important for the better understanding of cellular functioning. Meta-interactome addresses all possible interactions between evolutionarily conserved proteins. Meta-interactomics has enhanced the cumulative understanding of protein functions. In *Streptococcus pneumoniae* strain TIGR4, 2000 novel protein interactions were identified using Y2H, out of which 299 protein functions were predicted via meta-interactomics [34]. About 30% of the bacterial proteins and their functions are still unknown among which most of them are conserved in bacterial species. Meier et al. [35] identified putative role of 50 conserved proteins in *S. pneumoniae* strain

TIGR4 using powerful techniques viz microfluidic high-throughput assay technology and in vitro proteome-wide interaction screens. Caufield et al. [36] investigated 349 different bacterial species and strains for conserved interactions in different taxonomical groups and reported more than 52,000 unique PPIs.

E. coli has played a central role in the interpretation of underlying mechanism of basic cellular processes like metabolism, signaling, gene expression, and genome replication because of its comparative small genomic content. Several significant efforts have been made to study interactome of *E. coil* and its strains [37–41]. Rajagopala et al. [42] have reported Y2H screening of 3305 baits against 3606 preys. These results cover approximately 70% of mapped proteome of *E. coli*, identifying 2234 PPIs. Interactome databases of *E. coli* were developed in 2007 after compiling all analysis. Web link http://www.bacteriome.org, provides users access to information on PPIs, mining conserved pattern of gene, gene expression, and comparative genomics. This dataset is organized into three parts viz *functional dataset,* consisting of 3989 interactions from 1927 proteins, *core dataset* comprising high-quality experimental dataset of 4863 interactions from 1100 proteins, and *extended dataset* which contains 9860 interactions from 2131 proteins [43]. First protein interactomics data was reported for causal agent of gastric ulcer and gastric cancer [44], which covers 1500 binary protein interactions from 261 Y2H screens. Further experiments were performed using new approach, ORFeome-based proteome-wide Y2H screening which has resulted in 1515 PPIs, out of which 1461 interactions were new [45].

Bacterial interactomics research is not only limited to the cellular level but it has also successfully deciphered the bacterial microcompartments (MCPs). MCPs are well organized and bounded with proteins and involved in various cellular metabolic processes. The most complicated MCP is propanediol utilizing (Pdu) microcompartment involved in several cellular processes including degradation of 1,2-propanediol. Jorda et al. [46] have conducted PPIs via coevolution-based methods using pair-wise alignment. These results suggest that shell protein PduA serve as a universal hub for targeting various enzymes presenting special N-terminal extensions, namely Pdu C, D, E, L, and P. Also, it was revealed that a protein Pdu V, with unknown function in bacterial cell, showed remote similarity with Ras-like GTPase superfamily and may be located outside MCP where it interacts with protruding β-barrel of hexameric Pdu shell protein.

3.3 Eukaryotic interactomes

Although several attempts have been made to decode eukaryotic interactome yet, till date almost 90% of protein interactions are confined to yeast [47–49]. *C. elegans* is first multicellular organism whose genome was completely sequenced in 2012. Interaction of sperm membrane protein was established using split ubiquitin *m*embrane *y*east-*t*wo-

*h*ybrid (MYTH) system. Interactome data of SPE-38, a four-pass transmembrane protein with known roles in spermatogenesis, spermiogenesis, and fertilization was reported to interact with proteins essential for spermatogenesis and spermiogenesis [50].

3.4 Predicted interactomics

Diversified and complex cellular composition, costly experiments, and need of sophisticated instrumentation are important bottleneck in the development of different interactome databases. Plant genome-wide experimental approach-based interactome constriction has been limited to *A. thaliana* [51]. To circumvent posed hurdles, researchers have developed predicted interactome databases. These databases were developed by mining interlogs from different pre-developed interactome databases. Conserved interactions between a pair of proteins have interacting homologs in another organism. Online Predicted Human Interaction Database (OPHID; http://ophid.utoronto.ca) developed in 2005 was first predicted interactomics database. It was based on mining of interlogs in *Saccharomyces cerevisiae, C. elegans, D. melanogaster*, and *Mus musculus* with a total of 23,889 predicted interactions [52]. The proteome of human showed 31.9%, 39.7%, and 21.2% orthologs to *S. cerevisiae, D. melanogaster*, and *C. elegans*, respectively. Interactome for Arabidopsis is based on interlogs of yeast, nematode worm, fruit fly, and human. Results suggested that 1159 high-confidence; 5913 medium-confidence, and 12,907 low-confidence interactions were identified for 3617 conserved Arabidopsis proteins [53]. In the same series of database development, a Predicted Rice Interactome Network (PRIN) (http://bis.zju.edu.cn/prin/) was developed utilizing interactomic information from six model organisms *S. cerevisiae, C. elegans, D. melanogaster, H. sapiens, E. coli K12*, and *A. thaliana* [54]. PRIN database contains 76,585 nonredundant rice protein interaction pairs among 5049 rice protein interactions. Also, it revealed that the topology of the predicted rice protein interactions is more similar to *S. cerevisiae* (74%) than other five model organisms [54]. In 2013 two predicted interactomics databases were mapped for *X. oryzae* and *B. rapa*. The *X. oryzae* interactomics prediction was based on proteins, experimental interactome mapping, and protein structural interactome map (PSIMAP) from Protein Data Bank. Total 4538 proteins showed 26,932 possible interactions; out of 18,503 (PSIMAP), 3118 were PEIMAP and 8938 iPfam pairs. XooNET database (http://bioportal.kobic.kr/XooNET/) is placed in the public domain [55]. Ontogeny of *B. rapa* is similar to *A. thaliana* providing advantage of exploring Brassica interactome. Protein-protein interaction data of *A. thaliana* were extracted from three major databases viz BioGRID, IntAct, and TAIR as well as PPI prediction orthologs. The relationship between *B. rapa* and *A. thaliana* was completed in three complementary ways: (i) ortholog predictions, (ii) identification of gene duplication based on synteny and collinearity, and (iii) BLAST sequence similarity search. Protein-protein interaction for maize (PPIM; http://comp-sysbio.org/ppim) and predicted tomato interactome resource (PTIR; http://bdg.hfut.edu.cn/ptir/index) are database for maize and tomato, respectively, developed using an approach similar to that of mining model interactomics. Extending this work, 10 random interactions were cross-confirmed by Y2H screening and/or a bimolecular fluorescence complementation (BiFC) assay. It showed 357,946 nonredundant PPIs among 10,626 proteins including 12,291 high-confidence, 226,553 medium-confidence, and 119,102 low-confidence interactions. Interactome in various organisms are listed in Table 2.

4 In vivo pan-interactome mapping

The techniques involved in the study of interactomics can be broadly categorized as follows:

Label-free techniques comprising Y2H; surface plasmon resonance (SPR); biolayer interferometery (BLI), and nanotechnology-based biosensors.

Label-based technology includes fluorescent and radioisotopes label-based techniques.

4.1 Label-free technologies

The label-free technologies consist of Y2H, SPR, BLI, and nanotechnology-based biosensors developed using the physical properties of the sample viz mass, dielectric constant, etc. The major advantage of label-free technique is that it is a time-saving technique, no modification is required in the query molecule and the interacting molecules, and pre-information about the biochemical properties of the query molecules is not required. Moreover, it provides real time kinetics to study small molecule interactions. However, label-free methods are costlier techniques.

4.1.1 Y2H system

Y2H system is one of the most classical and effective way to identify PPI in yeast [117]. Later the technique was modified to detect protein-DNA interactions and DNA-DNA interactions. Hosts were also modified in addition to yeast [118, 119]. Snider et al. [120] modified the classical Y2H to detect the interaction between membrane proteins (Table 3). These modifications enhanced the potential of the technique to several folds. Most of the interactome databases developed so far were based on Y2H system; *E. coli; bacteriophage lambda* [122], *E. coli bacteriophage T7* [56], Chandipura virus [26], *HEV* [107], *C. elegans* [89], and *H. sapiens* [75]. The limitation of this system is that the subcellular location of the query proteins cannot be determined.

RNA-protein interactions are of great importance for RNA function. RNA interactome is characterized by the capture of polyadenylated RNAs, usually mRNAs, using $oligo_{(dT)}$-coated beads [123, 124] and linking 5-ethynyluridine (EU)-labeled RNAs and biotin using the click reaction for interactome of newly transcribed RNAs [125]. Click

Table 2 Published interactomes of various organisms

	Published interactome network maps	References
Bacteriophage	*Escherichia coli bacteriophage λ*	[24]
	Escherichia coli bacteriophage T7	[56]
	Streptococcus pneumoniae bacteriophage Dp-1	[57]
	Streptococcus pneumoniae bacteriophage Cp-1	[58]
Prokaryotes	PPI map of *Helicobacter pylori*	[44]
	Second-generation PPI network of *Helicobacter pylori*	[45]
	Helicobacter pylori	[59]
	Proteome-wide protein interaction map for *Campylobacter jejuni*	[60]
	Salmonella enterica (SalmoNet)	[61]
	Protein Interactome of *Streptococcus pneumoniae* and Bacterial Meta-interactomes	[34]
	Escherichia coli	[62]
	Binary PPI of *Escherichia coli*	[42]
	Bacterial Interactome (Desulfovibrio vulgaris and *E. coli*)	[41]
	Small RNA interactome of pathogenic *E. coli*	[63]
	FtsZ-ZapA-ZapB interactome of *E. coli*	[39]
	PPIs of nitrogen-fixing bacterium *Mesorhizobium loti*	[64]
	PPI *Synechocystis sp.*	[65]
	Binary protein interactome of *Treponema pallidum*	[66]
	Mycobacterium tuberculosis	[67]
	Mycoplasma genitalium	[68]
	Synechocystis sp. PCC6803	[65]
	Staphylococcus aureus (MRSA)	[69]
	Xanthomonas oryzae predicted interactome	[70]
Human	First human Interactome	[71]
	Interactome networks and human disease	[72]
	Studied temporal changes in human Interactome	[73]
	CoFrac 12	[74]
	HI-II-14	[75]
	BioPlex	[76]
	QUBIC	[77]
	CoFrac 15	[78]
	IID (integrated interactions database) providing tissue specific PPIs for model organisms and human	[79]
	Cross-species interactome mapping of yeast and human	[80]
	Proteome-Scale Human Interactomics	[81]
	Mitochondrial protein interactome	[82]
	Human Reference Protein Interactome Mapping Project (HuRI)	http://interactome.baderlab.org/

Table 2 Published interactomes of various organisms—cont'd

	Published interactome network maps	References
Yeast	Yeast protein interactome	[83]
	In vivo map of yeast protein Interactome	[84]
	In vivo map of yeast Interactome	[85]
	Yeast interactome network maps	[86]
	Cross species interactome map of *S. pombe* & *S. cerevisiae*	[87]
	Yeast Interactome	[88]
	IID (integrated interactions database) providing tissue specific PPIs for model organisms (yeast) and human	[79]
C. elegans	Map of interactome network metazoan (Worm Interactome5 or WI5)	[89]
	Systematic interactome mapping and genetic perturbation analysis of TGF-β Signaling network	[90]
	Worm Interactome version 8 (WI8)	[91]
	Interactome network for proteins involved in early embryonic cell divisions	[92]
	Worm SH3 interactome	[93]
	IID (integrated interactions database) providing tissue specific PPIs for model organisms and human	[79]
	Protein interactome mapping	[94]
Drosophila	Protein interaction map for fly proteome	[95]
	Drosophila Protein interaction Map (DPiM)	[96]
	IID (integrated interactions database) providing tissue specific PPIs for model organisms (fly) and human	[79]
	Drosophila Interactome Map—FlyBi Project	http://flybi.hms.harvard.edu/
Plants	Predicted interactome for Arabidopsis	[53]
	Arabidopsis Interactome Mapping Consortium	[97]
	Barley (*Hordeum vulgare*) interactome	[98]
	Oryza sativa predicted interactome	[54]
	Rice mitogen-activated protein kinase interactome	[99]
	Functional Gene Networks of soybean (SoyFGNs)	[100]
	Plant transcription factor Interactome	[101]
	Mapping genome-wide transcription factor-binding sites using DAP-seq	[102]
	Brassica rapa	[103]
Pathogen-host	Epstein-Barr virus (EBV)-human interactome map	[104], [25]
	Human varicella zoster virus (VZV)	[105]
	Chandipura virus	[26]
	Human-HCV interactions	[106]
	Hepatitis C virus (HPC)	[28]
	Hepatitis E virus (HEV)	[107]
	Herpes simplex virus 1 (HSV-1)	[25]
	Kaposi's sarcoma-associated herpesvirus (KSHV)	[107a]

Continued

Table 2 Published interactomes of various organisms—cont'd

	Published interactome network maps	References
	Murine cytomegalovirus (mCMV)	[25]
	Plant (Arabidopsis)-pathogen	[108]
	Viral-human interactome	[109]
	PPI maps of rice sheath blight by *Rhizoctonia solani*	[110]
	Global Dengue Virus NS1interactome	[111]
	Influenza A viruses (IAV)-human	[112]
	Human-bacterial interactomes	[113], [114]
	Pathogen contact point in host protein-protein Interactome	[115]
	ZikaBase (ZIKV-Human Interactome Map)	[116]

reaction refers to quick and sensitive detection of EU-labeled cellular RNA by using a copper (I)-catalyzed cycloaddition reaction with fluorescent azides, followed by microscopic imaging [126].

4.1.2 Surface plasmon resonance

The SPR is an optical effect utilized to measure quantitative equilibrium and kinetic parameters due to the interaction between various molecules in real-time without the use of labeled probes with great accuracy. In this method ligands are immobilized on the monolayer of carboxymethylated dextran attached to gold surface by covalent bonding. Recent findings have shown promising outcomes from the technique [127, 128].

4.2 Conventional label-based detection technologies

Various label-based detection methods evolved in last few decades to establish potential interactions. The major label-based detection techniques used in interactome mapping are Nematic Protein Organization Technique [129]; ellipsometry, split lactamase/split galactosidase [130], split YFP [131], split luciferase, FRET [132]/bioluminescence resonance energy transfer (BRET) [133].

4.3 Novel detection techniques for protein microarrays

Microarrays or biochip or DNA chip is one of the most effective techniques used for expression profiling of multiples genes. Microarray technology was also used for the development of biological networks for protein-DNA interactions [62], lectin-glycan recognition, viral kinase assay [134], PPI [135, 136], protein-DNA interaction [137], kinase assay [138], and protein-RNA interaction [139].

Table 3 Comparison of techniques used for PPIs studies [121]

Techniques	Principle	Interactions detected	Advantage	Disadvantage
Y2H	Two functional domains of transcription factor (TF), binding domain (BD) and activation domain (AD), are physically separated and fused to candidate proteins. Interaction between candidate proteins fused to AD and BD allows both domains to function together as TF and direct the expression of reporter gene (GFP, LacZ, etc.)	– Between two proteins; – Protein-nucleic acid; – Small molecule-target protein (using Y3H)	– Simple; – Well established; – Cost effective; – Scalable; – Effective for large-scale screening studies and for specific PPIs; – Being *in vivo* assay, it avoids cell lysis artefacts; – Best suited for detection of binary interactions	– Poor expression and/or lack of essential PTMs, cofactors, etc. may limit detection of some extra-organismal PPIs – Not suitable for membrane proteins as interacting proteins must enter nucleus to bring expression of reporter transcript – High false-positive due to non-specific interactions lead by overexpression of candidate proteins – Indirect readout prevent spatial or temporal analysis of PPIs
Membrane yeast-two-hybrid (MYTH)	Relies on split-ubiquitin approach where N-terminus (Nub) and C-terminus (Cub) fragments of ubiquitin protein are fused to prey and bait, respectively. Cub conjugated to an artificial TF is fused to cytosolic terminus of membrane protein (bait), while Nub is fused to membrane-associated or soluble	– Interactions of membrane proteins with membrane associated or soluble proteins	– Simple, – Cost effective; – Scalable in low- and high-throughput – Does not require any specialized equipment – Being *in vivo* assay allows study of full-length membrane proteins interactions in membrane environment	– PPI is adversely affected due to improper expression, modification and interaction of non-native proteins in yeast – Overexpression of candidate protein may lead to high false positives – Only applicable to membrane proteins having at least one cytosolic terminus accessible to cytosolic deubiquitinating enzymes

Continued

Table 3 Comparison of techniques used for PPIs studies [121]—cont'd

Techniques	Principle	Interactions detected	Advantage	Disadvantage
	potential interaction partners (preys). Bait-and-prey interaction brings Nub and Cub in close proximity forming pseudoubiquitin protein, recognizable by cellular deubiquitinating enzymes that cleave Cub-linked TF which on release is translocated into nucleus to activate expression of reporter gene		– Best suited for detection of binary interactions	– Soluble 'bait' protein should either be exceptionally large or anchored to intracellular structures to prevent diffusion of bait-TF fusion conjugate into nucleus preventing interaction-independent activation of reporter system – Indirect readout prevent spatial or temporal analysis of PPIs
Luminescence-based mammalian interactome mapping (LUMIER)	Co-immunoprecipitation based method where one protein is fused to *Renilla* luciferase and other to affinity tags (FLAG, HA, protein A). Such fusion constructs overexpress both proteins. Cells overexpressing proteins then are lysed, PPI complex is purified by immunoprecipition using antibody against affinity tag and interaction of fusion proteins then is measured by measuring	– Protein-protein interaction	– Easy to perform – Can be used in high throughput – Does not require specialized equipment except cell culturing reagents and instrument to measure bioluminescence – Can be used in various cell lines, facilitating PPIs for an organism in appropriate *ex vivo* format – Pertinent for studying binary interactions	– Requires lysis of cells prior to immunoprecipitation which can lead to disruption of weak and transient PPIs – Introduce artefacts as non-interacting proteins may come close together – May destabilize proteins as a result might expose concealed, non-native-binding surfaces – This assay should be optimized properly to normalize differences in transfection efficiency and expression hence minimizing noise signal

	luciferase activity which lowers due to interaction with affinity tagged fusion proteins			– Not suitable for studying spatial changes in PPIs with time or in response variable environments – Not compatible with full-length transmembrane proteins – Not suitable for spatial or temporal analysis of PPIs
Mammalian protein-protein interaction trap (MAPPIT)	A two-hybrid technique, used or mammalian cell lines, is based on cytokine signal transduction. Here, bait protein is fused to C-terminus of cytokine receptor lacking STAT3 binding, which otherwise is essential for signal transduction. Prey proteins are fused to receptor fragments containing functional STAT3 recruitment sites. When bait-and-prey proteins interact it results in a functional receptor which on stimulation with cytokine ligand activates STAT3 molecules via JAK activity and on entering nucleus activates reporter (luciferase) transcription under control of STAT3-responsive promoter	– Mammalian PPIs – Protein-small molecule interaction	– Mammalian PPIs – Enables high-throughput library and array screening – Easy procedure – Does not need specialized equipment except essential reagents for cell culture and instrument to record bioluminescence/fluorescence – Well suited for study of binary interactions – Its variants (three hybrid trap, reverse MAPPIT) can be used for screening for small-molecule	

Continued

Table 3 Comparison of techniques used for PPIs studies [121]—cont'd

Techniques	Principle	Interactions detected	Advantage	Disadvantage
Kinase substrate sensor (KISS)	Mammalian two-hybrid method to study PPIs, where bait protein is fused to kinase domain of TYK2 and prey proteins fused to a gp130 cytokine receptor fragment carrying TYK2 substrate motifs. Bait-and-prey interaction leads to phosphorylation of gp130 by bait fused TYK2 which in turn dock and activate STAT3 allowing STAT3 dimers to enter the nucleus and activate transcription of reporter gene (luciferase) under control of STAT3-responsive promoter	– Mammalian PPIs	– Allows assess PPIs directly in live mammalian cells – Sensitive detection of dynamic changes imposed by physiological or pharmacological challenges – Effective for membrane and cytosolic proteins – Best suited for study of binary interactions	– Indirect readout prevent spatial or temporal analysis of PPIs – Method is dependent on endogenous STAT3 and is not suitable to study PPIs proteins affecting STAT3 signaling
Bimolecular fluorescence complementation (BiFC)	Based on division of fluorescent protein (YFP) into two non-fluorescent segments that are used to bait-and-prey proteins. Bait-and-prey interaction brings both non-fluorescent segments in close proximity and	– Protein-protein interactions	– Direct visualization of PPIs in live cells allows subcellular localization of PPIs – Highly sensitive method, allows detection of low expressed proteins; weak and transient interactions	– Not suitable to study dynamics or real-time changes of PPIs due to slow reconstitution of fluorescent protein – Need of functionally active fusion proteins – False-positive fluorescent signals may arise due non-specific interactions

	forms a fluorescent complex that can be visualized by microscopy or flow cytometry		– Simple – Cost effective – Can be used for different organisms – Multiple PPIs can be visualized simultaneously in single cell by using combination of different fluorescent proteins as fusion – Best suited for detection of binary interactions	
Mammalian membrane two hybrid (MaMTH)	An *in vivo* technique for study of mammalian membrane PPIs is based on principle of split-ubiquitin, where candidate proteins are fused to Nub and Cub. Interaction of bait-and-prey proteins brings inactive Nub and Cub in close proximity leading to release of artificial TF to enter into nucleus and bringing expression of reporter (luciferase) transcript	– Interactions of full-length mammalian membrane proteins in with membrane associated or soluble proteins	– Cost effective – Highly scalable – Easily transferable almost all cell lines – Does not require any specialized reagent and equipment except reagents for cell culture reagents and instrument to record luciferase activity – Highly sensitive to measure transient or weak interactions and dynamic changes in PPIs imposed by intracellular stimulus – Best suited for study of binary interactions	– Functions when bait protein is membrane associated or associated intracellular structures, thus eliminating non-specific reporter gene activation – Cub should be fused to cytosolic terminus of protein and accessible to deubiquitinating enzymes for cleavage of TF from Cub – Not a well-suited method for spatial or real-time temporal analysis of PPIs

Continued

Table 3 Comparison of techniques used for PPIs studies [121]—cont'd

Techniques	Principle	Interactions detected	Advantage	Disadvantage
Fluorescence resonance energy transfer (FRET)	Relies on non-radiative transfer of excitation energy from donor fluorophore to a proximal acceptor molecule. Donor fluorescence is quenched and its fluorescence time is reduced by acceptor which increases acceptor fluorescence. For such quenching of fluorescence one candidate protein is fused to donor and other to an acceptor. Interaction of candidate proteins brings donor and acceptor fluorophores close to each other. Then excitation energy of donor is transferred to proximal acceptor producing detectable emission signal only for interacting candidate proteins	– Protein-protein interactions	– Allows real-time monitoring of PPIs including transient interactions – Can be directly used for live cells – Reversible fluorophore interaction allow monitoring of interaction dynamics	– Need of fluorescent fusion proteins and selection of fluorophore needs technical knowledge – Strong readout is dependent on close spatial proximity of fluorophores essential for energy transfer – Less sensitive than BiFC and BRET due to strong background signal – Not scalable to high-throughput screening of PPIs

Bioluminescence resonance energy transfer (BRET)	Developed to overcomes major limitation of FRET i.e. strong background signal. This method involves fusion of a protein of interest to *Renilla* luciferase (RLuc) energy donor, and interacting partner protein is fused to energy acceptor fluorescent protein (GFP or YFP). Interaction of candidate proteins bring donor and acceptor in close proximity allowing energy transfer and production of detectable fluorescent signal exhibiting PPIs	– Protein-protein interactions	– Allows real-time monitoring of PPIs including transient interactions – Can be directly used for live cells to – monitor cellular location of PPIs – Lower background makes it more sensitive than FRET	– Needs fusion proteins to work and BRET efficiency depends on spatial closeness of donor and acceptor for energy transfer – Signals are significantly weaker than those in FRET – Not scalable to high-throughput screening of PPIs
Affinity purification-mass spectrometry (AP-MS)	Bait protein is immobilized on to a solid support, i.e., agarose or magnetic beads, to capture target protein(s) from soluble phase. Affinity-purified captured proteins are digested with proteases (trypsin) producing peptides that are sub-fractionated by HPLC, ionized and detected using MS. AP-MS can	– Protein-protein interactions	– High-throughput screening method – Antibodies against endogenous, native bait proteins allow their purification in natural form without protein tagging and allowing simultaneous analysis of multiple isoforms of bait – Epitope tagging enables study of	– Needs cell lysis followed by affinity purification prevents detection of spatial or temporal and weak or transient PPIs – AP co-purifies abundant proteins as contaminants – Artefacts generated by exposure of proteins in cell lysate leading to illegitimate interactions or disruption interactions – High background may be produced due to improper

Continued

Table 3 Comparison of techniques used for PPIs studies [121]—cont'd

Techniques	Principle	Interactions detected	Advantage	Disadvantage
	be performed with endogenous, native bait proteins using antibodies raised against bait or with bait protein fused to epitope tag (TAP-, FLAG-, c-myc-, HA-, His-, protein A-, Strep-Tag)		proteins even in lack of their native antibodies	folding and mis-localization of proteins – Does not detect low expressing endogenous proteins – Data analysis of AP-MS requires expertise in MS and bioinformatics tools
Proximity-dependent biotin identification coupled to mass spectrometry (BioID-MS)	Bait protein is fused to prokaryotic biotin ligase (BirA) is expressed in cell. Proteins in close proximity to BioID fusion protein are biotinylated by BirA, allowing purification of such complexes by biotin-avidin/streptavidin affinity capture. Such purified biotinylated interaction partners of bait are then identified by MS	– Protein-protein interactions	– Library-independent method – Biotinylation of bait interacting proteins before cell lysis allows detection of interactions in natural cell environment – Stability of bait and prey and disruption of their interactions on cell lysis do not serve as limiting factors – Can detect weak or transient interactions – Detects low-abundant proteins more efficiently than AP-MS	– Fusion of BirA to bait protein significantly increases size of bait protein which can compromise targeting or function of bait – Low expressed interaction partners can lead to false negatives – Data analysis requires expertise in MS and bioinformatics

Proximity ligation assay (PLA)	*In situ* detection of PPIs in fixed cells/tissues using antibodies specific to two target protein Antibody specific to one protein is conjugated to Plus DNA oligonucleotide and that of other is conjugated with Minus DNA oligonucleotide. Plus and Minus DNA are complementary to DNA probe added at ligation step. Interaction of target proteins brings DNA nucleotides conjugated to respective antibodies in close proximity and serve as template for ligation of DNA fragments at ligation step and forming circular DNA molecule. Circular DNA is amplifies using rolling circle amplification primed by one of DNA oligonucleotide in proximity probe in presence of and forming	– Protein-protein interaction	– Detects and localizes PPIs with single molecule resolution – Can detect transient or weak interactions	– Dependent on activity and stability of enzymes (ligase and polymerase) – Expensive due to use of enzymes and target specific antibodies – Not suited for high-throughput PPI screening

Continued

Table 3 Comparison of techniques used for PPIs studies [121]—cont'd

Techniques	Principle	Interactions detected	Advantage	Disadvantage
	long DNA sequence physically linked to respective antibody and in turn to interacting target proteins. Newly synthesized DNA have various repetitive elements which on binding with fluorophore-labeled complementary oligonucleotide probes allow site-specific detection of interactions using fluorescence microscope			
Ligand-receptor capture—trifunctional chemoproteomics reagents (LRC-TriCEPS)	Relies on chemo-proteomics reagent containing three moieties, i.e. one binding to ligands of interest, second allows binding of glycosylated receptors on live cells, and a biotin tag. Reagent serves as a stable bridge that covalently links ligand of interest to carbohydrate groups on its receptor. After	– Receptor-ligand interactions	– Detects ligand-receptor interactions without genetic manipulations – Effectively detects transient surface interactions – Can identify cell surface-binding partners of various ligands viz peptides, proteins, viral particles, antibodies, and engineered affinity binders	– Efficient to identify N-glycoprotein receptors when glycans are sterically accessible – Coupling of ligands to TriCEPS reagent may alter its target-binding property

	treatment with TriCEPS, cells are lysed; tripsinized, and TriCEPS bound peptides are purified via biotin tag. Receptor peptides after separation from TriCEPS reagent are identified by MS			
Avidity-based extracellular interaction screen (AVEXIS)	Secreted recombinant bait-and-prey proteins are expressed in mammalian cell allowing essential post translational modification Biotinylated bait proteins are immobilized on streptavidin-coated solid phase and prey proteins are tagged with β-lactamase and a peptide sequence directing their pentamerization to increase prey concentration and assay sensitivity. Bait-and-prey interaction is then detection using ELISA-type method	– Novel extracellular receptor-ligand interactions	– Detects very weak interactions between membrane-embedded receptor proteins with low false-positives – High-throughput screening of extra-cellular interactions	– Limited to membrane proteins with self-contained extracellular domains – Method is not suitable for multi-pass membrane proteins and other proteins required to be embedded in plasma membrane for proper folding and functioning – Lengthy procedure – Expensive – Not well suited in detecting homophilic interactions – Artificial pentamerization hampers quantitation of strength of variable interactions

BD, DNA-binding domain; *AD*, transcriptional activation domain; *TF*, transcription factor; *PTM*, posttranslational modifications.

5 Applications

5.1 Role in disease diagnosis

Pathogenic microbes target a specialized host cell or tissue for propagation. Biochemistry of infected cells is altered to a great extent due to the expression of host defense gene products [140] and other specialized genes expressed only during or after pathogen invasion [141].

Complete genome sequences of microorganisms and human have allowed comparative interactome mapping for disease diagnosis. The Human Reference protein Interactome mapping project (HuRI) (http://interactome.baderlab.org/) has shown 74,820 possible interactions from 11,999 proteins. Results obtained from high-throughput Y2H were validated using two or more orthogonal assays and small-scale screening. Advanced host-pathogen interactome mapping facilitates the prediction of molecular and biochemical microenvironment of infected cells. This study revealed possible pathogen-associated genes, biochemical condition of cell, possible biochemical pathways during infection, and improves understanding of crosstalk among genes. Yeast interactome demonstrated that unlike nonessential genes, essential genes are highly connected. Differentially elevated genes identified via microarray during cancer were well connected than suppressed and randomly selected ones [142].

The study of network of genes involved in cancer showed that cancer proteins have almost double interaction partners than noncancer proteins. Clustering of network into overlapping subnetworks placed cancer proteins in larger clusters occupying more clusters than noncancer proteins [143].

Network analyses of herpes virus interactome were connected with networks of human proteins to simulate infection [144]. A network-based study performed to understand functions of proteins involved in Purkinje cell degeneration, identified 770 novel PPIs using stringent Y2H screening. It provided enhanced understanding of pathogenic mechanisms for a class of neurodegenerative disorders and identified candidate genes in inherited ataxias [145].

PPIs have been used to predict 300 candidate disease genes [146] and heredity disease genes [147]. Proteins encoded by genes mutated in inherited genetic disorders are likely to interact with proteins known to cause similar disorders, suggesting the existence of disease subnetworks. Human interaction map construction may facilitate an integrative systems biology approach for the elucidation of cellular networks that contribute to health and disease states [148].

Essential human genes encode hub proteins expressed in a wide range of tissues, suggesting their key role in human interactome. Goh et al. [149] reported that majority of the disease genes are nonessential and do not exhibit potential to encode hub proteins, and based on their expression patterns they occupy functional peripheral position in network.

Recent decades have witnessed exceptional growth in human-specific molecular interaction data, enabling us to understand the role of interconnected networks in human disease. PPI networks may contribute to the identification of new disease-associated gene targets for infectious disease, personalized medicine, and pharmacology [150].

The discovery of disease pathways by identifying disease-associated proteins may have potential to provide clinical insights into diagnosis, prognosis, and treatment of diseases. In silico methods provide better tools to discover PPI networks around known disease-associated proteins. Network connectivity alone cannot be enough for disease pathway discovery. Higher order network structures viz small subgraphs of the pathway may act as guide to the development of new methods [151].

5.2 Role in computational drug discovery

Molecular docking and scoring via computational approach provide the most promising route for drug designing and discovery. Computational drug designing and discovery is more economic, time saving, and effective over conventional technologies. The classical mechanism of computer-based drug designing methods masks the function of any individual proteins of interest. As a comprehensive approach, for the treatment of Parkinson and several inflammatory disorders, 238 drugs were used in 78 diseases showing that drugs can work in small network with genes present in related community [152]. The evolution of host-pathogen interactome database acts as a powerful tool in networking-based drug designing and discovery.

5.3 Role in identification of novel orphan gene in a pathway

Interactomics allows prediction of disease-associated pathways with complete information about related genes. Several attempts have been made for the screening of associated genes and other related biomolecules to discover disease-related pathways [153–155]. This method is not limited to disease-associated pathway but can be applied to signaling pathways [156, 157].

Zotenko et al. [158] developed a method to identify and represent overlapping protein complexes in interaction network. This method enabled to understand the decomposition of complexes, that is, transitions between functional groups and allowed tracking a protein's path through a cascade of functional groups.

Kelley et al. [159] aligned two PPI networks using a strategy that combined interaction topology and protein sequence similarity to identify conserved interaction pathways and complexes. It was used to align PPI networks of two distantly related species, *S. cerevisiae* and *Helicobacter pylori*. They reported that both species harbor a large complement of evolutionarily conserved pathways, and a large number of pathways appear to have duplicated and specialized within yeast.

Enormous PPI data available in databases are now being used for global alignment of multiple PPI networks of even unrelated species to deduce evolutionary relationships, trace novel pathway(s) involving orphan genes [160–165].

5.4 Role in understanding the parallel evolution of organisms

Comparative interactomics facilitates the study of evolution of any gene or set of genes in reference to preestablished functional information. In some cases there may not be sequence similarities but functional similarities may serve as evolutionarily conserved information [25]. The architecture of nuclear pore complex, regulating exchange of macromolecules between the nucleus and cytoplasm, has been best studied in yeast and human. It typically shows compositional similarities but display significant subcomplex structural variations [166].

5.5 Role in mutation studies

The sophisticated gene expression facilitates the cell to endure dynamic environmental and genetic conditions. The motifs prone to mutations can be easily mapped by interactomics. Signaling-regulatory Pathway INferencE (SPINE) framework explains activation/repression in gene expression profiles of multiple genes affected by knocking out of a gene. It permits the discovery of regulatory signaling networks that bridge random to affected genes on a web of network of physical interactions [12]. The Human Gene Mutation Database (HGMD; http://www.hgmd.org) collates all published gene lesions responsible for human inherited disease [167]. The mutation pattern proved to be useful for the better understanding of the evolution of respective species [168].

6 Conclusions and future perspectives

Pan-interactome mapping serve as a tool to identify the molecular networks in different physiological conditions of the cell. The prediction of unknown genes and their interactive partners can be mapped by different interactomics approaches. Modifications in sophisticated instrumentations took the interactomics research to next level. The combinations of two or more techniques provide precise understanding of pathway interactions. Different interactomics database improved the knowledge of host-parasite interaction mechanism and proved to be a milestone for better understanding of the biology of liaisons, thereby designing new approaches for more précised preventions. Moreover, deciphering the structure of infection network between the proteins and the host's cellular proteins generate new hypothesis to address the infections at molecular level to the advent of new drug discovery. The comparative interactome can also be used for exploring unknown pathway(s) or function(s) of a particular gene(s). In future it may be possible to unveil the whole cellular event of organisms by mapping interactome networks. Mutational analysis on a particular species unveils the function of all the co-expressing proteins in the absence of a single protein in any pathway. However, computational analysis needs a cross-examination through experiments or vice versa.

References

[1] N. Yadav, S.M.P. Khurana, D.K. Yadav, Plant secretomics: unique initiatives, in: D. Barh, M. Khan, E. Davies (Eds.), PlantOmics: The Omics of Plant Science, Springer, New Delhi, 2015, pp. 357–384, https://doi.org/10.1017/wsc.2018.33.

[2] X. Gao, U. Metzger, P. Panza, P. Mahalwar, S. Alsheimer, H. Geiger, et al., A floor-plate extracellular protein-protein interaction screen identifies Draxin as asectreted Nertin-1 antagonist, Cell Rep. 12 (2015) 694–708, https://doi.org/10.1016/j.celrep.2015.06.047.

[3] S. Yadav, D.K. Yadav, N. Yadav, S.M.P. Khurana, Plant glycomics. Advances and applications, in: D. Barh, M. Khan, E. Davies (Eds.), PlantOmics: The Omics of Plant Science, Springer, New Delhi, 2015, pp. 299–329, https://doi.org/10.1007/978-81-322-2172-2_10.

[4] P. Vechtova, J. Sterbova, J. Sterba, M. Vancova, R.O.M. Rego, M. Selinger, et al., A bite so sweet: the glycobiology interface of tick-host-pathogen interactions, Parasit. Vectors 11 (2018) 594–621, https://doi.org/10.1186/s13071-018-3062-7.

[5] J. Du, X. Ge, H. Dong, N. Zhang, L. Zhou, X. Guo, et al., The cellular interactome for glycoprotein 5 of the Chinese highly pathogenic porcine reproductive and respiratory syndrome virus, J. Integr. Agric. 15 (2016) 1833–1845, https://doi.org/10.1016/S2095-3119(15)61186-8.

[6] C. Sanchez, C. Lachaize, F. Janody, B. Bellon, L. Röder, J. Euzenat, et al., Grasping at molecular interactions and genetic networks in Drosophila melanogaster using FlyNets, an Internet database, Nucleic Acids Res. 27 (1) (1999) 89–94.

[7] X.-N. Shi, H. Li, H. Yao, X. Liu, L. Li, K.-S. Leung, et al., In silico identification and in vitro and in vivo validation of anti-psychotic drug fluspirilene as a potential CDK2 inhibitor and a candidate anti-cancer drug, PLoS One 10 (7) (2015) https://doi,org/10.1371/journal.pone.0132072.

[8] P. de los Reyes, F.J. Romero-Campero, M.T. Ruiz, J.M. Romero, F. Valverde, Evolution of daily gene co-expression patterns from algae to plants, Front. Plant Sci. 8 (2017) https://doi.org/10.3389/fpls.2017.01217.

[9] T. Ideker, O. Ozier, B. Schwikowski, A.F. Siegel, Discovering regulatory and signalling circuits in molecular interaction networks, Bioinformatics 18 (Suppl 1) (2002) S233–S240, https://doi.org/10.1093/bioinformatics/18.suppl_1.s.

[10] L. Cabusora, E. Sutton, A. Fulmer, C.V. Forst, Differential network expression during drug and stress response, Bioinformatics 21 (12) (2005) 2898–2905, https://doi.org/10.1093/bioinformatics/bti440.

[11] I. Ulitsky, R. Shamir, Identification of functional modules using network topology and high-throughput data, BMC Syst. Biol. 1 (1) (2007) 8, https://doi.org/10.1186/1752-0509-1-8.

[12] O. Ourfali, T. Shlomi, T. Ideker, E. Ruppin, R. Sharan, SPINE: a framework for signaling-regulatory pathway inference from cause-effect experiments, Bioinformatics 23 (13) (2007) i359–i366, https://doi.org/10.1093/bioinformatics/btm170.

[13] D.K. Yadav, N. Yadav, S. Yadav, S. Haque, N. Tuteja, An insight into fusion technology aiding efficient recombinant protein production for functional proteomics, Arch. Biochem. Biophys. 612 (2016) 57–77.

[14] H. Choi, T. Glatter, M. Gstaiger, A.I. Nesvizhskii, SAINT-MS1: protein-protein interaction scoring using label-free intensity data in affinity purification-mass spectrometry experiments, J. Proteome Res. 11 (2012) 2619–2624, https://doi.org/10.1021/pr201185r.

[15] H. Choi, G. Liu, M. Tyers, A.-C. Gingras, A.I. Nesvizhskii, Analyzing protein-protein interactions from affinity purification-mass spectrometry data with SAINT, Curr. Protoc. Bioinformatics (2012) https://doi.org/10.1002/0471250953.bi0815s39 Chapter 8: Unit 8.15.

[16] G. Teo, G. Liu, J.P. Zhang, A.I. Nesvizhskii, A.-C. Gingras, H. Choi, SAINTexpress: improvements and additional features in Significance Analysis of INTeractome for AP-MS data, J. Proteome 100 (2013) 37–43, https://doi.org/10.1016/j.jprot.2013.10.023.

[17] G.C. Teo, H. Koh, D. Fermin, J.-P. Lambert, J.D. Knight, A.-C. Gingras, et al., SAINTq: scoring protein-protein interactions in affinity purification-mass spectrometry experiments with fragment or peptide intensity data, Proteomics 16 (15-16) (2016) 2238–2245, https://doi.org/10.1002/pmic.201500499.

[18] M.F. Carazzolle, L.M. de Carvalho, H.H. Slepicka, R.O. Vidal, G.A. Pereira, J. Kobarg, et al., IIS—Integrated Interactome System: a web-based platform for the annotation, analysis and visualization of protein-metabolite-gene-drug interactions by integrating a variety of data sources and tools, PLoS One 9 (6) (2014) https://doi.org/10.1371/journal.pone.0100385.

[19] S.A. Lee, C.H. Chan, T.C. Chen, C.Y. Yang, K.C. Huang, C.H. Tsai, et al., POINeT: protein interactome with sub-network analysis and hub prioritization, BMC Bioinform. 10 (2009) 114, https://doi.org/10.1186/1471-2105-10-114.

[20] P. Shannon, A. Markiel, O. Ozier, N.S. Baliga, J.T. Wang, D. Ramage, et al., Cytoscape: a software environment for integrated models of biomolecular interaction networks, Genome Res. 13 (11) (2003) 2498–2504, https://doi.org/10.1101/gr.1239303.

[21] M.J. Cowley, M. Pinese, K.S. Kassahn, N. Waddell, J.V. Pearson, S.M. Grimmond, et al., PINA v2.0: mining interactome modules, Nucleic Acids Res. 40 (D1) (2012) D862–D865, https://doi.org/10.1093/nar/gkr967.

[22] M.J. Meyer, J.F. Beltran, S. Liang, R. Fragoza, A. Rumack, J. Liang, et al., Interactome INSIDER: a structural interactome browser for genomic studies, Nat. Methods 15 (2018) 107–114.

[23] H.B. Catabia, C. Smith, J. Ordovas, Inter-tools: a toolkit for interactome analysis, bioRxiv (2017) 150706, https://doi.org/10.1101/150706.

[24] S.V. Rajagopala, S. Casjens, P. Uetz, The protein interaction map of bacteriophage lambda, BMC Microbiol. 11 (1) (2011) 213, https://doi.org/10.1186/1471-2180-11-213.

[25] E. Fossum, C.C. Friedel, S.V. Rajagopala, B. Titz, A. Baiker, T. Schmidt, et al., Evolutionarily conserved herpesviral protein interaction networks, PLoS Pathog. 5 (9) (2009) https://doi.org/10.1371/journal.ppat.1000570.

[26] K. Kumar, J. Rana, R. Sreejith, R. Gabrani, S.K. Sharma, A. Gupta, et al., Intraviral protein interactions of Chandipura virus, Arch. Virol. 157 (10) (2012) 1949–1957, https://doi.org/10.1007/s00705-012-1389-5.

[27] I. García-Dorival, W. Wu, S.D. Armstrong, J.N. Barr, M.W. Carroll, R. Hewson, et al., Elucidation of the cellular interactome of Ebola virus nucleoprotein and identification of therapeutic targets, J. Proteome Res. 15 (12) (2016) 4290–4303, https://doi.org/10.1021/acs.jproteome.6b00337.

[28] N. Hagen, K. Bayer, K. Rösch, M. Schindler, The intraviral protein interaction network of hepatitis C virus, Mol. Cell. Proteomics 13 (7) (2014) 1676–1689, https://doi.org/10.1074/mcp.m113.036301.

[29] Z. Gao, J. Hu, Y. Liang, Q. Yang, K. Yan, D. Liu, et al., Generation and comprehensive analysis of host cell interactome of the PA protein of the highly pathogenic H5N1 avian influenza virus in mammalian cells, Front. Microbiol. 8 (2017) https://doi.org/10.3389/fmicb.2017.00739.

[30] I.P. Grégoire, C. Rabourdin-Combe, M. Faure, Autophagy and RNA virus interactomes reveal IRGM as a common target, Autophagy 8 (7) (2012) 1136–1137, https://doi.org/10.4161/auto.20339.

[31] E. Coyaud, C. Ranadheera, D.T. Cheng, J. Goncalves, B. Dyakov, E. Laurent, et al., Global interactomics uncovers extensive organellar targeting by Zika virus, Mol. Cell. Proteomics (2018) https://doi.org/10.1074/mcp.tir118.000800.

[32] L. Martinez-Gil, N.M. Vera-Velasco, I. Mingarro, Exploring the human-Nipah virus protein-protein interactome, J. Virol. 91 (23) (2017) https://doi.org/10.1128/jvi.01461-17.

[33] R.A. Flynn, L. Martin, R.C. Spitale, B.T. Do, S.M. Sagan, B. Zarnegar, et al., Dissecting noncoding and pathogen RNA–protein interactomes, RNA 21 (1) (2014) 135–143, https://doi.org/10.1261/rna.047803.114.

[34] S. Wuchty, S.V. Rajagopala, S.M. Blazie, J.R. Parrish, S. Khuri, R.L. Finley, The protein interactome of Streptococcus pneumoniae and bacterial meta-interactomes improve function predictions, mSystems 2 (3) (2017)https://doi.org/10.1128/msystems.00019-17.

[35] M. Meier, R.V. Sit, S.R. Quake, Proteome-wide protein interaction measurements of bacterial proteins of unknown function, Proc. Natl. Acad. Sci. 110 (2) (2012) 477–482, https://doi.org/10.1073/pnas.1210634110.

[36] J.H. Caufield, C. Wimble, S. Shary, S. Wuchty, P. Uetz, Bacterial protein meta-interactomes predict cross-species interactions and protein function, BMC Bioinform. 18 (1) (2017) https://doi.org/10.1186/s12859-017-1585-0.

[37] M. Arifuzzaman, M. Maeda, A. Itoh, K. Nishikata, C. Takita, R. Saito, et al., Large-scale identification of protein-protein interaction of Escherichia coli K-12, Genome Res. 16 (5) (2006) 686–691, https://doi.org/10.1101/gr.4527806.

[38] G. Butland, J.M. Peregrín-Alvarez, J. Li, W. Yang, X. Yang, V. Canadien, et al., Interaction network containing conserved and essential protein complexes in Escherichia coli, Nature 433 (7025) (2005) 531–537, https://doi.org/10.1038/nature03239.

[39] E. Galli, K. Gerdes, FtsZ-ZapA-ZapB interactome of Escherichia coli, J. Bacteriol. 194 (2) (2012) 292–302.

[40] S. Hu, Z. Xie, A. Onishi, X. Yu, L. Jiang, J. Lin, et al., Profiling the human protein-DNA interactome reveals ERK2 as a transcriptional repressor of interferon signaling, Cell 139 (3) (2009) 610–622, https://doi.org/10.1016/j.cell.2009.08.037.

[41] M. Shatsky, S. Allen, B.L. Gold, N.L. Liu, T.R. Juba, S.A. Reveco, et al., Bacterial interactomes: interacting protein partners share similar function and are validated in independent assays more frequently than previously reported, Mol. Cell. Proteomics 15 (5) (2016) 1539–1555, https://doi.org/10.1074/mcp.m115.054692.

[42] S.V. Rajagopala, P. Sikorski, A. Kumar, R. Mosca, J. Vlasblom, R. Arnold, et al., The binary protein-protein interaction landscape of Escherichia coli, Nat. Biotechnol. 32 (3) (2014) 285–290, https://doi.org/10.1038/nbt.2831.

[43] C. Su, J.M. Peregrin-Alvarez, G. Butland, S. Phanse, V. Fong, A. Emili, et al., Bacteriome.org an integrated protein interaction database for E. coli, Nucleic Acids Res. 36 (2007) D632–D636, https://doi.org/10.1093/nar/gkm807.

[44] J.-C. Rain, L. Selig, H. De Reuse, V. Battaglia, C. Reverdy, S. Simon, et al., The protein–protein interaction map of Helicobacter pylori, Nature 409 (6817) (2001) 211–215, https://doi.org/10.1038/35051615.

[45] R. Häuser, A. Ceol, S.V. Rajagopala, R. Mosca, G. Siszler, N. Wermke, et al., A second-generation protein protein interaction network of Helicobacter pylori, Mol. Cell. Proteomics 13 (5) (2014) 1318–1329, https://doi.org/10.1074/mcp.o113.033571.

[46] J. Jorda, Y. Liu, T.A. Bobik, T.O. Yeates, Exploring bacterial organelle interactomes: a model of the protein-protein interaction network in the Pdu microcompartment, PLoS Comput. Biol. 11 (2) (2015) https://doi.org/10.1371/journal.pcbi.1004067.

[47] N.J. Krogan, G. Cagney, H. Yu, G. Zhong, X. Guo, A. Ignatchenko, et al., Global landscape of protein complexes in the yeast Saccharomyces cerevisiae, Nature 440 (7084) (2006) 637–643, https://doi.org/10.1038/nature04670.

[48] B. Schwikowski, P. Uetz, S. Fields, A network of protein-protein interactions in yeast, Nat. Biotechnol. 18 (12) (2000) 1257–1261, https://doi.org/10.1038/82360. PMID 11101803.

[49] P. Uetz, L. Giot, G. Cagney, T.A. Mansfield, R.S. Judson, J.R. Knight, et al., A comprehensive analysis of protein–protein interactions in Saccharomyces cerevisiae, Nature 403 (6770) (2000) 623–627, https://doi.org/10.1038/35001009.

[50] M.R. Marcello, M. Druzhinina, A. Singson, Caenorhabditis elegans sperm membrane protein interactome, Biol. Reprod. 98 (6) (2018) 776–783, https://doi.org/10.1093/biolre/ioy055.

[51] S.Y. Rhee, W. Beavis, T.Z. Berardini, G. Chen, D. Dixon, A. Doyle, et al., The Arabidopsis Information Resource (TAIR): a model organism database providing a centralized, curated gateway to Arabidopsis biology, research materials and community, Nucleic Acids Res. 31 (1) (2003) 224–228, https://doi.org/10.1093/nar/gkg076.

[52] K.R. Brown, I. Jurisica, Online predicted human interaction database, Bioinformatics 21 (9) (2005) 2076–2082, https://doi.org/10.1093/bioinformatics/bti273.

[53] J. Geisler-Lee, N. O'Toole, R. Ammar, N.J. Provart, A.H. Millar, M. Geisler, A predicted interactome for Arabidopsis, Plant Physiol. 145 (2) (2007) 317–329, https://doi.org/10.1104/pp.107.103465.

[54] H. Gu, P. Zhu, Y. Jiao, Y. Meng, M. Chen, PRIN: a predicted rice interactome network, BMC Bioinform. 12 (2011) 161, https://doi.org/10.1186/1471-2105-12-161.

[55] J.-G. Kim, D. Park, B.-C. Kim, S.-W. Cho, Y. Kim, Y.-J. Park, et al., Predicting the interactome of Xanthomonas oryzae pathovar oryzae for target selection and DB service, BMC Bioinform. 9 (1) (2008) 41, https://doi.org/10.1186/1471-2105-9-41.

[56] P.L. Bartel, J.A. Roecklein, D. SenGupta, S. Fields, A protein linkage map of Escherichia coli bacteriophage T7, Nat. Genet. 12 (1) (1996) 72–77, https://doi.org/10.1038/ng0196-72.

[57] M. Sabri, R. Hauser, M. Ouellette, J. Liu, M. Dehbi, G. Moeck, et al., Genome annotation and intraviral interactome for the Streptococcus pneumoniae virulent phage Dp-1, J. Bacteriol. 193 (2) (2010) 551–562, https://doi.org/10.1128/jb.01117-10.

[58] R. Hauser, M. Sabri, S. Moineau, P. Uetz, The proteome and interactome of Streptococcus pneumoniae phage Cp-1, J. Bacteriol. 193 (12) (2011) 3135–3138, https://doi.org/10.1128/jb.01481-10.

[59] J. Yue, W. Xu, R. Ban, S. Huang, M. Miao, X. Tang, et al., PTIR: predicted tomato interactome resource, Sci. Rep. 6 (1) (2016) https://doi.org/10.1038/srep25047.

[60] J.R. Parrish, J. Yu, G. Liu, J.A. Hines, J.E. Chan, B.A. Mangiola, et al., A proteome-wide protein interaction map for *Campylobacter jejuni*, Genome Biol. 8 (7) (2007) R130.

[61] A. Métris, P. Sudhakar, D. Fazekas, A. Demeter, E. Ari, P. Branchu, et al., SalmoNet, an integrated network of ten *Salmonella enterica* strains reveals common and distinct pathways to host adaptation, NPJ Syst. Biol. Appl. 3 (2017) 31.

[62] P. Hu, S.C. Janga, M. Babu, J.J. Díaz-Mejía, G. Butland, W. Yang, et al., Global functional atlas of Escherichia coli encompassing previously uncharacterized proteins, PLoS Biol. 7 (4) (2009)https://doi.org/10.1371/journal.pbio.1000096.

[63] S.A. Waters, S.P. McAteer, G. Kudla, I. Pang, N.P. Deshpande, T.G. Amos, et al., Small RNA interactome of pathogenic E. coli revealed through crosslinking of RNase E, EMBO J. 36 (2016) 374–387.

[64] Y. Shimoda, S. Shinpo, M. Kohara, Y. Nakamura, S. Tabata, S. Sato, A large scale analysis of protein-protein interactions in the nitrogen-fixing bacterium *Mesorhizobium loti*, DNA Res. 15 (1) (2008) 13–23, https://doi.org/10.1093/dnares/dsm028.

[65] S. Sato, Y. Shimoda, A. Muraki, M. Kohara, Y. Nakamura, S. Tabata, A large-scale protein protein interaction analysis in Synechocystis sp. PCC6803, DNA Res. 14 (5) (2007) 207–216, https://doi.org/10.1093/dnares/dsm021.

[66] B. Titz, S.V. Rajagopala, J. Goll, R. Hauser, M.T. McKevitt, T. Palzkill, et al., The binary protein interactome of *Treponema pallidum*—the syphilis spirochete, PLoS One 3 (5) (2008) e2292, https://doi.org/10.1371/journal.pone.0002292.

[67] Y. Wang, T. Cui, C. Zhang, M. Yang, Y. Huang, W. Li, et al., Global protein – protein interaction network in the human pathogen Mycobacterium tuberculosis H37Rv, J. Proteome Res. 9 (12) (2010) 6665–6677, https://doi.org/10.1021/pr100808.

[68] S. Kuhner, V. van Noort, M.J. Betts, A. Leo-Macias, C. Batisse, M. Rode, et al., Proteome organization in a genome-reduced bacterium, Science 326 (5957) (2009) 1235–1240, https://doi.org/10.1126/science.1176343.

[69] A. Cherkasov, M. Hsing, R. Zoraghi, L.J. Foster, R.H. See, N. Stoynov, et al., Mapping the protein interaction network in methicillin-resistant Staphylococcus aureus, J. Proteome Res. 10 (3) (2011) 1139–1150, https://doi.org/10.1021/pr100918u.

[70] J. Guo, H. Li, J.-W. Chang, Y. Lei, S. Li, L.-L. Chen, Prediction and characterization of protein–protein interaction network in Xanthomonas oryzae pv. oryzae PXO99A, Res. Microbiol. 164 (10) (2013) 1035–1044, https://doi.org/10.1016/j.resmic.2013.09.001.

[71] J.F. Rual, K. Venkatesan, T. Hao, T. Hirozane-Kishikawa, A. Dricot, N. Li, et al., Towards a proteome-scale map of the human protein-protein interaction network, Nature 437 (7062) (2005) 1173–1178.

[72] M. Vidal, M.E. Cusick, A.L. Barabasi, Interactome networks and human disease, Cell 144 (2011) 986–998.

[73] A.R. Kristensen, J. Gsponer, L.J. Foster, A high-throughput approach for measuring temporal changes in the interactome, Nat. Methods 9 (2012) 907–909.

[74] P.C. Havugimana, G.T. Hart, T. Nepusz, H. Yang, A.L. Turnisky, Z. Li, et al., A census of human soluble protein complexes, Cell 150 (2012) 1068–1081.

[75] T. Rolland, M. Taşan, B. Charloteaux, S.J. Pevzner, Q. Zhong, N. Sahni, et al., A proteome-scale map of the human interactome network, Cell 159 (5) (2014) 1212–1226, https://doi.org/10.1016/j.cell.2014.10.050.

[76] E.L. Huttlin, L. Ting, R.J. Bruckner, F. Gebreab, M.P. Gygi, J. Szpyt, et al., The BioPlex network: a systematic exploration of the human interactome, Cell 162 (2) (2015) 425–440.

[77] M.Y. Hein, N.C. Hubner, I. Poser, J. Cox, N. Nagaraj, Y. Toyoda, et al., A human interactome in three quantitative dimensions organized by stoichiometries and abundances, Cell 163 (3) (2015) 712–723.

[78] C. Wan, B. Borgeson, S. Phanse, F. Tu, K. Drew, G. Clark, et al., Panorama of ancient metazoan macromolecular complexes, Nature 525 (7569) (2015) 339–344.

[79] M. Kotlyar, C. Pastrello, N. Sheahan, I. Jurisica, Integrated interactions database: tissue-specific view of the human and model organism interactomes, Nucleic Acids Res. 44 (2016) D536–D541.

[80] T.V. Vo, J. Das, M.J. Meyer, N.A. Cordero, N. Akturk, X. Wei, et al., A proteome-wide fission yeast interactome reveals network evolution principles from yeasts to human, Cell 164 (2016) 310–323.

[81] K. Luck, G.M. Sheynkman, I. Zhang, M. Vidal, Proteome-scale human interactomics, Trends Biochem. Sci. 42 (5) (2017) 342–354.

[82] D.K. Schweppe, J.D. Chavez, C.F. Lee, A. Caudal, S.E. Kruse, R. Stuppard, et al., Mitochondrial protein interactome elucidated by chemical cross-linking mass spectrometry, Proc. Natl. Acad. Sci. 114 (7) (2017) 1732–1737, https://doi.org/10.1073/pnas.1617220114.

[83] T. Ito, T. Chiba, R. Ozawa, M. Yoshida, M. Hattori, Y.A. Sakaki, Comprehensive two-hybrid analysis to explore the yeast protein interactome, Proc. Natl. Acad. Sci. U. S. A. 98 (8) (2001) 4569–4574, https://doi.org/10.1073/pnas.061034498.

[84] K. Tarassov, V. Messier, C.R. Landry, S. Radinovic, M.M. Serna Molina, I. Shames, et al., An in vivo map of the yeast protein interactome, Science 320 (2008) 1465–1470.

[85] J. Kast, Mapping connection for life: an in vivo map of the yeast interactome, HFSP J. 2 (2008) 244–250.

[86] H. Yu, P. Braun, M.A. Yildirim, I. Lemmens, K. Venkatesan, J. Sahalie, et al., High-quality binary protein interaction map of the yeast interactome network, Science 322 (5898) (2008) 104–110, https://doi.org/10.1126/science.1158684.

[87] J. Das, T.V. Vo, X. Wei, J.C. Mellor, V. Tong, A.G. Degatano, et al., Cross-species protein interactome mapping reveals species-specific wiring of stress response pathways, Sci. Signal. 6 (276) (2013) ra38.

[88] V. Janjić, R. Sharan, N. Pržulj, Modelling the yeast interactome, Sci. Rep. 4 (2014) 4273.

[89] S. Li, C.M. Armstrong, N. Bertin, H. Ge, S. Milstein, Boxem, et al., A map of the interactome network of the metazoan *C. elegans*, Science 303 (2004) 540–543.

[90] M. Tewari, P.J. Hu, J.S. Ahn, N. Ayivi-Guedehoussou, P.-O. Vidalain, S. Li, et al., Systematic interactome mapping and genetic perturbation analysis of a *C. elegans* TGF beta signaling network, Mol. Cell 13 (2004) 469–482.

[91] N. Simonis, J.F. Rual, A.R. Carvunis, M. Tasan, I. Lemmons, T. Hirozane-Kishikawa, et al., Empirically controlled mapping of the *Caenorhabditis elegans* protein-protein interactome network, Nat. Methods 6 (2009) 47–54.

[92] M. Boxem, Z. Maliga, N. Klitgord, N. Li, I. Lemmens, M. Mana, et al., A protein domain-based interactome network for C. elegans early embryogenesis, Cell 134 (3) (2008) 534–545.

[93] X. Xin, D. Gfeller, J. Cheng, R. Tonikian, L. Sun, A. Guo, et al., SH3 interactome conserves general function over specific form, Mol. Syst. Biol. 9 (2013) 652.

[94] S. Remmelzwaal, M. Boxem, Protein interactome mapping in *Caenorhabditis elegans*, Curr. Opin. Syst. Biol. 13 (2019) 1–9.

[95] L. Giot, J.S. Bader, C. Brouwer, A. Chaudhuri, B. Kuang, Y. Li, et al., A protein interaction map of Drosophila melanogaster, Science 302 (5651) (2003) 1727–1736.

[96] K.G. Guruharsha, R.A. Obar, J. Mintseris, K. Aishwarya, R.T. Krishnan, K. Vijayraghavan, S. Artavanis-Tsakonas, Drosophila protein interaction map (DPiM): a paradigm for metazoan protein complex interactions, Fly 6 (4) (2012) 246–253.

[97] Arabidopsis Interactome Mapping Consortium, Evidence for network evolution in an Arabidopsis interactome map, Science (New York, N.Y.) 333 (6042) (2011) 601–607.

[98] P.J. Schoonheim, H. Veiga, C. Pereira Dda, G. Friso, K.J. van Wijk, A.H. de Boer, A comprehensive analysis of the 14-3-3 interactome in barley leaves using a complementary proteomics and two-hybrid approach, Plant Physiol. 143 (2) (2007) 670–683.

[99] R. Singh, M.O. Lee, J.E. Lee, J. Choi, J.H. Park, E.H. Kim, et al., Rice mitogen-activated protein kinase interactome analysis using the yeast two-hybrid system, Plant Physiol. 160 (1) (2012) 477–487.

[100] Y. Xu, M. Guo, Q. Zou, X. Liu, C. Wang, Y. Liu, System-level insights into the cellular interactome of a non-model organism: inferring, modelling and analysing functional gene network of soybean (Glycine max), PLoS One 9 (2014).
[101] J. Yazaki, M. Galli, A.Y. Kim, K. Nito, F. Aleman, K.N. Chang, et al., Mapping transcription factor interactome networks using HaloTag protein arrays, PNAS 113 (29) (2016) E4238–E4247.
[102] A. Bartlett, R.C. O'Malley, S.C. Huang, M. Galli, J.R. Nery, A. Gallavotti, J.R. Ecker, Mapping genome-wide transcription-factor binding sites using DAP-seq, Nat. Protoc. 12 (8) (2017) 1659–1672.
[103] J. Yang, K. Osman, M. Iqbal, D.J. Stekel, Z. Luo, S.J. Armstrong, et al., Inferring the Brassica rapa interactome using protein–protein interaction data from Arabidopsis thaliana, Front. Plant Sci. 3 (2013) https://doi.org/10.3389/fpls.2012.00297.
[104] M.A. Calderwood, K. Venkatesan, L. Xing, M.R. Chase, A. Vazquez, A.M. Holthaus, et al., Epstein–Barr virus and virus human protein interaction maps, PNAS 104 (2007) 7606–7611.
[105] T. Stellberger, R. Häuser, A. Baiker, V.R. Pothineni, J. Haas, P. Uetz, Improving the yeast two-hybrid system with permutated fusions proteins: the Varicella Zoster Virus interactome, Proteome Sci. 8 (1) (2010) 8, https://doi.org/10.1186/1477-5956-8-8.
[106] Y. Han, J. Niu, D. Wang, Y. Li, Hepatitis C virus protein interaction network analysis based on hepatocellular carcinoma, PLoS One 11 (4) (2016) https://doi.org/10.1371/journal.pone.0153882.
[107] A. Osterman, T. Stellberger, A. Gebhardt, M. Kurz, C.C. Friedel, P. Uetz, et al., The Hepatitis E virus intraviral interactome, Sci. Rep. 5 (1) (2015) https://doi.org/10.1038/srep13872.
[107a] K. Heinzelmann, B.A. Scholz, A. Nowak, E. Fossum, E. Kremmer, J. Haas, et al., Kaposi's sarcoma-associated herpesvirus viral interferon regulatory factor 4 (vIRF4/K10) is a novel interaction partner of CSL/CBF1, the major downstream effector of notch signaling, J. Virol. 84 (23) (2010) 12255–12264, https://doi.org/10.1128/jvi.01484-10.
[108] M.S. Mukhtar, A.R. Carvunis, M. Dreze, P. Epple, J. Steinbrenner, J. Moore, et al., Independently evolved virulence effectors converge onto hubs in a plant immune system network, Science 333 (6042) (2011) 596–601, https://doi.org/10.1126/science.1203659.
[109] A. Segura-Cabrera, C.A. García-Pérez, X. Guo, M.A. Rodríguez-Pérez, A viral-human interactome based on structural motif-domain interactions captures the human infectome, PLoS One 8 (8) (2013).
[110] D. Lei, R. Lin, C. Yin, P. Li, A. Zheng, Global protein–protein interaction network of rice sheath blight pathogen, J. Proteome Res. 13 (7) (2014) 3277–3293.
[111] M.L. Hafirassou, L. Meertens, C. Umana-Diaz, A. Labeau, O. Dejarnac, L. Bonnet-Madin, et al., A global interactome map of the dengue virus NS1 identifies virus restriction and dependency host factors, Cell Rep. 21 (2017) 3900–3913.
[112] L. Wang, B. Fu, W. Li, G. Patil, L. Liu, M.E. Dorf, S. Li, Comparative influenza protein interactomes identify the role of plakophilin 2 in virus restriction, Nat. Commun. 8 (2017) 13876, https://doi.org/10.1038/ncomms13876.
[113] A. Pan, C. Lahiri, A. Rajendiran, B. Shanmugham, Computational analysis of protein interaction networks for infectious diseases, Brief. Bioinform. 17 (2016) 517–526.
[114] N. Crua Asensio, E. Munoz Giner, N.S. de Groot, M. Torrent Burgas, Centrality in the host-pathogen interactome is associated with pathogen fitness during infection, Nat. Commun. 8 (2017) 14092.
[115] H. Ahmed, T.C. Howton, Y. Sun, N. Weinberger, Y. Belkhadir, M.S. Mukhtar, Network biology discovers pathogen contact points in host protein-protein interactomes, Nat. Commun. 9 (2018) 2312.
[116] S. Gurumayum, R. Brahma, L.D. Naorem, M. Muthaiyan, J. Gopal, A. Venkatesan, ZikaBase: an integrated ZIKV- Human Interactome Map database, Virology 514 (2018) 203–210.
[117] S. Fields, O.K. Song, A novel genetic system to detect protein-protein interactions, Nature (London) 340 (1989) 245–246.
[118] L.B. Hays, Y.-S.A. Chen, J.C. Hu, Two-hybrid system for characterization of protein-protein interactions in E. coli, BioTechniques 29 (2) (2000) 288–296, https://doi.org/10.2144/00292st04.
[119] Y. Luo, A. Batalao, H. Zhou, L. Zhu, Mammalian two-hybrid system: a complementary approach to the yeast two-hybrid system, BioTechniques 22 (2) (1997) 350–352, https://doi.org/10.2144/97222pf02.

[120] J. Snider, S. Kittanakom, J. Curak, I. Stagljar, Split-ubiquitin based membrane yeast two-hybrid (MYTH) system: a powerful tool for identifying protein-protein interactions, J. Vis. Exp. 36 (2010) https://doi.org/10.3791/1698.
[121] J. Snider, M. Kotlyar, P. Saraon, Z. Yao, I. Jurisica, I. Stagljar, Fundamentals of protein interaction network mapping, Mol. Syst. Biol. 11 (2015) 848.
[122] S. Blasche, S. Wuchty, S.V. Rajagopala, P. Uetz, The protein interaction network of bacteriophage lambda with its host, Escherichia coli, J. Virol. 87 (23) (2013) 12745–12755, https://doi.org/10.1128/jvi.02495-13.
[123] A.G. Baltz, et al., The mRNA-bound proteome and its global occupancy profile on protein-coding transcripts, Mol. Cell 46 (2012) 674–690.
[124] A. Castello, B. Fischer, K. Eichelbaum, R. Horos, B.M. Beckmann, C. Strein, et al., Insights into RNA biology from an atlas of mammalian mRNA-binding proteins, Cell 149 (2012) 1393–1406.
[125] X. Bao, X. Guo, M. Yin, M. Tariq, Y. Lai, S. Kanwal, J. Zhou, N. Li, Y. Lv, C. Pulido-Quetglas, et al., Capturing the interactome of newly transcribed RNA, Nat. Methods 15 (2018) 213.
[126] C.Y. Jao, A. Salic, Exploring RNA transcription and turnover in vivo by using click chemistry, Proc. Natl. Acad. Sci. U. S. A. 105 (2008) 15779–15784.
[127] A. Florinskaya, P. Ershov, Y. Mezentsev, L. Kaluzhskiy, E. Yablokov, A. Medvedev, et al., SPR biosensors in direct molecular fishing: implications for protein interactomics, Sensors 18 (5) (2018) 1616, https://doi.org/10.3390/s18051616.
[128] S. Zhao, M. Yang, W. Zhou, B. Zhang, Z. Cheng, J. Huang, et al., Kinetic and high-throughput profiling of epigenetic interactions by 3D-carbene chip-based surface plasmon resonance imaging technology, Proc. Natl. Acad. Sci. 114 (35) (2017) E7245–E7254, https://doi.org/10.1073/pnas.1704155114.
[129] L. Wang, P. Eftekhari, D. Schachner, I.D. Ignatova, V. Palme, N. Schilcher, et al., Novel interactomics approach identifies ABCA1 as direct target of evodiamine, which increases macrophage cholesterol efflux, Sci. Rep. 8 (1) (2018) https://doi.org/10.1038/s41598-018-29281-1.
[130] A. Galarneau, M. Primeau, L.-E. Trudeau, S.W. Michnick, β-Lactamase protein fragment complementation assays as in vivo and in vitro sensors of protein–protein interactions, Nat. Biotechnol. 20 (6) (2002) 619–622, https://doi.org/10.1038/nbt0602-619.
[131] K.E. Miller, Y. Kim, W.-K. Huh, H.-O. Park, Bimolecular fluorescence complementation (bifc) analysis: advances and recent applications for genome-wide interaction studies, J. Mol. Biol. 427 (11) (2015) 2039–2055, https://doi.org/10.1016/j.jmb.2015.03.005.
[132] X. You, A.W. Nguyen, A. Jabaiah, M.A. Sheff, K.S. Thorn, P.S. Daugherty, Intracellular protein interaction mapping with FRET hybrids, Proc. Natl. Acad. Sci. 103 (49) (2006) 18458–18463, https://doi.org/10.1073/pnas.0605422103.
[133] S.W. Gersting, A.S. Lotz-Havla, A.C. Muntau, Bioluminescence resonance energy transfer: an emerging tool for the detection of protein-protein interaction in living cells, Methods Mol. Biol. 815 (2012) 253–263.
[134] J.-Y. Lu, Y.-Y. Lin, J.-C. Sheu, J.-T. Wu, F.-J. Lee, Y. Chen, et al., Acetylation of yeast AMPK controls intrinsic aging independently of caloric restriction, Cell 146 (6) (2011) 969–979, https://doi.org/10.1016/j.cell.2011.07.044.
[135] E.S. Johnson, Protein modification by SUMO, Annu. Rev. Biochem. 73 (1) (2004) 355–382, https://doi.org/10.1146/annurev.biochem.73.011303.0.
[136] H. Zhu, S. Hu, G. Jona, X. Zhu, N. Kreiswirth, B.M. Willey, et al., Severe acute respiratory syndrome diagnostics using a coronavirus protein microarray, Proc. Natl. Acad. Sci. 103 (11) (2006) 4011–4016, https://doi.org/10.1073/pnas.0510921103.
[137] J.S. Jeong, L. Jiang, E. Albino, J. Marrero, H.S. Rho, J. Hu, et al., Rapid identification of monospecific monoclonal antibodies using a human proteome microarray, Mol. Cell. Proteomics 11 (6) (2012) https://doi.org/10.1074/mcp.o111.016253.
[138] Y. Lin, J. Lu, J. Zhang, W. Walter, W. Dang, J. Wan, et al., Protein acetylation microarray reveals that NuA4 controls key metabolic target regulating gluconeogenesis, Cell 136 (6) (2009) 1073–1084, https://doi.org/10.1016/j.cell.2009.01.033.
[139] J. Huang, H. Zhu, S.J. Haggarty, D.R. Spring, H. Hwang, F. Jin, et al., Finding new components of the target of rapamycin (TOR) signaling network through chemical genetics and proteome chips, Proc. Natl. Acad. Sci. 101 (47) (2004) 16594–16599, https://doi.org/10.1073/pnas.0407117101.

[140] O. Visvikis, N. Ihuegbu, S.A. Labed, L.G. Luhachack, A.-M.F. Alves, A.C. Wollenberg, et al., Innate host defense requires TFEB-mediated transcription of cytoprotective and antimicrobial genes, Immunity 40 (6) (2014) 896–909, https://doi.org/10.1016/j.immuni.2014.05.002.
[141] I. Mitsuhara, T. Iwai, S. Seo, Y. Yanagawa, H. Kawahigasi, S. Hirose, et al., Characteristic expression of twelve rice PR1 family genes in response to pathogen infection, wounding, and defense-related signal compounds (121/180), Mol. Gen. Genomics. 279 (4) (2008) 415–427, https://doi.org/10.1007/s00438-008-0322-9.
[142] S. Wachi, K. Yoneda, R. Wu, Interactome-transcriptome analysis reveals the high centrality of genes differentially expressed in lung cancer tissues, Bioinformatics 21 (23) (2005) 4205–4208, https://doi.org/10.1093/bioinformatics/bti688.
[143] P.F. Jonsson, P.A. Bates, Global topological features of cancer proteins in the human interactome, Bioinformatics 22 (18) (2006) 2291–2297, https://doi.org/10.1093/bioinformatics/btl390.
[144] P. Uetz, Y.-A. Dong, C. Zeretzke, C. Atzler, A. Baiker, B. Berger, et al., Herpes viral protein networks and their interaction with the human proteome, Science 311 (5758) (2006) 239–242, https://doi.org/10.1126/science.1116804.
[145] J. Lim, T. Hao, C. Shaw, A.J. Patel, G. Szabó, J.-F. Rual, et al., A protein–protein interaction network for human inherited ataxias and disorders of purkinje cell degeneration, Cell 125 (4) (2006) 801–814, https://doi.org/10.1016/j.cell.2006.03.032.
[146] M. Oti, B. Snel, M.A. Huynen, H.G. Brunner, Predicting disease genes using protein-protein interactions, J. Med. Genet. 43 (8) (2006) 691–698, https://doi.org/10.1136/jmg.2006.041376.
[147] J. Xu, Y. Li, Discovering disease-genes by topological features in human protein–protein interaction network, Bioinformatics 22 (22) (2006) 2800–2805, https://doi.org/10.1093/bioinformatics/btl467.
[148] T.K. Gandhi, J. Zhong, S. Mathivanan, L. Karthick, K.N. Chandrika, S.S. Mohan, et al., Analysis of the human protein interactome and comparison with yeast, worm and fly interaction datasets, Nat. Genet. 38 (3) (2006) 285–293, https://doi.org/10.1038/ng1747.
[149] K.-I. Goh, M.E. Cusick, D. Valle, B. Childs, M. Vidal, A.-L. Barabasi, The human disease network, Proc. Natl. Acad. Sci. 104 (21) (2007) 8685–8690, https://doi.org/10.1073/pnas.0701361104.
[150] T. Ideker, R. Sharan, Protein networks in disease, Genome Res. 18 (4) (2008) 644–652, https://doi.org/10.1101/gr.071852.107.
[151] M. Agrawal, M. Zitnik, J. Leskovec, Large-scale analysis of disease pathways in the human interactome, in: Pacific Symposium on Biocomputing. Pacific Symposium on Biocomputing, vol. 23, 2018, pp. 111–122.
[152] E. Guney, J. Menche, M. Vidal, A.-L. Barábasi, Network-based in silico drug efficacy screening, Nat. Commun. 7 (2016) 10331, https://doi.org/10.1038/ncomms10331.
[153] M. Gustafsson, C.E. Nestor, H. Zhang, A.-L. Barabási, S. Baranzini, S. Brunak, et al., Modules, networks and systems medicine for understanding disease and aiding diagnosis, Genome Medicine 6 (10) (2014)https://doi.org/10.1186/s13073-014-0082-6.
[154] J. Pinero, N. Queralt-Rosinach, A. Bravo, J. Deu-Pons, A. Bauer-Mehren, M. Baron, et al., DisGeNET: a discovery platform for the dynamical exploration of human diseases and their genes, Database 2015 (2015) bav028, https://doi.org/10.1093/database/bav028.
[155] M.D. Ritchie, E.R. Holzinger, R. Li, S.A. Pendergrass, D. Kim, Methods of integrating data to uncover genotype–phenotype interactions, Nat. Rev. Genet. 16 (2) (2015) 85–97, https://doi.org/10.1038/nrg3868.
[156] E. Banks, E. Nabieva, R. Peterson, M. Singh, NetGrep: fast network schema searches in interactomes, Genome Biol. 9 (9) (2008) R138, https://doi.org/10.1186/gb-2008-9-9-r138.
[157] M. Steffen, A. Petti, J. Aach, P. D'haeseleer, G. Church, Automated modelling of signal transduction networks, BMC Bioinform. 3 (1) (2002) 34, https://doi.org/10.1186/1471-2105-3-34.
[158] E. Zotenko, K.S. Guimaraes, R. Jothi, T.M. Przytycka, Decomposition of overlapping protein complexes: a graph theoretical method for analyzing static and dynamic protein associations, Algorithms Mol Biol. 1 (2006) 7, https://doi.org/10.1186/1748-7188-1-7.
[159] B.P. Kelley, R. Sharan, R.M. Karp, T. Sittler, D.E. Root, B.R. Stockwell, et al., Conserved pathways within bacteria and yeast as revealed by global protein network alignment, Proc. Natl. Acad. Sci. U. S. A. 100 (20) (2003) 11394–11399, https://doi.org/10.1073/pnas.1534710.

[160] J. Flannick, A. Novak, B.S. Srinivasan, H.H. McAdams, S. Batzogloul, Graemlin: general and robust alignment of multiple large interaction networks, Genome Res. 16 (9) (2006) 1169–1181, https://doi.org/10.1101/gr.5235706.

[161] J. Gao, B. Song, W. Ke, X. Hu, Balanceali: multiple ppi network alignment with balanced high coverage and consistency, IEEE Trans. Nanobiosci. 16 (5) (2017) 333–340.

[162] S. Hashemifar, Q. Huang, J. Xu, Joint alignment of multiple protein–protein interaction networks via convex optimization, J. Comput. Biol. 23 (2016) 903–911.

[163] R. Sharan, T. Ideker, Modeling cellular machinery through biological network comparison, Nat. Biotechnol. 24 (4) (2006) 427–433, https://doi.org/10.1038/nbt1196.

[164] R. Sharan, S. Suthram, R.M. Kelley, T. Kuhn, S. McCuine, P. Uetz, et al., Conserved patterns of protein interaction in multiple species, Proc. Natl. Acad. Sci. 102 (6) (2005) 1974–1979.

[165] R. Singh, J. Xu, B. Berger, Global alignment of multiple protein interaction networks with application to functional orthology detection, Proc. Natl. Acad. Sci. 105 (35) (2008) 12763–12768, https://doi.org/10.1073/pnas.0806627105.

[166] S.O. Obado, M.C. Field, M.P. Rout, Comparative interactomics provides evidence for functional specialization of the nuclear pore complex, Nucleus 8 (4) (2017) 340–352, https://doi.org/10.1080/19491034.2017.1313936.

[167] P.D. Stenson, M. Mort, E.V. Ball, K. Evans, M. Hayden, S. Heywood, et al., The Human Gene Mutation Database: towards a comprehensive repository of inherited mutation data for medical research, genetic diagnosis and next-generation sequencing studies, Hum. Genet. 136 (6) (2017) 665–677, https://doi.org/10.1007/s00439-017-1779-6.

[168] F. Cheng, P. Jia, Q. Wang, C.-C. Lin, W.-H. Li, Z. Zhao, Studying tumorigenesis through network evolution and somatic mutational perturbations in the cancer interactome, Mol. Biol. Evol. 31 (8) (2014) 2156–2169, https://doi.org/10.1093/molbev/msu167.

Index

Note: Page numbers followed by *f* indicate figures and *t* indicate tables.

D

E

F

G

N

O

P

Q

R

Y

Z

FSC
www.fsc.org
MIX
Papier | Fördert gute Waldnutzung
FSC® C083411

Kolibri 360 GmbH
Ferdinand-Jühlke-Straße 7
99095 Erfurt, Deutschland
produktsicherheit@kolibri360.de